Improving water management in agriculture

Irrigation and food production

It is widely recognised that agriculture is a significant contributor to global warming and climate change. Agriculture needs to reduce its environmental impact and adapt to current climate change whilst still feeding a growing population, i.e. become more 'climate-smart'. Burleigh Dodds Science Publishing is playing its part in achieving this by bringing together key research on making the production of the world's most important crops and livestock products more sustainable.

Based on extensive research, our publications specifically target the challenge of climate-smart agriculture. In this way we are using 'smart publishing' to help achieve climate-smart agriculture.

Burleigh Dodds Science Publishing is an independent and innovative publisher delivering high quality customer-focused agricultural science content in both print and online formats for the academic and research communities. Our aim is to build a foundation of knowledge on which researchers can build to meet the challenge of climate-smart agriculture.

For more information about Burleigh Dodds Science Publishing simply call us on +44 (0) 1223 839365, email info@bdspublishing.com or alternatively please visit our website at www.bdspublishing.com.

Related titles:

Water management for sustainable agriculture
Print (ISBN 978-1-78676-176-7); Online (ISBN 978-1-78676-178-1, 978-1-78676-179-8)

Precision agriculture for sustainability
Print (ISBN 978-1-78676-204-7); Online (ISBN 978-1-78676-206-1, 978-1-78676-207-8)

Advances in Conservation Agriculture Volume 1: Systems and Science
Print (ISBN 978-1-78676-264-1); Online (ISBN 978-1-78676-266-5, 978-1-78676-267-2)

Advances in Conservation Agriculture Volume 2: Practice and Benefits
Print (ISBN 978-1-78676-268-9); Online (ISBN 978-1-78676-270-2, 978-1-78676-271-9)

Advances in Conservation Agriculture Volume 3: Adoption and Spread
Print (ISBN 978-1-78676-475-1); Online (ISBN 978-1-78676-477-5, 978-1-78676-478-2)

Climate change and agriculture
Print (ISBN 978-1-78676-320-4); Online (ISBN 978-1-78676-322-8, 978-1-78676-323-5)

Chapters are available individually from our online bookshop:
https://shop.bdspublishing.com

climate•SMART•publishing
IN AGRICULTURAL SCIENCE

BURLEIGH DODDS SERIES IN AGRICULTURAL SCIENCE
NUMBER 138

Improving water management in agriculture

Irrigation and food production

Edited by Professor Jerry W. Knox, Cranfield University, UK

burleigh dodds
SCIENCE PUBLISHING

Published by Burleigh Dodds Science Publishing Limited
82 High Street, Sawston, Cambridge CB22 3HJ, UK
www.bdspublishing.com

Burleigh Dodds Science Publishing, 1518 Walnut Street, Suite 900, Philadelphia, PA 19102-3406, USA

First published 2024 by Burleigh Dodds Science Publishing Limited

Library of Congress Control Number: 2023952208

British Library Cataloguing in Publication Data
A catalogue record for this book is available from the British Library

ISBN 978-1-80146-274-7 (print)
ISBN 978-1-80146-276-1 (PDF)
ISBN 978-1-80146-275-4 (ePub)
ISSN 2059-6936 (print)
ISSN 2059-6944 (online)

DOI: 10.19103/AS.2023.0123

Typeset by Deanta Global Publishing Services, Dublin, Ireland

Contents

Series list

Title	Series number
Achieving sustainable cultivation of maize - Vol 1 From improved varieties to local applications Edited by: Dr Dave Watson, CGIAR Maize Research Program Manager, CIMMYT, Mexico	001
Achieving sustainable cultivation of maize - Vol 2 Cultivation techniques, pest and disease control Edited by: Dr Dave Watson, CGIAR Maize Research Program Manager, CIMMYT, Mexico	002
Achieving sustainable cultivation of rice - Vol 1 Breeding for higher yield and quality Edited by: Prof. Takuji Sasaki, Tokyo University of Agriculture, Japan	003
Achieving sustainable cultivation of rice - Vol 2 Cultivation, pest and disease management Edited by: Prof. Takuji Sasaki, Tokyo University of Agriculture, Japan	004
Achieving sustainable cultivation of wheat - Vol 1 Breeding, quality traits, pests and diseases Edited by: Prof. Peter Langridge, The University of Adelaide, Australia	005
Achieving sustainable cultivation of wheat - Vol 2 Cultivation techniques Edited by: Prof. Peter Langridge, The University of Adelaide, Australia	006
Achieving sustainable cultivation of tomatoes Edited by: Dr Autar Mattoo, USDA-ARS, USA and Prof. Avtar Handa, Purdue University, USA	007
Achieving sustainable production of milk - Vol 1 Milk composition, genetics and breeding Edited by: Dr Nico van Belzen, International Dairy Federation (IDF), Belgium	008
Achieving sustainable production of milk - Vol 2 Safety, quality and sustainability Edited by: Dr Nico van Belzen, International Dairy Federation (IDF), Belgium	009
Achieving sustainable production of milk - Vol 3 Dairy herd management and welfare Edited by: Prof. John Webster, University of Bristol, UK	010
Ensuring safety and quality in the production of beef - Vol 1 Safety Edited by: Prof. Gary Acuff, Texas A&M University, USA and Prof. James Dickson, Iowa State University, USA	011
Ensuring safety and quality in the production of beef - Vol 2 Quality Edited by: Prof. Michael Dikeman, Kansas State University, USA	012
Achieving sustainable production of poultry meat - Vol 1 Safety, quality and sustainability Edited by: Prof. Steven C. Ricke, University of Arkansas, USA	013
Achieving sustainable production of poultry meat - Vol 2 Breeding and nutrition Edited by: Prof. Todd Applegate, University of Georgia, USA	014

Part 1

Current and future water demands for agriculture

Chapter 1

Reconciling water needs for food, the environment, energy and society – a quadrilemma facing the future of global irrigated agriculture

Jerry W. Knox, Cranfield University, UK*

1 Introduction

Although most agriculture globally is rainfed, irrigated agriculture is responsible for a significant proportion of overall food production, particularly in low- and middle-income countries (LMICs), where it typically accounts for ca. 20% by area but 40% of production (UN World Water Development Report, 2003). However, in terms of total freshwater abstractions, irrigated agriculture constitutes the largest use globally (1810 km^3), accounting for >70% of all freshwater withdrawals (FAO Aquastat, 2016; Springmann et al., 2018).

Given this high dependence, water allocations for irrigated agriculture are understandably facing increasing scrutiny due to the negative environmental impacts of over-abstraction from both surface and groundwater resources

* Corresponding author: j.knox@cranfield.ac.uk

http://dx.doi.org/10.19103/AS.2023.0123.01

on aquatic-dependent ecosystems and rising tensions over freshwater use between competing sectors including domestic water supply, energy generation and industry. A changing climate threatens to exacerbate the situation with increased climate extremes and drought risks reducing available water supplies and their reliability. There are also concerns regarding the contribution of agriculture to transgression of planetary boundaries, including land and water resources, biosphere integrity and nitrogen flows, with a so-called technological–cultural 'U-turn' being required to reduce the sector's environmental footprint whilst meeting future food demands and staying within these four planetary boundaries (Gerten et al., 2020). Current and future agricultural production is therefore vulnerable to a mix of interconnected water, climate, environmental and regulatory-related risks. These could also threaten or undermine the productivity of both rain-fed and irrigated, which would inevitably impact on food markets, trade flows, and food security with widespread regional and global repercussions.

Improving agricultural water management is widely viewed by many observers, including those from the scientific, governmental, agrifood industry and practitioner communities, as being a critical and essential prerequisite to achieving a more environmentally sustainable, resilient and productive agro-food sector (OECD, 2020). However, many fundamental questions remain unanswered, including reconciling the needs of water for food production with those of the environment and for energy and society – a wicked quadrilemma facing the future of global irrigated agriculture. What needs to be done, and by whom, and how can we bring together innovations in engineering, technology, management, science and governance to deliver much-needed solutions to improve agricultural water resources management, allocation and efficiency?

These are the pertinent priorities and questions that constitute the focus of this timely and policy-relevant book. It is divided into four main sections, with each covering related topics linked to:

- water resources allocation for agriculture – recognising the need for improved data and monitoring to support demand forecasting;
- improving on-farm agricultural practices through agronomy, plant science, soils, drainage and energy management;
- improving irrigation water resources management – understanding 'efficiency', supporting farmer-led collaboration, water sharing and benchmarking performance; and
- future climate and water-related climate risks – the role of precision irrigation, management and technology to address these challenges.

A brief overview of the topics and issues covered by each chapter in each section is provided as follows.

2 Allocating water resources for agriculture: recognising the need for improved data and monitoring to support demand forecasting

Allocating freshwater for irrigated agriculture is an essential component in meeting the food demands of a burgeoning global population, in supporting rural livelihoods in many LMICs where agriculture is the cornerstone of their economy and in buffering the impacts of rainfall deficits, drought risks and climate extremes on crop productivity (yield). Indeed, irrigated agriculture is, on average, at least twice as productive per unit of land compared to rainfed agriculture, thereby facilitating greater intensification and crop diversification (World Bank, 2024). However, water resources in many countries where irrigated agriculture is practiced are under severe pressure from multiple externalities, with a changing climate threatening to further destabilise current supply-demand imbalances.

This situation will be exacerbated by population growth which is expected to increase to over 10 billion by 2050 (coupled with increased urbanisation) and the need to increase food demands not only for the rising population but also to provide for increased consumption of calories and more complex diets, a direct response to socio-economic development. Indeed, it is estimated that agricultural production will need to increase by ca. 70% by 2050, with much of this demand expecting to be met from irrigated cropping. Many proponents would ask, 'Is there a pending crisis ahead?' with food production needing to increase by ca. 50% by 2050 to meet the demand of the world's growing population, rising from 1810 $km^3 year^{-1}$ to 2987 $km^3 year^{-1}$ by 2050. Blue water withdrawals for agriculture will also need to increase by ca. 65% if we are to continue producing and consuming food as we do today (FAO, 2017; Springmann et al., 2018).

Unfortunately, in many regions where irrigation demand is concentrated (Figure 1), high demand also coincides with countries where water resources are highly stressed, and where climate change is expected to have the greatest negative impacts on freshwater availability due to reduced rainfall and stream flows (Figure 2). Although there is expected to be an overall increase in global water resource availability, there is projected to be reduced surface and groundwater in most dry subtropical regions. The increase in water resources availability is mainly in the high latitudes, and the proportional changes are up to three times greater for runoff than for precipitation (Figure 2) (Jiménez Cisneros et al., 2014; IPCC Fifth Assessment Report, 2014).

Given these externalities on both water supplies and demands for agriculture, improved monitoring at much finer temporal and spatial scales will be necessary to support better water resources decision-making and policy development. This topic is the focus of Foster et al. (2024) who highlight that

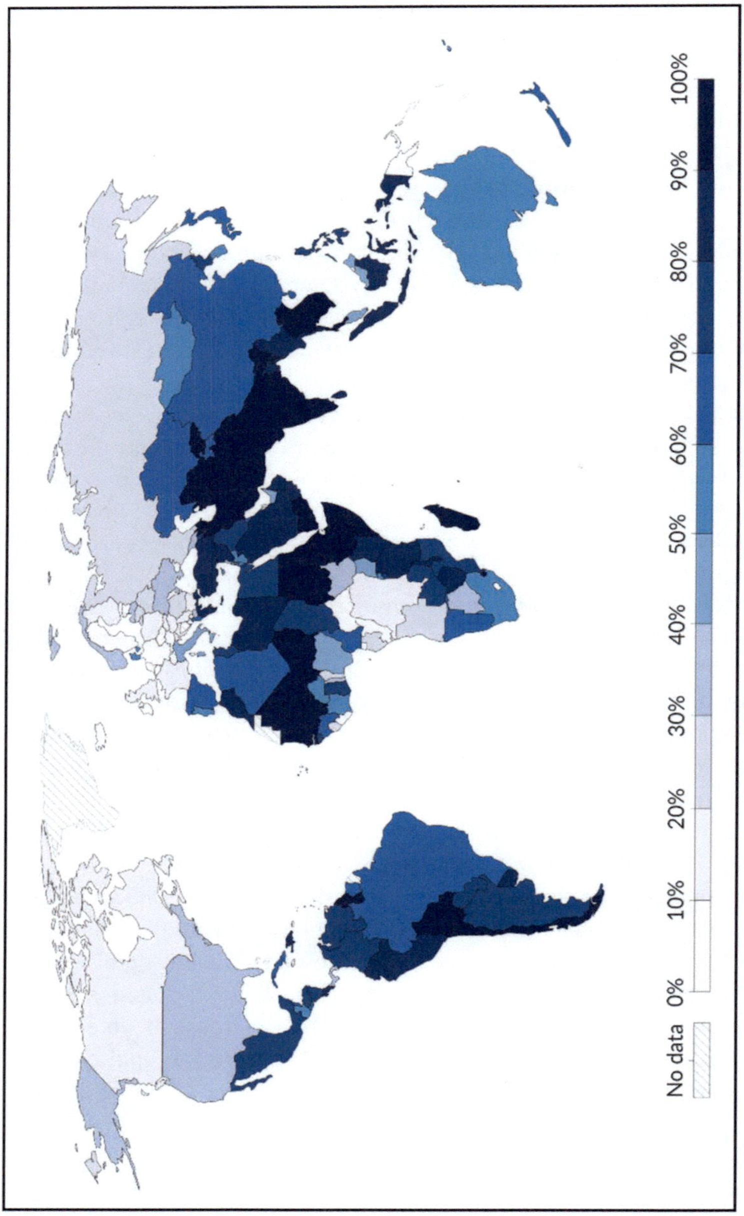

Figure 1 Water withdrawals for agriculture as a proportion (%) of total freshwater use. Source: FAO via World Bank (2019).

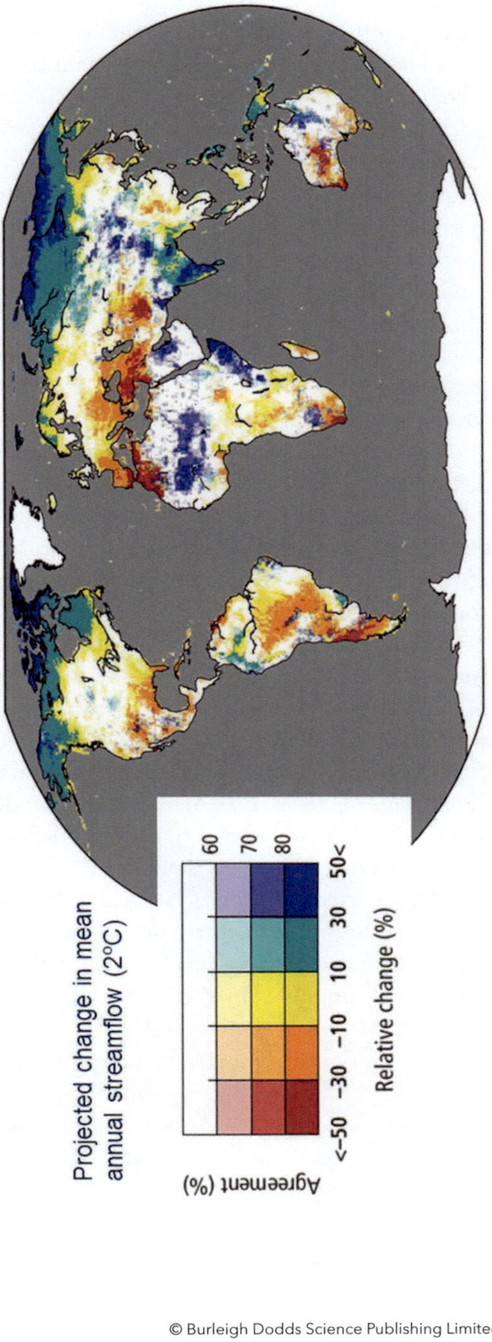

Figure 2 Projected changes in mean annual streamflow. Source: Jiménez Cisneros et al. (2014).

despite the importance of water for agriculture many countries face serious gaps in their ability to identify and accurately quantify where, when, how much and for what purposes water is withdrawn or abstracted by farmers (Foster et al., 2020). Indeed, in many countries, there are no digital records of the historical volumes of water abstracted, the sources used, their timing and peak rates of abstraction. The policy implications arising from this lack of monitoring are gradually being recognised, with unauthorised water use estimated to account for up to 30–50% of total abstractions and a key contributor to global water scarcity pressures (Felbab-Brown, 2017; Loch et al., 2020). In their chapter, Foster et al. (2024) critically review the evidence on the technical, financial, social and political barriers that need to be overcome to achieve more effective monitoring of agricultural water use, with a specific focus on abstraction licenses (water rights or permits) and quantifying volumetric withdrawals. They argue that having a better understanding and data to assess the nature and composition of abstractions will enable more accurate catchment or regional water accounting to be undertaken, will inform the implementation of appropriate abstraction management control regulations and help plan investments to reduce water resources risks and future water scarcity facing agriculture. Finally, Foster et al. (2024) provide insights into how the role of applied remote sensing and satellite imagery together with smart metering and telemetry systems can provide near real-time monitoring of agricultural abstractions and crop water use.

The importance of the need for improved monitoring of water use is particularly apparent in the context of demand forecasting and water policy planning to assess how future agricultural abstractions might change in response to, for example, population growth, socio-economic development, technology implementation or demands for greater environmental regulation. In their chapter, Amarasinghe et al. (2024) explain how deriving reliable estimates of future agricultural water demand are complex and challenging, due to the many externalities that impact on demand. They also argue how the situation is exacerbated by inadequate and unreliable data to assess historical trends and the impacts of exogenous factors such as climate change on the 'drivers of change', reinforcing the need for improved monitoring as described by Foster et al. (2024). Amarasinghe et al. (2024) also discuss 'uncertainty' in future demand forecasts and how it can impact water policy planning and resource development which typically requires a long-term strategic outlook, especially in the agricultural sector which is currently experiencing significant and rapid transformation which itself will impact on future water demand.

Other important issues raised include the need to address the challenges of estimating water demand at the sub-national level, which again is contingent on having access to granular data from local-scale water monitoring. With a changing climate, there will also be a need for better information on dietary and food consumption patterns, changing levels of income, national levels of self-sufficiency and targets for environmental protection to support accurate agricultural demand projections at the local level. Finally, Amarasinghe et al. (2024) emphasise the importance of linking demand forecasting to climate adaptation, since the uptake of most adaptation options and responses must be at the individual farmer or smallholder level or community level. Here, providing insights from projection studies will provide valuable evidence to build resilience to emergent climate and water-related risks.

3 Improving on-farm agricultural practices through agronomy, plant science, soils, drainage and energy management

More widespread monitoring of water abstractions and use at a multiscalar level will need to be accompanied by improvements in crop agronomy, in-field soil and water management (scheduling), coupled with plant science developments to reduce abiotic impacts. In addition, there will be a need to move away from fossil fuels to renewable energies including solar for irrigation pumping to reduce greenhouse gas (GHG) emissions and to transition agriculture to Net Zero (netzeroclimate, 2024). Improving the efficiency of water use in agriculture, which itself is an emotive topic, is discussed by Lankford and McCartney (2024) and will also depend on matching improvements in water supply (off-farm distribution systems) with appropriate incentives for on-farm investments to improve soil and water management. This will require more dynamic water delivery systems to provide water 'on-demand' as well as the use of advanced precision irrigation technologies (Hedley et al., 2014), including, for example, in situ soil moisture sensors, telemetry systems and satellite estimates of actual evapotranspiration to improve water productivity (World Bank, 2024). It will also need more concerted efforts to manage irrigation and drainage systems in a more integrated way, balancing water deficits with surpluses through greater dependence on rainwater harvesting and on-farm water storage. Similarly, we need to revisit soil and water management practices and policy, viewing them as part of a connected system rather than as discrete components (or silos) in the plant–soil–water continuum. These fundamental aspects form the second important theme in this book.

In their chapter, Haruna et al. (2024) focus on the range of agronomic and soil management practices such as tillage, cover cropping, crop rotation, and animal grazing that can enhance water infiltration and soil water retention. They stress the critical significance of soil water management and how fundamentally important it is for agronomic and environmental sustainability and that improved system management will become increasingly pertinent in a rapidly changing global climate. However, Haruna et al. (2024) highlight there can often be disagreement in evidence reported in the literature on the remedial actions required. Management practices that influence soil bulk density, the distribution of pore sizes, infiltration rates and soil moisture retention in the root zone, and saturated hydraulic conductivity can all vary depending on the intensity of management practices, differences in local agroclimatic conditions, soil characteristics including soil texture and local topography. They propose that, by practicing conservation tillage, with an emphasis on low or no till, rather than conventional tillage, soil water retention can be improved. Other important factors to consider also include the selection of appropriate crop cover species that are suitable for local soil and agroclimatic conditions, improving residue retention in the topsoil and practicing less-intensive (longer) grazing and longer crop rotations.

In the past, managing irrigation was usually considered in conjunction with the importance of drainage. However, in recent decades, the focus and hence investments in irrigation have often far exceeded or overshadowed equivalent investments in drainage, despite its importance for crop and soil water management. As a consequence, Ritzema (2024) reports that salinity and waterlogging problems in Egypt, which has one of the largest areas of agricultural drainage globally, now affect ca. 10–16% of the total irrigated area. Ritzema (2024) stresses that the role of drainage is changing markedly, largely in response to increased climate variability with greater rainfall uncertainty, but also due to changes in agricultural land use, with a shift from its single-purpose use for controlling waterlogging and/or salinity to becoming an essential component of integrated water resources management. Current advances in drainage design and management are now focussing more on improved control, reducing the environmental impacts of drainage as well as utilising new materials for drainage construction.

Drainage research is also focussing more on adapting institutional and governance arrangements for drainage water management to encourage greater farmer engagement and ownership. Ritzema (2024) argues that most historical 'top-down' approaches to drainage design and management had limited consideration of the local conditions and environmental issues and farmer needs, with a primary focus being more on the technical (infrastructural) aspects, and much less attention given to organizational (institutional) aspects. Although there has been a shift to a more 'service-oriented' approach since the 1990s, this has not gained sufficient traction in practice. Therefore, to ensure the benefits of drainage are realised in the future, Ritzema (2024) proposes three elements. Firstly, a need to balance 'top-down' against 'bottom-up' approaches with greater attention to farmer participation that considers societal choices. This will also require decentralized management with active farmer participation throughout the design, implementation, operational and management (O&M) phases. Secondly, there is a need to move away from a 'one-size-fits-all' or standard approach to one where there is greater flexibility which takes into account local conditions and farmer preferences. However, the challenge here is in reconciling one farmer's individual need for drainage, which varies from field to field to that of drainage needs at the farm, district or scheme level where collective action is needed. Finally, there is a pressing need for a greater focus on capacity building and development through multi-sector collaboration (Burns et al., 2015; Ghumman et al., 2012). In this context, Ritzema (2024) emphasises the need for integrating applied research with training and advisory services. Evidence from Egypt, India and Pakistan shows just how beneficial such an approach can be in effecting real change.

Whilst healthy soils and effective drainage help underpin on-farm crop production, improving the sustainability of agricultural systems will also require much greater attention to energy use in irrigation management, particularly given the spiralling costs of energy for pumping and government and societal demands for reducing GHG emissions from agriculture. In their chapter, Rodríguez Díaz et al. (2024) focus on the challenges of improving energy use and management in irrigated agriculture, highlighting how irrigation district modernisation in Spain (and other countries in the Mediterranean region), have focussed on replacing traditional open channel distribution networks with pressurised piped systems. These have dramatically changed the balance between water and energy use, but often with unintended consequences. In most cases, whilst water use efficiency (WUE) has increased markedly, so too has energy consumption, leading to increases in water supply costs and, in many cases, significant reductions in farm profits.

To address these imbalances, Rodríguez Díaz et al. (2024) set out the measures available to achieve significant energy savings in pressurised water supply networks (involving alternate management strategies coupled with improvements in hydraulic infrastructure) and the potential role of renewable energy (including solar photovoltaic and micro-hydraulic energy) which can reduce energy costs and the carbon footprint of agricultural production. This chapter also highlights how energy-saving measures can be based on deep learning from big data and artificial intelligence to enable demand forecasts at both district and field scales. However, one of the main limitations is the amount of data required for calibration, which requires extensive monitoring from near real-time sensors in-field. As the cost for advanced wireless sensor networks declines, more farm businesses should start to implement such technologies. Compelling evidence is provided with examples from irrigation districts in Spain.

Modern irrigated agriculture relying on pressurised water delivery and application has become inextricably dependent on energy, with initiatives to improve 'efficiency' often being more focussed on energy rather than water, given the unit cost differences. However, with concerns about the environmental impacts of over-abstraction and rising CO_2 emissions from agriculture. it becomes imperative to optimise both water and energy components with the water–energy–food (WEF) nexus. Some measures have required changes in network management, including sectorisation, critical point control and optimising pump station design and management. However, the optimal application of these measures in an integrated manner is complex and requires methods that take account of the large number of variables involved.

Switching to renewable energies also represents a major opportunity to provide clean energy at low cost, particularly from photovoltaic solar energy given the high levels of solar radiation in many semi-arid environments where irrigated agriculture is concentrated. Rodríguez Díaz et al. (2024) describe how early installations were relatively small-scale but are now being implemented across much larger irrigated areas in countries such as Spain, often with high-power solar plants. Selling excess energy produced from these plants into the national grid is also changing the economics of irrigation technologies, with investments becoming more viable and turning irrigation districts into renewable energy providers, as well as centres for food crop production.

Continuing the theme of renewable energy, in their chapter, Watto et al. (2024) focus on how recent developments in harnessing solar energy have transformed solar-powered irrigation systems (SPISs) to become a cost-effective, reliable and environmentally sustainable alternative to conventional fossil fuel-dependent energy-based irrigation systems. With reference to research from Pakistan, they review the latest developments and future prospects for SPIS, including smart solar irrigation systems and real-time water monitoring. Future advances including more advanced automation of water delivery and control systems might encourage wider uptake and adoption of SPIS. However, despite the significant potential opportunity for SPIS uptake, one barrier remains. Without effective regulatory control and market-based mechanisms, unlimited access to solar energy could encourage extensive over-exploitation of groundwater and lead to rural and social inequalities, unless the sustainability of groundwater resources is controlled by appropriate water abstraction policy and regulation.

In this final section, Dodd et al. (2024) focus on the critical issue of crop physiological responses to irrigation and their plant regulatory mechanisms. They highlight the research challenges that farmers are understandably risk-averse to growing crops with less irrigation (particularly as competition for 'blue water' increases and demands to restore environmental flows increases). This will require the irrigation research community to develop more innovative techniques to maintain and/or boost productivity whilst conserving limited water resources. However, developing so-called 'deficit irrigation' strategies (that are both suitable to specific crop types and local agroclimatic conditions) requires a fundamental understanding of the regulating mechanisms and relative sensitivity of different yield-determining physiological processes to soil and plant water deficits.

Monitoring and measuring these aspects is complex too. Dodd et al. (2024) present two contrasting case studies where varying either the timing (using regulated deficit irrigation, RDI) or placement (partial rootzone drying, PRD) of irrigation can enhance crop WUE. Their research demonstrates how restricting water to citrus crops at certain stages of fruit development, where the crop water status is allowed to fluctuate between defined thresholds, allows water savings to be made but without penalties to fruit growth being incurred. The use of PRD in tomato crops has also been shown to stimulate root-to-shoot signalling to increase WUE, whilst maintaining root and fruit growth at the expense of shoot growth. Dodd et al. (2024) argue that, under conditions of increasing water scarcity and resource competition, growers will increasingly adopt such techniques to maximise yield and crop quality. Understanding the sensitivity of physiological responses to changes in soil water status creates not only opportunities to reduce economic losses but also to potentially increase profits. However, Dodd et al. (2024) also emphasise the need for greater standardisation and refinement of RDI and PRD strategies through further research to account for contrasting soil and agroclimatic conditions. Technology adoption driven by precision agriculture will also result in deploying research instruments such as thermal imaging cameras in-field to help farmers detect crop canopy changes and support dynamic approaches to irrigation scheduling.

4 Improving irrigation water resources management: understanding 'efficiency', supporting farmer-led collaboration, water sharing and benchmarking performance

The World Bank (2024) recently identified a number of practical challenges hampering progress in agricultural water management. The main constraints included inadequate policies to enable change, a lack of governance and institutional under-performance and limitations in the provision of appropriate credit and finance. In addition, public and private institutions (which encompassed agricultural, environmental and water ministries, river basin authorities, irrigation agencies, water user associations and farmer organizations) were also reported to lack the enabling environment and capacities to effectively deliver on their expected functions. For example, basin authorities were found to have limited ability to enforce water allocations and to convene stakeholders to engage in multi-sector collaboration. Institutions responsible for designing and implementing irrigation development projects often focus too much on capital-intensive large-scale schemes which rely on public sector-based approaches rather than developing opportunities for small-scale private financing and irrigation management. Farmers and their local-scale organisations are also subjected to highly distorted incentive frameworks in terms of water pricing and agricultural support policies, which further hamper positive progress (World Bank, 2024).

Other challenges relate to the ongoing financial support for O&M of irrigation and drainage projects. Whilst inadequate management plays a major part in the generally low performance of many irrigation projects, a lack of sufficient funding to maintain systems inevitably results in their declining performance over time and subsequent need for rehabilitation. This results in the so-called build-neglect-rehabilitate-neglect cycle. Driving the changes needed to improve water resources allocation and management for irrigated agriculture will require concerted action across various fronts, ranging from supporting increased farmer-led collaboration to understanding how 'efficiency' needs to be interpreted across multiple scales, to implementing water trading (or sharing) and to assessing system performance through benchmarking.

With a specific focus on sub-Saharan Africa (SSA), Woodhouse (2024) considers the advances needed to support farmer-led irrigation development (FLID). Drawing on evidence from two decades of research, Woodhouse (2024) shows how the poor performance of many formal irrigation schemes in SSA has distracted attention away from the widespread adoption of irrigation by small-scale farmers. This so-called FLID has involved a variety of different techniques, including 'hill-furrow' irrigation using streams in hilly areas and the use of small motor pumps for community irrigation. Farmers have drawn on different sources of information to drive change including gaining knowledge from equipment suppliers, government and non-government technical services, and other farmers. Migrants have also been an important source of knowledge in sharing innovations that have been implemented elsewhere. Interestingly, Woodhouse (2024) suggests that, in aggregate, these small-scale farmer initiatives do appear to support much larger irrigated areas than those of formal schemes but are under-represented in official statistics that tend to focus on formal schemes characterised by permanent infrastructure.

FLID is being driven by expanding local and regional food markets as a result of urbanisation and population growth. Crop sales by households using irrigation developed by such farmer initiatives are reported to be five- to ten-fold higher than those from non-irrigating households, thereby offering a major opportunity for transformation of agricultural supply chains. This, in turn, is attracting international funding interest from agencies supporting the provision of solar pumping equipment. However, as such initiatives multiply, there will be policy and societal challenges over inequitable allocation and competition for limited water supplies, particularly between individual irrigators and between irrigation and other users, including water allocations for urban development, for hydropower and for restoring environmental flows. Woodhouse (2024) stresses that research is now urgently needed to understand both the processes of technological change being implemented by small-scale farmers and not only the scope for improved productivity but also the routes by which these interventions can be supported by scientists and policy-makers.

For FLID to be sustainable and to endure over the long term, Woodhouse (2024) identifies three salient areas for research. First, there is a need to better understand the medium-term consequences of FLID in terms of socio-economic impact, and who are the 'winners' and 'losers' and what should be done to ensure more equitable outcomes for rural communities. Secondly, how can institutions and governance be aligned to meet the organisational and regulatory challenges of small-scale irrigation. Finally, there is a question over to what extent the development of new water storage infrastructure might be needed to support future FLID initiatives and to ensure sustainable and reliable water supplies, against a potential backdrop of increasing competition, reducing water access and reliability.

One of the major contested debates over recent decades has been the misunderstanding (pitfalls and promises) of irrigation efficiency as a means to curb the rising demand for water to support the expansion of irrigated agriculture (Lankford et al., 2020). Farmers usually seek to use their water resources in the most efficient way within the constraints of their conveyance, scheduling and application systems (i.e. minimising non-beneficial losses and limiting any deterioration in water quality) and using their allocated water in the most agronomic and economically effectively way (maximising productive output, including both yield and quality). However, there is widespread evidence of the generally low levels of water efficiency and water productivity being practiced and achieved within irrigated agriculture globally. In their chapter, Lankford and McCartney (2024) turn their attention to managing the irrigation efficiency paradox in order to 'free' water for the environment. The hypothesis is that reducing irrigated agricultural water consumption will 'save' water, thereby enabling saved water to be reallocated for use by other competing sectors, including the environment to meet aquatic and ecological flow needs.

However, achieving this in practice is much harder and more complex as the local context is always specific and multi-scalar, different actors have different interests and objectives, the hydrological processes and their interactions with land use and water resources management in the catchment need to be fully understood, and there needs to be the relevant data available to order to complete water accounting in a robust way. Lankford and McCartney (2024) highlight that, without such a comprehensive approach, attempts to 'save water' within agriculture often result in no observed positive effect, or a reduction in crop production and, in some cases, paradoxically, lead to an increase in water consumption. Assuming allocations are not reduced, any 'saved water' achieved by farmers would be used to expand production. To overcome this challenge, Lankford and McCartney (2024) propose an 'irrigation savings allocation framework' (ISAF) which considers the agronomic crop water requirements and field applications, through intra-system and then system withdrawals of water. Their ISAF approach considers the irrigation consumption 'rebound effect' and reuse to reduce depletion at the river basin scale and aims to track the 'freed up' water to ensure it can be reallocated to other sectors and/or the environment.

The importance of transparent water accounting, considering scale impacts, as well as the need for appropriate institutional frameworks and governance to manage water allocations through water policy and legal frameworks, is also highlighted. Unfortunately, to date, exemplars of the successful transfer of real water savings to the environment remain limited due to different accounting methods, stakeholder disagreements over system boundaries and a lack of monitoring data to inform system accounting. It is hoped that the ISAF approach coupled with other initiatives will refocus attention on the 'irrigation efficiency paradox', helping to guide more informed policy dialogue, leading to more flexible mechanisms for abstraction management and reallocation. In many cases, it is often outdated and inflexible legal frameworks for water that no longer align with contemporary water resource challenges that are constraining innovative thinking in water management (e.g. supporting water sharing) and stifling attempts to foster multi-sectoral collaboration.

In their chapter, Haensch and Wheeler (2024) focus on the fundamentals and developments in water sharing and water trading, with reference to evidence and experiences from Australia. Although both approaches (but notably water trading) have been implemented in many countries for some time, water sharing is gaining particular interest, given its informal and farmer-led control (e.g. Chengot et al., 2021). Haensch and Wheeler (2024) stress there still remains a need for ongoing dialogue and debate to increase awareness, adoption and implementation of sharing and trading to learn lessons and to avoid potential adverse effects. They identified three key areas for attention, including building institutional and regulatory capacity, the integration of relevant actors and stakeholders into the process, and improved monitoring and control systems to facilitate trading or sharing. Their research also highlights the importance of regulatory frameworks to ensure that sustainable outcomes are achieved from both formal and informal water sharing and water trading to limit negative or unintended impacts, including increased agricultural abstractions, which is a concern from the regulatory authority's perspective when unused or underused licensed water is activated. Facilitating water sharing and trading also requires access to high-quality monitoring data on abstractions in both spatial and temporal dimensions, reemphasising the need for improved approaches not only to agricultural monitoring (e.g. using remote sensing to track consumption) but also to support regulatory enforcement (Foster et al., 2024).

Research described earlier by Rodríguez Díaz et al. (2024) focused on the impacts of irrigation modernisation on energy use in Mediterranean agriculture. To complement this, Burton et al. (2024) address the key challenges facing the modernisation of medium and large irrigation schemes in India and how to formulate effective programmes for change. This Indian perspective provides important and valuable insights of relevance to many other countries given that the externalities and 'drivers for change' in many ways are similar. These include concerns regarding the impacts of climate change on crop production and water resources availability, the impacts of population growth on water and food demands, projected reductions in available water supplies, changing demographics (increasing urbanisation) and diets (increasing and more diversified food demands), the competing demands for water, and the challenges in attaining Sustainable Development Goals (SDGs). In India and elsewhere, there have also been positive enablers of change to support modernisation with improved communications, technology and a willingness from farming communities to embrace change and to transform their agricultural systems through better management, technology and innovation.

Burton et al. (2024) describe how a national scale, multi-disciplinary program to modernise irrigation and drainage schemes across India has been implemented, including the need for policy and institutional reforms, water resources identification, agricultural reforms and upgrading of pathways for knowledge transfer and information exchange. Their analysis recognises that the institutional and management dimensions are often the most challenging aspects, as they take time, commitment and resources but are essential if modernised schemes are to be sustainable in the long run. The importance of scheme management, water user associations and the need for farmer engagement in the process are emphasised; farmers are more likely to accept and internalise the changes required by modernisation and make the process successful and sustainable if discussions and decisions are conducted in a transparent and collaborative manner. The authors therefore propose that, rather than adopting conventional approaches to irrigation modernisation that typically follows an existing blueprint, instead a 'cafeteria' approach is recommended, applying technology and support that is most appropriate to each specific scheme. Particularly important is the need to adopt a phased approach, the consideration of farmers' ability and willingness to change and achieving progressive capacity development of key personnel involved in the modernisation process.

5 Future climate and water-related risks: the role of precision irrigation, management and technology to address the nexus challenges of agriculture

Globally, irrigation accounts for >70% of total freshwater withdrawals and >90% of total consumptive water use (Siebert et al., 2010). However, allocations of water for agriculture and the need to increase irrigation efficiency are under increased scrutiny, driven by public and regulatory concerns regarding the consequences of over-abstraction on the environment and water resources sustainability (Knox et al., 2018; Lankford et al., 2020). Rising competition for water resources and demands for greater environmental protection means that irrigated production is under severe pressure to improve WUE and reduce its environmental footprint. The major yield benefits of irrigation over the last few decades have often been overshadowed by adverse effects on natural resources, threatening its long-term productive potential. For example, land degradation and large-scale salinisation of irrigated areas are often linked to irrigation development, and numerous examples in the scientific literature describe the negative impacts of unsustainable abstraction for irrigation on the aquatic environment (e.g. Ketchum et al., 2023).

The importance of water for irrigated agriculture and its contribution to national economies in the Mediterranean region is well known and documented (e.g. Daccache et al., 2012), but irrigated agriculture is one of the sectors that is most vulnerable to water scarcity. In their chapter, Ortuño et al. (2024) describe the development of a decision support tool termed PRECIMED, which aims to support farmers in raising the efficiency of both irrigation and fertiliser management. The PRECIMED DSS offers end users a platform through which they can deploy and integrate a range of IoT devices and web services from different sources, with advanced agronomic services. The outputs from the system were validated in four different Mediterranean countries, including Spain, Algeria, Tunisia and Greece, under both greenhouse (protected cropping) and open-field scale cropping systems.

Ortuño et al. (2024) recognise that DSS development and implementation is a time-consuming, resource-intensive and data-demanding activity and requires continual development for future sustainability. It also requires extensive ongoing farmer engagement and support to realise the benefits of the DSS for on-farm decision-making. New technologies will need to be integrated over time, including, for example, from satellites, updating hardware designs to optimise energy consumption, configuring IoT devices with sufficient computing power to support embedded AI algorithms and incorporating technologies (e.g. blockchain) to secure data. They also consider how the DSS platform will need to evolve to meet future farmer needs and what business model options based on the cloud software as a service (SaaS) distribution model (e.g. Freemium, pay-as-you-go, subscription or tiered fixed fee) will be most suitable for commercial exploitation.

As the costs for water, fertilisers and energy continue to rise, the development of DSS such as PRECIMED will also support greater interest and implementation of precision irrigation technologies to improve water and nutrient use efficiencies, to reduce the environmental impacts of irrigation and to maximise yields and crop quality. Daccache et al. (2024) describe the challenges and approaches for integrating both biophysical and ballistics-based water distribution models to assess the agronomic and environmental impacts of precision irrigation (PI). Whilst PI offers scope to save water, improve yields and support the sustainable intensification of agriculture, it could also help reduce environmental impacts of food production linked to nutrient leaching and GHGs.

Daccache et al. (2024) note that, although many modelling studies have been undertaken to assess the potential benefits of precision agriculture, most have ignored the fundamentally important impacts that in-field soil heterogeneities and irrigation non-uniformity have on crop growth and productivity. The different approaches (both deterministic and empirical) that have been developed to simulate irrigation water distribution are explained and a modelling framework is then developed that couples a dynamic ballistic model with a crop growth model (Aquacrop). The application of this integrated modelling approach is described with reference to onions grown in the UK, to quantify the agronomic and environmental impacts of irrigation heterogeneity on crop yield, nutrient uptake and deep drainage. Changing irrigation equipment and management practices and different operational modes (conventional versus PI) on crop production are demonstrated. The need for further research is highlighted, specifically incorporating farm economic models to assess the financial impacts of different irrigation management strategies on production (to inform investment in PI), to evaluate the use of remote sensed crop cover data for improving model calibration and to assess how soil monitoring techniques such as electromagnetic induction (EMI) scanning might help define soil and irrigation management zones. Daccache et al. (2024) also consider how such modelling frameworks can support training and knowledge transfer to farmers to highlight the agronomic, water resource and environmental benefits associated with precision irrigation.

Climate change is a serious threat to crop productivity in regions that are already food insecure (Knox et al., 2012a). One of the major challenges facing the future sustainability of irrigated agriculture will be coping with the direct and indirect impacts of increased climate variability due to a changing climate and the increased frequency, magnitude and duration of droughts. These issues are discussed in detail by Ray-Shyan Wu and Hussain (2024) in their chapter on managing the impacts of climate change, droughts and water scarcity on agriculture. They highlight various strategies available, including adaptation, mitigation, improved water management, the importance of good governance and policy and international cooperation. Indeed, a recent UN study has estimated that crop yields could decline by up to 30% in some regions due to drought and that, by 2025, nearly 2 billion people will be living in countries or regions with absolute water scarcity.

Although various adaptation strategies have been proposed, in reality their potential effectiveness is not well understood. Whilst adaptation strategies such as improved crop rotations, conservation tillage and water storage might help make agricultural systems more resilient to changes in water availability, they may not be sufficient to fully address the impacts of recurrent droughts and water scarcity. This is particularly the case for regions or countries with limited capacity to adapt. For example, SSA is often cited as one of the most vulnerable regions (Slingo et al., 2005) since a significant portion of its national economies are dependent on agriculture (Schlenker and Lobell, 2010), and most of its available water resources (85%) are used for agriculture (Downing et al., 1997). Farming techniques are also relatively primitive, the majority of the continent is arid and the smallholder systems that dominate the agricultural landscape have very limited capacity to adapt (Müller et al., 2011). Governments, international agencies and the private sector will also have an increasingly important role to play in supporting smallholders, by providing financial resources, technical expertise and policy support and access to information, training and technology to promote sustainable and climate-resilient agricultural practices.

It is apposite that the final chapter in this book by Melo et al. (2024) brings together the WEF nexus to evaluate their interdependencies and trade-offs. The impacts of food crop production on our water environment are tightly coupled and depend on a range of factors including cropping patterns, agricultural practices, agroclimate and the status of water resources (Hess et al., 2014). There is, therefore, a need for integrated approaches that consider water, energy, food and environment components from both governance and management perspectives (Mahlknecht et al., 2020). Considering that projected global demands for food, water and energy by 2030 are expected to increase by 35%, 40% and 50%, respectively, the interlinkages between these discrete sectors require an integrated approach to ensure that both future food and water security, and sustainable agricultural and energy production needs are both balanced and sustainable (Momblanch et al., 2019).

Melo et al. (2024) recognise the extensive literature published on the WEF nexus and therefore aim to illustrate the WEF concept through selected examples of each element (water - energy, energy - food, and food - water) using three Andean country case studies to analyse the trade-offs between the elements that arise when designing sustainable development strategies. Two case studies use a water–energy–food–environment modelling toolkit

(Correa-Cano et al., 2022) which is applied to agricultural development in Peru and Ecuador. The third case study demonstrates a water–food–energy model that links water used in agriculture with hydroelectricity to evaluate climate change adaptation strategies. Their research also highlights the benefits of a nexus approach to support more integrated and holistic frameworks to inform water research and policy, and the importance of considering legal aspects, knowledge and communication, policy integration and governance.

6 Conclusions

6.1 Role of irrigation for crop production and food security

Over the last five decades, major changes in agricultural land management have underpinned substantial increases in crop productivity. Prior to 1960, most gains were attributed to increases in agricultural land area. But most yield gains (ca. 70%) have arisen mainly from so-called Green Revolution technologies, including the introduction of high-yielding cultivars, higher levels of inputs and dependence on chemical fertilisers, pesticides, farm mechanisation, and, importantly, the expansion of irrigation (Foley et al., 2005). For example, in the UK, average yields in 1948 were ca. 3 t ha^{-1}, today they are around 8 t ha^{-1} and often near to 10 t ha^{-1}, especially where supplemental irrigation is applied. Consequently there have been significant increases in global fertiliser use and the total areas irrigated.

Although development has slowed in recent years, irrigated croplands have nevertheless increased from ca. 100 M ha in 1950 to ca. 367 M ha in 2018 (Lal, 2009; Siebert et al., 2013b; Meier et al., 2018). They are also reported to be roughly twice as productive as non-irrigated or rainfed croplands. In developing countries, irrigated areas typically account for 20% of the total area but provide 40% of production (and 60% of cereal production) (UN World Water Development Report, 2003). With more than a third of the world's food being produced on irrigated croplands (Goodwin and O'Connell, 2008), global food supply, in terms of quantity, quality and diversity, continues to be sustained by irrigation. However, meeting future food demands will be increasingly challenging, given the rising global population with changing dietary needs and competing demands for freshwater resources, especially in arid and semi-arid regions, where irrigated production is concentrated.

6.2 Have we passed the planetary boundary for freshwater?

The concept of planetary boundaries (PBs) aimed to define the 'safe operating space for humanity' by identifying the environmental processes that regulate Earth system resilience, allowing for non-human uses of water including environmental flows (Rockström et al., 2009). A number of planetary boundaries have been defined, including the maximum amount of freshwater that can be appropriated for use by society. This original threshold was initially set at 4000 km^3 $year^{-1}$ for the global consumption of blue water abstracted from surface and groundwater sources, including reservoir storage. However, further analysis that accounted for environmental flow requirements to maintain aquatic dependent ecosystems in a more spatially explicit manner proposed a PB water threshold of 2800 km^3 $year^{-1}$ with an uncertainty range of 1100–4500 km^3 $year^{-1}$ (Gerten et al., 2013). Exceeding the upper threshold was reported by Rockström et al. (2009) to significantly increase the risk of a collapse in terrestrial and aquatic ecosystems.

However, current blue water consumption (1700–2270 km^3 $year^{-1}$) already exceeds the lower end of this planetary boundary range (Gerten et al., 2013). The situation is exacerbated when we also consider the additional future food demand required to meet a growing global population. Indeed the FAO (2017) reported that food production must increase by ca. 50% by 2050 to meet the demand of the world's growing population. To achieve this, blue water withdrawals for agriculture would need to increase by +65% if we were to continue producing and consuming food as we do today (rising from a present estimate of 1810 km^3 $year^{-1}$ to 2987 km^3 $year^{-1}$ by 2050) (Rockström et al., 2003; Springmann et al., 2018). Collectively, these analyses raise a number of fundamentally important questions. These include

- 'Have we already passed the planetary boundary for freshwater?
- 'How and where will our future food demand be met?
- as Hess (2020) stated, 'How can we avoid eating ourselves out of water?'

Over a decade ago, Fereres et al. (2011) posed a similar question. Before we ask, 'Will there be enough food to feed a growing population?' maybe we should ask, 'Will there be enough water to grow that food?' Global freshwater withdrawals are perilously close to the planetary boundary.

However, recent research by Gerten et al. (2020) offers some hope. They reported that a transformation towards more sustainable production and consumption patterns could support >10 billion people within four planetary boundaries (biosphere integrity, land-system change, freshwater use, nitrogen flows), but there would need to be significant changes in current societal and water management practices. Given that, at present, nearly half of our global food production is transgressing these planetary boundaries, the prerequisites for achieving water-environmental-food sustainability will require a spatial redistribution of cropland areas, significant improvements in water (irrigation) and nutrient management, coupled with reducing food waste and modifying diets (Gerten et al., 2020).

As a society, we will need to increase food production with less water and less impact, recognise that (over the next 25 years) demand for water for food will increase and water supplies for food production will decrease, the availability of sustainable freshwater will be a significant constraint on future global food production, and we cannot continue to produce and consume food in the same way that we have done in the past. What is certain is that we will need another 'Green Revolution' but one that happens in half the timeframe and one that is underpinned by 'green' technology, management and investment principles. As Fereres et al. (2011) stated, given the future uncertainties in both water availability and reliability and its implications for agricultural production and food security, research efforts will need to be escalated to unprecedented levels in the near future, as it is evident that time is running out. Reconciling the water needs for food, the environment, energy and society remains a significant but surmountable challenge – it is not only a quadrilemma facing the future of global irrigated agriculture but also an opportunity for change.

7 References

Amarasinghe, U. A., Sikka, A. and Amaranath, G. (2024). Forecasting future global agricultural water demand: current projections and future challenges. In: *Improving Water Management in Agriculture: Irrigation and Food Production*. Burleigh Dodds Publishing, UK.

Burns, M. J., Wallis, E. and Matic, V. (2015). Building capacity in low-impact drainage management through research collaboration. *Freshwater Science* 34(3), 1176–1185.

Burton, M., Shivakumar, M. G., Sen, R. and Palanisami, K. (2024). Modernizing major and medium irrigation schemes in India: formulating a program of change. In: *Improving Water Management in Agriculture: Irrigation and Food Production*. Burleigh Dodds Publishing, UK.

Chengot, R., Knox, J. W. and Holman, I. P. (2021). Evaluating the feasibility of water sharing as a drought risk management tool for irrigated agriculture. *Sustainability* 13(3), 1–16. https://doi.org/10.3390/su13031456.

Cisneros, J. B. E., Oki, T., Arnell, N. W., Benito, G., Cogley, J. W., Döll, P., Jiang, T. and Mwakalila, S. S. (2014). Freshwater resources. In: Field C. B., Barros V. R., Dokken D. J., Mach K. J., Mastrandrea M. D., Bilir T. E., Chatterjee M., Ebi K. L., Estrada Y. O., Genova R. C., Girma B., Kissel E. S., Levy A. N., MacCracken S., Mastrandrea P. R., and White L. (Eds.), *Climate Change 2014 – Impacts, Adaptation and Vulnerability: Part A: Global and Sectoral Aspects. Contribution of Working Group II to the Fifth Assessment Report of the Intergovernmental Panel on Climate Change,* Cambridge University Press, pp. 229–269. https://doi.org/10.1017/CBO9781107415379.008.

Correa-Cano, M. E., Salmoral, G., Rey, D., Knox, J. W., Graves, A., Melo, O., Foster, W., Naranjo, L., Zegarra, E., Johnson, C., Viteri-Salazar, O. and Yan, X. (2022). A novel modelling toolkit for unpacking the water-energy-food-environment (WEFE) nexus of agricultural development. *Renewable and Sustainable Energy Reviews* 159, 112182. https://doi.org/10.1016/j.rser.2022.112182.

Daccache, A., Ciurana, J. S., Rodriguez Diaz, J. A. and Knox, J. W. (2014). Water and energy footprint of irrigated agriculture in the Mediterranean region. *Environmental Research Letters* 9(12), 124014.

Daccache, A., Knox, J. W. and Hess, T. M. (2024). Integrating biophysical and hydraulic models to assess the agronomic and environmental impacts of precision irrigation. In: *Improving Water Management in Agriculture: Irrigation and Food Production*. Burleigh Dodds Publishing, UK.

Dodd, I. C., Huntenburg, K. and Pérez-Pérez, J. G. (2024). Crop physiological responses to irrigation and regulatory mechanisms. In: *Improving Water Management in Agriculture: Irrigation and Food Production*. Burleigh Dodds Publishing, UK.

Downing, T. E., Ringuis, L., Hulme, M. and Waughray, D. (1997). Adapting to climate change in Africa. *Mitigation and Adaptation Strategies for Global Change* 2(1), 19–44.

FAO AQUASTAT (2016). *Food and Agriculture Organisation (FAO) Global Information System on Water and Agriculture*. Available at: https://www.fao.org/aquastat/en/databases/.

FAO (2017). *The Future of Food and Agriculture – Trends and Challenges*. FAO, Rome. Available at: http://www.fao.org/3/a-i6583e.pdf.

Felbab-Brown, V. (2017). *Water Theft and Water Smuggling: Growing Problem or Tempest in a Teapot?* Brookings Institution, Washington, D.C.

Fereres, E., Orgaz, F. and Gonzalez-Dugo, V. (2011). Reflections on food security under water scarcity. *Journal of Experimental Botany*. https://doi.org/10.1093/jxb/err165.

Foley, J. A., Defries, R., Asner, G. P., Barford, C., Bonan, G., Carpenter, S. R., Chapin, F. S., Coe, M. T., Daily, G. C., Gibbs, H. K., Helkowski, J. H., Holloway, T., Howard, E. A., Kucharik, C. J., Monfreda, C., Patz, J. A., Prentice, I. C., Ramankutty, N. and Snyder, P. K. (2005). Global consequences of land use. *Science* 309(5734), 570–574.

Foster, T., Mieno, T. and Brozović, N. (2020). Satellite-based monitoring of irrigation water use: assessing measurement errors and their implications for agricultural water management policy. *Water Resources Research* 56(11), e2020WR028378.

Foster, T., Pérez-Blanco, C. D. and Schmidt, G. (2024). Monitoring agricultural water use: challenges and solutions for sustainable water management. In: *Improving Water Management in Agriculture: Irrigation and Food Production*. Burleigh Dodds Publishing, UK.

Gerten, D., Hoff, H., Rockström, J., Jägermeyr, J., Kummu, M. and Pastor, A. V. (2013). Towards a revised planetary boundary for consumptive freshwater use: role of environmental flow requirements. *Current Opinion in Environmental Sustainability* 5(6), 551–558. https://doi.org/10.1016/j.cosust.2013.11.001.

Gerten, D., Heck, V., Jägermeyr, J., Bodirsky, B. L., Fetzer, I., Jalava, M., Kummu, M., Lucht, W., Rockström, J., Schaphoff, S. and Schellnhuber, H. J. (2020). Feeding ten billion people is possible within four terrestrial planetary boundaries. *Nature Sustainability* 3(3), 200–208. https://doi.org/10.1038/s41893-019-0465-1.

Ghumman, A. R., Ghazaw, Y. M., Hashmi, H. N., Kamal, M. A. and Niazi, M. F. (2012). Environmental and socio-economic impacts of pipe drainage in Pakistan. *Environmental Monitoring and Assessment* 184(3), 1671–1681.

Goodwin, I. and O'Connell, M. G. (2008). The future of irrigated production horticulture – world and Australian perspective. *Acta Horticulturae* 792, 449–458.

Haensch, J. and Wheeler, S. (2024). Developments in water sharing and water trading. In: *Improving Water Management in Agriculture: Irrigation and Food Production*. Burleigh Dodds Publishing, UK.

Haruna, S. I. and Anderson, S. H. (2024). Agronomic practices for optimising soil water retention. In: *Improving Water Management in Agriculture: Irrigation and Food Production*. Burleigh Dodds Publishing, UK.

Hedley, C. B., Knox, J. W., Raine, S. R. and Smith, R. (2014). Water: advanced irrigation technologies. In: Van Alfen, N. (Editor-in-Chief). *Encyclopaedia of Agriculture and Food Systems* (vol. 5). Elsevier, San Diego, pp. 378–406.

Hess, T., Aldaya, M., Fawell, J., Franceschini, H., Ober, E., Schaub, R. and Schulze-Aurich, J. (2014). Understanding the impact of crop and food production on the water environment–using sugar as a model. *Journal of the Science of Food and Agriculture* 94(1). https://doi.org/10.1002/jsfa.6369.

Hess, T. M. (2021). How can we avoid eating ourselves out of water? *Nature Food* 2(4), 225. https://doi.org/10.1038/s43016-021-00258-0. PMID: 37118461.

Ketchum, D., Hoylman, Z.H., Huntington, J. et al. (2023) Irrigation intensification impacts sustainability of streamflow in the Western United States. *Commun Earth Environ* 4, 479. https://doi.org/10.1038/s43247-023-01152-2.

IPCC (2014) *Climate Change 2014: Synthesis Report. Contribution of Working Groups I, II and III to the Fifth Assessment Report of the Intergovernmental Panel on Climate Change* [Core Writing Team, Pachauri, R. K. and Meyer, L. A. (Eds)]. IPCC, Geneva, 151 pp.

Knox, J. W., Hess, T. M., Daccache, A. and Wheeler, T. (2012a). Climate change impacts on crop productivity in Africa and South Asia. *Environmental Research Letters* 7(3), 034032.

Knox, J. W., Kay, M. G. and Weatherhead, E. K. (2012b). Water regulation, crop production and agricultural water management – understanding farmer perspectives on irrigation efficiency. *Agricultural Water Management* 108, 3–8.

Knox, J. W., Haro-Monteagudo, D., Hess, T. M. and Morris, J. (2018). Identifying trade-offs and reconciling competing demands for water: integrating agriculture into a robust decision-making framework. *Earth's Future* 6(10), 1457–1470.

Lal, R. (2009). Soils and world food security. *Soil and Tillage Research* 102(1), 1–4.

Lankford, B., Closas, A., Dalton, J., López Gunn, E., Hess, T., Knox, J. W., van der Kooij, S., Lautze, J., Molden, D., Orr, S., Pittock, J., Richter, B., Riddell, P. J., Scott, C. A., Venot, J., Vos, J. and Zwarteveen, M. (2020). A scale-based framework to understand the promises, pitfalls and paradoxes of irrigation efficiency to meet major water challenges. *Global Environmental Change*. https://doi.org/10.1016/j.gloenvcha.2020.102182.

Lankford, B. and McCartney, M. (2024). Managing the irrigation efficiency paradox to "free" water for the environment. In: *Improving Water Management in Agriculture: Irrigation and Food Production*. Burleigh Dodds Publishing, UK.

Loch, A., Pérez-Blanco, C. D., Carmody, E., Felbab-Brown, V., Adamson, D. and Seidl, C. (2020). Grand theft water and the calculus of compliance. *Nature Sustainability* 3(12), 1012–1018.

Mahlknecht, J., González-Bravo, R. and Loge, F. J. (2020). Water-energy-food security: a Nexus perspective of the current situation in Latin America and the Caribbean. *Energy* 194, 116824. https://doi.org/10.1016/j.energy.2019.116824.

Meier, J., Zabel, F. and Mauser, W. (2018). A global approach to estimate irrigated areas – a comparison between different data and statistics. *Hydrology and Earth System Sciences* 22(2), 1119–1133. https://doi.org/10.5194/hess-22-1119-2018.

Melo, O., Vendrell, S. and Foster, W. (2024). The water-energy-food nexus (WEF). In: *Improving Water Management in Agriculture: Irrigation and Food Production*. Burleigh Dodds Publishing, UK.

Momblanch, A., Papadimitriou, L., Jain, S. K., Kulkarni, A., Ojha, C. S. P., Adeloye, A. J. and Holman, I. P. (2019). Untangling the water-food-energy-environment nexus for global change adaptation in a complex Himalayan water resource system. *Science of the Total Environment* 655, 35–47.

Monaghan, J. M., Daccache, A., Vickers, L. H., Hess, T. M., Weatherhead, E. K., Grove, I. G. and Knox, J. W. (2013). More 'crop per drop' – constraints and opportunities for precision irrigation in European agriculture. *Journal of the Science of Food and Agriculture* 93(5), 977–980.

Müller, C., Cramer, W., Hare, W. L. and Lotze-Campen, H. (2011). Climate change risks for African agriculture. *Proceedings of the National Academy of Sciences of the United States of America* 108(11), 4313–4315.

Netzeroclimate (2024). *Land Use and Agriculture*. Available at: https://netzeroclimate.org/sectors/agriculture/#:~:text=Achieving%20net%20zero%20agriculture%20requires,about%20net%20zero%20agriculture%20quickly.

OECD/FAO (2020). *OECD-FAO Agricultural Outlook 2020-2029*. OECD Publishing, Paris/FAO, Rome. https://doi.org/10.1787/1112c23b-en.

Ortuño, m. F., Skarmeta, A., Katsoulas, N., Semiani, M., Boukadi, K., Parra, A., Gómez-Bellot, M. J., Mora, M., Marín, R., Faliagka, S., Papanastasiou, D. K., Belkhiri, F. E., Dalila, S., Grati, R., Bouaziz, B. and Alarcón, J. J. (2024). PRECIMED: development of a decision support system (DSS) for precision irrigation in Mediterranean agriculture. In: *Improving Water Management in Agriculture: Irrigation and Food Production*. Burleigh Dodds Publishing, UK.

Ritzema, H. (2024). Advances in drainage design and management for irrigated agriculture. In: *Improving Water Management in Agriculture: Irrigation and Food Production*. Burleigh Dodds Publishing, UK.

Rockström, J. (2003). Water for food and nature in drought-prone tropics: vapour shift in rain-fed agriculture. *Philosophical Transactions of the Royal Society of London. Series B, Biological Sciences* 358(1440), 1997–2009.

Rockstrom, J., Steffen, W., Noone, K., Persson, Å., Chapin, F. S. I. III, Lambin, E., Lenton, T. M., Scheffer, M., Folke, C., Schellnhuber, H. J., Nykvist, B., de Wit, C. A., Hughes, T., van der Leeuw, S., Rodhe, H., Sörlin, S., Snyder, P. K., Costanza, R., Svedin, U., Falkenmark, M., Karlberg, L., Corell, R. W., Fabry, V. J., Hansen, J., Walker, B., Liverman, D., Richardson, K., Crutzen, P. and Foley, J. (2009). Planetary boundaries: exploring the safe operating space for humanity. *Ecology and Society* 14(2), 32. Available at: http://www.ecologyandsociety.org/vol14/iss2/art32/.

Rodríguez Díaz, J. A., Mérida García, A., González Perea, R. and Fernández García, I. (2024). Managing energy demands in irrigated agriculture. In: *Improving Water Management in Agriculture: Irrigation and Food Production*. Burleigh Dodds Publishing, UK.

Schlenker, W. and Lobell, D. B. (2010). Robust negative impacts of climate change on African agriculture. *Environmental Research Letters* 5(1). https://doi.org/10.1088/1748-9326/5/1/014010.

Siebert, S., Henrich, V., Frenken, K. and Burke, J. (2013a). Update of the digital global map of irrigation areas to version 5. https://doi.org/10.13140/2.1.2660.6728.

Siebert, S., Kummu, M., Porkka, M., Döll, P., Ramankutty, N. and Scanlon, R. (2013b). A global dataset of the extent of irrigated land from 1900 to 2005. *Hydrology and Earth System Sciences Discussions* 11, 13207–13258. https://doi.org/10.5194/hessd-11-13207-2014.

Slingo, J. M., Challinor, A. J., Hoskins, B. J. and Wheeler, T. R. (2005). Introduction: food crops in a changing climate. *Philosophical Transactions of the Royal Society of London. Series B, Biological Sciences* 360(1463), 1983–1989.

Springmann, M., Clark, M., Mason-D'Croz, D., Wiebe, K., Bodirsky, B. L., Lassaletta, L., de Vries, W., Vermeulen, S. J., Herrero, M., Carlson, K. M., Jonell, M., Troell, M., DeClerck, F., Gordon, L. J., Zurayk, R., Scarborough, P., Rayner, M., Loken, B., Fanzo, J., Godfray, H. C. J., Tilman, D., Rockström, J. and Willett, W. (2018). Options for keeping the food system within environmental limits. *Nature* 562(7728), 519–525.

Thenkabail, P. S., Knox, J. W., Ozdogan, M., Gumma, M. K., Congalton, R., Wu, Z., You, S., Milesi, C., Giri, C., Nagler, P., Finkral, A., Marshall, M. and Mariotto, I. (2012). Assessing future risks to agricultural productivity, water resources and food security: how can remote sensing help? *Photogrammetric Engineering and Remote Sensing*, 773–782.

Tilman, D. (1999). Global environmental impacts of agricultural expansion: the need for sustainable and efficient practices. *Proceedings of the National Academy of Sciences of the United States of America* 25; 96(11), 5995–6000.

UN World Water Development Report (2003). *Water for People, Water for Life*. Available at: https://www.unwater.org/publications/un-world-water-development-report-2003.

Watto, M. A., Shabir, M., Sher, A. and Ali Khan, M. A. (2024). Solar powered irrigation systems: recent developments and future trends. In: *Improving Water Management in Agriculture: Irrigation and Food Production*. Burleigh Dodds Publishing, UK.

Woodhouse, P. (2024). Advances in farmer-led irrigation development in Africa. In: *Improving Water Management in Agriculture: Irrigation and Food Production*. Burleigh Dodds Publishing, UK.

World Bank (2019). *Water Withdrawals for Agriculture as a Proportion (%) of Total Freshwater Use*. Available at: https://data.worldbank.org/indicator/ER.H2O.FWAG.ZS.

World Bank (2024). *Water in Agriculture*. Available at: https://www.worldbank.org/en/topic/water-in-agriculture.

Wu, R. S. and Hussain, F. (2024). Managing climate change, droughts and water scarcity affecting agriculture.

Chapter 2

Monitoring agricultural water use: challenges and solutions for sustainable water management

T. Foster, University of Manchester, UK; C. D. Pérez-Blanco, Universidad de Salamanca, Spain; and G. Schmidt, Fresh Thoughts Consulting, Austria*

1 Introduction

Water scarcity is a growing challenge facing many areas of the world, negatively impacting food security, economic development, human and ecosystem health (Dolan et al., 2021). As the largest sectoral user of water in most water-scarce regions, agriculture is central to efforts to address the environmental and socio-economic challenges posed by water scarcity. Agricultural productivity is heavily dependent on access to reliable and affordable water supplies for irrigation. This dependence is expected to grow further in the coming decades as demands increase to intensify food production, and warming temperatures simultaneously drive higher crop water requirements (Rosa, 2022). In parallel, there is also a growing recognition that continued unimpeded growth in agricultural water abstractions would have severe negative consequences for freshwater-dependent ecosystems and other human uses of water such as urban supplies, industry, rainfed agriculture and freshwater fisheries (McDermid et al., 2021; Schmitt et al., 2022; van Maanen et al., 2022).

* Corresponding author: timothy.foster@manchester.ac.uk.

http://dx.doi.org/10.19103/AS.2023.0123.26

Much of the academic and policy debates around agricultural water use have historically focussed on how to mitigate water scarcity and conflicts through improved management. At the farm level, narratives have centred around goals of improving agricultural water productivity to achieve 'more crop per drop' (Giordano et al., 2021). Proposed solutions include innovations in heuristics and technologies to better target how irrigation is applied and scheduled (McCarthy et al., 2020; Kelly et al., 2023), optimising crop choices (Beyer et al., 2022; Richter et al., 2023) or via development of more drought tolerant cultivars (Passioura, 2006). At larger catchment or landscape scales, extensive work has evaluated policies and strategies to sustainably and equitably manage trade-offs between agricultural and non-agricultural (environmental and human) water needs. These include the design of demand management instruments (including caps and economic instruments, such as water markets or charges) that reallocate water among alternative uses (Rey et al., 2019; Deines et al., 2021; Koebele et al., 2022), along with supply enhancement policies that rely on green or grey infrastructures to expand the water supply base available to both agricultural and non-agricultural users (van Vliet et al., 2021; Schmitt et al., 2022; Sun et al., 2023).

Central to all efforts to ensure food security, sustainability and economic efficiency under conditions of water scarcity is the ability to spatially and temporally monitor agricultural water use accurately and reliably (Foster et al., 2020). Information about where and how much water is used by agriculture is essential to accurately assess the sector's contribution to a region's water budget, to design or enforce policies to achieve more sustainable levels of agricultural abstractions, to guide or incentivise farmers on ways to improve on-farm water management practices and productivity and to identify and plan investments to mitigate agricultural water insecurity. More broadly, monitoring is also essential to help ensure that policy interventions do not backfire in ways that ultimately end up further exacerbating scarcity and inter-sectoral conflicts over water. For example, shifts towards more efficient irrigation technologies have often been assumed by planners and policymakers to reduce water use when, in fact, monitoring reveals that they more commonly have had no impact on consumptive use or have led to increased water use through expansion of irrigated lands and/or switching to more water-intensive crops (Pérez-Blanco et al., 2021).

While the value of monitoring agricultural water use is clearly apparent, attention given to this topic in policy and research has grown mostly in the last decade. Recent research has shown that even countries with highly developed systems of environmental regulation still lack robust and actionable monitoring of agricultural water use, while other regions collect little or no data at all on agricultural water use (Foster et al., 2020). The implications of these monitoring gaps for water management and policy are increasingly being recognised and highlighted. For example, monitoring gaps have been cited as a key enabler of illegal or unauthorised water use, which by some estimates account for up to 30–50% of total abstractions and are a key contributor to water scarcity pressures globally (Felbab-Brown, 2017; Loch et al., 2020). A lack of monitoring also limits the evidence base and can favour the consolidation of ill-informed systems of beliefs (e.g. that water availability can be increased through improvements to irrigation efficiency – Lankford et al., 2020) or failure to learn from past policies (e.g. failures of large-scale irrigation infrastructure development projects – Higginbottom et al., 2021). The need to strengthen monitoring of agricultural water use is also being acknowledged by governments and the agricultural industry, with growing interest in the potential for innovative monitoring solutions such as satellite remote sensing (Massari et al., 2021) and smart metering (Ouvrard et al., 2023) to support agricultural water management and stewardship.

In this chapter, we contribute to debates around agricultural water use management by reviewing the features, implementation and innovations of irrigation monitoring systems and approaches globally. The following sections of this chapter focus on reviewing historic approaches for identifying and quantifying agricultural water use, followed by a critical discussion of the opportunities provided by recent innovations in satellite-based monitoring approaches. We then discuss the need to think beyond improvements in data collection and reporting to consider how agricultural water use policies are enacted and enforced, if improvements in sustainability and efficiency of water use are to be achieved. The chapter draws on the authors' extensive experience in different irrigated regions across Europe, North America and other parts of the world, drawing recommendations about the opportunities and challenges for strengthening agricultural water use monitoring and accounting globally.

2 Identifying and detecting where water is abstracted

An essential starting point for monitoring agricultural water use is knowledge and intelligence about the locations of agricultural abstractions, including where rights to water exist and how these change and evolve over time. Knowledge of where irrigation abstractions are permitted is an essential part of any water management system, both for understanding how agriculture is influencing catchment water budgets and for enforcing demand management policies and instruments.

To this end, many countries require farmers to register and obtain permission for abstraction before water can be taken for irrigation. Typically, this process involves a farmer making an application to the relevant environmental regulator or agency to obtain a permit or licence for a water right before a new abstraction point is developed and made operational by a farmer. Based on local rules and environmental assessments (e.g. to determine that the abstraction will not negatively impact ecosystems or infringe existing abstraction rights of other users), an abstraction permit or right may then be granted. Abstraction licences may include stipulations about the water source, where, when and/or how much (including peak daily, monthly or annual rates) water can be abstracted. These conditions can be specified as hard limits on the rate or cumulative amount of abstraction in a given year or may be linked to the hydrologic condition of the water source. For example, in the UK many agricultural licences specify the maximum volume of water that can be abstracted each year while also curtailing abstraction in the event that streamflow or groundwater levels drop below minimum thresholds, known as 'Hands off Flows' or 'Hands off Levels' (Erfani et al., 2015). Alternatively, licences may instead restrict farmers' water use based on average levels of abstraction over multiple years, allowing irrigators to adjust water use from 1 year to the next in line with climate conditions and crop water needs. For example, in the Po River Basin in northern Italy, the majority of water licences specify only the average water use restriction (Santato et al., 2016), with similar rules also common in parts of the United States such as the states of Nebraska and Kansas (Young et al., 2021), where groundwater is the primary source of water for irrigation.

While abstraction registries are, on paper, an essential starting point for effective monitoring, the quality and completeness of official irrigation abstraction registries are highly variable in reality. Large data gaps often exist in abstraction registries even in countries with strong environmental compliance systems and laws. Farmers, for example, may decide to develop an abstraction point without registering or applying for an official licence to bypass costs associated with registration or subsequent fees for water access, where procedures are overly onerous or bureaucratic, or where moratoriums exist on registering new licences for water abstraction (ECA, 2021). These problems are especially pronounced for groundwater abstractions where the infrastructure (typically a pump and well) required to access water for irrigation is privately owned (Foster et al., 2020; Molle and Closas, 2020) as opposed to surface water irrigation that often depends in part on public or shared infrastructure such as canal networks. Indeed, challenges of identifying and deterring illegal groundwater well development have been widely reported around the world, including in countries such as Spain (Novo et al., 2015), Australia (Holley et al., 2020) and Mexico (Hoogesteger and Wester, 2017), to name only a select few examples.

The likelihood of abstractions not being officially registered is, in part, influenced by the extent to which a farmer expects non-compliance with permitting or registration requirements to be punished or penalised. Where consequences are expected to be below the returns from non-compliance, there are greater risks of unauthorised abstractions being developed. Over time, this may, in turn, serve to further delegitimise or undermine compliance with abstraction registration rules. This is especially common in countries or regions with weak environmental protection and governance systems. In countries such as India and Pakistan, for example, private groundwater well drilling for irrigation is now the de facto standard practice, with little expectation that these wells will be officially registered or permitted, even though this may still be required on paper (Molle and Closas, 2020b). However, issues of non-compliance with abstraction permitting are also common in countries where strong environmental protections exist but are inconsistently enforced. For example, in Doñana in Southern Spain, 2000 cases of illegal water abstractions were reported in the period 2003–2020. Yet, despite the scale of illegal use and its substantial human and ecological impacts, official data report 28 guilty verdicts for water theft equating to a prosecution rate of 2.2% (Loch et al., 2020) – a rate that is not unique to Doñana or Spain (Schmidt et al., 2020), highlighting the often limited costs of non-compliance with abstraction registration rules for farmers.

Nonetheless, it is important to recognise that incomplete abstraction registries are not purely due to farmers deliberately seeking to bypass official rules. Data gaps also reflect challenges and deficiencies in the institutions and processes of governance of agricultural water use. A common challenge is that responsibility for maintaining abstraction records often falls to different agencies depending on the source of water or when an abstraction point was first developed. In other cases, abstraction rights may be recorded, but critical information may often fail to be captured (e.g. coordinates of an abstraction point), making it hard to track and monitor that water user in the future. This is especially common for older water rights for which requirements to register an abstraction may not have existed at the time the abstraction was originally developed, or for which records may only exist in paper form. In Spain, the 1985 Water Act aimed at transitioning from private to public ownership of groundwater, and created a 'Public Water Registry' and a 'Catalogue of Private Waters' to keep track of the ownership and water withdrawn. Yet this database is far from complete almost 40 years after its creation – and thus the characteristics of some formal groundwater withdrawals are still essentially unknown (De Stefano et al., 2015). Similar challenges have been reported in California in the United States, where the implementation of a new Sustainable Groundwater Management Act (SGMA) has been complicated by an antiquated system of paper-based water rights (Kiparsky et al., 2021), which is difficult or impossible to access and use for water management decision-making.

This logistical challenge of maintaining accurate and complete abstraction records and registries is compounded by financial pressures faced by many regulatory agencies around the world, many of whom have experienced declining funding levels over recent years and decades. In the UK, for example, the Environment Agency, the regulatory authority responsible for abstraction licensing and management, has experienced budgets and staffing reductions of 80–90% in the past decade (NAO, 2022). Similarly, in the EU, over two-thirds of Member States cite lack of finance and staffing resources as a major barrier to the implementation of water resource management plans, including systems of effective abstraction control and monitoring (EC, 2021). However, inadequate registries are also linked to institutional rigidities and often high transaction costs of reforming abstraction licensing. Further to the abatement costs faced by users who are constrained to relinquish (part of) their use, water allocation reforms also face non-trivial transaction costs related to the implementation, evaluation and monitoring of public policy programmes, which can often represent up to 30% of the total costs of the water allocation reform (Loch and Gregg, 2018). Lobbying, information gaps and asymmetries and policy dynamics can result in institutional rigidities that increase transaction costs and make policy reform expensive or even not feasible.

Finally, it is important to note that gaps in registries may occur intentionally due to legal exemptions, which mean that not all abstraction points are officially recorded and tracked. Exemptions are typically conditional on the intended volume of abstraction, purpose of water use, or, in some cases, even based on the type of infrastructure used to abstract water (Santato et al., 2016; Molle and Closas, 2020a; Benson et al., 2022). These exemptions are often applied to limit burden on regulators and abstractors. While exemptions often mean that individual abstractions are small in nature – for example, in the UK, no licence is required if irrigation abstractions are less than 20 m^3/day (Benson et al., 2022) – collectively, these users can amount to substantial volumes of agricultural water demand with significant potential impacts on water availability for licensed irrigators, other users and the environment. The lack of need to register these types of abstractions also makes it hard to verify that actual abstraction practices remain within the bounds of exemption rules. For example, an irrigator, who, on paper, qualifies for an exemption to registration is, therefore, unlikely to have their abstraction rates monitored or metered to verify they do not exceed prescribed limits specified by the exemption policy.

3 Quantifying volumes and purposes of water use

Knowledge of where agricultural abstractions exist on its own is insufficient to support effective and sustainable management of agricultural water use. Additional information is also required about the volumes, application methods (e.g. irrigation technologies used), purposes (e.g. types of crops irrigated, and whether grown outdoors or under protected conditions such as polytunnels or glasshouses), locations and associated costs and revenues of water use by farmers in order to support the design and implementation of regulatory and non-regulatory water management policies. For example, without knowledge of the quantities of water abstracted, it is impossible to assess and regulate levels of agricultural water demand to balance competing demands for scarce water. Similarly, without information on the expected returns from illegal water abstraction – which depend heavily on the volume of extraction and how water was used – it is challenging to design penalties that effectively deter this practice.

Monitoring of agricultural water use is typically achieved via on-farm metering, in the form of a flow meter attached to the point of water abstraction. However, where metering exists, there are large differences in how data are collected and verified in different regions around the world. Most commonly, meter records are self-reported manually by farmers (Foster et al., 2020). This approach is relatively low cost for regulators whose staff and budgets are often stretched. However, self-reporting also poses risks to the quality and completeness of monitoring of agricultural water use (ECA, 2021). For example, farmers may be incentivised to misreport or fail to report water use and other required information if the likelihood of spot checks and penalties are low. Incidents of meter manipulation (e.g. reversing meters) and tampering (Greiner et al., 2016; Al Naber and Molle, 2017; Sidhu et al., 2020; Molle and Closas, 2021) to falsify reporting of water use will also further degrade the quality of data collected from meters. Genuine or accidental reporting errors may also occur, remaining uncorrected if regulators lack capacity to properly scrutinise and follow up on potentially erroneous or spurious water use reports by farmers. Where meter records do exist, it is common for data to include records that appear spurious or outlier records, or for records to be incomplete or missing critical information for some abstraction rights even where otherwise robust metering systems are in place (Foster et al., 2019; Filippelli et al., 2022).

While metering of agricultural water use is accepted to be the gold standard for monitoring, levels of adoption and implementation of meters remain far from universal. There are still regions with significant water scarcity problems linked to agricultural abstractions where there remains no active metering (smart or otherwise) or reporting (self-reported or otherwise) of the amounts of water being abstracted by farmers for irrigation. In Middle East and North African countries, land ownership is typically associated with the right to abstract groundwater, and several small and large water users have historically abstracted the resource without metering or formal reporting (Molle and Closas, 2021). Lebanon's National Water Sector Strategy, for example, approved in 2012, established the need to adopt demand-based policies such as metering and volumetric charging, but none of these policies have yet been implemented. Similar data gaps are also observed in countries such as Australia, where a report by the Murray–Darling Basin Commission found as little as 25% of total abstractions are metered in some parts of the basin (Hanemann and Young, 2020), while in the United States, there is little official metering of groundwater abstractions in states such as California and Texas with large irrigated agricultural sectors (Foster et al., 2020).

The reasons for a lack of adoption of meters are complex and reflect a range of technical, economic and socio-political barriers to implementation. First, installing and maintaining meters, including ensuring that these systems are not tampered with by farmers, is often a very expensive and time-consuming task for regulators whose resources are already stretched (Foster et al., 2020; Molle and Closas, 2021). For example, Kinzelbach et al. (2022) highlight the challenge facing a country such as China where metering would entail installing and maintaining infrastructure on millions of individual abstraction points. Financial and logistical challenges to the implementation of metering are compounded by the fact that agricultural water use is typically composed of large numbers of individual abstractors, spread over wide and sometimes difficult-to-access rural areas. Metering has also frequently been opposed by farmers seeking to limit regulation or surveillance by the government. This opposition is often tolerated, or in some cases even enabled and supported, for political reasons or by businesses dependent on unmonitored and unregulated water use. All of this makes the implementation of metering an often slow and laborious process, with the end result being that agricultural abstractions may often go unmetered until impacts on water resources become so severe that public pressure forces an increased focus on improving monitoring and data collection. The case of California in the United States is an exemplar of this pattern, where water metering has only recently received substantive focus in policy despite decades of prolonged unsustainable groundwater abstraction (Griggs, 2020; Bruno et al., 2022).

Metering alone - even when implemented reliably and robustly - is only sufficient to determine the volume of water abstracted. Collection and documenting of additional information about features such as the purpose or location of use is extremely uncommon, despite the fact that these may be important to assess compliance with the terms of a farmer's water right. In Spain, for example, it has been observed that farmers may sometimes use legally defined water rights to irrigate land in different locations than for which their permit or licence was not initially specified (Bea Martínez et al., 2021), creating challenges for tracking if and where water is being used legally and illegally for irrigation purposes. Similar challenges are also common in the context of water sharing and trading systems. In some regions, neighbouring farmers may share or pool water supplies between co-located farms as part of collective responses to drought and water scarcity. This makes attributions of metered volumes to specific users or land parcels challenging without additional data on water sharing practices (Chengot et al., 2021). Formal trading systems often specify that trading of abstraction licences must not alter levels of consumptive water use or return flows (Brozović and Young, 2014) – yet both of these require additional data (e.g. on crop type or irrigation technology) that go beyond the information provided by flow metering alone.

4 Remote sensing – a tool to plug agricultural monitoring data gaps?

The previous sections of this chapter have highlighted that significant gaps remain in understanding of where, how much and for what purposes water is used by farmers. It has been shown that many of these gaps are caused by technical, financial, social and political challenges of implementing and maintaining reliable *in situ* systems of agricultural water use monitoring. Given these challenges, there is growing interest in the potential of new technological solutions to help address data gaps in a cost-effective and scalable manner. In particular, the potential of using satellite remote sensing to strengthen monitoring of agricultural water use has received growing attention from policymakers and researchers. This focus has been driven in large part by the explosion of increasingly high-resolution and sophisticated satellite imagery and algorithms over recent years, which now offer the potential to monitor agricultural water and land use at unprecedented resolutions and scales.

Remote sensing can potentially support the monitoring of agricultural water use in two main ways. First, analysis of satellite imagery can be used to detect where land is likely to have been irrigated or not (Deines et al., 2019; Dari et al., 2021; Higginbottom et al., 2023). If these estimates can be compared and contrasted with records of officially permitted abstractions, satellite data could therefore be used to identify errors or omissions in existing official abstraction registries or to pinpoint locations of potential illegal water use by farmers. Second, satellite data can also be used to generate estimates of rates of agricultural water use as a substitute or complement for *in situ* metering of abstractions (Foster et al., 2020; Massari et al., 2021). The method through which water use is quantified varies depending on the type of satellite sensor. For example, thermal-infrared satellite such as the Landsat can be used to quantify crop water use through the solution of a surface energy balance (Vogels et al., 2020; Brookfield et al., 2023), radar sensors such as Sentinel-1 satellite seek to estimate irrigation water use events through the detection of soil moisture anomalies (Modanesi et al., 2022; Dari et al., 2023), while optical sensors such as Sentinel-2 can support crop water budget estimates through monitoring of crop growth over the growing season (Foster et al., 2019; Garrido-Rubio et al., 2020).

Operational products and tools for satellite-based irrigation monitoring have recently been developed in several regions around the world. For example, in the United States, the recently launched OpenET platform led by NASA operationalises satellite estimates of agricultural water use from six evapotranspiration models at field scales across the western United States (Melton et al., 2022). In Spain, the Mancha Oriental Central Irrigation Water Board (in Spanish: *Junta de Comunidades de Regantes de la Mancha Oriental*) has been using remote sensing techniques since 1996 for the management and monitoring of irrigation uses in its area, and since 2017, all these monitoring data are available free and open access via the SPIDERwebGIS system (http://maps.spiderwebgis.org/webgis/), allowing different users to study the temporal sequences of the different products generated, including water use (Garrido-Rubio et al., 2020). It is important to acknowledge, however, that such tools and platforms are not the norm globally. In general, developing and implementing satellite monitoring approaches requires substantial applied remote sensing and GIS technical skills, time and finance, which must be considered in the context of the declining resources often available to many water managers and regulators.

Satellite data and tools such as those described earlier undoubtedly have the potential to be a transformative solution for how agricultural water use is monitored and managed (Massari et al., 2021). However, it is important to recognise that satellite-based approaches will not be a perfect solution to regulatory challenges of identifying and tracking agricultural water use. One challenge is that errors in satellite-based estimates of agricultural water use can often be large, especially when being used to estimate rates of crop water use at field or farm scales (Foster et al., 2020). Even detecting locations of irrigated fields can be fraught with challenges, especially in regions with heterogeneous cropping patterns, high levels of cloud cover, or in humid and temperate regions where supplemental irrigation is small relative to rainfall during the growing season (Ozdogan et al., 2010; Xie et al., 2021). Moreover, some crops cannot be observed via remote sensing, notably greenhouses where incentives for illegal water use are often particularly high due to the high return they deliver to the farmer (Caparrós-Martínez et al., 2020). These uncertainties and omissions pose challenges for trust and acceptance of satellite-based monitoring, although evidence from countries where remote sensing use for monitoring is established within water management practice – such as Spain (Garrido-Rubio et al., 2020) – suggests that the perception of satellite estimates as 'unbiased' may help adoption and acceptance by irrigators.

In addition, a further challenge is that the majority of satellite-based approaches provide estimates of consumptive water use (i.e. crop evapotranspiration). However, this emphasis creates a disconnect with the framing of abstraction rules and regulations, which often define rights or regulations based on the volumes or depths of water applied or abstracted by a farmer. Data derived satellite imagery may therefore have to be transformed into estimates of applied or abstracted water. This will introduce further uncertainties regarding irrigation efficiency and farmers irrigation management practices (Brookfield et al., 2023), with potential negative impacts for the effectiveness of policies based on these derived estimates (Foster et al., 2020). Satellite data alone are also not able to determine the source of water used by crops, which is often an important requirement for management and policy. This presents further challenges in regions where farmers use irrigation to supplement rainfall, where multiple water rights serve individual fields or farms, or where different sources of water (e.g. groundwater and surface water) are used conjunctively but regulated independently.

Beyond these challenges, advances in satellite technologies do not remove the need for investment in *in situ* monitoring. The estimation of crop water use from satellite imagery also depends on a range of additional data, including meteorological, soil and land cover datasets that are often derived from *in situ* monitoring. New data-driven satellite models of agricultural water use (Majumdar et al., 2020; Filippelli et al., 2022; Wei et al., 2022) may also require significant observations of actual water use or abstraction from *in situ* monitoring to train algorithms so that they can be applied robustly to new locations. Furthermore, knowledge of the location of agricultural water abstractions and rights is also essential for satellite estimates to be used operationally in management and policy, yet as previously noted these data may not always be present. For example, using satellite data to distinguish legal from illegal farm water use can only be successful if satellite data advances occur alongside efforts to address gaps and deficiencies in abstraction registries and reporting discussed previously. Here, digitalisation of water rights presents opportunities, as shown by the approach of the Spanish Guadalquivir, Ebro and Duero River Basin authorities which have included in their public geographical information system details about corresponding water rights, water sources and associated plots (GHC, 2023).

Given the various constraints around satellite-based monitoring of agricultural water use in policy and management, there is arguably a need to focus on how improvements in both approaches (satellite and *in situ*) to monitoring can be leveraged to complement each other rather than satellite data being seen as a replacement for *in situ* monitoring (Foster et al., 2020). For example, remote sensing could offer a powerful tool to target limited resources or staff conducting spot checks of low-cost self-reported water use estimates by helping to identify possible locations to check for illegal or unauthorised abstractions. Similarly, remote sensing could be used as a low-cost tool for verifying abstraction levels in areas where the costs – environmental and economic – of failing to detect non-compliance are low, with metering prioritised in sensitive areas so that the quality of meter data that are collected can be maximised.

5 Beyond improved data to improved management

The previous sections of this chapter have highlighted the difficulties faced by regulators and water managers with regard to permitting and quantification of water abstraction by farmers, along with the potential for new technologies such as remote sensing to strengthen monitoring of agricultural water use. However, improved knowledge about where, how much and for what purposes water is being abstracted does not automatically mean that agricultural water use will be managed more effectively, or that abstraction rules and regulations will be implemented as intended. For example, unregulated irrigation abstractions in water-scarce regions have been visible and acknowledged for many years (Stava, 1946; Baird and Walters, 2020; Loch et al., 2020; Baird et al., 2021; Donna and Espín-Sánchez, 2021), but this has not necessarily led to substantial changes in the way those abstractions are managed in these regions. Moreover, given the technical, logistical and economic challenges of monitoring large numbers of agricultural abstraction points distributed over large areas, no monitoring system will ever be perfect. For these reasons, improvements in monitoring systems and data – whether through greater *in situ* data collection and/or derived from remote sensing and other technologies – must be accompanied by improvements to the way data are used in policy to control and enforce water use policies.

From a sustainability and equity perspective, one of the main concerns around gaps in agricultural water use monitoring is the potential to allow or encourage illegal or unregulated water use by farmers. In the absence of perfect monitoring, efforts must be made to strengthen deterrents and penalties when farmers are found to have violated extraction rules or limits. For example, fines for illegal abstraction are often small in comparison to financial gains (Greiner et al., 2016; Baird et al., 2021). Moreover, when illegal or unlicensed abstractions are detected, evidence indicates that legal prosecutions are frequently never fully enacted by regulators (Loch et al., 2020; ECA, 2021). Strengthening the severity or likelihood of these penalties being implemented could help deter farmers from attempting to violate permitting or abstraction regulations (Castilla-Rho et al., 2019). Time lags also matter in this context. Long delays in the implementation of enforcement action reduce deterrents to non-compliance, especially in a drought year where the value of water is high and/or where the perceived likelihood of detection is low. For example, Loch et al. (2020) document a case in the Doñana area in Southern Spain 77 illegal wells were detected in 2012 and an administrative procedure opened in 2014. This led to a sanction that was appealed to the Andalusian High Court of Justice, which finally ratified the sanctions in 2017–2018. This 6-year period from theft to implementation of sanctions provided a substantial opportunity for the additional income obtained via illegal water abstraction to be leveraged to generate further financial returns, substantially reducing the severity of the original sanction.

Efforts to strengthen enforcement and compliance will only be achievable if, in parallel, progress is also made to tackle the transaction costs that often constrain reforms to institutions and policies governing agricultural water use. Larger fines, for example, are politically costly; while supporting or siding with farmers, for example, suggesting an amnesty and regularisation of informal uses, can be seen as politically expedient, especially at times of elections or where powerful farm lobby groups exist (Rijsberman, 2008; Grafton and Williams, 2020; Loch et al., 2020; Schmidt et al., 2020; Jacoby et al., 2021). Implementing penalties can also be a cumbersome and time-consuming process for regulators. This is in part due to the significant under-resourcing of environmental compliance and regulation in many countries, along with the significant difficulties often faced in being able to detect and verify the occurrence of unregulated or unpermitted water use. In some cases, this can extend to directly limit the ability of regulators to monitor and carry out enforcement actions, as shown by examples in Donaña in southern Spain, where lobbying has led some local municipalities to appeal against some fines imposed on irrigators for illegal abstraction or even pay fines through public money where water theft has occurred on land publicly owned and leased by the government (Loch et al., 2020).

Water scarcity and volatility in water supplies – driven in part by climate change – are likely to increase the need for and economic value of water for agriculture in many locations around the world (D'Odorico et al., 2020; Balasubramanya et al., 2022). Importantly, this may create increased incentives for non-compliance with abstraction regulations by farmers, especially those operating in regions where monitoring is patchy or limited, and/or where there is limited support for farmers to mitigate the physical and financial impacts of water scarcity through other means. Given the infeasibility of perfect monitoring in practice, it is essential that efforts are therefore also made to address and mitigate the root causes and consequences of water scarcity for farmers. This could include measures such as insurance systems to financially protect farmers in the event of water allocations being curtailed in drought years, avoiding the need to abstract illegally to protect against economic and financial losses. Currently, most insurance schemes provide cover against drought primarily for rainfed crops (Abdi et al., 2022). Those which do provide cover for irrigated crops typically cover losses not related to water scarcity or include stipulations that farmers must fully utilise all possible water sources (e.g. the US Federal crop insurance scheme – Suchato et al., 2022), which could incentivise inefficient or unregulated use of water. However, research has shown that provision of financial safety nets, such as insurance, could provide an important mechanism to protect irrigated agriculture against risks posed by physical and regulatory water scarcity (Gómez-Limón, 2020; Agudo-Domínguez et al., 2022), which may, in turn, help to incentivise compliance with abstraction regulations and restrictions.

Alongside support for dealing with the consequences of water scarcity, there are opportunities to strengthen resilience through interventions that aim to enhance supply reliability under drought and climate change. In some regions, this will require a greater focus on recognising the importance of water for agriculture – and of agriculture for food supply and economies – through changes to prioritisation rules around who gets water in times of scarcity. For example, in the UK, agriculture is often the first sector to experience curtailments to water supply during droughts (Knox et al., 2018). Financial or technical support could also be used to enhance on-farm water supply and reduce the need for irrigation abstraction at times of scarcity or drought. Examples of interventions that could be used include the development of new physical storage infrastructure, such as on-farm reservoirs to smooth out hydrologic variability (Rey et al., 2017), nature-based solutions such as improved soil management or use of cover crops (Wens et al., 2020; Nouri et al., 2021) to retain water in soils in order to limit supplemental irrigation abstraction needs, or exploitation of non-conventional water sources such as treated wastewater for irrigation (Ricart et al., 2021). However, care must be taken that such measures do not exacerbate existing scarcity issues. For example, on-farm reservoir construction has been a source of controversy during recent droughts in Europe and elsewhere (Perin et al., 2021; Brown et al., 2022) due to the perception of withholding water from other downstream uses. Storage systems may themselves also be vulnerable to climate change (Ghansah et al., 2022; Malerba et al., 2022), while also creating challenges for regulation when monitoring systems and infrastructure are insufficient to reliably and robustly track when and where reservoirs are actually being filled.

Beyond these interventions, there are also ways that the wider food supply chain and agribusiness market could more directly incentivise greater monitoring, collection and reporting of water use data. One promising area would be to more closely link farm subsidies and other forms of financial support, such as payments linked to product or supply chain certification, to requirements for farms and farmers to demonstrate robust accounting of water use data and compliance with water use regulations. Verifying the legality of water use and evidencing good water stewardship by growers is an increasingly relevant element for the certification of products, especially for water-intensive crops such as fresh fruits and vegetables. For example, the latest certification standards by GLOBAL G.A.P.'s Farm Sustainability Assessment (FSA) as well as the Sustainable Program for Irrigation and Groundwater Use (SPRING), the Alliance for Water Stewardship, Rainforest Alliance, and organic certification labels such as Naturland or CAAE have developed criteria and rules to validate the legal use of water at the farm level (Schmidt et al., 2020). However, some of these standards and frameworks still face challenges to widespread implementation gaps, particularly linked to the lack of baseline water metering and monitoring and the difficulties in linking monitoring data to specific users or plots in some farming systems.

Finally, on a broader level, there is a need to move beyond thinking about monitoring and the collection of agricultural water use data as a tool purely for regulation and demand management. Greater efforts should be made by both the public and private sectors to support farmers in leveraging monitoring data to enable positive changes in water management and stewardship that reduce exposure to scarcity and improve the resilience of connected human-water systems. For example, at present, farmers' main incentive for monitoring and reporting water use is compliance with regulations. However, the value of water use data is arguably much greater than this. Opportunities exist for the agricultural sector to learn from monitoring by other sectors (e.g. domestic water use) and for other resources (e.g. energy), where monitoring of usage is used to enable better demand management. For example, greater sharing of agricultural water use data could be used to drive innovations in benchmarking (addressing performance 'gaps') and decision support that help farmers learn and identify opportunities and pathways to improve their water management, in ways that, in turn, help mitigate the rising pressures posed by water scarcity and drought.

6 Conclusion

Agriculture is central to debates around how to manage scarce and limited freshwater resources around the world. Achieving sustainable and efficient water management in agriculture is predicated on the ability to monitor agricultural water use effectively and reliably over space and time. However, as we have shown in this chapter, the majority of regions around the world lack effective monitoring of where, when, how much and for what purposes water is withdrawn or abstracted by farmers. These gaps are the result of a complex interplay of technical, financial, social and political barriers, which limit the ability to conduct and maintain robust systems of accounting for agricultural water use. New technologies such as satellite remote sensing provide a possible pathway to plug gaps in traditional *in situ* monitoring approaches, but, as we discuss, should not be seen as panacea solutions due to the limitations and uncertainties inherent in remote monitoring of agricultural water use. We argue that there is a need for greater focus on leveraging complementarily of different monitoring approaches and enabling actors to support sustainable, efficient and equitable agricultural water management. Beyond this, we identify a need to focus not just on improving data but also on improving how data are used in policy and practice, without which new technical advances in monitoring are unlikely to yield necessary improvements in agricultural water management policy and practice on the ground.

7 Where to look for further information

The following articles provide a good overview of the subject:

- Foster, T., Mieno, T. and Brozović, N. (2020). Satellite-based monitoring of irrigation water use: Assessing measurement errors and their implications for agricultural water management policy. *Water Resources Research*, 56(11), e2020WR028378.
- Massari, C., Modanesi, S., Dari, J., Gruber, A., De Lannoy, G. J., Girotto, M., Quintana-Seguí, P., Le Page, M., Jarlan, L., Zribi, M. and Ouaadi, N. (2021). A review of irrigation information retrievals from space and their utility for users. *Remote Sensing*, 13(20), p. 4112.
- Schmidt, G., De Stefano, L., Bea, M., Carmody, E., van Dyk, G., Fernández-Lop, A., Fuentelsaz, F., Hatcher, C., Hernandez, E., O'Donnell, E. and Rouillard, J. (2020). *How to tackle illegal water abstractions: taking stock of experience.*

8 References

Abdi, M. J., Raffar, N., Zulkafli, Z., Nurulhuda, K., Rehan, B. M., Muharam, F. M., Khosim, N. A. and Tangang, F. (2022). Index-based insurance and hydroclimatic risk management in agriculture: a systematic review of index selection and yield-index modelling methods. *International Journal of Disaster Risk Reduction* 67, 102653.

Agudo-Domínguez, A., Pérez-Blanco, C. D., Gil-García, L., Ortega, J. A. and Dasgupta, S. (2022). Climate-sensitive hydrological drought insurance for irrigated agriculture under deep uncertainty. Insightful results from the Cega River Basin in Spain. *Agricultural Water Management* 274, 107938.

Al Naber, M. and Molle, F. (2017). Controlling groundwater over abstraction: state policies vs local practices in the Jordan highlands. *Water Policy* 19(4), 692–708.

Baird, A. and Walters, R. (2020). Water theft through the ages: insights for green criminology. *Critical Criminology* 28(3), 371–388.

Baird, A., Walters, R. and White, R. (2021). Water theft maleficence in Australia. *International Journal for Crime, Justice and Social Democracy* 10(1), 83–97.

Balasubramanya, S., Brozović, N., Fishman, R., Lele, S. and Wang, J. (2022). Managing irrigation under increasing water scarcity. *Agricultural Economics* 53(6), 976–984.

Bea Martínez, M., Fernández Lop, A., Gil, T., Seiz Puyuelo, R. and Cols, R. (2021). *El Robo del Agua: Cuatro Ejemplos Flagrantes del Saqueo Hídrico en España*. WWF, Madrid, Spain.

Benson, D., Cook, H., Yasir Ak, M. and Demirbilek, B. (2022). England and Wales: countering 'unsustainable abstraction' with the catchment-based approach. In: Rouillard, J., Babbitt, C., Challies, E. and Rinaudo, J.-D. (Eds), *Water Resources Allocation and Agriculture: Transitioning from Open to Regulated Access*. IWA Publishing: London, UK.

Beyer, R. M., Hua, F., Martin, P. A., Manica, A. and Rademacher, T. (2022). Relocating croplands could drastically reduce the environmental impacts of global food production. *Communications Earth and Environment* 3(1), 49.

Brookfield, A. E., Zipper, S., Kendall, A. D., Ajami, H. and Deines, J. M. (2023). Estimating groundwater pumping for irrigation: a method comparison. *Ground Water* 62, 15-33.

Brown, P., Colloff, M. J., Slattery, M., Johnson, W. and Guarino, F. (2022). An unsustainable level of take: on-farm storages and floodplain water harvesting in the northern Murray–Darling Basin, Australia. *Australasian Journal of Water Resources* 26(1), 43–58.

Brozović, N. and Young, R. (2014). Design and implementation of markets for groundwater pumping rights. In: Easter, K. W. and Huang, Q. (Eds.), *Water Markets for the 21st Century: What Have We Learned?* pp. 283–303. Springer, Dordrecht.

Bruno, E. M., Hagerty, N. and Wardle, A. R. (2022). The political economy of groundwater management: descriptive evidence from California. In: Libecap, G. D. and Dinar, A. (Eds) *American Agriculture, Water Resources, and Climate Change*. University of Chicago Press, In press, Chicago, USA.

Caparrós-Martínez, J. L., Rueda-Lópe, N., Milán-García, J. and de Pablo Valenciano, J. (2020). Public policies for sustainability and water security: the case of Almeria (Spain). *Global Ecology and Conservation* 23, e01037.

Castilla-Rho, J. C., Rojas, R., Andersen, M. S., Holley, C. and Mariethoz, G. (2019). Sustainable groundwater management: how long and what will it take? *Global Environmental Change* 58, 101972.

Chengot, R., Knox, J. W. and Holman, I. P. (2021). Evaluating the feasibility of water sharing as a drought risk management tool for irrigated agriculture. *Sustainability* 13(3), 1456.

Dari, J., Quintana-Seguí, P., Escorihuela, M. J., Stefan, V., Brocca, L. and Morbidelli, R. (2021). Detecting and mapping irrigated areas in a Mediterranean environment by using remote sensing soil moisture and a land surface model. *Journal of Hydrology* 596, 126129.

Dari, J., Brocca, L., Modanesi, S., Massari, C., Tarpanelli, A., Barbetta, S., Quast, R., Vreugdenhil, M., Freeman, V., Barella-Ortiz, A., Quintana-Seguí, P., Bretreger, D. and Volden, E. (2023). Regional data sets of high-resolution (1 and 6 km) irrigation estimates from space. *Earth System Science Data* 15(4), 1555–1575.

Deines, J. M., Kendall, A. D., Crowley, M. A., Rapp, J., Cardille, J. A. and Hyndman, D. W. (2019). Mapping three decades of annual irrigation across the US high plains aquifer using Landsat and Google Earth Engine. *Remote Sensing of Environment* 233, 111400.

Deines, J. M., Kendall, A. D., Butler, J. J., Basso, B. and Hyndman, D. W. (2021). Combining remote sensing and crop models to assess the sustainability of stakeholder-driven groundwater management in the US high plains aquifer. *Water Resources Research* 57(3), WR027756.

De Stefano, L., Fornés, J. M., López-Geta, J. A. and Villarroya, F. (2015). Groundwater use in Spain: an overview in light of the EU water framework directive. *International Journal of Water Resources Development* 31(4), 640–656.

D'Odorico, P., Chiarelli, D. D., Rosa, L., Bini, A., Zilberman, D. and Rulli, M. C. (2020). The global value of water in agriculture. *Proceedings of the National Academy of Sciences of the United States of America* 117(36), 21985–21993.

Dolan, F., Lamontagne, J., Link, R., Hejazi, M., Reed, P. and Edmonds, J. (2021). Evaluating the economic impact of water scarcity in a changing world. *Nature Communications* 12(1), 1915.

Donna, J. D. and Espín-Sánchez, J. A. (2021). Water theft as social insurance: south-eastern Spain, 1851–1948. *The Economic History Review* 74(3), 721–753.

Erfani, T., Binions, O. and Harou, J. J. (2015). Protecting environmental flows through enhanced water licensing and water markets. *Hydrology and Earth System Sciences* 19(2), 675–689.

European Commission (EC). (2021). *European Overview - Implementation of Planned Programmes of Measures and New Priority Substances*. European Commission, Brussels.

European Court of Auditors (ECA). (2021). Sustainable water use in agriculture: CAP funds more likely to promote greater rather than more efficient water use. Special Report 20/2021. ECA, Luxembourg.

Felbab-Brown, V. (2017). *Water Theft and Water Smuggling: Growing Problem or Tempest in a Teapot?* Brookings Institution, United States, Washington, D.C.

Filippelli, S. K., Sloggy, M. R., Vogeler, J. C., Manning, D. T., Goemans, C. and Senay, G. B. (2022). Remote sensing of field-scale irrigation withdrawals in the central Ogallala aquifer region. *Agricultural Water Management* 271, 107764.

Foster, T., Gonçalves, I. Z., Campos, I., Neale, C. M. U. and Brozović, N. (2019). Assessing landscape scale heterogeneity in irrigation water use with remote sensing and in situ monitoring. *Environmental Research Letters* 14(2), 024004.

Foster, T., Mieno, T. and Brozović, N. (2020). Satellite-based monitoring of irrigation water use: assessing measurement errors and their implications for agricultural water management policy. *Water Resources Research* 56(11), WR028378.

Ghansah, B., Foster, T., Higginbottom, T. P., Adhikari, R. and Zwart, S. J. (2022). Monitoring spatial-temporal variations of surface areas of small reservoirs in Ghana's Upper East Region using Sentinel-2 satellite imagery and machine learning. *Physics and Chemistry of the Earth, Parts A/B/C* 125, 103082.

Giordano, M., Scheierling, S. M., Tréguer, D. O., Turral, H. and McCornick, P. G. (2021). Moving beyond 'more crop per drop': insights from two decades of research on agricultural water productivity. *International Journal of Water Resources Development* 37(1), 137–161.

Garrido-Rubio, J., Calera, A., Arellano, I., Belmonte, M., Fraile, L., Ortega, T., Bravo, R. and González-Piqueras, J. (2020). Evaluation of remote sensing-based irrigation water accounting at river basin district management scale. *Remote Sensing* 12(19), 3187.

Gómez-Limón, J. A. (2020). Hydrological drought insurance for irrigated agriculture in southern Spain. *Agricultural Water Management* 240, 106271.

Grafton, R. Q. and Williams, J. (2020). Rent-seeking behaviour and regulatory capture in the Murray-Darling Basin, Australia. *International Journal of Water Resources Development* 36(2–3), 484–504.

Greiner, R., Fernandes, L., McCartney, F. and Durante, J. (2016). Reasons why some irrigation water users fail to comply with water use regulations: a case study from Queensland, Australia. *Land Use Policy* 51, 26–40.

Griggs, B. W. (2020). Reaching consensus about conservation: high plains lessons for California's sustainable groundwater management act. *The University of the Pacific Law Review* 52, 495.

Guadalquivir Hydrographic Confederation (GHC). (2023). Hydrographic demarcations of Guadalquivir, Ceuta and Melilla. Available at: https://idechg.chguadalquivir.es/nodo/index.html.

Hanemann, M. and Young, M. (2020). Water rights reform and water marketing: Australia vs the US West. *Oxford Review of Economic Policy* 36(1), 108–131.

Higginbottom, T. P., Adhikari, R., Dimova, R., Redicker, S. and Foster, T. (2021). Performance of large-scale irrigation projects in sub-Saharan Africa. *Nature Sustainability* 4(6), 501–508.

Higginbottom, T. P., Adhikari, R. and Foster, T. (2023). Rapid expansion of irrigated agriculture in the Senegal River Valley following the 2008 food price crisis. *Environmental Research Letters* 18(1), 014037.

Holley, C., Mutongwizo, T., Pucci, S., Castilla-Rho, J. and Sinclair, D. (2020). *Groundwater Regulation, Compliance and Enforcement: Insights on Regulators, Regulated Actors and Frameworks in New South Wales, Australia*. Springer International Publishing, pp. 411–433.

Hoogesteger, J. and Wester, P. (2017). Regulating groundwater use: the challenges of policy implementation in Guanajuato, Central Mexico. *Environmental Science and Policy* 77, 107–113.

Jacoby, H. G., Mansuri, G. and Fatima, F. (2021). Decentralizing corruption: irrigation reform in Pakistan. *Journal of Public Economics* 202, 104499.

Kelly, T. D., Foster, T. and Schultz, D. M. (2023). Assessing the value of adapting irrigation strategies within the season. *Agricultural Water Management* 275, 107986.

Kinzelbach, W., Wang, H., Li, Y., Wang, L. and Li, N. (2022). *Groundwater Overexploitation in the North China Plain: A Path to Sustainability*. Springer Nature, Singapore. p. 157.

Kiparsky, M., Miller, K., Roos-Collins, R., Roos-Collins, E. and Rademacher, D. (2021). *Piloting a Water Rights Information System for California*. UC Berkeley Center for Law, Energy and the Environment, Berkeley, CA.

Knox, J. W., Haro-Monteagudo, D., Hess, T. M. and Morris, J. (2018). Identifying trade-offs and reconciling competing demands for water: integrating agriculture into a robust decision-making framework. *Earth's Future* 6(10), 1457–1470.

Koebele, E. A., Singletary, L., Hockaday, S. E. and Ormerod, K. J. (2022). A role for water markets in enhancing water security in the western United States?: lessons from the Walker River Basin. *Water Policy* 24(11), 1757–1771.

Lankford, B., Closas, A., Dalton, J., Gunn, E. L., Hess, T., Knox, J. W., van der Kooij, S., Lautze, J., Molden, D., Orr, S., Pittock, J., Richter, B., Riddell, P. J., Scott, C. A., Venot, J., Vos, J. and Zwarteveen, M. (2020). A scale-based framework to understand the promises, pitfalls and paradoxes of irrigation efficiency to meet major water challenges. *Global Environmental Change* 65, 102182.

Loch, A. and Gregg, D. (2018). Salinity management in the Murray-Darling Basin: a transaction cost study. *Water Resources Research* 54(11), 8813–8827.

Loch, A., Pérez-Blanco, C. D., Carmody, E., Felbab-Brown, V., Adamson, D. and Seidl, C. (2020). Grand theft water and the calculus of compliance. *Nature Sustainability* 3(12), 1012–1018.

Majumdar, S., Smith, R., Butler Jr., J. J. and Lakshmi, V. (2020). Groundwater withdrawal prediction using integrated multitemporal remote sensing data sets and machine learning. *Water Resources Research* 56(11), WR028059.

Malerba, M. E., Wright, N. and Macreadie, P. I. (2022). Australian farm dams are becoming less reliable water sources under climate change. *Science of the Total Environment* 829, 154360.

Massari, C., Modanesi, S., Dari, J., Gruber, A., De Lannoy, G. J. M., Girotto, M., Quintana-Seguí, P., Le Page, M., Jarlan, L., Zribi, M., Ouaadi, N., Vreugdenhil, M., Zappa, L., Dorigo, W., Wagner, W., Brombacher, J., Pelgrum, H., Jaquot, P., Freeman, V., Volden, E., Fernandez Prieto, D., Tarpanelli, A., Barbetta, S. and Brocca, L. (2021). A review of irrigation information retrievals from space and their utility for users. *Remote Sensing* 13(20), 4112.

McCarthy, B., Anex, R., Wang, Y., Kendall, A. D., Anctil, A., Haacker, E. M. K. and Hyndman, D. W. (2020). Trends in water use, energy consumption, and carbon emissions from irrigation: role of shifting technologies and energy sources. *Environmental Science and Technology* 54(23), 15329–15337.

McDermid, S. S., Mahmood, R., Hayes, M. J., Bell, J. E. and Lieberman, Z. (2021). Minimizing trade-offs for sustainable irrigation. *Nature Geoscience* 14(10), 706–709.

Melton, F. S., Huntington, J., Grimm, R., Herring, J., Hall, M., Rollison, D., Erickson, T., Allen, R., Anderson, M., Fisher, J. B., Kilic, A., Senay, G. B., Volk, J., Hain, C., Johnson, L., Ruhoff, A., Blankenau, P., Bromley, M., Carrara, W., Daudert, B., Doherty, C., Dunkerly, C., Friedrichs, M., Guzman, A., Halverson, G., Hansen, J., Harding, J., Kang, Y., Ketchum, D., Minor, B., Morton, C., Ortega-Salazar, S., Ott, T., Ozdogan, M., ReVelle, P. M., Schull, M., Wang, C., Yang, Y. and Anderson, R. G. (2022). OpenET: filling a critical data gap in water management for the western United States. *JAWRA Journal of the American Water Resources Association* 58(6), 971–994.

Modanesi, S., Massari, C., Bechtold, M., Lievens, H., Tarpanelli, A., Brocca, L., Zappa, L. and De Lannoy, G. J. M. (2022). Challenges and benefits of quantifying irrigation through the assimilation of Sentinel-1 backscatter observations into Noah-MP. *Hydrology and Earth System Sciences* 26(18), 4685–4706.

Molle, F. and Closas, A. (2020a). Groundwater licensing and its challenges. *Hydrogeology Journal* 28(6), 1961–1974.

Molle, F. and Closas, A. (2020b). Why is state-centered groundwater governance largely ineffective? A review. *Wiley Interdisciplinary Reviews: Water* 7(1), e1395.

Molle, F. and Closas, A. (2021). Groundwater metering: revisiting a ubiquitous 'best practice'. *Hydrogeology Journal* 29(5), 1857–1870.

National Audit Office UK (NAO). (2022). *Environmental Compliance and Enforcement*. Department for Environment, Food, and Rural Affairs (Defra), London.

Nouri, A., Yoder, D. C., Raji, M., Ceylan, S., Jagadamma, S., Lee, J., Walker, F. R., Yin, X., Fitzpatrick, J., Trexler, B., Arelli, P. and Saxton, A. M. (2021). Conservation agriculture increases the soil resilience and cotton yield stability in climate extremes of the southeast US. *Communications Earth and Environment* 2(1), 155.

Novo, P., Dumont, A., Willaarts, B. A. and López-Gunn, E. (2015). More cash and jobs per illegal drop? The legal and illegal water footprint of the Western Mancha Aquifer (Spain). *Environmental Science and Policy* 51, 256–266.

Ouvrard, B., Préget, R., Reynaud, A. and Tuffery, L. (2023). Nudging and subsidising farmers to foster smart water meter adoption. *European Review of Agricultural Economics* 50(3), 1178–1226.

Ozdogan, M., Yang, Y., Allez, G. and Cervantes, C. (2010). Remote sensing of irrigated agriculture: opportunities and challenges. *Remote Sensing* 2(9), 2274–2304.

Passioura, J. (2006). Increasing crop productivity when water is scarce—from breeding to field management. *Agricultural Water Management* 80(1–3), 176–196.

Pérez-Blanco, C. D., Loch, A., Ward, F., Perry, C. and Adamson, D. (2021). Agricultural water saving through technologies: a zombie idea. *Environmental Research Letters* 16(11), 114032.

Perin, V., Tulbure, M. G., Gaines, M. D., Reba, M. L. and Yaeger, M. A. (2021). On-farm reservoir monitoring using Landsat inundation datasets. *Agricultural Water Management* 246, 106694.

Rey, D., Pérez-Blanco, C. D., Escriva-Bou, A., Girard, C. and Veldkamp, T. I. E. (2019). Role of economic instruments in water allocation reform: lessons from Europe. *International Journal of Water Resources Development* 35(2), 206–239.

Ricart, S., Villar-Navascués, R. A., Hernández-Hernández, M., Rico-Amorós, A. M., Olcina-Cantos, J. and Moltó-Mantero, E. (2021). Extending natural limits to address water scarcity? The role of non-conventional water fluxes in climate change adaptation capacity: a review. *Sustainability* 13(5), 2473.

Richter, B. D., Ao, Y., Lamsal, G., Wei, D., Amaya, M., Marston, L. and Davis, K. F. (2023). Alleviating water scarcity by optimizing crop mixes. *Nature Water*, 1, 1035–1047.

Rijsberman, F. R. (2008). Water for food: corruption in irrigation systems. Global Corruption Report 2008, 67.

Rosa, L. (2022). Adapting agriculture to climate change via sustainable irrigation: biophysical potentials and feedbacks. *Environmental Research Letters* 17(6), 063008.

Santato, S., Mysiak, J. and Pérez-Blanco, C. D. (2016). The water abstraction license regime in Italy: a case for reform? *Water* 8(3), 103.

Schmidt, G., De Stefano, L., Bea, M., Carmody, E., van Dyk, G., Fernández-Lop, A., Fuentelsaz, F., Hatcher, C., Hernandez, E., O'Donnell, E. and Rouillard, J. (2020). *How to Tackle Illegal Water Abstractions: Taking Stock of Experience and Lessons Learned*. Fundacion Botin, Madrid, Spain.

Schmitt, R. J. P., Rosa, L. and Daily, G. C. (2022). Global expansion of sustainable irrigation limited by water storage. *Proceedings of the National Academy of Sciences of the United States of America* 119(47), e2214291119.

Sidhu, B. S., Kandlikar, M. and Ramankutty, N. (2020). Power tariffs for groundwater irrigation in India: a comparative analysis of the environmental, equity, and economic tradeoffs. *World Development* 128, 104836.

Stava, W. (1946). Detecting illegitimate uses of water. *Journal AWWA* 38(1), 53–56.

Suchato, P., Mieno, T., Schoengold, K. and Foster, T. (2022). The potential for moral hazard behavior in irrigation decisions under crop insurance. *Agricultural Economics* 53(2), 257–273.

Sun, S., Tang, Q., Konar, M., Fang, C., Liu, H., Liu, X. and Fu, G. (2023). Water transfer infrastructure buffers water scarcity risks to supply chains. *Water Research* 229, 119442.

van Maanen, N., Andrijevic, M., Lejeune, Q., Rosa, L., Lissner, T. and Schleussner, C. F. (2022). Accounting for socioeconomic constraints in sustainable irrigation expansion assessments. *Environmental Research Letters* 17(7), 075004.

Van Vliet, M. T. H., Jones, E. R., Flörke, M., Franssen, W. H. P., Hanasaki, N., Wada, Y. and Yearsley, J. R. (2021). Global water scarcity including surface water quality and expansions of clean water technologies. *Environmental Research Letters* 16(2), 024020.

Vogels, M. F. A., de Jong, S. M., Sterk, G., Wanders, N., Bierkens, M. F. P. and Addink, E. A. (2020). An object-based image analysis approach to assess irrigation-water consumption from MODIS products in Ethiopia. *International Journal of Applied Earth Observation and Geoinformation* 88, 102067.

Wens, M., Veldkamp, T. I. E., Mwangi, M., Johnson, J. M., Lasage, R., Haer, T. and Aerts, J. C. J. H. (2020). Simulating small-scale agricultural adaptation decisions in response to drought risk: an empirical agent-based model for semi-arid Kenya. *Frontiers in Water* 2, 15.

Xie, Y., Gibbs, H. K. and Lark, T. J. (2021). Landsat-based Irrigation Dataset (LANID): 30 m resolution maps of irrigation distribution, frequency, and change for the US, 1997–2017. *Earth System Science Data* 13(12), 5689–5710.

Young, R., Foster, T., Mieno, T., Valocchi, A. and Brozović, N. (2021). Hydrologic-economic trade-offs in groundwater allocation policy design. *Water Resources Research* 57(1), WR027941.

Chapter 3

Forecasting future global agricultural water demand: current projections and future challenges

Upali A. Amarasinghe and Giriraj Amaranath, International Water Management Institute, Sri Lanka; and Alok Sikka, International Water Management Institute, India*

1 Introduction

Future agricultural water demand estimates are essential for water policy planning and water infrastructure development and management. The agricultural sector currently accounts for about 69% of the total freshwater withdrawals (UN, 2021). However, the total water withdrawal has decreased since the 1990s, but the share of agricultural consumptive water use (CWU) (or evapotranspiration) in the total process depletion is substantially higher (~90%) (FAO, 2022). But reliable estimates of future water demand are increasingly complex and challenging (Amarasinghe and Smakhtin, 2014). The complexity is due to integrating many interdependent drivers that affect demand, such as changing land-use patterns, technological advances, and climate change. This situation is exacerbated due to inadequate and unreliable data to assess historical trends and exogenous factors such as climate change that abruptly affect the drivers. Indeed, many past studies have not internalized climate change drivers in their future demand projections. One reason for this lacuna was the unavailability of details on climate change drivers at the local scale. The other was the inherent uncertainties associated with climate change

* Corresponding author: u.amarasinghe@cgiar.org.

http://dx.doi.org/10.19103/AS.2023.0123.02

projections. The major objective of this chapter is to discuss current projections of agricultural water demand and the challenges in estimating them in the future.

Why projections? A forecast predicts a future using historical trends. Uncertainty is a key factor in long-term forecasts. Short-term (months to few years) predictions have narrower confidence limits or lower uncertainty, while medium- to long-term forecasts have wider confidence limits or higher uncertainty (Chambers et al., 1971). The short-term forecasts in the water sector, especially in the domestic sector, are generally more accurate than their long-term forecasts (Vijai and Sivakumar, 2018; Pacchin et al., 2019). However, water policy planning and resource development require a long-term strategic outlook, especially for the agricultural sector which is experiencing significant transformation (25–50 years).

Surface water resource development also has a high development cost. The cost of the inaccuracy of forecasts of agricultural water demand, whether on the higher or lower side, could be very high when developing large water infrastructure such as reservoirs or distribution networks. Using upper limits of forecasts with high uncertainties could incur substantial socio-economic, financial, and environmental costs. Conversely, developments using the lower limits of forecasts with high uncertainty may not be adequate to meet the future demand, giving rise to water scarcity, food insecurity, conflicts, and unrest, and would also increase the risk of failure of existing infrastructure due to over-utilization. Climate change and other man-made risks could aggravate insecurity. Therefore, existing long-term water demand estimates are projections that use the current water use pattern as a baseline and consider alternative scenarios of potential actions/decisions before the projection endpoint for estimating the demand. This chapter focuses primarily on agricultural water demand projections.

The two terms that often describe water demand are water withdrawals and CWU. However, these terms have different meanings. Withdrawals represent the volume of water that is diverted or withdrawn from surface or groundwater sources to meet the CWU. The CWU is the part permanently removed from the water source as evapotranspiration (ET) and is not available for further use (see Perry (2007) for the nomenclature proposed for irrigation). The return flows of water withdrawn from a field could be recycled several times downstream and/or on the same farm, depending on the situation. Thus, the withdrawals and the CWU significantly differ in different hydrological units, from field to irrigation systems to river basins. Therefore, the two terms, withdrawals and CWU need to be carefully defined and distinguished while discussing the water demand for different spatial units.

In 2018, the global agricultural water withdrawals was estimated to be 2882 billion cubic meters (BCM), but the CWU was about 1487 BCM or about 52%

of the total agricultural water withdrawals (FAO, 2022). India, one of the largest water users, reported total water withdrawals for agriculture as being 688 BCM in 2018, but the CWU was 370 BCM, about 53% of the total water withdrawals (FAO, 2022). These estimates show that India, with some of the largest irrigation infrastructure systems, is as inefficient as global estimates for water delivery. The water use efficiency (WUE) in agriculture is the ratio of depletion (i.e. the water lost from the system as evapotranspiration) to agricultural withdrawals. The depletion includes both process and non-process evapotranspiration. The WUEs are often claimed to be very low in irrigation systems. The official records of WUE of an irrigation system, i.e. the ratio of irrigation CWU and the releases from the reservoir, are often quoted as 35–45% (GOI, 2013). Low WUE at the field level does not necessarily mean that it is low at a larger hydrological unit, such as the command area of an irrigation system or at the river basin scale (Perry et al., 2009). The actual CWU in some water-scarce irrigation systems could be almost equal to the total water released from the reservoir due to the reuse of return flows (Amarasinghe et al., 2021; UN Water, 2021). However, there is currently no information available on the widespread reuse of return flows within and outside the command areas for beneficial purposes. But return flows that flow downstream always have multiple benefits such as for ecosystems and supporting environmental flows, if not for crop production.

Globally, water withdrawals have increased by 600% over the last 100 years (UN, 2021). They have increased on average by 2.7% per annum between 1992 and 2002, 0.55% per annum between 2002 and 2012, and 0.06% per annum until 2018 (FAO, 2022). Total agricultural water withdrawals have increased by 3.8%, 0.94%, and 0.29% annually in the above three periods. In India, the growth of agricultural water withdrawals in 1992–2002 and 2002–11 was 1.70% and 1.74%, respectively, with no significant change since then. But irrigation withdrawals are still increasing in many other countries or regions with potential for irrigation expansion. For example, in sub-Saharan Africa (SSA), total agricultural withdrawals have increased by 4.8%, 0.83%, and 0.87% over these three periods. Within the SSA, Ethiopia, Ghana, Kenya, Mali, Niger, Nigeria, and Zimbabwe had substantially higher growth rates since 2010. Indonesia in South East Asia has also had high irrigation expansion rates lately.

Globally, the domestic and industrial demands are expected to grow faster than agriculture, more than doubling by 2050 (CA, 2007; De Fraiture and Wichelns, 2010). But agricultural water demand will still be the dominant sector, accounting for 58% of total water demand by 2050, compared to 75% in 2000. Projections under different scenarios show that the demand for total irrigation water withdrawals will increase by 5–57% between 2000 and 2050 (Alexandratos and Bruinsma, 2012; CA, 2007; De Fraiture and Wichelns, 2010; Raskin et al., 1998; Wada et al., 2016). As per CA (2007) projections, much of the increase in irrigation demand will be in Asia and Africa. South and East Asia will account for 29% and 27% of the additional demand for water withdrawals by 2050, and Africa's share will be *c*9%. Latin America, Asia, and Africa will also require a substantial increase in water withdrawals to meet their growing CWU demands.

Population growth, socio-economic development, changing lifestyles, and dietary consumption patterns are major drivers of increasing agricultural water demand. The World Water Development Report (UN, 2018) stated that more than 6 billion people will experience clean, fresh water scarcities by 2050. But this may be under-estimated due to the underrated effects of the linkages of water scarcities, unequal growth, and access and needs (Boretti and Rosa, 2019). This situation could be more precarious than many of the human impacts projected by the other global scenarios by not considering local needs, accessibility, and changes. Moreover, climate change could exacerbate the situation in many places with higher inter- and intra-seasonal irrigation requirements, i.e. higher CWU, due to temperature increases. In addition, these conditions need to be internalized into agricultural demand projections for designing useful adaptation strategies by the public and private sectors.

The outline of this chapter is as follows. Section 2 provides an account of the current state of water withdrawals and CWU in different regions and trends from past projections. In Section 3, we discuss the current state of the water demand drivers and the difficulties of setting targets for the relevant parameters. Section 4 concludes the chapter with policy implications.

2 Agricultural withdrawals: current status and past demand projections

2.1 Agricultural water withdrawals

The agricultural sector still accounts for most of the total water withdrawals (about 72%). The FAO AQUASTAT database shows that the agricultural share has slightly increased (by about 3%) between 2002 and 2017. This share increase was due to increased water withdrawals in South and Southeast Asia, West Asia and North Africa (WANA), sub-Saharan Africa (SSA), and Latin America (Table 1). The ten largest water users, India, China, Indonesia, USA, Pakistan, Iran, Vietnam, Philippines, Mexico, and Egypt, share 68% of the global water withdrawals. All countries except the USA have increased their agricultural water supplies – Indonesia by 79%, and India, Philippines, Mexico, and Vietnam by more than 10%. Even China had a slight increase of about 4%. Regionally, Latin America had the highest growth of water withdrawals over this period.

Data on irrigation requirements, i.e. CWU data for most countries, are available only after 2007 (FAO, 2022). In 2007, the global irrigation requirement

Table 1 Agricultural water withdrawals across regions (in BCM)

Region/country	2002	2007	2012	2017	Growth (%) 2002–17
World	2578	2704	2832	2883	11.8
South Asia	903	977	1019	1019	12.8
South East Asia	715	758	821	838	17.2
WANA	220	226	234	243	10.5
SSA	54	57	59	62	14.0
Latin America	198	222	242	255	28.9
Others (developed)	489	464	456	467	−4.5
India	584	649	688	688	17.7
China	370	360	388	385	4.0
Indonesia	106	139	171	190	79.2
Pakistan	165	171	172	172	4.4
USA	195	185	176	176	−9.5

Source: FAO-AQUASTAT (2022).

was estimated to be 1426 BCM, representing only 53% of the total water withdrawals for agriculture. Of this, 68% was in Southeast Asia and South Asia. The total irrigation requirement increased by 4% between 2007 and 2017. Much of the increase in irrigation requirement between 2002 and 2017 was in South Asia (29 BCM or 4%) and in the developed countries (16 BCM or 10%), followed by Latin America (10 BCM or 15%) and central Asia (9 BCM or 17%).

2.2 Total water demand projections

Pre- and post-1990 water demand projections used different approaches to demand estimation (Amarasinghe et al., 2014). The 1960s–1980s projections used constant norms, such as per capita domestic, industrial, or irrigation requirements, to estimate sectoral water use. Population growth was the main driver during that period. Post-1990 projections factored in many water demand 'drivers', including assumptions of technological advancement, economic growth and changing lifestyles and consumption patterns, water availability from different sources such as groundwater, water use efficiency, and the reuse of return flows. Yet, the post-1990 projections to 2025 also show substantial variations (Fig. 1), with total water demand decreasing by −0.02% to increasing by +35% from the 2000 baseline level of 3669 BCM. Unrealistic assumptions on the drivers for sustainable development targets are the reasons for negative water demand growth.

Most pre-1990 studies projected demand for years between 2000 and 2015. These demand projections significantly over-estimated the actual water

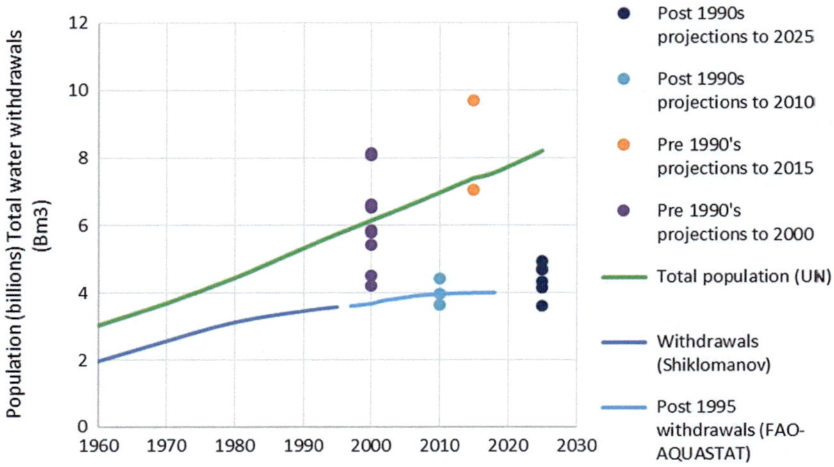

Figure 1 Estimates (1960–2017) and projections (2000–25) of population and total water withdrawals. Sources: Population estimates and projections are from the UN 2019 revision (UN, 2022). Water withdrawal estimates before 1995 are from Shiklonomav (1997), and after 1995 are from AQUASTAT (FAO, 2022); Pre- and post-1995 water demand projections are taken from Amarasinghe et al. (2014). *Note:* Amarasinghe et al. (2014) used FAO-AQUASTAT data up to 2002 for 149 countries. AQUASTAT data accessed for this chapter contain water withdrawal information from 180 countries.

withdrawals. The post-1990 projections for 2010 and 2025 show marked improvements, although most long-term projections to 2025 could still be larger than the trends that AQUASTAT data indicates. This is because the actual water withdrawals in recent years (2010–18), as shown by AQUASTAT, were more or less flat. It is likely that water withdrawals by 2025 to be at a similar or slightly higher level than 4016 BCM in 2018. This estimate is closer to the demand projections of the sustainable development scenario of Alcomo et al. (1998), Raskin et al. (1998), Seckler et al. (1998), and Rosegrant and Cai (1998). The post-1990s global projections may have helped countries to focus on demand-side management actions and driving efficiency improvements. The projection exercises also helped planners identify major drivers affecting future demand and where to prioritize investments. More informative projections at the sub-national level, including identifying limitations of further developments and setting targets on key drivers can be useful as the climate change impacts start to be felt, and planned adaptation strategies are implemented more at the local level.

Raskin et al. (1998) explored policy options to bend the curve toward sustainable development from the conventional approach. In the latter, Raskin et al. (1998) assumed that development follows the current trends of economic globalization, and developing countries will increase their push towards industrialization. The conventional or reference scenario assumed that between

1995 and 2050, the world population would grow by 50%, with average income increasing by 250% and economic output increasing fourfold. The agricultural land area would increase to reduce the forested area by 10% to meet the growing demand for food. The demand for water and energy (>125%) would increase substantially by more than 50% and 125%, respectively. The policy reforms included a set of institutional and technical instruments for attaining sustainable social and environmental objectives to bend the growth curve of the drivers identified earlier.

In Rosegrant et al. (2002), the sustainable scenario assumed a dramatic reduction in water consumption across all sectors from the business as usual (BAU) scenario, which projected an +8% total water withdrawal increase between 2000 and 2025. Increased basin efficiency and reduced water consumption coupled with decreased rice cropped areas were reported to decrease evapotranspiration demands and irrigation withdrawals. Increased efficiency and pricing of water would also reduce the demand in the industrial and service sectors, mainly in the developed world.

In Seckler et al. (1998), the scenarios which included water demand projections for 118 countries assumed that per capita demand and water use efficiency were the key drivers. In Seckler et al. (1998), high irrigation efficiency scenario assumed an increased irrigation efficiency with a ceiling of 70% for all countries. This ceiling allowed sufficient return flows from irrigation to be reused for irrigation and in other sectors.

These scenarios have similar projections because they assumed that technology use and institutional reforms would increase water use efficiency and reduce non-beneficial consumption (evaporation) through crop diversification. The discourses on emerging water scarcities during the 1990s, especially at the second World Water Forum in the Hague (Cosgrove and Rajisberman, 2000), may have triggered many governments and stakeholders to make course corrections and address the excessive water use patterns in the irrigation sector. For instance, increasing water use efficiency slogans of India (GOI, 2013) have contributed to bending the BAU trajectories resulting in lower withdrawal demands. At the same time, surface irrigation development over the last decades has become financially expensive with environmental costs (Dhawan, 1997; FAO, 1998; MA, 2005). Due to high development costs, many countries, such as those in Africa, have not kept pace with surface irrigation development with the demand increase (Inocencio et al., 2005; Lebdi, 2016).

It is unclear whether the projections for 2050 will follow actual water use patterns. Only a few global projections for 2050 are available (Raskin et al., 1998; CA, 2007; Wada et al., 2016). Figure 2 shows the projected changes from 2000 to 2050. The Comprehensive Assessment of Agriculture (CA) scenarios used 2000 data for the baseline. The bending of the curve of water withdrawals, which started around 2005, shows that the CA water withdrawals projections

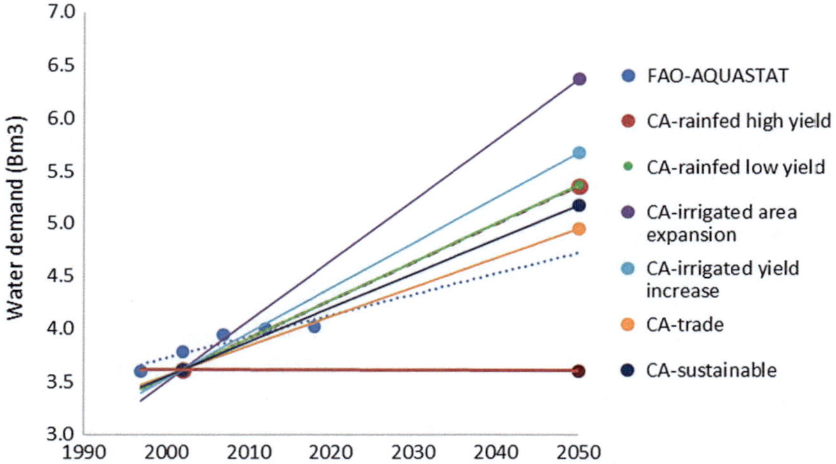

Figure 2 Reported total water demand projections from 1995 to 2050. Sources: *CA-*scenarios are from CA (2007); Raskin et al. (1998).

could be larger than the current trends of water withdrawals indicated in the FAO AQUASTAT data. The sustainable development or trade scenario projections are the closest to the AQUASTAT trends, which assumed rigid assumptions on trade between countries. In fact, with economic growth and uncertainties associated with climate change, many developing countries may enhance their trade of commodities, and with that virtual water trade. The present trends indicate that even the CA trade scenario over-estimates the 2050 demand.

Raskin et al. (1998) used 1995 as their baseline. The projection of the policy reform scenario, adjusted using 2011 population revision data, shows a drastic reduction in total withdrawal demands. The projected demand would not expect any increase from the present level. Given the increasing demand due to population growth and lifestyle changes and fast-changing dietary habits, it is unlikely such a dramatic reduction in water withdrawals will occur by 2050.

Over the last few decades, private investments in groundwater development have been a key strategy for underpinning irrigation expansion and productivity growth (Shah, 2010). Groundwater development has brought substantial benefits for rural development and income generation, but the environmental consequences of over-exploiting groundwater resources are already emerging in many places (Scott and Shah, 2004; Mohanavelu et al., 2021; Yang et al., 2018). Filling groundwater tables, land subsidence, secondary salinization of soil, drying up of streams and wetlands, and seawater intrusions are some of the emerging issues that need to be tackled. Moreover, there is no clear evidence on the extent of groundwater recharge by reservoirs and canal irrigation. Without these water storages, groundwater development in some areas to the present level would not have occurred

The impacts of these emerging issues and unknowns on water, food, and environmental security at finer spatial resolutions need to be better integrated with water supply and demand projections. These issues threaten not only the agricultural sector as a whole but also the domestic, industrial, and service sectors, which are the main drivers of economic growth in many developing countries at present.

3 Challenges of water demand projections

The prominent drivers influencing water demand projections are area expansion or cropping intensity increase (irrigated and rainfed), yield growth (irrigated and rainfed), and water productivity and efficiency improvements. There are exceptions, such as in India, with one of the largest irrigated areas in the world. In India, the net sown area has not increased over the last two decades, but the gross irrigated area has increased by 31% (GOI, 2023). Some of the other drivers influencing irrigation water demand are crop diversification, sources of water supply, return flows and reuse, changes in food consumption patterns, post-harvest losses, and food waste. Climate change is not an explicit factor in many water demand estimations. However, changes in rainfall, temperature, extreme heat, or rainfall events will impact on crop choices and growing periods, CWU, and crop production (Liu et al., 2013). Given the unreliability associated with data, changes in climate and economic conditions, technological growth, conflicts, and governance issues, many assumptions underlying future demand projections are unrealistic. In the case of climate change, it would be challenging to internalize the driver in water demand projections given the uncertainties associated in the climate change scenarios themselves, especially at local scales. This section explores some of the key drivers behind these projections, the issues that may distort the assumptions, and the challenges of setting the pathways for meeting future demand.

3.1 Food consumption patterns

Food consumption patterns, lifestyle changes, and changing dietary habits profoundly influence water use or water footprint (WFP) patterns (Kim et al., 2020; Vanham et al., 2018). A significant part of the energy supply from diets is from vegetable products. In 2018, this share was 88% of the total energy supply of 2960 kcal/day/person (FAO, 2022). Although declining lately, cereals still contribute to the largest energy supply per person at the global level – equating to *c*1310 kcal/day/person (or 43% of the total and 54% of the vegetable products consumed (Fig. 3).

Increasing energy supply from animal and other vegetable products (Fig. 4) has compensated for the reducing energy supply from cereals. With

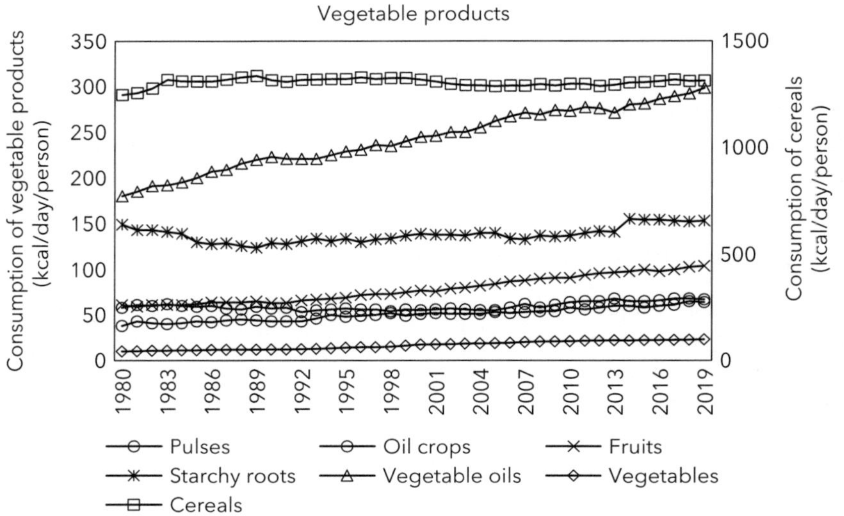

Figure 3 Total energy supply from vegetable products. Source: FAOSTAT (2022).

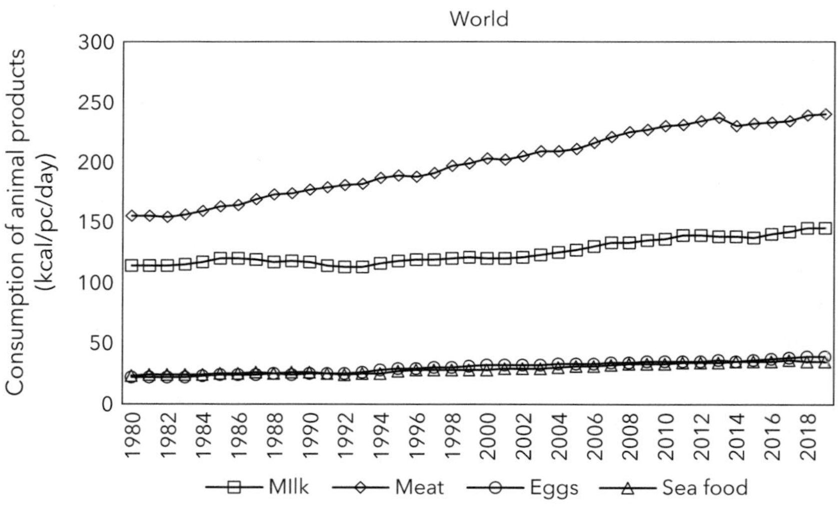

Figure 4 Total energy supply from animal products. Source: FAOSTAT in FAO (2022).

that, the total energy supply at the global level has increased by 18% from the 1980s level. Similar consumption patterns (Fig. 5) exist for energy supply in Asia and Africa, two regions with severe water-related issues. The most significant increases were in East Asia (45%), west Africa (37%, and SE Asia (35%), followed by South Asia (27%) and north Africa (22%). Middle and eastern Africa, with some of the lowest total calorific supplies (2189 kcal/person/day and 2259

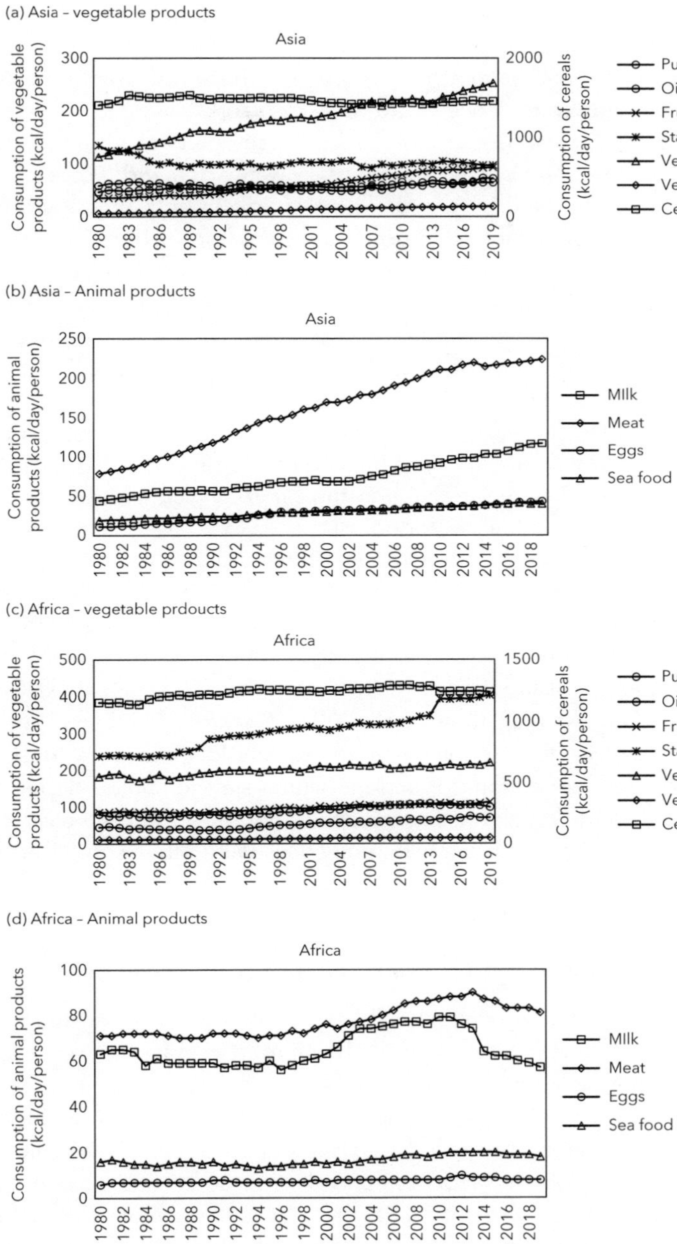

Figure 5 Vegetable and animal product calorie supply in Asia (a, b) and Africa (c, d). Source: FAOSTAT in FAO (2022).

kcal/person/day), also have similar increases in animal and some vegetable product consumption. Given the increasing physical water scarcities in Asia and precarious economic water scarcities in Africa, these consumption pattern changes will profoundly influence future water demand.

Large countries, such as India, China, and many parts of Africa will have a substantial sectoral and spatial variations in water demand. For example, in China, consumption patterns are changing rapidly in both urban and rural sectors (He et al., 2021). Animal product consumption in both sectors has increased. In contrast, food grain consumption has declined. Between 1981 and 2016, water requirements for the urban sector diets increased from 167 billion m^3 to 671 billion m^3. In contrast, water requirements for the rural sector decreased from 485 billion m^3 to 405 billion m^3 over the same period. Changes in both animal and cereals products have contributed to the overall water requirement. Meeting the increasing water requirements for animal products in both sectors in the future will be challenging. He et al. (2021) projected 24–41 BCM to meet dietary demands in the future, which is more than the recommended limits of China's water policy.

Unfortunately, detailed water requirements data for South Asia or African urban and rural sectors and diets are unavailable. However, similar consumption pattern changes within countries require more grains to feed both the population and for livestock. In India, dietary patterns are changing with the increasing consumption of animal products, especially milk (Golait and Pradhan, 2006; Amarasinghe et al., 2007). Changes in dietary patterns will also significantly impact on water requirements for producing food and feed grain demand in irrigation-dependent South Asia and rainfed-dependent Africa. Climate change-related disasters such as floods, droughts, heat waves, reduced rainy days, and high intense rainy spells will further complicate the challenge of projecting future water demands.

3.2 Rainfed agriculture and rainfall consumptive water use

Improvements in rainfed agriculture will be crucial in determining the future water demand for irrigation withdrawals. This importance is due to the large extent of rainfed agricultural land, low rainfed productivity, and the high cost of surface irrigation expansion for agricultural production. Of the arable and permanent cropped area, the rainfed area is about 98% in Africa, 75% in Latin America and the Caribbean, and 45% and 49% in South and Eastern Asia, respectively (FAO, 2022 AQUASTAT). Rainfed grains cover more than 80% of the world's cropland and produce 60% of grains (CA, 2007). Rainfed agriculture generally has a lower productivity compared to irrigated agriculture. There is thus substantial scope for reducing irrigation expansion by increasing rainfed yields. In SSA, rainfed grain yields range from 0.5 tons/ha to 2 tons/ha, averaging

1 ton/ha with substantial yield gaps (Rockström and Falkenmark, 2000; Wani et al., 2009). Current wheat yields in Africa are two to four times lower than their potential or achievable yield.

SSA and South Asia are regions that both have a large rainfed agriculture-dependent population and high levels of rural poverty. In SSA, rainfed yields are typically only 53% of irrigated yields (FAO, 2022). With the existing rainfed area, a 25–50% increase in rainfed yield can increase total production by about 20–48%. The latter estimate is more than sufficient to meet the increasing food demand by 2050, as projected by the Comprehensive Assessment of Agriculture (CA, 2007). The total grain demand (food, feed, and other uses) projection for 2050 in the SSA under the rainfed agriculture improvement scenario is about 150–200 kg/person/year – a 33% increase from the base year 2000. In South Asia, the rainfed yield is only 48% in the irrigated area. A 25–50% increase in rainfed yield can increase total production by 46–55%, which is more than sufficient to meet South Asia's projected food demand increase. The rainfed agriculture growth scenario includes investments in in-situ water, soil and nutrient management, and water harvesting and conservation. It also provides ex-situ water harvesting in farm fonds, sub-surface tanks, through bunds, and then supplementary irrigation through pumping. Providing 1000 m^3 of supplemental irrigation through fonds and groundwater, with appropriate soil and nutrient water management, can double the rainfed yield (CA, 2007). Sharma et al. (2010) showed that meeting critical water needs during the critical stage of crop growth can significantly enhance the rainfed crop yields in India. Amarasinghe et al. (2021) showed that farmers in a surface irrigation system in northern Sri Lanka managed rainfall (by delayed sowing) and used groundwater irrigation to cultivate 100% of the command area in the dry season. This increase was achieved even when there was water released from the reservoir. In some other rainfed areas, reliable rainfall forecasts can reduce even the supplemental irrigation provided to the maize crops. Rainfed agriculture has the potential to more than double the crop productivity in Africa and South Asia (Wani et al., 2009).

The high rainfed agricultural yield growth scenario projects irrigation withdrawal demand to increase only by a small margin (CA, 2007). However, if the high rainfed agriculture yield scenario fails, it requires expansion of the rainfed area and irrigation. This option is only feasible in large parts of the SSA, Latin America, and the Caribbean. Only 228 million ha is currently cropped in SSA, although more than 1030 Mha of land is available for expansion of rainfed agriculture. In Latin America, these areas are 203 Mha and 1066 Mha, respectively. The CA scenario of low rainfed yield projects an additional 400 Mha of rainfed area expansion by 2050. There is a potential for increasing irrigation in low-productivity rainfed lands in South Asia. In India, almost half of the net planted area is rainfed, but the potential for increasing supplemental irrigation depends on water availability in different river basins.

Expanding rainfed agriculture on new land always has environmental costs. It can reduce groundwater recharge and change surface runoff patterns, disturbing many livelihoods that depend on water downstream. These are unknowns and are difficult to integrate into global modeling. Increasing natural disaster risks affecting water demand and supply would be another missing element in global projections. In most climate change-related disasters, the impacts are local and can have short- to medium-term impacts on agriculture cropping patterns and productivity in many locations. Thus the envisaged increases in area and productivity, especially in rainfed agricultural growth, are uncertain. This uncertainty increases the inaccuracies associated with demand projections.

3.3 Irrigation growth

The surface or groundwater-dependent irrigation growth is also a key driver of agricultural water demand projections. Surface irrigation growth depends on the potential for increasing surface or groundwater irrigation supplies and increasing irrigated yields. Expanding irrigated areas depends on physical water scarcity, which is already a constraint in arid and semi-arid regions in southern Africa, north Africa, western Asia, and parts of Central Asia, India, China, Northern Americas, and Australia (CA, 2010). But many surface irrigation systems are also underperforming (Lie et al., 2020). Poor irrigation management and low irrigation water use efficiency are major performance constraints.

In many underperforming canal irrigation systems, water reuse can be used to expand the cropped area and/or increase crop yields. There is substantial scope for increasing CWU in canal command areas (CCA). Studies show that considerable groundwater irrigation already exists in the CCA and the buffer zone of CCA in water-scarce areas. A part of this groundwater is the recharge from canal irrigation return flows. For example, in the Sina medium irrigation system in Maharashtra, India, the reuse of return flows as groundwater helps 100% cropping intensity in the dry season, with substantially higher water use efficiency (Amarasinghe et al., 2021). In the Hurulu Wewa irrigation system in Sri Lanka, efficient use of rainfall and the groundwater recharged from the reservoir storage has helped farmers to achieve 200% cropping intensity when no water was available from the reservoir in drought years. Irrigation managers adjust the water release dates to use rainfall and storage efficiently. Minor changes in system management can therefore substantially improve canal irrigation performance.

Indeed, in the International Water Management Institute's (IWMI) comprehensive assessment (CA, 2010), irrigation growth scenario shows an optimistic pathway for increasing crop yield and meeting future water demand.

This scenario includes institutional reforms, resolving head-tail competition and conflicts, improving water allocation mechanisms, and motivating farmers and managers to enhance crop yield. These are less costly than large-scale infrastructural area expansion. With a slight increase in irrigated and rainfed cropped areas (about 10% and 12%, respectively), South Asia can become self-sufficient in grains, vegetables, roots, and tubers. East Asia has less potential for increasing already high crop yields.

The challenge for the irrigation sector is to design pathways for achieving these optimistic goals. The lack of reliable information in the canal irrigation systems is the main constraint. Many of the projections use official records for setting the baseline. However, many irrigation systems do not have correct irrigated area estimates. The analysis of land use and cropping patterns using remote sensing and GIS (RS/GIS) in the Sina irrigation system in Maharashtra showed substantial differences between the officially reported and RS/GIS based area estimates (Amarasinghe et al., 2021). The RS/GIS estimates show that the dry season irrigated area in the command area is as large as the command area. This was mainly achieved through conjunctive use of canal and groundwater in the command areas. If such under-estimating patterns affect the baseline assumptions, then the projections themselves could also be substantially under-estimated.

The assumption that groundwater is the source for irrigated area expansion in the long term is also questionable. In many areas in India, Pakistan and China, the groundwater tables are depleting rapidly (Shah, 2010; Wang et al., 2020; Qureshi, 2020), and many regions already have unsustainable levels of agricultural production. Moreover, many believe canal irrigation projects are poor investments, and groundwater with local recharge mechanisms can meet the water needs. But the emerging evidence indicates that the canal irrigation projects recharge groundwater and contribute to groundwater irrigation in many cropped areas outside the canal command areas (Aamarasinghe et al., 2021). Because of low and erratic rainfall patterns, cropping in many regions in the dry season could not be possible without such groundwater recharge. These interactions between surface irrigation and groundwater recharge need to be integrated into modeling to estimate the potential for increasing the groundwater irrigated area through conjunctive use of water. These are possible only through modeling at local scales.

3.4 Water productivity growth

Increasing water productivity can potentially reduce the demand, i.e. increasing the production per unit of water consumed or applied/withdrawn (physical water productivity) or the value of output per unit of water consumed or applied/withdrawn (economic water productivity) (Molden, 1998). Increasing physical water productivity (PWP) precedes area expansion when water is scarce. Physically, water scarcity conditions emerge when the potentially utilizable water supply is insufficient for further development. When basins reach physical water scarcity, the value of production or economic water productivity (EWP) takes precedence. This precedence is possible by crop or agricultural diversification, such as livestock or fisheries or integration with domestic use or adoption to integrated farming systems or multiple water use systems.

CA (2007) estimated that a 35% increase in water productivity could reduce additional crop consumption demand down from 80% to 20%. These estimates assume the fact that water productivity at the global and national levels is low. But significant variation exists at the sub-national level (Fig. 6). The potential for a marginal PWP (MWP) increase is very small in many irrigated areas. MWP is the increase in PWP for a unit increase CWU. This increase is mainly possible with better water management and agronomic practices, including deficit irrigation with precision and timely applications, and improved varieties. However, a high potential for increasing PWP exists in rainfed areas. Increasing land productivity in many rainfed areas by providing small supplemental irrigation with better agronomic practices can lead to substantial water productivity and production gains. These positive impacts in rainfed agriculture will reduce large increases in surface irrigation demand.

In water-scarce areas, improving EWP is a solution. This is proposed mainly with high-value agriculture, such as vegetables and fruits, and combining

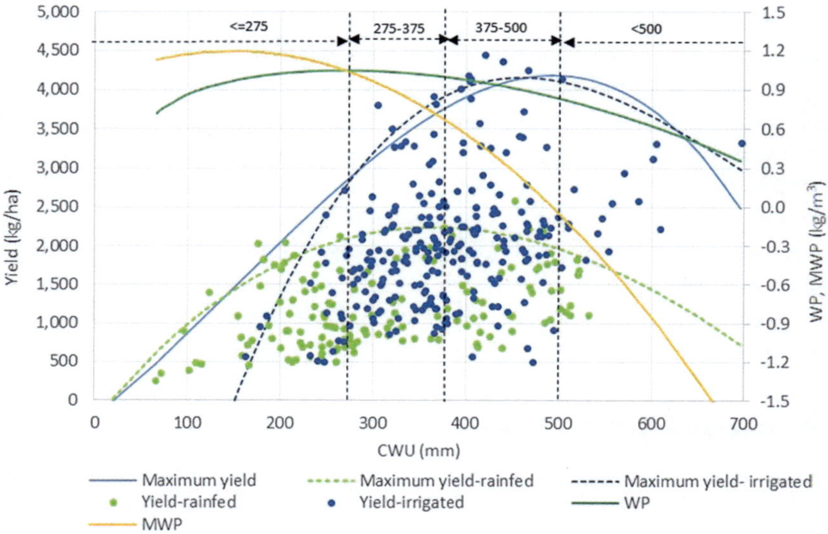

Figure 6 Yield, WP and MWP of cereal production in Indian districts. Source: Amarasinghe and Smakhtin (2014).

fisheries, and livestock. However, the potential for crop diversification to high-value crops depends on factors outside the water sector, including household food security, prices, and profitability, access to improved technology, climate information, markets and risks of investments, education and incentives, and institutional structures.

Despite the constraints and challenges, improving water productivity, PWP or EWP, could be the best way forward to address increasing water and food demand and reduce poverty. In many areas in Africa, Asia, and Latin America, with little water development can have significant gains in PWP and EWP of existing crops or new cropping or agricultural production systems. These developments require better storage information, including soil moisture, groundwater, tanks and surface, reservoirs, and wetlands. An increase or decrease in water storage at source can affect the storage of the others and risks development around them. Moreover, climate change impacts on storage is another uncertainty that may affect the farmer's decisions for investments in increasing PWP or EWP.

3.5 Increasing irrigation water use efficiency

Increasing irrigation water use efficiency (WUE), the ratio of crop evapotranspiration to irrigation supply, is proposed as a panacea for improving irrigation performance, saving water, and expanding the irrigated area. The general perception of canal irrigation systems is that they have low WUE and a large potential for saving water. This perception is correct at the field level, except when drip or sprinklers provide irrigation to crops. But much of the irrigation return flows either recharge groundwater or flow downstream, are captured by other users, or contribute to other uses. These aspects of return flows and reuses are major concerns for achieving water savings in irrigated agriculture and in assessing irrigation performance (Perry et al., 2009; Koech and Langat, 2018).

Improved irrigation technologies, such as drip irrigation, can potentially reduce non-beneficial evaporation. But the impact of drip irrigation on microclimate, such as humidity, around plants can have some impacts on transpiration, which is directly related to yield (Perry et al., 2009). Such yield impacts are less in tree crops such as fruits but could be higher in field crops.

Another major factor that affects surface irrigation performance is service delivery (Molden et al., 2007; Hasanain et al., 2013). Poor operation and maintenance have deteriorated the physical conditions of already existing assets beyond any meaningful operation contribution. Better asset management could significantly impact the WUE of existing infrastructure and returns on investments. Future projection studies should integrate these aspects in their estimates of future irrigation demand (or withdrawals). Finally, the information currently available on gravity irrigation statistics is insufficient for assessing the potential for increasing irrigation WUE (Kumar et al., 2022). Accurate statistics of irrigation water use are not possible to obtain without appropriate accounting. The increased use of advanced technologies, such as remote sensing will be essential for estimating cropped areas and irrigation depletion (Amarasinghe et al., 2021).

4 Conclusion

Increasing agricultural water demand will be inevitable with an increasing population, changing lifestyles and food consumption patterns. The past projections studies have already incorporated the changing food consumption patterns into water demand estimations at the national level. Future projection studies should consider changing food consumption patterns from the rural to urban sectors and at the sub-national level in estimating agricultural water demand. Moreover, the magnitude and the location that will increase the water demand depends on many other factors, including the current level of irrigated and rainfed areas, cropping patterns, land productivity, and water reuse. Impacts of climate change on crop water requirements in general and irrigation requirements, in particular vary with locations and over time, especially during different growth periods. The projections at the sub-national level should be further improved by incorporating climate change impacts on the demand drivers. This information has not been available for past projection studies and was not integrated due to large uncertainties associated with climate change projections.

Global projections of agricultural water demand capture the general trends of water demand. Many of these studies use national-level data and assumptions on critical parameters. The sub-national trends of rainfed agriculture, surface irrigation, and water reuse are important in assessing actual water demand. As such, global water withdrawal demand projections differ substantially from current use. The variations could be substantial at the sub-national national level. However, the global projection studies help generate awareness of issues and challenges. The planners may already have used these studies to decide on policies and investments to meet the demand. This may be the reason for the gaps in past projections and current water use. The ignorance of climate change in future demand projections would result in further untargeted investments and inefficiency of resource use.

There is much to do now to address the challenges of estimating water demand at the sub-national level. Global studies have given sufficient guidance for conducting detailed studies at the national and sub-national levels. Past studies have mainly used national-level data, and some have used non-comparable estimates of demand drivers. The challenge is to overcome acquiring accurate data and information required for the critical parameters that influence water demand at local level. Given that climate is rapidly changing, information such as food consumption patterns, food security targets, employment, income and poverty, cropping and water use patterns, water depletion, and rainfall and water supply variations are critical for accurate agricultural demand projections at the local level.

Future studies should also have more focus on crop models and seasonal forecast information. The projections can be better for agriculture and water management scenarios to help policymakers to plan water supply demands and needs across multiple levels. Future studies should consider water demands and supply across agriculture systems from rainfed, irrigated, and pastoral in the new economic or institutional regime by accounting for substitution and expansion effects and promoting system transformation approaches. Technology can help transform the global food production system and mitigate its impact on the climate and environment.

Finally, projection studies should assess total district demand to directly determine future variations in water demand due to climate change based on seasonal and peak weekly demands. The emerging technologies such as RS/GIS and their use in estimating local level water parameters would fill the gaps in many places. The implementation of adaptation options mostly occurs at the farmer and community level, and better guidance for them through projections studies are critical to overcome the emerging challenge that individual farmers, communities, and water managers are facing.

5 Acknowledgements

We would like to acknowledge funding from CGIAR ClimBeR: Building Systemic Resilience Against Climate Variability and Extremes. We also acknowledge the Indian Council of Agricultural Research for financial support for collecting data and conducting analysis for this paper.

6 References

Alexandratos, N. and Bruinsma, J. 2012. World agriculture towards 2030/2050: The 2012 revision. E.S.A. Working paper No. 12-03. Rome, Food and Agriculture Organization of the United Nations FAO. www.fao.org/docrep/016/ap106e/ap106e.pdf.

Amarasinghe, U. A., Shah, T. and Singh, O. P. 2007. *Changing Consumption Patterns: Implications on Food and Water Demand in India*. Colombo: International Water Management Institute (IWMI). 43p. (IWMI Research Report 119).

Amarasinghe, U. A. and Smakhtin, V. 2014a. Global Water Demand Projections: Past, Present, and Future. Colombo: International Water Management Institute (IWMI). 32p. (IWMI Report 156).

Amarasinghe, U. A. and Smakhtin, V. 2014b. Water productivity and water footprint: Misguided concepts or useful tools in water management and policy? *Water International* 39(7), 1000–1017, https://doi.org/10.1080/02508060.2015.986631.

Amarasinghe, U. A., Sikka, A., Mandave, V., Panda, R. K., Gorantiwar, S., Chandrasekharan, K. and Ambast, S. K. 2021. A re-look at canal irrigation system performance: A pilot study of the SINA Irrigation System in Maharashtra, India. *Water Policy* 23(1), 114–129. https://doi.org/10.2166/wp.2020.291.

Amarasinghe, U. A., Amarnath, G., Alahacoon, N., Aheeyar, M., Ghosh, S., Chandrasekharan, K. and Nakada, T. 2021. *Adaptation to Climate Variability in Sri Lanka: A Case Study of the Huruluwewa Irrigation System in the Dry Zone* (Vol. 200). Colombo: International Water Management Institute (IWMI).

Boretti, A. and Rosa, L. 2019. Reassessing the projections of the world water development report. *npj Clean Water* 2(1), 15. https://doi.org/10.1038/s41545-019-0039-9.

Burek, P., Satoh, Y., Fischer, G., Kahil, M. T., Scherzer, A., Tramberend, S., Nava, L. F., Wada, Y., Eisner, S., Flörke, M. and Hanasaki, N. 2016. Water futures and solution: Fast track initiative (final report). IIASA Working Paper. Laxenburg: International Institute for Applied Systems Analysis (IIASA).

Buzby, J. C., Farah-Wells, H. and Hyman, J. 2014. The estimated amount, value, and calories of postharvest food losses at the retail and consumer levels in the United States. *Economic Research Service, SDA*. Available at: https://www.ers.usda.gov/webdocs/publications/43833/43680_eib121.pdf.

CA (Comprehensive Assessment) 2007. *Water for Food Water for Life: A Comprehensive Assessment of Water Management in Agriculture*. London: EarthScan, and International Water Management Institute.

Chambers, J. C., Mullick, S. K. and Smith, D. D. 1971. *How to Choose the Right Forecasting Technique*. Cambridge, MA: Harvard University, Graduate School of Business Administration.

Cosgrove, W. and Rajisberman, F. 2000. *World Water Vision: Making Water Everybody's Business*. London: Routledge.

de Amorim, W. S., Valduga, I. B., Ribeiro, J. M. P., Guazzelli Williamson, V., Krauser, G. E., Magtoto, M. K. and de Andrade Guerra, J. B. S. O. 2018. The nexus between water, energy, and food in the context of the global risks: An analysis of the interactions between food, water, and energy security. *Environmental Impact Assessment Review* 72, 1–11. https://doi.org/10.1016/j.eiar.2018.05.002.

De Fraiture, C. and Wichelns, D. 2010. Satisfying future water demands for agriculture. *Agricultural Water Management* 97(4), 502–511.

Dhawan, B. D. 1997. Large-scale canal irrigation: How cost effective? *Economic and Political Weekly*, 32(52), A71–A77.

FAO (Food and Agricultural Organization) 2022. *AQUASTAT Database*. http://www.fao.org/nr/water/aquastat/data/ query/index.html?lang=en.

GOI (Government of India) 2013. *Twelfth Five Year Plan (2012–17): Faster, Sustainable, and More Inclusive Growth* (Vol. I). New Delhi: Planning Commission, Government of India. https://niti.gov.in/planningcommission.gov.in/docs/plans/planrel/fiveyr/12th/pdf/ 12fyp_vol1.pdf.

GOI (Government of India) 2023. Agricultural statistics at a glance 2021. https://eands.dacnet.nic.in/PDF/Agricultural%20Statistics%20at%20a%20Glance%20-%202021%20(English%20version).pdf.

Golait, R. and Pradhan, N. C. 2006. Changing food consumption pattern in rural India: Implication on food and nutrition security. *Ind. Jn. of Agri. Econ* 61(3), 374-388.

Hasanain, A., Ahmad, S., Mehmood, M. Z., Majeed, S. and Zinabou, G. 2013. Irrigation and water use efficiency in South Asia. https://www.gdn.int/sites/default/files/WP/WP68-SA-Irrigation-Water_64f.pdf.

Inocencio, A., Kikuchi, M., Merrey, D., Tonosaki, M., Maruyama, A., de Jong, I., Sally, H. and de Vries, F. P. 2005. *Lessons from Irrigation Investment Experiences: Cost-Reducing and Performance-Enhancing Options for sub-Saharan Africa*. Final Report Submitted to the African Development Bank. Battaramulla: International Water Management Institute.

Kim, B. F., Santo, R. E., Scatterday, A. P., Fry, J. P., Synk, C. M., Cebron, S. R., Mekonnen, M. M., Hoekstra, A. Y., De Pee, S., Bloem, M. W., Neff, R. A. and Nachman, K. E. 2020. Country-specific dietary shifts to mitigate climate and water crises. *Global Environmental Change* 62, 101926.

Koech, R. and Langat, P. 2018. Improving irrigation water use efficiency: A review of advances, challenges and opportunities in the Australian context. *Water* 10(12), 1771.

Kumar, M. D., Sahasranaman, M., Verma, M. S., Kumar, S. and Narayanamoorthy, A. 2022. Getting the irrigation statistics right. *International Journal of Water Resources Development* 38(3), 536–543.

Lebdi, F. 2016. Irrigation for agricultral transformation. Background paper for African transformation report 2016: Transforming Africa's agriculture. https://www.jica.go.jp/jica-ri/publication/booksandreports/l75nbg0000004aet-att/l75nbg0000004ai9.pdf.

Liu, J., Folberth, C., Yang, H., Röckström, J., Abbaspour, K. and Zehnder, A. J. 2013. A global and spatially explicit assessment of climate change impacts on crop production and consumptive water use. *PLoS ONE* 8(2), e57750.

MA (Millennium Ecosystem Assessment) 2005. *Ecosystems and Human Well-Being–Synthesis Report*. Washington, DC: Island Press.

Mohanavelu, A., Naganna, S. R. and Al-Ansari, N. 2021. Irrigation induced salinity and sodicity hazards on soil and groundwater: An overview of its causes, impacts and mitigation strategies. *Agriculture* 11(10), 983.

Molden, D., Burton, M. and Bos, M. G. 2007. Performance assessment, irrigation service delivery and poverty reduction: Benefits of improved system management. *The Journal of the International Commission on Irrigation and Drainage* 56(2–3), 307–320.

Pacchin, E., Gagliardi, F., Alvisi, S. and Franchini, M. 2019. A comparison of short-term water demand forecasting models. *Water Resources Management* 33(4), 1481–1497.

Perry, C. 2007. Efficient irrigation; inefficient communication; Flawed recommendations. *The Journal of the International Commission on Irrigation and Drainage* 56(4), 367–378.

Perry, C., Steduto, P., Allen, R. G. and Burt, C. M. 2009. Increasing productivity in irrigated agriculture: Agronomic constraints and hydrological realities. *Agricultural Water Management* 96(11), 1517–1524.

Perry, C., Steduto, P. and Karajeh, F. 2017. *Does Improved Irrigation Technology Save Water? A Review of Field Evidence*. Discussion Paper on Irrigation and Sustainable Water Resources Management in the Near East and North Africa. Food and Agriculture Organization of the United Nations.

Qureshi, A. S. 2020. Groundwater governance in Pakistan: From colossal development to neglected management. *Water* 12(11), 3017.

Raskin, P., Gallopin, G., Gutman, P., Hammond, A. and Swart, R. 1998. *Bending the Curve: Toward Global Sustainability* (Vol. 8). Stockholm: Stockholm Environment Institute 8.

Rinaudo, J.-D. 2015. Long-term water demand forecasting. Understanding and managing urban water in transition, pp. 239–268. ff10.1007/978-94-017-9801-3_11ff. ffhal-01183853.

Rockström, J. and Falkenmark, M. 2000. Semi arid crop production from a hydrological perspective: Gap between potential and actual yields. *Critical Reviews in Plant Sciences* 19(4), 319–346.

Rosegrant, M. W., Cai, X. and Cline, S. A. 2002. *World Water and Food to 2025: Dealing with Scarcity*. Washington, DC: International Food Policy Research Institute (IFPRI).

Scott, C. A. and Shah, T. 2004. Groundwater overdraft reduction through agricultural energy policy: Insights from India and Mexico. *International Journal of Water Resources Development* 20(2), 149–164.

Seckler, D., Amarasinghe, U., Molden, D., de Silva, R. and Barker, R. 1998. *World Water Demand and Supply, 1990 to 2025: Scenarios and Issues*. Colombo: International Irrigation Management Institute (IIMI). 47p. (IIMI Research Report 19).

Shah, T. 2010. *Taming the Anarchy: Groundwater Governance in South Asia*. Routledge.

Sharma, B. R., Rao, K. V., Vittal, K. P. R., Ramakrishna, Y. S. and Amarasinghe, U. 2010. Estimating the potential of rainfed agriculture in India: Prospects for water productivity improvements. *Agricultural Water Management* 97(1), 23–30. https://doi.org/10.1016/j.agwat.2009.08.002.

UN (United Nations) 2021. *The United Nations World Water Development Report 2021: Valuing Water*. Paris: UNESCO.

UN-Water 2021. *UN-Water Analytical Brief on Water Use Efficiency*. Geneva: UN-Water.

Vanham, D., Comero, S., Gawlik, B. M. and Bidoglio, G. 2018. The water footprint of different diets within European sub-national geographical entities. *Nature Sustainability* 1(9), 518–525. https://doi.org/10.1038/s41893-018-0133-x.

Wada, Y., Flörke, M., Hanasaki, N., Eisner, S., Fischer, G., Tramberend, S., Satoh, Y., Van Vliet, M. T. H., Yillia, P., Ringler, C. J. G. M. D., Burek, P. and Wiberg, D. 2016. Modeling global water use for the 21st century: The Water Futures and Solutions (WFaS) initiative and its approaches. *Geoscientific Model Development* 9(1), 175–222.

Wang, J., Jiang, Y., Wang, H., Huang, Q. and Deng, H. 2020. Groundwater irrigation and management in northern China: Status, trends, and challenges. *International Journal of Water Resources Development* 36(4), 670–696.

Wani, S. P., Rockström, J. and Oweis, T. Y. (Eds). 2009. *Rainfed Agriculture: Unlocking the Potential*. Wallingford: CABI; Patancheru, Andhra Pradesh, India: International Crops Research Institute for the Semi-Arid Tropics (ICRISAT); Colombo, Sri Lanka: International Water Management Institute (IWMI). 310p.(Comprehensive Assessment of Water Management in Agriculture Series 7).

Vijai, P. and Sivakumar, P. B. 2018. Performance comparison of techniques for water demand forecasting. *Procedia Computer Science* 143, 258–266. https://doi.org/10.1016/j.procs.2018.10.394.

Yang, W., Long, D., Wada, Y., Burek, P. and Pan, Y. 2018. Evaluation of groundwater recovery using GRACE observations and high-resolution community water model in the North China plain. In: *AGU Fall Meeting Abstracts* (Vol. 2018, pp. H13A-07).

Part 2

Improving agricultural water management practices

Chapter 4

Agronomic practices for optimising soil water retention

Samuel I. Haruna, Middle Tennessee State University, USA; and Stephen H. Anderson, University of Missouri, USA*

1 Introduction

An important ecological function of soils is their ability to retain nutrients, and especially water (Blum, 2005). Prevailing climatic conditions and soil management practices can significantly influence the ability of the soil to provide these benefits. Increasing variability of climatic conditions means that management practices need to be optimised to maintain, and possibly improve crop productivity for the increasing global human population. These practices include tillage, cover cropping, crop rotation, and livestock grazing.

Tillage involves the manipulation of the soil for seedbed preparation, mechanical weed termination, relieving compaction, and increasing water evaporation from fields in very wet conditions. There are different types of tillage, and their choice is based on several soil properties including soil wetness, temperature and structure, and susceptibility to compaction, crusting, and erosion (Logan et al., 1991). The choice can also be culturally motivated. Conventional tillage (CT) often involves the use of traditional (to that environment) and often more destructive equipment to partially or completely overturn the soil and expose buried soil materials. Conservation tillage often

* Corresponding author: AndersonS@missouri.edu.

http://dx.doi.org/10.19103/AS.2023.0123.04

requires that at least 30% of crop residues be left on the soil surface and this frequently results in reduced susceptibility to erosion and stratification of organic material in the topsoil horizon (Mathew et al., 2012).

Cover cropping systems place more emphasis on the ecological benefits of crops rather than the economic benefits, especially during periods when the soil would otherwise be left bare. These ecological benefits include bio-tillage (Zhang and Peng, 2021), weed suppression through biomass production (MacLaren et al., 2019) and allelopathy (Osipitan et al., 2019), carbon sequestration and soil organic carbon (SOC) improvement (Lal, 2015), nutrient optimisation through scavenging and recycling (Haruna and Nkongolo, 2020), improving the abundance and diversity of soil organisms (Thapa et al., 2021), and soil water conservation (Daigh et al., 2014).

Crop rotation involves growing a sequence of plants during the regular growing season with slightly more emphasis placed on their economic benefits rather than their ecological benefits. Although more recent iterations of crop rotation have demonstrated many ecological benefits (West and Post, 2002; Venter et al., 2016), potential yield increase is still a predominant reason for their adoption (Dury et al., 2012; Sumner, 2018). However, due to improvements with and more reliance on genetically modified crops, the adoption of crop rotation in most developed parts of the world has been declining in recent years (Fausti et al., 2018).

Livestock grazing practices such as rotational grazing are being implemented to reduce the environmental impact of the more intensive continuous grazing that was dominant for much of agricultural history. Rotational grazing can reduce some of the impacts of a near-ubiquitous intensive grazing management practice like loss of plant and soil organism diversity, disruption of ecosystem functions, and change in physical characteristics of the soil (Fleischner, 1994). More recently, reports have shown positive or neutral effects of rotational grazing on soil ecosystem functions and plant species richness (Fensham et al., 2014). As a result, there is a desire for rotational grazing to be used as a tool for soil conservation (Williamson et al., 2014).

One of the ways to improve soil conservation and improve plant water use efficiency is to optimise available soil water. This can be achieved by implementing practices that relieve natural- and anthropogenic-induced soil particle consolidation, improve SOC, increase soil porosity, enhance infiltration, improve water retention, and reduce surface runoff. Current research efforts in soil conservation are focused on identifying new practices or ways to improve present management practices to enhance crop productivity and environmental sustainability. This chapter reviews the literature on the influence of soil management practices on water infiltration and retention.

2 Agronomic practices

2.1 Tillage

Tillage is a broad term used to describe any form of seedbed preparation for crop production and has been used since intentional crop production began. Early civilisations (e.g. the ancient Egyptians, the Incas of the Andes of South America) as well as indigenous cultures around the world frequently used sticks to make holes in the ground, put seeds in the soil by hand, and then covered the seeds with the foot (Harlan, 1995). As such, early tillage was a means of planting seeds, probably to enhance seed-soil contact and because of the human power requirements.

As a result of this power requirement, the earliest forms of tillage were an inefficient use of time and energy, and only a small portion of land could be tilled. With increasing human population and a more sedentary lifestyle, animals were introduced to provide the required power and reduce the physical stress on the farmer. Even with a reduction in the physical stress, the producer still needed to walk about 35 km behind their plough for each hectare tilled (Harlan, 1995). Over time, the plough was mechanised, and this greatly increased the area that could be tilled while also reducing the physical stress on the producer. As a result, the use of tillage expanded to include relieving surface soil compaction, raising soil temperature in cold climates, increasing water evaporation from wet soils, and terminating weeds.

Even though this conventional form of crop production increased yields early on, incessant use led to some environmental challenges, most notably the 'Dust Bowl' which occurred on the American Great Plains in the early-mid 1930s. This led to research on 'conservation' and 'stubble mulch' tillage, both of which were forerunners of the current no-till (NT) practices (Sainju, 2020). The Conservation Technology Information Center (CTIC) defined conservation tillage as any form of tillage system in which at least 30% of the soil surface is covered with crop residue after planting to reduce soil erosion (CTIC, 2004). Further, Busari et al. (2015) defined conservation tillage systems as those that involve the management of soil cover through residue management achievable by practising reduced or no tillage with minimum soil disturbance. The minimum soil disturbance can improve soil biological activity, while the retained crop residue can protect the soil from the direct impact of raindrops, increase the roughness coefficient of the soil surface, and protect the soil surface from direct sunlight, helping to protect aggregate stability and reducing surface soil water evaporation, respectively. These practices can potentially reduce soil degradation while also improving soil productivity.

Edward Faulkner's book 'Plowman's Folly' (Faulkner, 1943) as well as the introduction of plant growth regulators developed during World War II has been credited for revolutionising the gradual shift from the conventional plough (Phillips and Phillips, 1984). During the mid-twentieth century, Moody et al. (1961), Free et al. (1963), and Triplett et al. (1963) were among the first researchers to report on NT crop production. Over the next decade, through research, education, and some government incentives, NT adoption in the USA increased steadily from 2.2 million ha in 1973, 4.8 million ha in 1983, 22 million ha in 2003, 38 million ha in 2012, and 42 million ha in 2017 (USDA-NASS, 2017).

In South America, NT experimentation began in Brazil in 1971 and has been gaining popularity over time. Currently, more than 50% of cropland in Brazil, Argentina, and Paraguay are managed with NT (Calegari et al., 2020). Compared to North and South America, NT adoption is still slow in Africa, Asia, and Europe, possibly because most farmers in these regions are small landholders depending on peer-to-peer extension activities (Thierfelder, 2020; Jayaraman et al., 2020; Stroud, 2020). Even though NT research and practice did not begin until the late twentieth century, NT adoption has grown rapidly in Australia. Currently, Australia is one of the world leaders in NT adoption, with about 67% of its cropland under NT management (Cornish et al., 2020).

2.2 Cover crops

Most land management practices involve a fallow period, usually between the harvest of one season's crops and the planting of the next. Left bare, the soils can be susceptible to soil loss from the action of water, wind, or both. As such, one of the main goals of current cover crop (CC) practices is soil conservation. They can also be used to prepare the soil before planting (Yunusa and Newton, 2003). Due to their benefits in reducing soil bulk density (Haruna et al., 2017; Adeli et al., 2020), increasing the proportion of transmission pores (Villamil et al., 2006), increasing water infiltration (Haruna et al., 2018a) and water retention (Hubbard et al., 2013; Haruna et al., 2022a), increasing transpiration of water from wet soils (Daigh et al., 2014), and increasing SOC content (Haruna, 2019) and microbial activity (Zhu et al., 2019), CC adoption has been rising gradually in recent decades (Kaye and Quemada, 2017). However, current adoption levels of CCs do not reflect the historical trend.

Historically, CCs were adopted into crop production systems (Reeves, 1994), first as green manures serving as a mulch and soil amendment, and then later for soil fertility improvement (Kasper and Singer, 2011). Leguminous CCs were mainly used for atmospheric N fixation until the mid-twentieth century when synthetic N fertilisers became more available in most of the developed parts of the world (MacRae and Mehuys, 1985). Currently, widespread adoption of CCs is limited due to the availability of synthetic fertilisers and potential management challenges for most first-time adopters. In recent decades, however, CC adoption is increasing gradually in some regions of the world.

During the late eighteenth and early nineteenth century in the USA, considerable soil loss was reported due to the abundance of fertile arable land and the associated soil stewardship illiteracy at that time (Langdale et al., 1991). Over the next century, poor management led to several soil erosion problems that culminated in the 'Dust Bowl'. The aftermath of this event was intensive research into soil loss mitigation, and this ushered in the establishment of several soil erosion demonstration plots and the development of the Universal Soil Loss Equation C (cover and management) and P (supporting practice) factors (Wischmeier and Smith, 1965). This focus on improving conservation management led to a considerable reduction in soil loss. For example, Bamett (1972) reported a 73% reduction in sedimentation rates in some northern Georgia reservoirs between 1939 and 1967. However, due to an increase in export opportunity for soybean (*Glycine max* L.) and wheat (*Triticum aestivum*) during the 1970s and 1980s, monocropping increased and CC adoption decreased (Follett and Stewart, 1985). Currently, due to increased research and education efforts by scientists and extension agents, and government incentives, CC usage has been increasing in the USA in the past decade. For example, in 2017, US farmers reported planting CCs on 6 million ha (3.9% of all croplands). This represented a 50% increase from the 4 million ha (2.6% of all croplands) planted to CCs in 2012 (Wallander et al., 2021).

In Europe, Smit et al. (2019) reported that the adoption of catch (CCs grown primarily to reduce soil erodibility and increase nutrient uptake and transfer to subsequent crops) and CCs based on the share of farmers adopting catch and CCs ranged from 12% in Spain to 46% in Romania, to 84% in France, and to 99% in the Netherlands. These authors noted that, except for the Netherlands (about 90%), adopters grow catch and CCs on a small share of their arable land (<20%). Most non-adopters in these regions cited lack of benefits, high cost and labour requirements, lack of awareness, unfavourable weather, and crop rotations as the major barriers to adoption and expressed willingness to adopt if additional subsidies were provided.

2.3 Crop rotation

The sequential cultivation of different crops in a specific order on the same field has been traced to ancient Greece as early as 6000 BCE when monoculture could no longer support the growing population (White, 1970). For most of documented agricultural history, some form of crop rotation has been practised. For example, a system in Africa employed a 36-year rotation where a finger millet (*Eleusine coracana* L.) is grown after a 35-year growth of woody shrubs and trees has been cut down. During the late seventeenth century, the 'Norfolk four-course rotation', was developed and this included wheat and turnip (*Brassica rapa* subsp. *rapa*) in the first two years, with barley (*Hordeum vulgare*) undersown with clover (*Trifolium repens* L.) and ryegrass (*Lolium perenne* L.) in the third year. The clover and ryegrass were grazed to cattle during the fourth year (Overton, 1989).

Early experiments at the Rothamsted experimental station in England demonstrated the usefulness of proper crop rotation cycles based on cultivated rows, close-growing grains, and sod-forming crops (Overton, 1989). This led to the widespread adoption of various rotation systems. One of the important advantages of crop rotation that played a significant role in their adoption was their flexibility; crops can be adjusted to many soil and environmental situations, changes can be made easily, and it favours the implementation of green manures.

Over the decades, economic conditions, government policies, and the reliance on genetically modified crops in various parts of the world have limited the adoption and continued practice of robust crop rotations (Fausti et al., 2018). Fausti et al. (2018) reported that increasing reliance on insecticides and herbicides for mitigating pests and weeds, respectively, corresponds with the reduction in the reliance on traditional crop rotation practices in the northern Great Plains and much of the Midwestern United States. Furthermore, Dill et al. (2008) reported that the use of glyphosate-resistant crops has grown to more than 74 million ha in 5 crop species in 13 countries. The largest areas of glyphosate-resistant crops, they argued, are soybean (54.2 million ha), corn (*Zea mays*) (13.2 million ha), cotton (*Gossypium* herbaceum L.) (5.1 million ha), canola (*B. napus*) (2.3 million ha), and alfalfa (*Medicago sativa*) (0.1 million ha) with the USA, Argentina, Brazil, and Canada having the largest plantings globally. If current trends continue, traditional crop rotations may continue to dwindle. However, the increasing adoption of CCs may be a way to increase cropping system diversity in the future.

2.4 Grazing practices

The early nomads first domesticated sheep and goats, around 7000 BCE, and are credited with the first grazing practice ever implemented. The gradual shift from a hunter-gathering to a more sedentary one led to the domestication of more animals, including cattle and sheep (Painter et al., 1993). Over several millennia, grazing has evolved to include conservation practices that maintain the sustainability of the system, animal, and soil health. The first step towards improving grazing practice was rotating pastures within the year or alternate years to give the forage time to recover (as compared to continuous grazing). The next step involved rotating the cattle more frequently while still allowing the forage time to recover. This was more management-intensive than the first step. Eventually, high-stock-density grazing was developed, which involves intensive grazing on pre-assigned portions of the field and rotating the cattle after a pre-determined period (Painter et al., 1993; Briske et al., 2011).

Livestock grazing on both public and private lands in the USA dates to the nineteenth-century westward expansion. Since this time, grazing has contributed to a lot of controversy over the management of public lands due to the degradation of rangelands and water resources. This led to the passage of the Taylor Grazing Act by the US Congress in 1934 in order to regulate rangelands and water resources (Ross, 1984). Despite this legislation, grazing practices have faced reduced adoption. For example, in 2015 alone, about 1.4 million ha of grazing land (out of a total of 314 million ha) was lost to excessive degradation (Cherney, 2015). Due to the environmental effects of intensive grazing, about 25% of grazing land in the USA need some conservation practices and structures to reduce soil erosion (USDA-NRCS).

In South America, the Food and Agricultural Organization (FAO, 2000) estimated that about 26% of cattle in this region are grazed on 600 million ha of grazing land. Compared to North America, the adoption of cattle grazing has been increasing at a rate of 1.6% annually over the past 3 decades and is projected to grow by 0.9% annually by 2030. The population of sheep, goats, and pigs on grazing land is projected to grow by 0.7%, 0.7%, and 0.9%, respectively, annually by 2030. Although these numbers represent a decline from previous decades, the adoption rate is expected to exceed those in the more developed countries by 2030.

Finally, the increasing variation in global climatic patterns and the increased focus on soil conservation in recent decades is responsible for the decline in the adoption of conservation grazing practices globally (Hyland et al., 2018; Schaak and Mußhoff, 2018). In later sections, the influence of grazing practices on water infiltration and retention will be discussed.

3 Soil water infiltration and retention

Water is an essential component of life, and in crop production, it is the most limiting factor. When water is applied to the soil surface, and depending on the hydraulic conductivity of the soil, the antecedent soil moisture content, and soil aggregation, part of that water can either enter the soil, remain ponded on the soil surface (if depressions are present), or runoff as overland flow. The movement of water from the atmosphere into the soil (infiltration) is inversely proportional to the runoff from the soil, so it is critically important for agronomic and environmental purposes. Furthermore, the rate of water infiltration governs the amount of water storage in the soil. The following subsections detail the factors influencing water infiltration and retention and their long-term benefits.

3.1 Factors influencing soil water infiltration and retention

The infiltration rate (volume flux of water entering through a unit surface area) of soil is dependent on soil structure, particle size distribution, antecedent soil moisture content, soil matric potential, as well as size, continuity, and stability of soil pores (Lal and Vandoren, 1990). Land management practices such as tillage, cover cropping, crop rotation, residue management and conservation practices including buffer strips can influence these factors. The arrangement of soil particles can increase or reduce the rate of soil water infiltration. These structures could be granular, platy, blocky, or elongated. Generally, water infiltration is higher in soils with granular and elongated structures as compared with platy structures with more tortuosity (Hillel, 1998). However, platy structures generally retain more water than any other soil structure because of the long, indirect route of water through these structures, and their larger surface area. Similarly, the rate of water infiltration is generally higher in soils with sandy and silty size particles compared with clay size particles (Hillel, 1998).

In nature, the soil is seldom devoid of any water molecules. The amount of water present in the soil before precipitation or irrigation is known as the antecedent soil moisture content. In soils with low antecedent soil moisture content, the rate of water infiltration is high, initially. Over time, the rate of infiltration then gradually decreases until it reaches a steady state, also known as the equilibrium infiltration rate. Soils with higher antecedent moisture content may be indicative of higher water retention (Hillel, 1998).

Soil pore sizes can be classified, based on their effective diameters, as macropores (>1000 μm), coarse mesopores (60–1000 μm), fine mesopores (10–60 μm), and micropores (<10 μm) (Anderson et al., 1990). A higher proportion of macropores and mesopores tend to increase water infiltration rates while a higher proportion of micropores increase soil water retention (Hillel, 1998). Besides pore size distribution, the connectivity of soil pores also significantly influences soil water infiltration. Interconnected and stable soil pores tend to increase water infiltration (Bhattacharyya et al., 2006).

Soil matric potential represents the water energy due to the interaction of water with soil particles through capillarity or adsorption and refers to the amount of energy that must be exerted in order to extract a molecule of water from the soil and is related to soil water content. At saturation (SAT), the matric potential of the soil is close to 0 kPa, while at permanent wilting point (PWP), the matric potential is close to −1500 kPa. Therefore, soils with lower matric potentials generally have higher water infiltration and retention rates (Swarowsky et al., 2011).

3.2 Long-term benefits of soil water infiltration and retention

Researchers have reported on the benefits of water infiltration and retention on crop productivity (Bharati et al., 2002; TerAvest et al., 2015; Adugna, 2016) and environmental sustainability (Yu et al., 2003; Cawson et al., 2016). Increased water infiltration and retention can reduce surface runoff and erosion (Abrol et al., 2016) and this can maintain the structure of surface soils which translates to further improvements in water infiltration and storage. These benefits can accrue over time and have been studied in long-term experimental research field trials such as the historic Sanborn Field in Columbia, Missouri, USA.

The Rotation Field, later renamed Sanborn Field in 1924, after the former Dean of the College of Agriculture J. W. Sanborn, was established in late 1888 as a research tool (Upchurch et al., 1985). It was established to study the effects of various rotation practices and constant tillage on soil fertility and to also evaluate the effects of various fertility practices (chemical fertilisers, manured and non-manured) on crops. Over time, research from the field proved to be important and the experiment continues to be maintained for current research purposes. For a full history of the Sanborn field, readers are directed to Upchurch et al. (1985) and Brown and Wyman (1989).

Anderson et al. (1990) evaluated the effects of a century of crop and land management on soil physical properties at Sanborn Field. These researchers reported that, although management practices may have subtle effects at the site annually, over time, these benefits accumulate and may reduce runoff and the potential for soil erosion. Similarly, Buyanovsky and Wagner (2019) reported that the long-term level of SOC resulted from annual inputs, which cumulatively benefits crop productivity and environmental sustainability. Further, Cetin (2015) conducted a study on the hydraulic properties at Sanborn Field 127 years after establishment and reported significant differences in soil water content between various land management practices [continuous corn (*Z. mays*), wheat, timothy (*Phleum pratense*), and a rotation of corn, wheat, and red clover (*Trifolium pratense*) with and without farmyard manure] at 0 (SAT), −33 [field capacity (FC)), and −1500 (PWP) kPa matric potentials as a result of long-term management.

In a study on the effect of long-term management practices on soil physical properties related to erodibility, Acikgoz et al. (2017) demonstrated that long-term perennial grass management reduced splash detachment by 75% compared with continuous corn management and concluded that long-term studies are important for the determination of the soil erosion possibilities of different management practices and for establishing conservation structures.

4 Agronomic practices and water infiltration

4.1 Tillage and water infiltration

Mechanically agitating the soil with various tillage equipment can break up soil clods and may temporarily increase water infiltration. Blanco-Canqui et al. (2017) evaluated the impact of moldboard plough on soil hydraulic properties and infiltration. They reported that, after 1 year, the cumulative 3 h infiltration was significantly higher under moldboard plough compared with NT. Similarly, Haruna et al. (2018a) reported that 3 months after moldboard plough tillage, cumulative infiltration was higher under tilled, compared with NT management. This was attributed to the lower antecedent water content induced by higher evaporation rates due to tillage.

Further, Lipiec et al. (2006) reported that ploughing to a depth of 0.2 m increased cumulative infiltration (after 3 h) by 61% compared with NT due to the temporary large flow-active pores generated by tillage. In an earlier study by Jones et al. (1994) right after tillage, cumulative infiltration after 2 h was 90% higher under a stubble mulch tillage compared with NT. These authors reported that the stubble mulch tillage destroyed the soil surface crust, lowered soil bulk density, and increased depression storage capacity.

Although CT systems have been shown to temporarily improve water infiltration, their contribution to sustainability should be evaluated in the longer term. Due to increased equipment traffic and the significant incorporation of surface residue deeper into the soil (in extreme cases), repeated tillage may lead to soil compaction and consequently reduced water infiltration over time. For example, Capowiez et al. (2009) and de Almeida et al. (2018) reported that cumulative infiltration was similar between tilled and NT fields. In fact, de Almeida et al. (2018) concluded that vegetation cover (depending on the type), rather than tillage systems, significantly influence water infiltration due to their ability to reduce rainfall energy and the sustainability of root-generated bio-pores.

The lack of mechanical soil agitation may benefit water infiltration over time due to reduced traffic, increased surface residue, and improved soil structure. Earlier studies by Azooz and Arshad (1996) and Shukla et al. (2003) showed that, due to reduced traffic, bulk density was significantly lower and cumulative water infiltration (after 3 h) was significantly higher under NT management compared with CT. More recently, Blanco-Canqui and Ruis (2018) reported that due to improvement in wet aggregate stability, NT management had 86% higher water infiltration compared with CT. In addition, Stone and Schlegel (2010) reported that the steady-state infiltration rate was 63% higher under NT management compared to CT in the West-Central Great Plains of the United States. These authors concluded that NT management may be more sustainable in this region compared with CT due to better aggregate stability, higher infiltration rate and the lack of reconsolidation during the growing season.

4.2 Cover crops and water infiltration

Cover crops can increase water infiltration through increased water transpiration, especially in areas with very heavy spring precipitation amounts (Haruna et al., 2018a). Additionally, the leaves of living plants and the residue of CCs can reduce the destructive kinetic energy of raindrops, and the roots of CCs can also increase soil macropores, all of which are beneficial for water infiltration. For example, Kemper and Derpsch (1981) reported that CCs (winter annual legume and rapeseed [*B. napus* L.]) increased water infiltration rate into an Alfisol in Brazil by 629% due to CC-induced bio-pores. Similarly, Wilson et al. (1982) reported that a Stylo (*Stylosanthes guianensis*) CC significantly increased macropores and water infiltration rate into an Alfisol in Nigeria by 13% compared with no CC management. Furthermore, Bruce et al. (1992) (CC = crimson clover; *Trifolium incarnatum*) and Chalise et al. (2018) (CC = rye; *Secale cereale* and hairy vetch; *Vicia villosa*) reported that CCs increased infiltration rate by 100% and 79%, respectively, compared with no CCs.

Besides improving the infiltration rate, CCs have also been reported to improve cumulative infiltration. For example, Folorunso et al. (1992) reported that bromegrass (*Bromus* spp.) and strawberry clover (*Trifolium fragiferum*) improved cumulative infiltration by 97% and 91%, respectively, compared with a fallow rotation. Similarly, Blanco-Canqui et al. (2011) reported that Sunn hemp (*Crotalaria juncea*) increased cumulative infiltration by 163% relative to no CC management. This was attributed to high earthworm population enhanced by NT and CC management. Additionally, Nouri et al. (2019) reported that NT vetch and wheat CCs improved cumulative water infiltration by 86% and 166%, respectively, compared with no CC management due to higher porosity generated by CC roots. Additionally, Haruna et al. (2022b) reported that multi-species CCs significantly increased cumulative infiltration by 52% prior to termination and by 68% two months after termination, compared with no CC management (Fig. 1). These authors attributed this difference to the CC residues reducing soil particle rearrangement by absorbing the kinetic energy of raindrops two months after termination.

Due to the ability of CCs to transpire excess water from the field, soil and climatic conditions are important considerations before the adoption of CCs. In very wet environments, CCs can help lengthen the growing season and this can lead to improved crop productivity and a reduction in soil erosion. However, this occurrence can be damaging in arid and semiarid regions and during periods

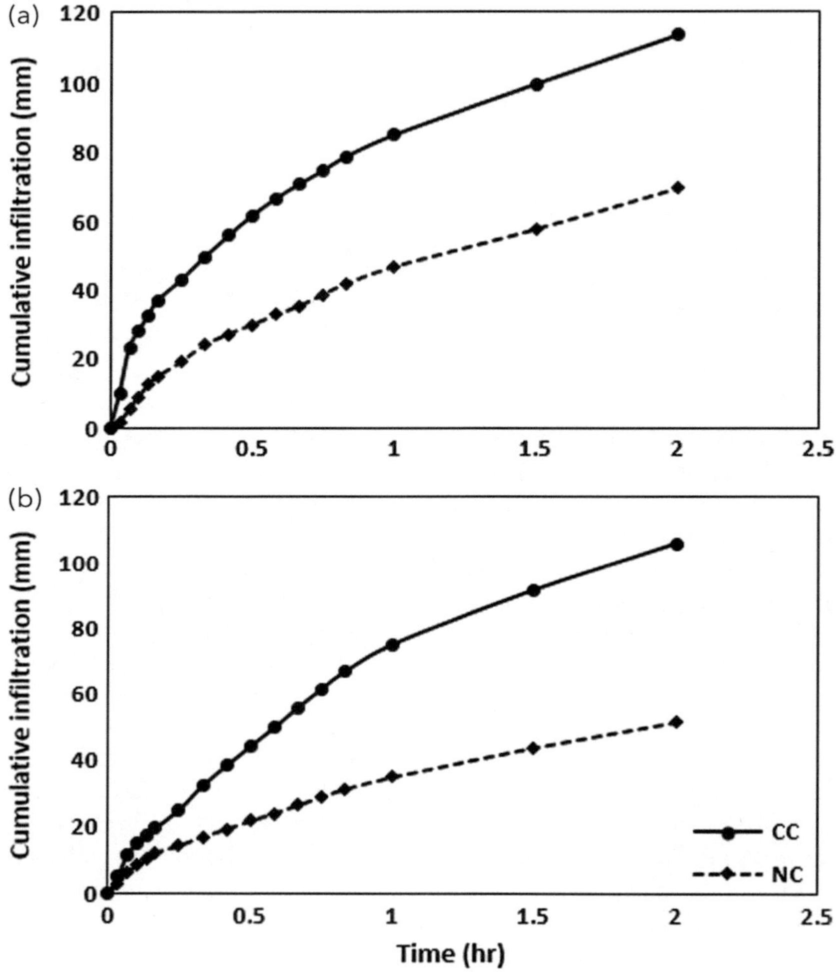

Figure 1 Cumulative infiltration between cover crop (CC) and no cover crop (NC) management (a) just prior to CC termination and (b) 2 months after CC termination. The 2-h cumulative infiltration for CC was 52% and 68% higher than NC in (a) and (b), respectively (redrawn from Haruna et al., 2022b).

of drought in humid and sub-humid regions. However, it has been reported that the differences in soil water content between CC and no CC management may not be significant in the long-term, to reduce crop productivity (Daigh et al., 2014). As such, appropriate CC species selection and proper CC termination timing can help maintain and improve overall crop productivity.

4.3 Crop rotation and water infiltration

Crop rotations are beneficial for biological pest (Rusch et al., 2013) and weed (Simić et al., 2016) control, increasing soil nutrient availability and turnover (Haruna and Nkongolo, 2020), and improving soil health indicators (Gozubuyuk et al., 2014). An important soil health indicator is SOC which can indirectly improve water infiltration. For example, a 2-year wheat–green gram (*Vigna radiata* L.) rotation study by Sarma et al. (2018) showed that SOC was significantly improved by crop rotation compared with a fallow system. Chan et al. (2001) reported that, after 4 years of lucerne (*Medicago sativa*)–consol lovegrass (*Eragrostis curvula*)–barrel medic (*Medicago truncatula*) rotation, water infiltration was significantly higher in the rotation compared with fallow due to 26% higher SOC. Similar findings were reported by Govaerts et al. (2006).

Crop rotations can also influence water infiltration directly. For example, Bruce et al. (1990) showed that soybean and grain sorghum (*Sorghum bicolor*) following winter wheat rotation improved water infiltration compared with no rotation. These researchers concluded that crop rotation effects can only be observed under NT and can be erased or modified by tillage, after 8 years of management. Further, Blair et al. (2006) conducted a long-term (1966-1999) experiment and reported that lucerne–clover–medic rotation resulted in 87% higher surface infiltration (10 mm water tension) compared with long fallow due to better aggregate stability and more continuous pores from the plant roots. Additionally, Govaerts et al. (2007) conducted a 12-year study on the effects of a monoculture and maize–wheat rotation on soil water infiltration and water content. They found that NT crop rotation significantly increased water infiltration and content by inducing a 'vertical' mulching effect compared with maize monoculture.

In a 33-year study, Blair and Crocker (2000) reported that lucerne–clover–wheat rotation significantly improved surface infiltration (at 30 mm and 40 mm water tensions) compared with fallow treatment. Conversely, Chang and Lindwall (1992) did not find any significant difference in water infiltration between winter wheat–barley rotation and fallow after 8 years. However, these researchers (Chang and Lindwall) reported a numerically higher infiltration rate under rotation treatment as compared with fallow.

Most of the currently available literature suggests that crop rotation alone (without other supporting practice such as tillage) that included three or more different crops tend to significantly improve water infiltration compared with rotations only including two crops (Basche and DeLonge, 2019). This could be due to the higher density of crop roots and more SOC generated over time by the many roots of these crops. Also, with the increasing availability of genetically modified crops that can provide some of the soil property improvement benefits of traditional crop rotations, the adoption of these rotation systems has been declining in recent decades. However, the increasing adoption of CCs is a method used by many farmers to rotate their cropping systems and improve crop productivity.

4.4 Grazing practices and water infiltration

Grazing timing, intensity, and frequency can influence water infiltration indirectly or directly. Intensive/conventional grazing can influence vegetation dynamics, and soil properties which may reduce water infiltration (Teague et al., 2011). For example, Sirimarco et al. (2018) reported that conventional grazing increased soil bulk density and reduced water infiltration. Further, intensive grazing can result in the early death of plants which reduces SOC accumulation and water infiltration (Teague et al., 2011). Similarly, Kumar et al. (2012) reported slightly better infiltration parameters under rotationally stocked pasture compared with continuously stocked pasture due to less animal density and traffic in the rotationally stocked pasture.

An alternative to intensive grazing is adaptive multi-paddock grazing (or adaptive grazing) which references natural rotational grazing dynamics by large herbivores in grassland environments (Hillenbrand et al., 2019). This system of conservative grazing is characterised by very short grazing periods at high stock densities followed by an extended 'rest' period for plant regeneration (Savory and Butterfield, 2016). Adaptive grazing has been reported to break up soil surface crusts, ease the integration of residues into the soil (Butterfield et al., 2019), and significantly increase water infiltration compared with intensive grazing (Hillenbrand et al., 2019). Additionally, Döbert et al. (2021) reported that adaptive grazing increased water infiltration by 42% compared with traditional (intensive) grazing practices in grassland soils of Canada. This increase in water infiltration was positively correlated with increased litter mass and lower bulk density.

In an earlier study at the High Plains Grasslands Research Station in Cheyenne, Wyoming, USA, Abdel-Magid et al. (1987) reported that heavy grazing (2.25 ha/steer) significantly increased soil bulk density due to the static loads of grazing animals and this led to a reduction in water infiltration compared with moderate (3.0 ha/steer) and light (5.25 ha/steer) grazing. In addition, Pluhar et al. (1987) showed that continuous grazing increased soil bulk density while rotational stock grazing increased infiltration rates and decreased sediment production rates.

An important consideration with grazing practices and water infiltration is the duration of the grazing management and what will happen upon ending this practice. Sharrow (2007) showed that grazing in a silvopasture significantly reduced soil porosity and water infiltration after 2 years compared with silviculture. However, these effects were quickly reversed upon cessation of grazing and had little detrimental long-term effect on silvopasture forage or tree production. Further, Bell et al. (2011) showed that although animal treading can increase soil bulk density and reduce water infiltration, most of these effects were restricted to the soil surface (0–0.1 m depth). When animal traffic was reduced or eliminated, most of these detrimental effects were short-lived due to amelioration through natural processes and were not enough to significantly reduce subsequent crop yields.

Since the detriments of temporary intensive grazing are limited to the soil surface, their effects may be too small in magnitude or depth to incur long-term yield penalties. Furthermore, the root morphology and density, and the season-dependent nature of most subsequent grain crops might further help to reduce any long-term effects of grazing. Therefore, to improve water infiltration, studies have shown that a reduction in the intensity, density, and frequency of grazing is desirable (Bell et al., 2011; Hunt et al., 2016).

5 Agronomic practices and water retention

5.1 Tillage and water retention

Crop productivity depends on the amount of water retained in soils and was one of the primary reasons for tillage: 'to bring soil moisture from the deeper depths to the soil surface' (Faulkner, 1943). Faulkner argued that the premise for tillage adoption and encouragement was anecdotal rather than scientific. Since then, there have been mixed results on the effects of tillage on water retention, with studies showing either an increase, no effect, or a decrease in water retention as a result of tillage.

In a study on the impact of different types of tillage on soil moisture dynamics, Slawinski et al. (2012) demonstrated that reduced tillage had 10% higher water retention at 0.05 m soil depth compared with traditional tillage and NT. At 0.15 m depth; however, traditional tillage had 16% higher water retention compared with reduced tillage and NT. More recently, Soltani et al. (2019) reported that soil water retention improved under CT compared with NT. This was attributed to the use of organic fertilisation in combination with CT. Obour et al. (2018) reported a similar result, showing a strong relation between SOC and the range of water contents for tillage.

Crittenden et al. (2015) evaluated the effects of tillage on soil physical properties at two sites with organic soils and reported that non-inversion tillage (tillage depth = 0.18–0.23 m) improved soil water retention compared with moldboard plough (tillage depth = 0.23–0.25 m) at the second site but had no effect at the first site. This was attributed to better aggregate stability (at the 0.10–0.20 m depth) under non-inversion tillage compared with moldboard plough tillage at both sites. Similarly, Blanco-Canqui et al. (2017) reported that there was no difference in water retention between moldboard plough, chisel tillage, and NT management, and this was similar to reports by Haruna et al. (2018b).

Despite these studies, other researchers have found that NT improves soil water retention compared with conventional and reduced tillage. For instance, Diiwu et al. (1998) found that available water content was higher under direct seeding compared with CT. However, in an earlier long-term (12 years) research, Hill (1990) reported that CT significantly increased the proportion of pore spaces available for water storage compared with NT; utilisation of total water resources was better under NT management because of lower soil water evaporation and higher water infiltration. Further, Huang et al. (2021) reported that NT soil, in the field and laboratory, had a higher water retention capacity and a relatively stable water retention curve. This was attributed to the higher rigidity of soil pore systems and the straw mulch on the soil surface in NT management. However, under field conditions, the water retention capacity of tilled soils (conventional and rotational tillage) increased gradually with wet-drying cycles while that of NT at the 0 to 0.05m depth decreased slightly. Similarly, Tuzzin de Moraes et al. (2016) reported that a decadal use of NT management improved soil physical quality and provided higher plant available water content in the soil between −10 kPa and −200 kPa matric potentials compared with conventional and minimum (yearly chisel) tillage.

In a study from two sites in the northern British Columbia region of Canada, Arshad et al. (1999) demonstrated that NT management resulted in 34% higher SOC and 55% higher water-stable aggregation compared with CT at the top 0.05 m of a silt loam soil. As such, NT management significantly increased soil water retention compared with CT without changing bulk density due to redistribution of pore sizes into small and large pores. Also, Bescansa et al. (2006) reported that soil organic matter was 13% higher (at 0 to 0.15m depth) under NT compared with moldboard plough tillage. These researchers further reported that, due to the reorganisation of pore sizes because of tillage, available water capacity was significantly higher under NT compared with moldboard tillage. Further, Schwen et al. (2011) demonstrated that NT management significantly improved pore connectivity and reduced pore tortuosity. This resulted in higher soil water retention under NT management compared with conventional and reduced tillage. Further, NT management resulted in a better temporal stability of the pore network and soil hydraulic properties. These results were attributed to better established soil structure and microbial activity under NT management.

Surface conditions can also determine how tillage behaves relative to water retention. For example, Lampurlanes et al. (2001) reported that root length density and volumetric water content were 5% (at 0–0.25 m depth) and 6% (at

0.25–0.50 m depth) higher under NT compared with minimum (field cultivator to a depth of 0.15 m) tillage and this led to higher yields under NT management. These authors concluded that surface conditions are of major importance in soil water content and are determined by differences in tillage types. They also surmised that NT was a better management practice for subsequent years of lower precipitation compared with minimum tillage.

Temporal variability can also determine the overall effects of tillage on soil water retention. Pena-Sancho et al. (2017) reported that, right after tillage, no significant differences were observed in bulk density and water content at saturation between moldboard plough tillage and NT. However, post-tillage rainfall significantly increased soil bulk density and reduced water content at saturation under CT compared with NT management. In fact, these authors reported that the biggest difference in water retention occurred right after the first rainfall events (>100 mm) following tillage.

Current literature is not conclusive on the exact effect of tillage on soil water retention. However, the predominant conclusions show that right after tillage, increased macroporosity leads to higher water retention under tillage systems. After a few rainfall events, soil consolidation tends to reverse this initial benefit. As such, NT management may be more beneficial to water retention in the long term. Further, conservation tillage tends to be more beneficial in the well-drained soils found in arid and semi-arid regions compared with poorly drained soils in humid and subhumid regions (Lampurlanes et al., 2001).

5.2 Cover crops and water retention

Soil water storage is dependent on soil texture and structure, and porosity, among other factors. Cover crop root-generated biopores can increase water retention at FC (Basche et al., 2016). Bilek (2007) reported that cereal rye CC increased volumetric water content at −10 kPa and −30 kPa water pressures by 4% and 5%, respectively, compared with no CC management. Since −10 kPa and −30 kPa water pressure corresponds to the presence of fine mesopores (10–30 μm empirical spherical diameter) (Kay, 1997), this suggests that CC roots can generate coarse mesopores which are instrumental in the retention of water at these pressures, as reported by Haruna et al. (2018b) (Fig. 2). Additionally, Hubbard et al. (2013) and Haruna and Anderson (2022) reported an 18% and 52% higher water retention at FC, respectively, under CC compared with no CC management. Further, Bilek (2007) and Villamil et al. (2006) both reported that CC management significantly increased soil organic matter and improved soil aggregation, which led to an increase in water retention at FC, compared with no CC management.

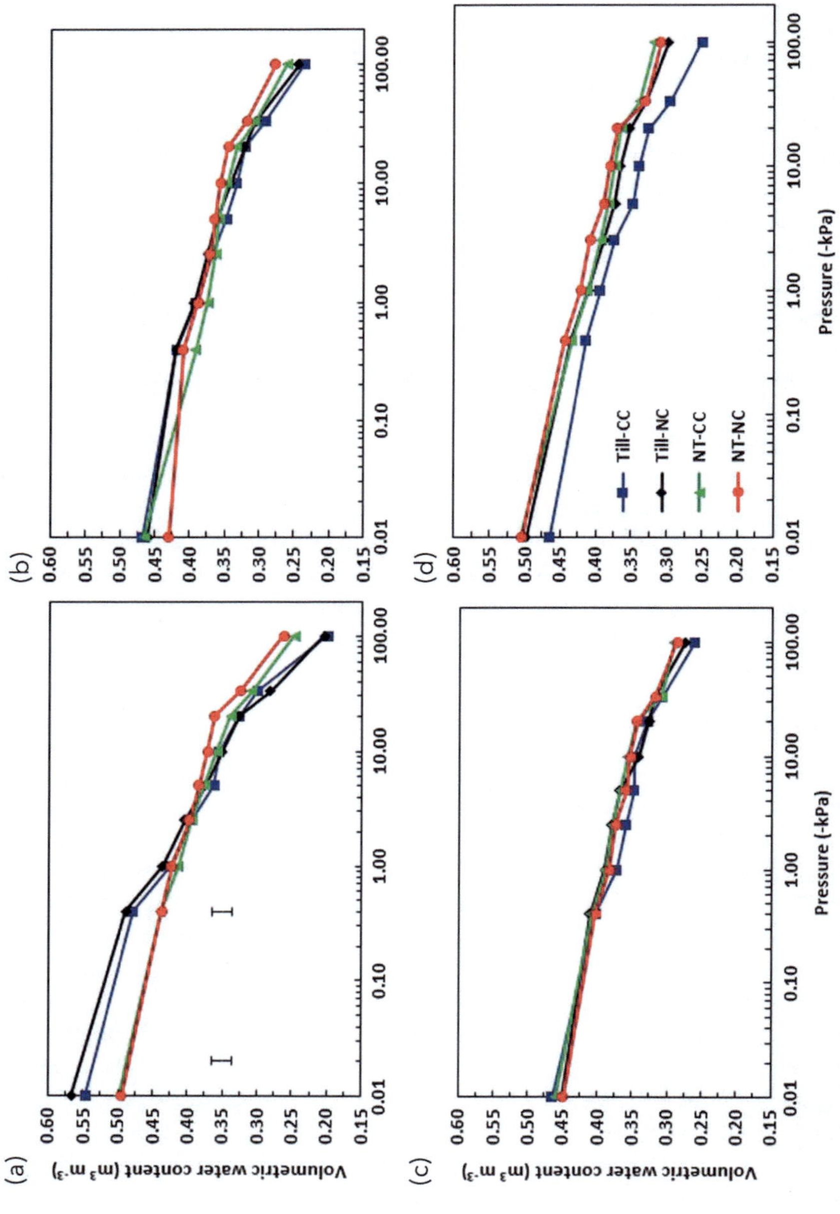

Figure 2 Soil water retention at selected depths; (a) 0–10 cm, (b), 10–20 cm, (c) 20–30 cm, and (d) 30–40 cm under different tillage and cover crop management 2 weeks after termination. Please note: Till-CC = tillage with cover crop, Till-NC = tillage with no cover crop, NT-CC = no-till with cover crop, and NT-NC = no-till with no cover crop. Note that statistical differences were only observed in (a) at 0 kPa and −0.4 kPa soil water pressures (redrawn from Haruna et al., 2018a).

In contrast, Ward et al. (2012) reported that CCs had no significant impact on soil water storage during summer and fall periods due to their little impact on total evaporation in a Mediterranean climate. However, these authors (Ward et al., 2012) reported that CCs occasionally provided short-term benefits right after rainfall. Similarly, other authors (Blanco-Canqui et al., 2011; Rorick and Kladivko, 2017; Nouri et al., 2019) showed that CCs had no significant impact on soil water retention at FC and wilting point, and thus, available water content was similar between CC and no CC plots.

These studies show that, despite the benefits of CCs on soil structure and water retention, climatic conditions may limit their water retention benefits. Since CCs can transpire excessive water from the field (Daigh et al., 2014), it might be detrimental for water storage in arid and semi-arid environments but beneficial in humid and sub-humid climates.

5.3 Crop rotation and water retention

With the current reduction in the practice of crop rotation in favour of various monocultures, an important consideration is their effect on soil water retention. Since SOC directly and indirectly influences soil water retention, Bullock (1992) reported that a 2-year corn-soybean rotation significantly improved SOC and soil water storage, compared with continuous corn, and this resulted in a 20% higher yield under the rotation system. In fact, these authors concluded that no amount of soil amendments can completely compensate for the difference in crop yield. López-Bellido et al. (1996), Karlen et al. (2006), and Zhao et al. (2020) reported similar findings. Furthermore, Tennakoon and Hulugalle (2006) reported that cotton–wheat rotation improved water availability and water use efficiency compared with monocropping in most years between 1997 and 2003.

A cotton-wheat rotation has also been reported to improve soil water storage, particularly during the early part of the growing season, compared with a cotton monoculture (Hulugalle et al., 2010). Bhattacharyya et al. (2006) evaluated the effect of tillage and crop rotations on pore size distribution and soil hydraulic conductivity in sandy clay loam soil and found that soybean–lentil and soybean–pea rotations resulted in better soil water retention compared with monocropping resulting from improved soil transmission properties caused by crop rotation.

Some of the improvements in water retention can be attributed to improvements in soil structure and weed suppression by crop rotation practices. For example, Chan and Heenan (1996) investigated the influence of crop rotation on soil structure and soil physical properties of an Oxic Paleustalf. These authors reported that crop rotation significantly increased volumetric water content at FC compared with monocropping. This was attributed to rotation-induced improvement in soil structure rather than an improvement in soil organic matter. Further, Weisberger et al. (2019) reported that crop rotations can reduce weed density by as much as 49% compared with continuous monocropping. This reduction in weed density can reduce the demand on soil water and may lead to increased water storage under these systems.

Conversely, some researchers either noticed that monoculture increased soil water storage or they did not notice any difference in water storage due to crop rotation. For example, Blanco-Canqui et al. (2010) reported that continuous wheat increased water retention by 10% between 0 kPa and −3 kPa matric potentials compared with sorghum-fallow rotation. This was attributed to higher SOC concentration under continuous wheat management due to continuous residue addition. On the other hand, Haruna and Nkongolo (2015) reported that a corn–soybean rotation did not significantly influence soil water content compared with either a corn or soybean monoculture. These contrasting reports could have resulted from the duration of the rotation. For example, Kong et al. (2019) reported that a 2-year wheat–soybean and wheat–corn rotation did not show any significant influence on soil water content compared with continuous monocropping, regardless of the higher SOC under rotation management. These researchers concluded that long-term practice of rotation may be needed to confirm their influence on soil water storage in dryland cropping systems.

5.4 Grazing practices and water retention

The influence of livestock grazing on water retention is dependent on several factors including climate and grazing intensity, among others. In a study on the effects of livestock grazing on the hydraulic properties of gypsic soils, Moret-Fernández et al. (2021) reported that, although no livestock trampling effects were observed on soil water retention characteristics, high grazing intensities decreased soil water content. This was attributed to reduced vegetative and litter cover under the conventionally grazed management which caused a reduction in soil porosity. They concluded that more conservative, moderate grazing practices are more acceptable management for soil conservation in semi-arid regions with high gypsum content. Similarly, Krümmelbein et al. (2009) showed that conventional grazing reduced soil pore volume and this led to a significant decrease in saturated hydraulic conductivity and its anisotropy and water retention, compared with ungrazed fields. However, their hydrological model suggested that the ungrazed sites had higher evapotranspiration due to wetter soil conditions, decreased runoff, and more dense vegetation and litter cover, compared with the grazed sites. This suggests that 'no grazing' increased water use efficiency as compared with grazing management.

In a study of the effects of grazing intensity on soil water regime in Inner Mongolia grassland, Lei et al. (2012) showed that heavy grazing (1.2 sheep ha^{-1} $year^{-1}$) reduced soil water content by 43–48% and plant available water by 46–61% compared with continuous grazing (2.0 sheep ha^{-1} $year^{-1}$) management. Interestingly, these researchers demonstrated that net water flux during the growing season was similar between the heavy and continuous grazing sites at the 0.05–0.20 m depths, but the heavy grazing site had significantly higher upward flux compared with the continuous grazing sites. This, they suggested, showed that water was stored in the soil under continuous grazing management but depleted by root uptake under the heavy grazing management. Further, Bonetti et al. (2019) reported that intensive grazing modified pore size distribution decreased volumetric water content between −1 kPa and −10 kPa soil water pressure and increased it between −10 kPa and −1500 kPa of soil water pressure. Wang et al. (2019) reported similar findings at the 0–30 cm depth. Additionally, Tobin et al. (2020) reported that ungrazed treatment significantly increased soil water retention at the 0–0.05 m depths compared with grazing and this was attributed to several factors: (1) reduction in surface pore volume by grazing, (2) higher SOC levels under ungrazed management which has a higher water retention capacity, and (3) destruction of vegetation coverage by grazing which can result in water and wind erosion.

Conversely, Teutscherova et al. (2021) reported that, as early as 9 months after adoption, intensive short-duration rotational grazing (4.2 livestock units ha^{-1}) had significantly lower bulk density and higher water retention capacity as compared with the control (1 livestock unit ha^{-1}). This was attributed to higher SOC stocks and improved soil aggregation, and higher macrofauna abundance which can improve soil structure through pedoturbation. Therefore, these authors concluded that short-duration grazing can potentially increase cattle production per unit area while also improving or maintaining soil quality parameters in the short term. Zúñiga et al. (2015) reported similar findings.

Finally, intensive grazing tends to decrease SOC (Raiesi and Riahi, 2014), increase the bulk density and reduce the porosity of topsoil layers (0 to 0.15m depth) (de Andrade Bonetti et al., 2019), and reduce soil water content (Wang et al., 2019). Moderate grazing has been reported to better conserve SOC in the topsoil layers (0–0.15 m) (Becker et al., 2022), favour seed dispersion, and reduce the isolation between patches of vegetation (Pueyo et al., 2008). Therefore, moderate grazing systems might have a more beneficial influence on the ecosystem and offer better opportunities to increase carbon sequestration into soils as compared with intensive grazing.

6 Conclusion

Soil water management is important for agronomic and environmental sustainability, and improved management will be more pertinent in a rapidly changing global climate. In agriculture, several practices have been implemented, primarily to improve soil tilth and overall productivity. These practices include various types of tillage, CCs and residue management, crop rotation, and animal grazing practices, among others. With increasing variability in climatic conditions, efforts in recent decades have been focused on understanding the influence of these practices on soil water infiltration and retention. Researchers have reported varying, and sometimes contrasting, results of these management practices on soil bulk density, pore size distributions, water infiltration and retention, and saturated hydraulic conductivity. Reasons for the varying results include intensity of some of these management practices, differences in climatic conditions, soil types, and topography. In general, these practices can be used to optimise soil water retention by practising conservation tillage (reduced and NT) rather than CT, appropriate CC species selection (with considerations for the prevailing climatic conditions) and proper termination timing, improved surface residue retention, and less intensive crop rotation and grazing practices.

7 Where to look for further information

Below is a recommended list of relevant books, book chapters, and peer-reviewed articles that provide further information on how management practices can be used improve soil physical and biological properties for water optimisation.

7.1 Books

- Dang, Y. P., Dalal, R. C. and Menzies, N. W. 2020. *No-Till Farming Systems for Sustainable Agriculture: Challenges and Opportunities*. Springer, Cham, Switzerland.
- Eilittä, M., Mureithi, J. and Derpsch, R. 2004. *Green Manure/Cover Crop Systems of Smallholder Farmers: Experiences from Tropical and Subtropical Regions*. Springer Netherlands, Dordrecht.
- Gruhn, P., Goletti, F. and Yudelman, M. 2000. *Integrated Nutrient Management, Soil Fertility, and Sustainable Agriculture: Current Issues and Future Challenges*. International Food Policy Research Institute.
- Hatfield, J. L. and Sauer, T. J. 2020. *Soil management: building a stable base for agriculture*. American Society of Agronomy and Soil Science Society of America, Madison.

- Liebman, M., Mohler, C. L. and Staver, C. P. 2001. *Ecological Management of Agricultural Weeds*. Cambridge University Press.
- Ouda, S., Zohry, A. E. H. and Noreldin, T. 2018. *Crop Rotation: an Approach to Secure Future Food*. Springer International Publishing.

7.2. Book chapters

- Hatfield, J. L. and Sauer, T. J. 2011. Emerging challenges in soil management. In: Hatfield, J. L. and Sauer, T. J. (Eds). *Soil Management: Building a Stable Base for Agriculture*. American Society of Agronomy and Soil Science Society of America, Madison. pp 391–394.
- Haruna, S. I. and Anderson, S. H. 2020. No-till farming systems for enhancing soil water storage. In: Dang, Y. P., Dalal, R. C. and Menzies, N. W. (Eds). *No-Till Farming Systems for Sustainable Agriculture*. Springer, Cham. pp 213–231.
- Kaspar, T. C., Singer, J. W., Hatfield, J. L. and Sauer, T. J. 2011. The use of cover crops to manage soil. In: Hatfield, J. L. and Sauer, T. J. (Eds). *Soil Management: Building a Stable Base for Agriculture*. American Society of Agronomy and Soil Science Society of America, Madison. pp 321–338.
- Taboada, M. A., Rubio, G. and Chaneton, E. J. 2011. Grazing impacts on soil physical, chemical, and ecological properties in forage production systems. In: Hatfield, J. L. and Sauer, T. J. (Eds). *Soil Management: Building a Stable Base for Agriculture*. American Society of Agronomy and Soil Science Society of America, Madison. pp 301–320.

7.3. Articles

- Byrnes, R. C., Eastburn, D. J., Tate, K. W. and Roche, L. M. 2018. A global meta-analysis of grazing impacts on soil health indicators. *Journal of Environmental Quality* 47(4):758–765. https://doi.org/10.2134/jeq2017.08.0313.
- Cercioglu, M., Anderson, S. H., Udawatta, R. P. and Haruna, S. I. 2018. Effects of cover crop and biofuel crop management on computed tomography-measured pore parameters. *Geoderma* 319:80–88. https://doi.org/10.1016/j.geoderma.2018.01.005.
- Congreves, K. A., Hayes, A., Verhallen, E. A. and Van Eerd, L. L. 2015 Long-term impact of tillage and crop rotation on soil health at four temperate agroecosystems. *Soil and Tillage Research* 152:17–28. https://doi.org/10.1016/j.still.2015.03.012.
- Dabney, S. M., Delgado, J. A. and Reeves, D. W. 2001. Using winter cover crops to improve soil and water quality. *Communications in Soil Science*

and Plant Analysis 32(7–8):1221–1250. https://doi.org/10.1081/CSS-100104110.

- Irmak, S., Sharma, V., Mohammed, A. T. and Djaman, K. 2018. Impacts of cover crops on soil physical properties: field capacity, permanent wilting point, soil-water holding capacity, bulk density, hydraulic conductivity, and infiltration. *Transactions of the ASABE* 61(4):1307–1321. https://doi.org/10.13031/trans.12700.
- Karlen, D. L., Cambardella, C. A., Kovar, J. L. and Colvin, T. S. 2013. Soil quality response to long- term tillage and crop rotation practices. *Soil and Tillage Research* 133:54–64.https://doi.org/10.1016/j.still.2013.05.013.
- Kumar, S., Kadono, A., Lal, R. and Dick, W. 2012. Long-term no-till impacts on organic carbon and properties of two contrasting soils and corn yields in Ohio. *Soil Science Society of America Journal* 76(5):1798–1809. https://doi.org/10.2136/sssaj2012.0055.
- McDaniel, M. D., Tiemann, L. K. and Grandy, A. S. 2014. Does agricultural crop diversity enhance soil microbial biomass and organic matter dynamics? A meta-analysis. *Ecological Applications* 24(3):560–570. https://doi.org/10.1890/13-0616.1.
- Meyer, N., Bergez, J., Constantin, J., Belleville, P. and Justes, E. 2020. Cover crops reduce drainage but not always soil water content due to interactions between rainfall distribution and management. *Agricultural Water Management* 231:105998. https://doi.org/10.1016/j.agwat.2019.105998.
- Norkaew, S., Miles, R. J., Brandt, D. K. and Anderson, S. H. 2019. Effects of 130 years of selected cropping management systems on soil health properties for Sanborn field. *Soil Science Society of America Journal* 83(5):1479–1490. https://doi.org/10.2136/sssaj2019.03.0086.
- Unger, P. W. and Vigil, M. F. 1998. Cover crop effects on soil water relationships. *Journal of Soil and Water Conservation* 53:200–207.
- Wander, M. M., Cihacek, L. J., Coyne, M., Drijber, R. A., Grossman, J. M., Gutknecht, J. L. M., Horwath, W. R., Jagadamma, S., Olk, D. C., Ruark, M., Snapp, S. S., Tiemann, L. K., Weil, R. and Turco, R. F. 2019. Developments in Agricultural Soil Quality and Health: Reflections by the Research Committee on Soil Organic Matter Management. *Frontiers in Environmental Science* 7. https://doi.org/10.3389/fenvs.2019.00109.

8 References

Abdel-Magid, A. H., Schuman, G. E. and Hart, R. H. 1987. Soil bulk density and water infiltration as affected by grazing systems. Rangeland ecology & management. *Journal of Range Management Archives* 40:307–309.

Abrol, V. and Ben-Hur, M., Verheijen, F. G., Martins, J. J., Keizer, M. A., Tenaw, H., Tchehansky, L., Graber, E. R.. 2016. Biochar effects on soil water infiltration and erosion under seal formation conditions: rainfall simulation experiment. *Journal of Soils and Sediments*. 16:2709–2719. https://doi.org/10.1007/s11368-016-1448-8.

Acikgoz, S., Anderson, S. H., Gantzer, C. J., Thompson, A. L. and Miles, R. J. 2017. 125 years of soil and crop management on Sanborn field: effects on soil physical properties related to soil erodibility. *Soil Science* 182(5):172–180. https://doi.org/10.1097/SS.0000000000000207.

Adeli, A., Brooks, J. P., Read, J. J., Feng, G., Miles, D., Shankle, M. W., Barksdale, N. and Jenkin, J. N. 2020. Management strategies on an upland soil for improving soil properties. *Communications in Soil Science and Plant Analysis* 1:413–429. https://doi.org/10.1080/00103624. 2019.1709490.

Adugna, G. 2016. A review on impact of compost on soil properties, water use and crop productivity. *Academic Research Journal of Agricultural Science and Research* 4:93–104. https://doi.org/10.14662/ARJASR2016.010.

Anderson, S. H., C. J., GantzerJ. R., and Brown 1990. Soil physical properties after 100 years of continuous cultivation. *Journal of Soil and Water Conservation* 45:117–121.

Arshad, M. A., Franzluebbers, A. J. and Azooz, R. H. 1999. Components of surface soil structure under conventional and no-tillage in northwestern Canada. *Soil and Tillage Research* 53(1):41–47. https://doi.org/10.1016/S0167-1987(99)00075-6.

Azooz, R. H. and Arshad, M. A. 1996. Soil infiltration and hydraulic conductivity under long-term no-tillage and conventional tillage systems. *Canadian Journal of Soil Science* 76(2):143–152. https://doi.org/10.4141/cjss96-021.

Bamett, A. P. 1972. Agriculture and a quality environment. *Journal of Soil and Water Conservation* 27:104–108.

Basche, A. D. and DeLonge, M. S. 2019. Comparing infiltration rates in soils managed with conventional and alternative farming methods: A meta-analysis. *PLoS ONE* 14(9):e0215702. https://doi.org/10.1371/journal.pone.0215702.

Basche, A. D., Kaspar, T. C., Archontoulis, S. V., Jaynes, D. B., Sauer, T. J., Parkin, T. B. and Miguez, F. E. 2016. Soil water improvements with the long-term use of winter rye cover crop. *Agricultural Water Management* 172:40–50. https://doi.org/10.1016/j.agwat.2016.04.006.

Becker, A. E., Horowitz, L. S., Ruark, M. D. and Jackson, R. D. 2022. Surface-soil carbon stocks under well-managed grazed pasture than row crops. *Soil Science Society of America Journal*. in press. https://doi.org/10.1002/saj2.20388.

Bell, L. W., Kirkegaard, J. A., Swan, A., Hunt, J. R., Huth, N. I. and Fettell, N. A. 2011. Impacts of soil damage by grazing livestock on crop productivity. *Soil and Tillage Research* 113(1):19–29. https://doi.org/10.1016/j.still.2011.02.003.

Bescansa, P., Imaz, M. J., Virto, I., Enrique, A. and Hoogmoed, W. B. 2006. Soil water retention as affected by tillage and residue management in semiarid Spain. *Soil and Tillage Research* 87(1):19–27. https://doi.org/10.1016/j.still.2005.02.028.

Bharati, L., Lee, K.-H., Isenhart, T. M. and Schultz, R. C. 2002. Soil-water infiltration under crops, pasture, and established riparian buffer in Midwestern USA. *Agroforestry Systems* 56(3):249–257. https://doi.org/10.1023/A:1021344807285.

Bhattacharyya, R., Prakash, V., Kundu, S. and Gupta, H. S. 2006. Effect of tillage and crop rotations on pore size distribution and soil hydraulic conductivity in sandy clay loam soil of the Indian Himalayas. *Soil and Tillage Research* 86(2):129–140. https://doi.org/10.1016/j.still.2005.02.018.

Bilek, M. 2007. Winter annual rye cover crops in no-till grain crop rotations: impacts on soil physical properties and organic matter. Master of Science thesis. University of Maryland. Available at: https://drum.lib.umd.edu/handle/1903/7268.

Blair, N. and Crocker, G. J. 2000. Crop rotation effects on soil carbon and physical fertility of two Australian soils. *Soil Research* 38(1):71–84. https://doi.org/10.1071/SR99064.

Blair, N., Faulkner, R. D., Till, A. R. and Crocker, G. J. 2006. Long-term management impacts on soil C, N and physical fertility: Part III: Tamworth. *Soil and Tillage Research* 91(1–2):48–56. https://doi.org/10.1016/j.still.2005.11.003.

Blanco-Canqui, H. and Ruis, S. J. 2018. No-tillage and soil physical environment. *Geoderma* 326:164–200. https://doi.org/10.1016/j.geoderma.2018.03.011.

Blanco-Canqui, H., Stone, L. R. and Stahlman, P. W. 2010. Soil response to long-term cropping systems on an Argiustoll in the central Great Plains. *Soil Science Society of America Journal* 74(2):602–611. https://doi.org/10.2136/sssaj2009.0214.

Blanco-Canqui, H., Mikha, M. M., Presley, D. R. and Claassen, M. M. 2011. Addition of cover crops enhances no-till potential for improving soil physical properties. *Soil Science Society of America Journal* 75(4):1471–1482. https://doi.org/10.2136/sssaj2010.0430.

Blanco-Canqui, H., Wienhold, B. J., Jin, V. L., Schmer, M. R. and Kibet, L. C. 2017. Long-term tillage impact on soil hydraulic properties. *Soil and Tillage Research* 170:38–42. https://doi.org/10.1016/j.still.2017.03.001.

Blum, W. E. H. 2005. Functions of soil for society and the environment. *Reviews in Environmental Science and Bio/Technology* 4(3):75–79. https://doi.org/10.1007/s11157-005-2236-x.

Briske, D. D., Sayre, N. F., Huntsinger, L., Fernandez-Gimenez, M., Budd, B. and Derner, J. D. 2011. Origin, persistence, and resolution of the rotational grazing debate: integrating human dimensions into rangeland research. *Rangeland Ecology and Management* 64(4):325–334. https://doi.org/10.2111/REM-D-10-00084.1.

Brown, J. R. and Wyman, G. W. 1989. Sanborn Field: an overview. In: Proceedings of the Sanborn Field Centennial: A Celebration of 100 Years of Research. pp 53–63.

Bruce, R. R., Langdale, G. W. and Dillard, A. L. 1990. Tillage and crop rotation effect on characteristics of a sandy surface soil. *Soil Science Society of America Journal* 54(6):1744–1747. https://doi.org/10.2136/sssaj1990.03615995005400060039x.

Bruce, R. R., Langdale, G. W., West, L. T. and Miller, W. P. 1992. Soil surface modification by biomass inputs affecting rainfall infiltration. *Soil Science Society of America Journal* 56(5):1614–1620. https://doi.org/10.2136/sssaj1992.03615995005600050046x.

Bullock, D. G. 1992. Crop rotation. *Critical Reviews in Plant Sciences* 11(4):309–326. https://doi.org/10.1080/07352689209382349.

Busari, M. A., Kukal, S. S., Kaur, A., Bhatt, R. and Dulazi, A. A. 2015. Conservation tillage impact on soil, crop and the environment. *International Soil and Water Conservation Research* 3(2):119–129. https://doi.org/10.1016/j.iswcr.2015.05.002.

Butterfield, J., Bingham, S. and Savory, A. 2019. *Holistic Management Handbook: Regenerating Your Land and Growing Your Profits*. Island Press, Washington, DC. pp 57–104.

Buyanovsky, G. A. and Wagner, G. H. 2019. Crop residue input to soil organic matter on Sanborn Field. In: *Soil Organic Matter in Temperate Agroecosystems*. CRC Press, Boca Raton. pp 73–83.

Calegari, A., de Araujo, A. G., Tiecher, T., Bartz, M. L. C., Lanillo, R. F., dos Santos, D. R., Capandeguy, F., Zamora, J. H., Jump, J. R. B., Moriya, K. and Dabalá, L. 2020. No-till farming systems for sustainable agriculture in South America. In: Dang, Y. P., Dalal, R. C. and Menzies, N. W. (Eds). *No-Till Farming Systems for Sustainable Agriculture*. Springer, Cham. pp 533–565.

Capowiez, Y., Cadoux, S., Bouchant, P., Ruy, S., Roger-Estrade, J., Richard, G. and Boizard, H. 2009. The effect of tillage type and cropping system on earthworm communities, macroporosity and water infiltration. *Soil and Tillage Research* 105(2):209–216. https://doi.org/10.1016/j.still.2009.09.002.

Cawson, J. G., Nyman, P., Smith, H. G., Lane, P. N. J. and Sheridan, G. J. 2016. How soil temperatures during prescribed burning affect soil water repellency, infiltration and erosion. *Geoderma* 278:12–22. https://doi.org/10.1016/j.geoderma.2016.05.002.

Cetin, E. 2015. *Soil and Crop Management Influences on Soil Physical Quality of Sanborn Field*. University of Missouri-Columbia.

Chalise, K. S., Singh, S., Wegner, B. R., Kumar, S., Pérez-Gutiérrez, J. D., Osborne, S. L., Nleya, T., Guzman, J. and Rohila, J. S. 2018. Cover crops and returning residue impact on soil organic carbon, bulk density, penetration resistance, water retention, infiltration, and soybean yield. *Agronomy Journal* 111:99–108. https://doi.org/10.2134/agronj2018.03.0213.

Chan, K. Y. and Heenan, D. P. 1996. The influence of crop rotation on soil structure and soil physical properties under conventional tillage. *Soil and Tillage Research* 37(2-3):113–125. https://doi.org/10.1016/0167-1987(96)01008-2.

Chan, K. Y., Bowman, A. and Oates, A. 2001. Oxidizable organic carbon fractions and soil quality changes in an oxic Paleustalf under different pasture leys. *Soil Science* 166(1):61–67.

Chang, C. and Lindwall, C. W. 1992- 1987(92)90051-C. Effects of tillage and crop rotation on physical properties of a loam soil. *Soil and Tillage Research* 22:383–389. https://doi.org/10.1016/0167.

Cherney, J. H. 2015. Forage crop systems: challenges, opportunities, and technology transfer. In: Proceedings of the 17th Australian Society of Agronomy Conference. pp 20–24.

Conservation Technology Information Center 2004. *National Crop Residue Management Survey*. NACD's Conservation Technology Information Center.

Cornish, P. S., J. N., Tullberg, D., Lemerle and Flower, K. 2020. No-till farming systems in Australia. In: Dang, Y. P., Dalal, R. C. and Menzies, N. W. (Eds). *No-Till Farming Systems for Sustainable Agriculture*. Springer, Cham. pp 511–531.

Crittenden, S. J., Poot, N., Heinen, M., van Balen, D. J. M. and Pulleman, M. M. 2015. Soil physical quality in contrasting tillage systems in organic and conventional farming. *Soil and Tillage Research* 154:136–144. https://doi.org/10.1016/j.still.2015.06.018.

Daigh, A. L., Helmers, M. J., Kladivko, E., Zhou, X., Goeken, R., Cavdini, J., Barker, D. and Sawyer, J. 2014. Soil water during the drought of 2012 as affected by rye cover crops in fields in Iowa and Indiana. *Journal of Soil and Water Conservation* 69(6):564–573. https: //doi.org/10.2489/jswc.69.6.564.

de Almeida, W. S., Panachuki, E., de Oliveira, P. T. S., da Silva Menezes, R., Sobrinho, T. A. and de Carvalho, D. F. 2018. Effect of soil tillage and vegetal cover on soil water infiltration. *Soil and Tillage Research* 175:130–138. https://doi.org/10.1016/j.still.2017.07.009.

de Andrade Bonetti, J., Anghinoni, I., Ivonir Gubiani, P., Cecagno, D. and de Moraes, M. T. 2019. Impact of a long-term crop-livestock system on the physical and hydraulic properties of an oxisol. *Soil and Tillage Research* 186:280–291. https://doi.org/10.1016/j.still.2018.11.003.

Diiwu, J. Y., Rudra, R. P., Dickinson, W. T. and Wall, G. J. 1998. Tillage and heterogeneity effects on the performance of soil water characteristic models. *Journal of Agricultural Engineering Research* 71(3):307–313. https://doi.org/10.1006/jaer.1998.0358.

Dill, G. M., Cajacob, C. A. and Padgette, S. R. 2008. Glyphosate-resistant crops: adoption, use and future considerations. *Pest Management Science* 64(4):326–331. https://doi.org/10.1002/ps.1501.

Döbert, T. F., Bork, E. W., Apfelbaum, S., Carlyle, C. N., Chang, S. X., Khatri-Chhetri, U., Silva Sobrinho, L., Thompson, R. and Boyce, M. S. 2021. Adaptive multi-paddock grazing improves water infiltration in Canadian grassland soils. *Geoderma* 401:115314. https://doi.org/10.1016/j.geoderma.2021.115314.

Dury, J., Schaller, N., Garcia, F., Reynaud, A. and Bergez, J. E. 2012. Models to support cropping plan and crop rotation decisions. A review. *Agronomy for Sustainable Development* 32(2):567–580. https://doi.org/10.1007/s13593-011-0037-x.

Faulkner, E. H. 1943. *Plowman's Folly (vol. 04)(S603), F3*. University of Oklahoma Press, Norman.

Fausti, S., Kolady, D. E., Van der Sluis, E., Lundgren, J. and Qasmi, B. A. 2018. Extensive usage of insecticide and changing crop rotation patterns: A South Dakota case study. *PLoS ONE* 13(11):e0208222. https://doi.org/10.1371/journal.pone.0208222.

Fensham, R. J., Silcock, J. L. and Firn, J. 2014. Managed livestock grazing is compatible with the maintenance of plant diversity in semidesert grasslands. *Ecological Applications* 24(3):503–517. https://doi.org/10.1890/13-0492.1.

Fleischner, T. L. 1994. Ecological costs of livestock grazing in western North America. *Conservation Biology* 8(3):629–644. https://doi.org/10.1046/j.1523-1739.1994.08030629.x.

Food and Agriculture Organization (FAO) 2000. Latin America and the Caribbean – farming systems and poverty. Available at: https://www.fao.org/3/y1860e/y1860e09.htm. (Verified August 10, 2022).

Follett, R. F. and Stewart, B. A. (Eds.) 1985. Soil erosion and crop productivity. *American Society of Agronomy*. Madison Book Company, WI. pp 533. Madison.

Folorunso, O. A., Rolston, D. E., Prichard, T. and Loui, D. T. 1992. Soil surface strength and infiltration rate as affected by winter cover crops. *Soil Technology* 5(3):189–197. https://doi.org/10.1016/0933-3630(92)90021-R.

Free, G. R., Fertig, S. N. and Bay, C. E. 1963. Zero tillage for corn following sod 1. *Agronomy Journal* 55(2):207–208.

Gan, L., Peng, X., Peth, S. and Horn, R. 2012. Effects of grazing intensity on soil water regime and flux in Inner Mongolia grassland, China. *Pedosphere* 22(2):165–177. https://doi.org/10.1016/S1002-0160(12)60003-4.

Govaerts, B., Sayre, K. D. and Deckers, J. 2006. A minimum data set for soil quality assessment of wheat and maize cropping in the highlands of Mexico. *Soil and Tillage Research* 87(2):163–174. https://doi.org/10.1016/j.still.2005.03.005.

Govaerts, B., Fuentes, M., Mezzalama, M., Nicol, J. M., Deckers, J., Etchevers, J. D., Figueroa-Sandoval, B. and Sayre, K. D. 2007. Infiltration, soil moisture, root rot and nematode populations after 12 years of different tillage, residue and crop rotation managements. *Soil and Tillage Research* 94(1):209–219. https://doi.org/10.1016/j.still.2006.07.013.

Gozubuyuk, Z., Sahin, U., Ozturk, I., Celik, A. and Adiguzel, M. C. 2014. Tillage effects on certain physical and hydraulic properties of a loamy soil under a crop rotation in a semi-arid region with a cool climate. *CATENA* 118:195–205. https://doi.org/10.1016/j.catena.2014.01.006.

Harlan, J. R. 1995. *The Living Fields: Our Agricultural Heritage*. Cambridge University Press. pp 1–25.

Haruna, S. I. 2019. Influence of winter wheat on soil thermal properties of a Paleudalf. *International Agrophysics* 33(3):389–395. https: //doi.org/10.31545/intagr/110850.

Haruna, S. I. and Anderson, S. H. 2022. Influence of no-till cover crop management on soil thermal properties. *Soil Research* 60(6):580–589. https://doi.org/10.1071/SR21197.

Haruna, S. I. and Nkongolo, N. V. 2015. Effects of tillage, rotation and cover crop on the physical properties of a silt-loam soil. *International Agrophysics* 29(2):137–145. https://doi.org/10.1515/intag-2015-0030.

Haruna, S. I. and Nkongolo, N. V. 2020. Influence of cover crop, tillage, and crop rotation management on soil nutrients. *Agriculture* 10(6):225. https://doi.org/10.3390/agriculture10060225.

Haruna, S. I., Anderson, S. H., Nkongolo, N. V., Reinbott, T. and Zaibon, S. 2017. Soil thermal properties influenced by perennial biofuel and cover crop management. *Soil Science Society of America Journal* 81(5):1147–1156. https://doi.org/10.2136/sssaj2016.10.0345.

Haruna, S. I., Nkongolo, N. V., Anderson, S. H., Eivazi, F. and Zaibon, S. 2018a. In situ infiltration as influenced by cover crops and tillage management. *Journal of Soil and Water Conservation* 73(2):164–172. https://doi.org/10.2489/jswc.73.2.164.

Haruna, S. I., S. H., Anderson, S. H., Nkongolo, N. V. and Zaibon, S. 2018b. Soil hydraulic properties: influence of tillage and cover crops. *Pedosphere* 28:430–442. https://doi.org/10.1016/S1002-0160(17)60387-4.

Haruna, S. I., Ritchey, E., Mosley, C. and Ku, S. 2022a. Effects of cover crops on soil hydraulic properties during commodity growing season. *Soil Use and Management* 00:1–14. https://doi.org/10.1111/sum.12803.

Haruna, S. I., R. C., Eichas, O. M., Peters, A. C., Farmer, D. Q., Lackey, N. A. and Slone, N. A. 2022b. In situ water infiltration: influence of cover crops after growth termination. *Soil Science Society of America Journal* 00:1–12. https://doi.org/10.1002/saj2.20390.

Hill, R. L. 1990. Long-term conventional and no-tillage effects on selected soil physical properties. *Soil Science Society of America Journal* 54(1):161–166. https://doi.org/10.2136/sssaj1990.03615995005400010025x.

Hillel, D. 1998. *Environmental Soil Physics: Fundamentals, Applications, and Environmental Considerations*. Elsevier, The Netherlands. pp 112–114.

Hillenbrand, M., Thompson, R., Wang, F., Apfelbaum, S. and Teague, R. 2019. Impacts of holistic planned grazing with bison compared to continuous grazing with cattle in South Dakota shortgrass prairie. *Agriculture, Ecosystems and Environment* 279:156–168. https://doi.org/10.1016/j.agee.2019.02.005.

Huang, X., Wang, H., Zhang, M., Horn, R. and Ren, T. 2021. Soil water retention dynamics in a mollisol during a maize growing season under contrasting tillage systems. *Soil and Tillage Research* 209:104953. https://doi.org/10.1016/j.still.2021.104953.

Hubbard, R. K., Strickland, T. C. and Phatak, S. 2013. Effects of cover crop systems on soil physical properties and carbon/nitrogen relationships in the coastal plain of southeastern USA. *Soil and Tillage Research* 126:276–283. https://doi.org/10.1016/j.still.2012.07.009.

Hulugalle, N. R., Weaver, T. B. and Finlay, L. A. 2010. Soil water storage and drainage under cotton-based cropping systems in a furrow-irrigated vertisol. *Agricultural Water Management* 97(10):1703–1710. https://doi.org/10.1016/j.agwat.2010.06.001.

Hunt, J. R., Swan, A. D., Fettell, N. A., Breust, P. D., Menz, I. D., Peoples, M. B. and Kirkegaard, J. A. 2016. Sheep grazing on crop residues do not reduce crop yields in no-till, controlled traffic farming systems in an equi-seasonal rainfall environment. *Field Crops Research* 196:22–32. https://doi.org/10.1016/j.fcr.2016.05.012.

Hyland, J. J., Heanue, K., McKillop, J. and Micha, E. 2018. Factors influencing dairy farmers' adoption of best management grazing practices. *Land Use Policy* 78:562–571. https://doi.org/10.1016/j.landusepol.2018.07.006.

Jayaraman, S., Naorem, A., Lal, R., R. C., Dalal and Patra, A. K. 2020. No-till farming systems in South Asia. In: Dang, Y. P., Dalal, R. C. and Menzies, N. W. (Eds). *No-Till Farming Systems for Sustainable Agriculture*. Springer, Cham. pp 459–476.

Jones, O. R., Hauser, V. L. and Popham, T. W. 1994. No-tillage effects on infiltration, runoff, and water conservation on dryland. *Transactions of the ASAE* 37(2):473–479.

Karlen, D. L., Hurley, E. G., Andrews, S. S., Cambardella, C. A., Meek, D. W., Duffy, M. D. and Mallarino, A. P. 2006. Crop rotation effects on soil quality at three northern corn/soybean belt locations. *Agronomy Journal* 98(3):484–495. https://doi.org/10.2134/agronj2005.0098.

Kasper, T. C. and Singer, J. W. 2011. The use of cover crops to manage soil. In: Hartfield, J. L. and Sauer, T. J. (Eds). *Soil Management: Building a Stable Base for Agriculture*. ASA and SSSA. Madison Book Company, WI. pp 321–337. Madison, Wisconsin, USA.

Kay, B. D. 1997. Soil structure and organic carbon. In: Lal, R., Kimble, J. M., Follet, R. F. and Stewart, B. A. (Eds). *Soil Processes and the Carbon Cycle*. CRC Press, Boca Raton. pp 169–198.

Kaye, J. P. and Quemada, M. 2017. Using cover crops to mitigate and adapt to climate change. A review. *Agronomy for Sustainable Development* 37:4. https://doi.org/10.1007/s13593-016-0410- x.

Kemper, B. and Derpsch, R. 1981. Results of studies made in 1978 and 1979 to control erosion by cover crops and no-tillage techniques in Parana, Brazil. *Soil and Tillage Research* 1:253–267. https://doi.org/10.1016/0167-1987(80)90028-8.

Kong, D., Liu, N., Wang, W., Akhtar, K., Li, N., Ren, G., Feng, Y. and Yang, G. 2019 Soil respiration from fields under three crop rotation treatments and three straw retention treatments. *PLoS ONE* 14(9):e0219253. https://doi.org/10.1371/journal.pone.0219253.

Krümmelbein, J., Peth, S., Zhao, Y. and Horn, R. 2009. Grazing-induced alterations of soil hydraulic properties and functions in Inner Mongolia, PR China. *Journal of Plant Nutrition and Soil Science* 172(6):769–776. https://doi.org/10.1002/jpln.200800218.

Kumar, S., Anderson, S. H., Udawatta, R. P. and Kallenbach, R. L. 2012. Water infiltration influenced by agroforestry and grass buffers for a grazed pasture system. *Agroforestry Systems* 84(3):325–335. https://doi.org/10.1007/s10457-011-9474-4.

Lal, R. 2015. Cover cropping and the "4 per Thousand" proposal. *Journal of Soil and Water Conservation* 70(6):141A–141A. https://doi.org/10.2489/jswc.70.6.141A.

Lal, R. and Vandoren, D. M. 1990 Influence of 25 years of continuous corn production by three tillage methods on water infiltration for two soils in Ohio. *Soil and Tillage Research* 16(1–2):71–84.

Lampurlanés, J., Angás, P. and Cantero-Martínez, C. 2001. Root growth, soil water content and yield of barley under different tillage systems on two soils in semiarid conditions. *Field Crops Research* 69(1):27–40. https://doi.org/10.1016/S0378-4290(00)00130-1.

Langdale, G. W., Blevins, R. L., Karlen, D. L., McCool, D. K., Nearing, M. A. and Skidmore, E. L.. and Williams, J. R. 1991. Cover crop effects on soil erosion by wind and water. *Cover Crops for Clean Water*: 15–22.

Lipiec, J., Kuś, J., Słowińska-Jurkiewicz, A. and Nosalewicz, A. 2006. Soil porosity and water infiltration as influenced by tillage methods. *Soil and Tillage Research* 89(2):210–220. https://doi.org/10.1016/j.still.2005.07.012.

Logan, T. J., Lal, R. and Dick, W. A. 1991- 1987(91)90042-V. Tillage systems and soil properties in North America. *Soil and Tillage Research* 20:241–270. https://doi.org/10.1016/0167.

López-Bellido, L., Fuentes, M., Castillo, J. E., López-Garrido, F. J. and Fernández, E. J. 1996. Long-term tillage, crop rotation, and nitrogen fertilizer effects on wheat yield under rainfed Mediterranean conditions. *Agronomy Journal* 88(5):783–791. https://doi.org/10.2134/agronj1996.00021962008800050016x.

MacLaren, C., Swanepoel, P., Bennett, J., Wright, J. and Dehnen-Schmutz, K. 2019. Cover crop biomass production is more important than diversity for weed suppression. *Crop Science* 59(2):733–748. https://doi.org/10.2135/cropsci2018.05.0329.

MacRae, R. J., G. R. and Mehuys 1985. The effect of green manuring on the physical properties of temperate-area soils. In: *Advances in Soil Science* Stewart, B. A. (Ed.). Springer, New York. pp 71–94.

Mathew, R. P., Feng, Y., Githinji, L., Ankumah, R. and Balkcom, K. S. 2012. Impact of no-tillage and conventional tillage systems on soil microbial communities. *Applied and Environmental Soil Science* 2012:1–10. https://doi.org/10.1155/2012/548620.

Moody, J. E., Shear, G. M. and Jones, J. N., Jr. 1961. Growing corn without tillage. *Soil Science Society of America Proceedings* 25:516–517.

Moret-Fernández, D., Arroyo, A. I., Herrero, J., Barrantes, O., Alados, C. L. and Pueyo, Y. 2021. Livestock grazing effect on the hydraulic properties of gypseous soils in a Mediterranean region. *CATENA* 207:105697. https://doi.org/10.1016/j.catena.2021.105697.

Nouri, A., Lee, J., Yin, X., Tyler, D. D. and Saxton, A. M. 2019. Thirty-four years of no-tillage and cover crops improve soil quality and increase cotton yield in alfisols, southeastern USA. *Geoderma* 337:998–1008. https://doi.org/10.1016/j.geoderma.2018.10.016.

Obour, P. B., Jensen, J. L., Lamandé, M., Watts, C. W. and Munkholm, L. J. 2018. Soil organic matter widens the range of water contents for tillage. *Soil and Tillage Research* 182:57–65. https://doi.org/10.1016/j.still.2018.05.001.

Osipitan, O. A., Dille, J. A., Assefa, Y., Radicetti, E., Ayeni, A. and Knezevic, S. Z. 2019. Impact of cover crop management on level of weed suppression: a meta-analysis. *Crop Science* 59(3):833–842. https://doi.org/10.2135/cropsci2018.09.0589.

Overton, M. 1989. Agricultural revolution? England, 1540–1850. In: *New Directions in Economic and Social History*. Palgrave Macmillan, London. pp 9–21.

Painter, E. L., Detling, J. K. and Steingraeber, D. A. 1993 Plant morphology and grazing history. *Vegetatio* 106(1):37–62. https://doi.org/10.1007/BF00044857.

Pena-Sancho, C., Lopez, M. V., Garcia, R. and Moret-Fernandez, D. 2017. Effects of tillage on the soil water retention curves during a fallow period of a semiarid dryland. *Soil Research* 55(2):114–123. https://doi.org/10.1071/SR15305.

Phillips, E. R. and Phillips, S. H. 1984. *No-Tillage Agriculture, Principles and Practices*. Van Nostrand Reinhold Co., New York.

Pluhar, J. J., Knight, R. W. and Heitschmidt, R. K. 1987. Infiltration rates and sediment production as influenced by grazing systems in the Texas Rolling Plains. Rangeland ecology & management. *Journal of Range Management Archives* 40:240–243.

Pueyo, Y., Alados, C. L., Barrantes, O., Komac, B. and Rietkerk, M. 2008. Differences in gypsum plant communities associated with habitat fragmentation and livestock grazing. *Ecological Applications* 18(4):954–964. https://doi.org/10.1890/07-1770.1.

Raiesi, F. and Riahi, M. 2014. The influence of grazing exclosure on soil C stocks and dynamics, and ecological indicators in upland arid and semi-arid rangelands. *Ecological Indicators* 41:145–154. https://doi.org/10.1016/j.ecolind.2014.01.040.

Reeves, D. 1994. Cover crops and rotations. In: Hartfield, J. L. and Stewart, B. A. (Eds). *Crop Residue Management*. Lewis Publishers, Boca Raton. pp 125–172.

Rorick, J. D. and Kladivko, E. J. 2017. Cereal rye cover crop effects on soil carbon and physical properties in southeastern Indiana. *Journal of Soil and Water Conservation* 72(3):260–265. https://doi.org/10.2489/jswc.72.3.260.

Ross, J. V. 1984. Managing the public range-lands: 50 years since the Taylor Grazing Act. *Rangelands Archives* 6:147–151.

Rusch, A., Bommarco, R., Jonsson, M., Smith, H. G. and Ekbom, B. 2013. Flow and stability of natural pest control services depend on complexity and crop rotation at the landscape scale. *Journal of Applied Ecology* 50(2):345–354. https://doi.org/10.1111/1365-2664.12055.

Sainju, U. M. 2020. No-till farming systems in North America. In: Dang, Y. P., Dalal, R. C. and Menzies, N. W. (Eds). *No-Till Farming Systems for Sustainable Agriculture*. Springer, Cham. pp 587–599.

Sarma, B., Farooq, M., Gogoi, N., Borkotoki, B., Kataki, R. and Garg, A. 2018. Soil organic carbon dynamics in wheat-Green gram crop rotation amended with vermicompost and biochar in combination with inorganic fertilizers: A comparative study. *Journal of Cleaner Production* 201:471–480. https://doi.org/10.1016/j.jclepro.2018.08.004.

Savory, A., J. and Butterfield 2016. *Holistic Management: a Commonsense Revolution to Restore Our Environment*. Island Press, Washington, DC. Pp 412–470.

Schaak, H. and Mußhoff, O. 2018. Understanding the adoption of grazing practices in German dairy farming. *Agricultural Systems* 165:230–239. https://doi.org/10.1016/j.agsy.2018.06.015.

Schwen, A., Bodner, G., Scholl, P., Buchan, G. D. and Loiskandl, W. 2011. Temporal dynamics of soil hydraulic properties and the water-conducting porosity under different tillage. *Soil and Tillage Research* 113(2):89–98. https://doi.org/10.1016/j.still.2011.02.005.

Sharrow, S. H. 2007. Soil compaction by grazing livestock in silvopastures as evidenced by changes in soil physical properties. *Agroforestry Systems* 71(3):215–223. https://doi.org/10.1007/s10457-007-9083-4.

Shukla, M. K., Lal, R., Owens, L. B. and Unkefer, P. 2003. Land use and management impacts on structure and infiltration characteristics of soils in the North Appalachian region of Ohio. *Soil Science* 168(3):167–177. https://doi.org/10.1097/01.ss.0000058889.60072.aa.

Simić, M., I., Spasojević, I., Kovacević, D., Brankov, M. and Dragicević, V. 2016. Crop rotation influence on annual and perennial weed control and maize productivity. *Romanian Agricultural Research* 33:125–133.

Sirimarco, X., Barral, M. P., Villarino, S. H. and Laterra, P. 2018. Water regulation by grasslands: A global meta-analysis. *Ecohydrology* 11(4):e1934. https://doi.org/10.1002/eco.1934.

Sławiński, C., Cymerman, J., Witkowska-Walczak, B. and Lamorski, K. 2012. Impact of diverse tillage on soil moisture dynamics. *International Agrophysics* 26(3):301–309.

Smit, B., Janssens, B., Haagsma, W., Hennen, W., Adrados, J. L., Kathage, J. and Domínguez, I. P. 2019. *Adoption of Cover Crops for Climate Change Mitigation in the EU*. Publications Office of the European Union, Luxembourg.

Soltani, I., Fouad, Y., Michot, D., Pichelin, P. and Cudennec, C. 2019. Relevance of a near infrared spectral index for assessing tillage and fertilization effects on soil water retention. *Soil and Tillage Research* 194:104345. https://doi.org/10.1016/j.still.2019.104345.

Stone, L. R. and Schlegel, A. J. 2010. Tillage and crop rotation phase effects on soil physical properties in the west-central Great Plains. *Agronomy Journal* 102(2):483–491. https://doi.org/10.2134/agronj2009.0123.

Stroud, J. L. 2020. No-till farming systems in Europe. In: Dang, Y. P., Dalal, R. C. and Menzies, N. W. (Eds). *No-Till Farming Systems for Sustainable Agriculture*. Springer, Cham. pp 567–585.

Sumner, D. R. 2018. Crop rotation and plant productivity. In: *CRC handbook of agricultural productivity*. CRC Press. pp 273–314. Boca Raton, USA.

Swarowsky, A., Dahlgren, R. A., Tate, K. W., Hopmans, J. W. and O'Geen, A. T. 2011. Catchment-scale soil water dynamics in a Mediterranean-type oak woodland. *Vadose Zone Journal* 10(3):800–815. https://doi.org/10.2136/vzj2010.0126.

Teague, W. R., Dowhower, S. L., Baker, S. A., Haile, N., DeLaune, P. B. and Conover, D. M. 2011. Grazing management impacts on vegetation, soil biota and soil chemical, physical and hydrological properties in tall grass prairie. *Agriculture, Ecosystems and Environment* 141(3–4):310–322. https://doi.org/10.1016/j.agee.2011.03.009.

Tennakoon, S. B. and Hulugalle, N. R. 2006. Impact of crop rotation and minimum tillage on water use efficiency of irrigated cotton in a vertisol. *Irrigation Science* 25(1):45–52. https://doi.org/10.1007/s00271-006-0033-0.

TerAvest, D., Carpenter-Boggs, L., Thierfelder, C. and Reganold, J. P. 2015. Crop production and soil water management in conservation agriculture, no-till, and conventional tillage systems in Malawi. *Agriculture, Ecosystems and Environment* 212:285–296. https://doi.org/10.1016/j.agee.2015.07.011.

Teutscherová, N., Vázquez, E., Sotelo, M., Villegas, D., Velásquez, N., Baquero, D., Pulleman, M. and Arango, J. 2021. Intensive short-duration rotational grazing is associated with improved soil quality within one year after establishment in Colombia. *Applied Soil Ecology* 159:103835. https://doi.org/10.1016/j.apsoil.2020.103835.

Thapa, V. R., Ghimire, R., Acosta-Martínez, V., Marsalis, M. A. and Schipanski, M. E. 2021. Cover crop biomass and species composition affect soil microbial community structure and enzyme activities in semiarid cropping systems. *Applied Soil Ecology* 157:103735. https://doi.org/10.1016/j.apsoil.2020.103735.

Thierfelder, C. 2020. No-till farming systems in Southern Africa. In: Dang, Y. P., Dalal, R. C. and Menzies, N. W. (Eds). *No-Till Farming Systems for Sustainable Agriculture*. Springer, Cham. pp 493–510.

Tobin, C., Singh, S., Kumar, S., Wang, T. and Sexton, P. 2020. Demonstrating short-term impacts of grazing and cover crops on soil health and economic benefits in an integrated crop-livestock system in South Dakota. *Open Journal of Soil Science* 10(3):109–136. https://doi.org/10.4236/ojss.2020.103006.

Triplett, Jr., G. B., Johnson, W. H. and Van Doren, D. M., Jr. 1963. Performance of two experimental planters for no-tillage corn culture 1. *Agronomy Journal* 55(4):408–409.

Tuzzin de Moraes, M., Debiasi, H., Carlesso, R., Cezar Franchini, J., Rodrigues da Silva, V. and Bonini da Luz, F. 2016. Soil physical quality on tillage and cropping systems after two decades in the subtropical region of Brazil. *Soil and Tillage Research* 155:351–362. https://doi.org/10.1016/j.still.2015.07.015.

Upchurch, W. J., Kinder, R. J., Brown, J. R. and Wagner, G. H. 1985. *Sanborn Field: Historical Perspective*. pp 1–52. University of Missouri located in Columbia, USA.

USDA National Agricultural Statistics Service (USDA-NASS) 2017. Available at: https://www.nass.usda.gov/Publications/AgCensus/2017/index.php.

Venter, Z. S., Jacobs, K. and Hawkins, H. 2016. The impact of crop rotation on soil microbial diversity: A meta-analysis. *Pedobiologia* 59(4):215–223. https://doi.org/10.1016/j.pedobi.2016.04.001.

Villamil, M. B., Bollero, G. A., Darmody, R. G., Simmons, F. W. and Bullock, D. G. 2006. No-till corn/soybean systems including winter cover crops. *Soil Science Society of America Journal* 70(6):1936–1944. https://doi.org/10.2136/sssaj2005.0350.

Wallander, S., D., Smith, M., Bowman, R. and Claassen 2021. Cover crop trends, programs, and practices in the United States. US Department of Agriculture-Economic Research service. *Economic Information Bulletin No. 222:1-33*. Available at: https://www.ers.usda.gov/webdocs/publications/100551/eib-222.pdf?v=9246. (Verified 02/18/2022).

Wang, S., Fan, J., Li, Y., Wu, D., Zhang, Y. and Huang, L. 2019. Dynamic response of water retention to grazing activity on grassland over the Three River Headwaters region. *Agriculture, Ecosystems and Environment* 286:106662. https://doi.org/10.1016/j.agee.2019.106662.

Ward, P. R., Flower, K. C., Cordingley, N., Weeks, C. and Micin, S. F. 2012. Soil water balance with cover crops and conservation agriculture in a Mediterranean climate. *Field Crops Research* 132:33–39. https://doi.org/10.1016/j.fcr.2011.10.017.

Weisberger, D., Nichols, V. and Liebman, M. 2019. Does diversifying crop rotations suppress weeds? A meta-analysis. *PLoS ONE* 14(7):e0219847. https://doi.org/10.1371/journal.pone.0219847.

West, T. O. and Post, W. M. 2002. Soil organic carbon sequestration rates by tillage and crop rotation: a global data analysis. *Soil Science Society of America Journal* 66(6):1930–1946. https://doi.org/10.2136/sssaj2002.1930.

White, K. D. 1970. Fallowing, crop rotation, and crop yields in Roman times. *Agricultural History* 44:281–290.

Williamson, G. J., Murphy, B. P. and Bowman, D. M. J. S. 2014. Cattle grazing does not reduce fire severity in eucalypt forests and woodlands of the Australian Alps. *Austral Ecology* 39(4):462–468. https://doi.org/10.1111/aec.12104.

Wilson, G. F., Lal, R. and Okigbo, B. N. 1982. Effects of cover crops on soil structure and on yield of subsequent arable crops grown under strip tillage on an eroded alfisol. *Soil and Tillage Research* 2(3):233–250. https://doi.org/10.1016/0167-1987(82)90013-7.

Wischmeier, W. H. and Smith, D. D. 1965. *Rainfall-Erosion Losses from Cropland East of the Rocky Mountains. Agronomy Handbook* (vol. 282). United States Department of Agriculture, Washington, DC.

Yu, J., Lei, T., Shainberg, I., Mamedov, A. I. and Levy, G. J. 2003. Infiltration and erosion in soils treated with dry PAM and gypsum. *Soil Science Society of America Journal* 67(2):630–636. https://doi.org/10.2136/sssaj2003.6300.

Yunusa, I. A. M., P. J. and Newton 2003. Plants for amelioration of subsoil constraints and hydrological control: the primer-plant concept. *Plant and Soil* 257:261–281.https://doi.org/10.1023/A: 1027381329549.

Zhang, Z. and Peng, X. 2021. Bio-tillage: a new perspective for sustainable agriculture. *Soil and Tillage Research* 206:104844. https://doi.org/10.1016/j.still.2020.104844.

Zhao, J., Yang, Y., Zhang, K., Jeong, J., Zeng, Z. and Zang, H. 2020. Does crop rotation yield more in China? A meta-analysis. *Field Crops Research* 245:107659. https://doi.org/10.1016/j.fcr.2019.107659.

Zhu, X., Hu, Y., Wang, W. and Wu, D. 2019. Earthworms promote the accumulation of maize root-derived carbon in a black soil of Northwest China, especially in soil from long-term no-till. *Geoderma* 340:124–132. https://doi.org/10.1016/j.geoderma.2019.01. 003.

Zúñiga, F., Ivelic-Sáez, J., López, I., Huygens, D. and Dörner, F. J. 2015. Temporal dynamics of the physical quality of an andisol under a grazing system subjected to different pasture improvement strategies. *Soil and Tillage Research* 145:233–241. https://doi.org/10.1016/j.still.2014.09.014.

Chapter 5

Advances in drainage design and management for irrigated agriculture

Henk Ritzema[1], Wageningen University, The Netherlands

1 Introduction

At present, about 233 Mha (or 14%) of the arable and permanently cropped area worldwide are irrigated (International Commission on Irrigation and Drainage, 2022), contributing as much as 40% of the gross agricultural output (Faures et al., 2007). However, despite the importance of drainage as a component of agricultural water management (Fig. 1), only about 22% of these irrigated lands are drained (Schultz et al., 2007). As a result, salinity and waterlogging problems affect about 10–16% of these areas because the natural drainage is not sufficient for controlling soil salinity levels (Ritzema, 2016). In humid regions, drainage is needed to control soil water for better aeration, higher temperatures and easier workability; by contrast in arid and semi-arid regions, its primary function is to prevent irrigation-induced waterlogging and salinization of the soil (Ritzema, 2018). In regions with prevailing dry and high evaporative conditions, increased salt concentrations and river depletion have become two inevitable collaterals of irrigated crop production. Salinity build-up is a slow process, so farmers, engineers and government departments all see irrigation as a need for today and salinization as a problem of tomorrow.

Drainage therefore has a lower priority than other agricultural activities such as irrigation and other on-farm management practices. This is due in part to drainage being at the end of the pipeline: to clean up the 'mess' other

[1] Corresponding author: henk.ritzema@wur.nl.

http://dx.doi.org/10.19103/AS.2023.0123.05

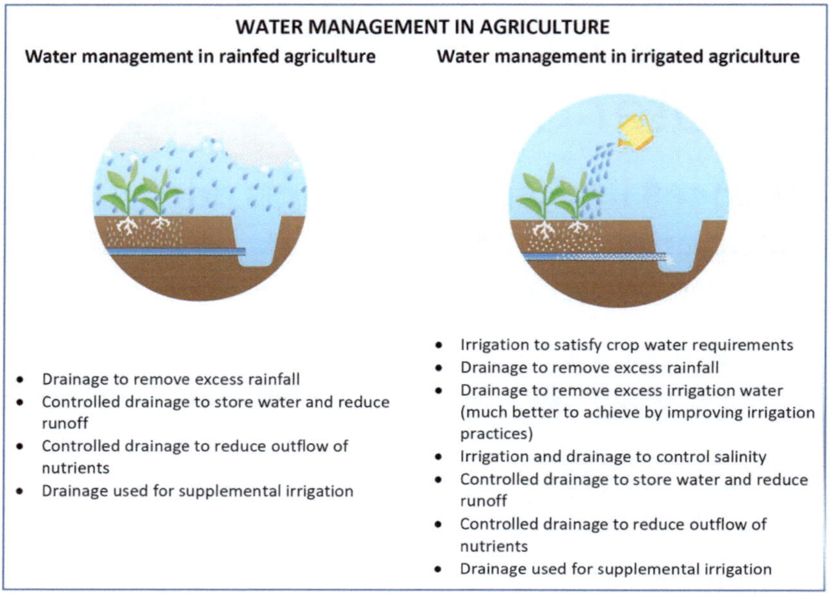

Figure 1 The role of drainage in agricultural water management is more complex in irrigated agriculture than in rainfed agriculture.

activities leave behind like salts brought in by irrigation water and residues of fertilizers and pesticides. Most people do not like to be reminded of this and therefore ignore it (Scheuman, 1997). On top of this, irrigation needs are changing triggered by land use changes. In the past, large-scale irrigation systems were built to supply water to farmers for a limited number of crops, mainly irrigated by surface water diverted from rivers, streams or lakes. In recent decades, there has been a gradual change in land use: urbanization and non-agricultural uses, including ecosystem services, are increasing (Ritzema et al., 2016). This has changed the demand for water (De Fraiture et al., 2010). Global climate change may further exacerbate the pressure on supply and demand through changing temperatures and long-term variation in annual precipitation amounts and regional distribution patterns. In addition to the changing climate, cropping patterns are diversifying and field irrigation methods are changing. Irrigation and drainage water demands are no longer homogeneous because surface water is often supplemented with water from other sources: groundwater extraction, (treated) wastewater and/or the reuse of drainage water (Singh, 2014). Although irrigation system design and management have evolved tremendously over the past few decades, the design methodology and management criteria for subsurface drainage (SSD) have not changed much in the last 50 years (Ayars and Evans, 2015). The aforementioned developments have already led to certain changes in drainage water management; however, greater awareness of the entire system is needed. For example, recent approaches that include multiple water services have repercussions for the hydraulic design of irrigation and drainage systems (Renault et al., 2007). To optimize water use and control salinity, it is important to match rainfall, irrigation and drainage (Van Hoorn and Van Alphen, 2006). In both rainfed and irrigated agriculture, the only agronomically significant criterion for installing drainage is to maximize the commercial crop yield (and sometimes crop quality). This makes the need for drainage complex as the direct effects of drainage, i.e. controlling the water table and discharge, are mainly determined by the hydrological conditions, the hydraulic properties of the soil and the physical characteristics of the drainage system (Oosterbaan, 1988). Higher crop yields and improved agricultural management, such as accessibility and nutrient management, are indirect effects. Salinity control is also an indirect effect of drainage; thus water and salt balance analyses are needed to check whether the salt build-up in the root zone stays within acceptable limits (Ayers and Westcot, 1994) and to develop irrigation and drainage management strategies to satisfy the leaching requirements. For these analyses, the crop tolerance to waterlogging and salinity needs to be known.

Artificial drainage is needed if the natural hydrological and soil conditions cannot cope with these leaching requirements. Traditional surface irrigation methods, such as basin and furrow irrigation, have field application efficiencies that are generally lower than the leaching requirement; but modern irrigation methods, including sprinkler or drip, can have field application efficiencies that are higher than the required leaching (Brouwer et al., 1989). In light of all the changes in water demand, supply and use, the role of (subsurface) drainage in arid and semi-arid regions has changed from a single-purpose measure for controlling waterlogging and/or salinity to an essential element of integrated water management under multiple land use scenarios (Schultz et al., 2007). To move towards more reasonable sustainability, drainage must be given its appropriate role in agricultural water management (Ritzema, 2016; Ayars and Evans, 2015). The advances in drainage design and management to achieve this change are discussed in the following sections.

2 Advances in drainage design and management for irrigated agriculture

To give drainage its appropriate role in agricultural water management in both rainfed and irrigated agriculture, the advances in drainage design and management are now focussing on the following issues:

- More control in the drainage system to match irrigation and drainage needs;
- Reducing the environmental effects of drainage;

- Reuse of drainage water;
- Safe disposal of the drainage effluent;
- Drainage materials; and
- Organization of drainage water management.

2.1 More control in the drainage system to match irrigation and drainage needs

Drainage systems that were traditionally built for rapid removal of all excess water have to be transformed into systems that can better control water levels (Vlotman et al., 2007). The ongoing changes in land use and climate change require a fundamental shift towards a more adaptive and participatory form of drainage (Ritzema and van Loon-Steensma, 2018). This requires more control in the drainage system – instead of free outflow (FD) conditions, the focus has shifted to controlled drainage in a three-step approach of decreasing priority: (i) after heavy rainfall or excessive irrigation to retain excess water in the field by storing the water on the soil surface and in the soil profile, (ii) increasing the storage capacity in the drainage system and (iii) enabling reuse or controlled removal (Fig. 2).

The second and third steps, storage of excess water in the field and controlled reuse/removal, build on experiences with controlled drainage from, among other places, the Netherlands, the United States, Egypt and India (Vlotman and Jansen, 2003). Controlled drainage is a system for physically restricting discharge with adjustable gates installed at the outlet of the drain with in-line control structures. The aim is to control water level during the growing season (Fig. 3) (Nasralla, 2019). Controlled drainage is the most promising practice to increase crop production and irrigation efficiency and at the same time to reduce the total salt and nutrient load in the drainage water. Controlled drainage shows promise as a tool to improve the balance between various types of land use not only between different types of agricultural use but also between agriculture and nature, an often-delicate balance (Ritzema et al., 2016). Studies in both semi-arid and humid areas demonstrate that it is possible to actively manage SSD systems to reduce flow and store water for crop water use (Ayars and Evans, 2015; El-Atfy et al., 1991; Ritzema et al., 2008a).

The challenge is to optimize the depth of the water table between the various growing seasons or during the growing season. For example, a 4-year study conducted in Sari, North Iran, shows that controlled SSD can enable the cultivation of winter crops in the consolidated paddy fields (Mehdi et al., 2015). In a study to optimize the water table management during the growing season conducted in two fields (about 32 ha), located in the Moghan plain, North Iran, three options for controlled drainage were tested (Jouni et al., 2018): subsurface

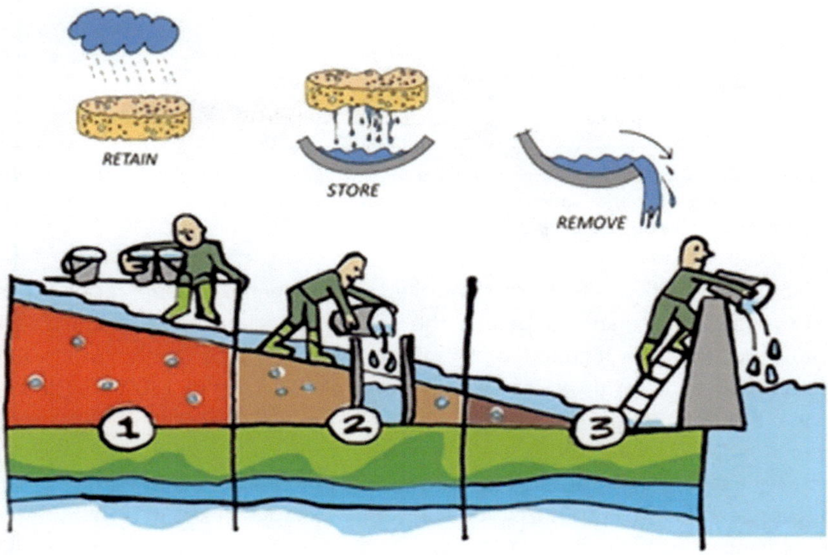

Figure 2 The focus in agricultural water management has shifted from increasing drainage intensities to retain (1), store (2) and reuse (3) of the drainage water. Source: Adapted from: Ritzema and Stuyt (2015).

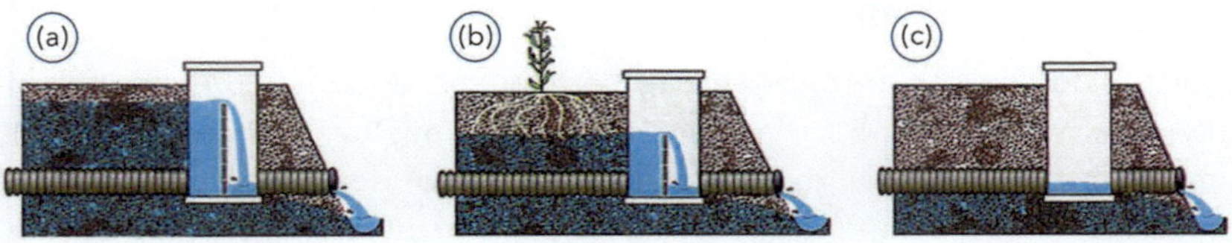

Figure 3 Examples of options that farmers have for manipulating the drainage base in a controlled drainage system to promote desirable impacts at the farm level: (a) high water table during the initial stage of crop development; (b) optimum water table during crop development and mid- and late season; and (c) low water table during harvest and after harvest. Source: Adapted from: Poole et al. (2018).

drains at 2 m with FD, controlled drainage at 70 cm below the soil surface (CD70) and controlled drainage with a varying depth depending on the crop stage (CDch) (Fig. 4). The highest crop yields were found for CDch followed by CD70, both significantly higher compared to the free drainage (FD) (Table 1). The same trends were observed for water use efficiency (WUE). Controlled drainage reduced not only SSD rates but also nitrate and phosphorous losses. Soil salinity decreased in FD but slightly increased in the CD treatments. Thus, although controlled drainage clearly has advantages above FD practices, to optimize CD management options, more research is needed on the long-term effects of controlled drainage on soil salinity.

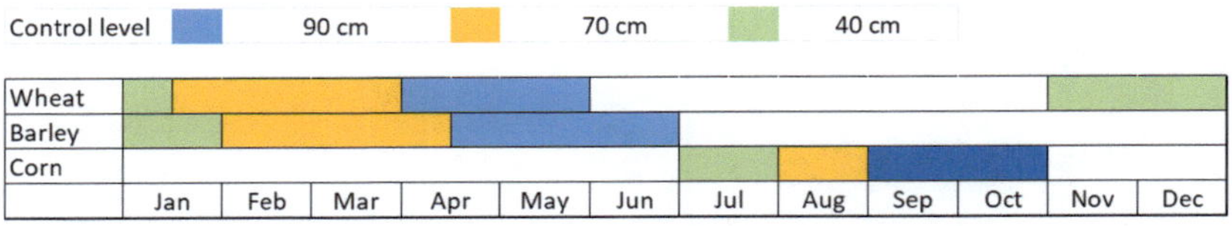

Figure 4 Example of controlled drainage with a varying depth during the growing seasons of wheat, barley and corn in two fields in the Moghan plain, North Iran. Source: Adapted from: Jouni et al. (2018).

Similar positive effects of controlled drainage have been reported for crop yields in maize and soybean in the US and Canada. For example, Ng et al. (2002) observed an increase of 64% in maize yield for a controlled drained system compared with a free draining system. Borin et al. (2001) observed higher plant biomass in crops grown under controlled rather than conventional drainage. Mejia et al. (2000) reported higher crop yield (maize and soybean) with controlled drainage. Additional examples of similar results can be found in Noory and Liaghat (2009), Drury et al. (2009), Skaggs et al. (2012), Nash et al. (2015) and Sunohara et al. (2015), among others. These studies showed that controlled drainage can contribute to increased productivity in this important agricultural region.

2.2 Reuse of drainage water

Drainage outflow can be reused to supplement freshwater resources but usually only after mixing (Fig. 5). There are various options for this: it can be done directly in the field, through the irrigation system or downstream in the basin. The challenge is the need to improve the quality of the drainage water before it can be reused (Tanji and Kielen, 2002). There are a number of ways to improve the quality of the drainage, for example, preventive methods by legislation to restrict the use of fertilizer and/or manure at farm level (Ritzema, 2007a) or to let farmers pay for the quality of their drainage effluent or with subsidies and advisory services to reduce losses of nitrogen (Christianson et al., 2013) or to treat the polluted drainage water, for example, through desalination by using reverse osmosis membrane technology (Thompson et al., 2013; Uléna et al., 2012). The challenge is to find the most appropriate method or combination of methods for achieving the target quality. A complicating factor is that reuse is often done unofficially, for example farmers pumping directly from their field drains, but there is a need for regulation at the project or basin scale. To adapt or implement national reuse strategies at the local level a participatory approach in defining goals by merging local visions, and expert judgement increases its functionality and robustness (Jonsson et al., 2011; Speelman et al., 2011). Whatever approach is chosen, continuous monitoring of both the

Table 1 Crop yield, water use efficiency (WUE), drain discharge (q) and nitrogen (N) and phosphorous (P) losses for the three treatments: free drainage (FD), controlled drainage at 0.70 (CD70) and controlled drainage with varying depths depending on crop stage (CDch)

Crop	Treatment	Crop yield		WUE		*q*		N-losses		P-losses	
		(t/ha)		**(kg/m³)**		**(mm/d)**		**(kg/ha)**		**(g/ha)**	
Wheat	FD	5.1	-	0.54	-	2.10	-	0.24	-	9.14	-
	CD70	6.4	(27%)	0.68	(26%)	1.16	(−45%)	0.15	(−38%)	5.69	(−38%)
	CDch	7.2	(41%)	0.76	(41%)	1.41	(−33%)	0.18	(−25%)	6.21	(−32%)
Barley	FD	4.3	-	0.52	-	2.60	-	0.33	-	13.7	-
	CD70	5.3	(23%)	0.62	(19%)	1.17	(−55%)	0.17	(−48%)	8.4	(−39%)
	CDch	5.8	(34%)	0.69	(33%)	1.43	(−45%)	0.19	(−42%)	9.6	(−30%)
Maize	FD	49.1	-	1.18	-	5.10	-	0.63	-	46.8	-
	CD70	58.6	(19%)	1.53	(30%)	2.50	(−51%)	0.31	(−51%)	28.5	(−39%)
	CDch	61.4	(25%)	1.7	(44%)	2.86	(−44%)	0.34	(−46%)	30.1	(−36%)

Source: Jouni et al. (2018).

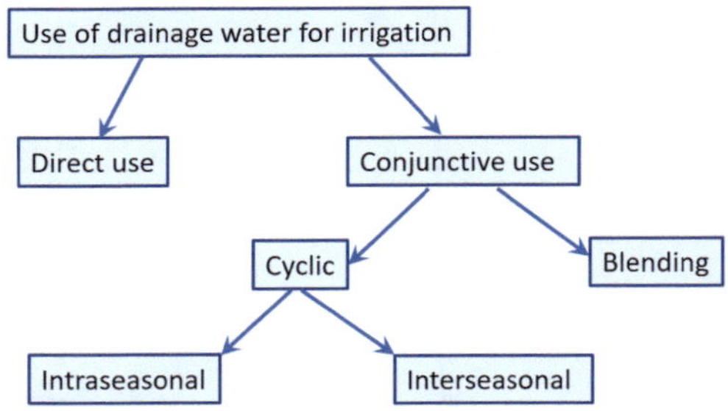

Figure 5 Reuse of drainage water for irrigation. Source: Adapted from: Tanji and Kielen, (2002).

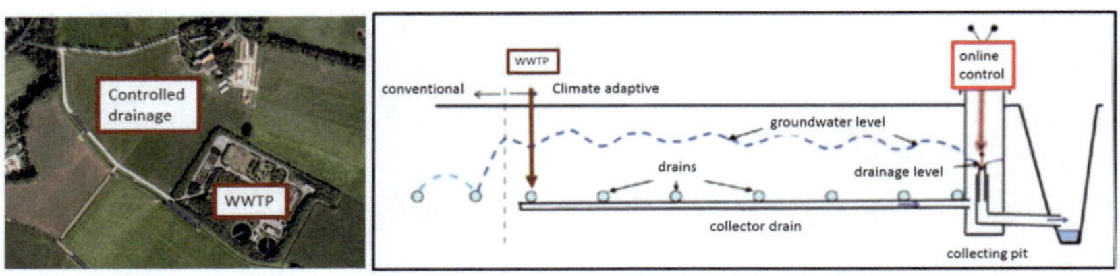

Figure 6 Controlled drainage and sub-irrigation using (partly) treated wastewater from a municipality Waste Water Treatment Plant (WWTP) in Haaksbergen, The Netherlands (https://www.knowh2o.nl/en/).

quantity and quality of the water is needed to ensure good-quality irrigation water (El-Quosy, 1989).

2.3 Drainage to enable the use of poor-quality irrigation water

Experiments in a greenhouse in Sistan, South Iran, showed that brackish or even saline irrigation water can be used for irrigation of salt-tolerant crops, such as *Samphire*, if drainage is used in combination with soil amendments, for example composted cow manure, to reclaim salt-affected soils (Kalanaki et al., 2022). Another option is the direct use of treated wastewater in a controlled SSD system (Fig. 6) (Bartholomeus et al., 2016). The most important advantages are (i) better control over the soil moisture regime and therefore better growth conditions; (ii) lower irrigation water requirements (water saving) and no direct evaporation losses; and (iii) a reduced load on surface water with fertilizers and anthropogenic contaminants, in particular residues of medicines and plant protection products. Less effluent is discharged into the surface water and/or indirectly because the effluent only reaches the surface water after it has flowed through the soil. There are also risks and challenges: (i) groundwater contamination with various micro-contaminants including pharmaceutical by-products is an important potential risk and (ii) the extent to which and how residues of drugs and pathogens/bacteria/viruses spread during sub-irrigation.

Note, however, that if there is no sub-irrigation with wastewater, then direct irrigation with the surface water will also have these problems because, in that case, the wastewater effluent is discharged into the surface water.

Long-term field experiments (1995–2013) in north-eastern Italy indicate that the benefits of combining controlled drainage and sub-irrigation are especially beneficial in years with wet springs followed by summer droughts. Sub-irrigation via controlled drainage helped to achieve higher yields when soil moisture content was declining due to prolonged dry periods (Tolomio and Borin, 2019). It should be remembered, however, that successful sub-irrigation using controlled drainage is only possible under the following conditions (Bos and Wolters, 2006) including (i) a flat soil surface, (ii) small water losses to the underground and adjacent areas and (iii) permeable soils (also when always saturated or at field capacity).

2.4 Safe disposal of drainage effluent

Controlled drainage and reuse can reduce outflow, but in the end, the salts brought in by the irrigation water, together with the residues of fertilizers, pesticides and wastewater coming from non-agricultural sources, require safe disposal (Ritzema and Braun, 2006). This is a final challenge to drainage being able to have its place as part of integrated and sustainable water management. Evaporation ponds can be used for salt disposal, for example, in Pakistan, where these ponds are located in the desert area outside the irrigated plain in southeast Punjab. However, they are not very effective as only 13% of the salts are deposited in these ponds (Quereshi and Sarwar, 2009). Discharging the salts into lakes and/or rivers that eventually reach the sea or ocean seems more logical, but the consequences of these higher salt concentrations, not to mention the other (toxic) elements dissolved in the water, require serious consideration. If the river or lake in which the drainage effluent is discharged cannot cope with the salt-laden drainage water, then a separate outfall drain directly to a safe outlet, usually the sea, has to be constructed. Two of the best-known outfall drains, especially created for the disposal of highly saline drainage water, are the Left Bank Outfall Drain in Pakistan and the Third River in Iraq. The Left Bank Outfall Drain has been constructed to drain approximately 0.5 million ha in the Sindh Province of Pakistan (McCready, 1987). The disposal of the drainage effluent back into the Indus River, or one of its branches, is unacceptable because of the high salinity levels: the effluent from SSD can vary from 4.7 to 15 dS/m and that from tube wells can be twice as saline. Disposal into the river would result in excessively high salinity levels and would make downstream use for irrigation impossible. In Iraq, the Third River, which was completed in 1993, acts as an outfall drain for the area between the Euphrates and the Tigris. But even disposal to a 'safe' outlet (the sea) can have environmental effects. In the northwest European countries, the excessive use of manure, fertilizers and pesticides in agriculture has resulted in the eutrophication of the North Sea and Baltic Sea caused by the leaching of minerals from these agricultural lands (Skogena et al., 2014). Similarly, in the USA, the nitrate-nitrogen losses from subsurface tile drain systems, that were installed in about 40% of the farmlands in the US Midwest, have now created a hypoxic zone in the Gulf of Mexico (Anderson and Wall, 2006).

Another situation where reuse of drainage water at the field level is needed is when the drainage effluent is so contaminated that it is forbidden to discharge the polluted drainage effluent back to the river. For example, in the western San Joaquin Valley, California, USA, the Integrated on-Farm Drainage Management (IFDM) system was developed to dispose of selenium (Se)-contaminated agricultural drainage water through the sequential reuse of that drainage water to grow crops with increased tolerance to salinity (Lin et al., 2002). The concept of the IFDM system involves several stages (Fig. 7), which are as follows:

- Stage 1: the use of freshwater irrigation (i.e. with low salinity and Se-concentration) to grow salt-sensitive crops, like vegetables;
- Stage 2: reuse of the resulting drainage water, which of course has a higher salinity than the fresh irrigation water to irrigate more salt-tolerant crops, like cotton, alfalfa, sugar beet and grasses;
- Stage 3: the drainage water from the salt-tolerant cropping area, with a still higher salt concentration, is then used to irrigate even more salt-tolerant trees or grasses;

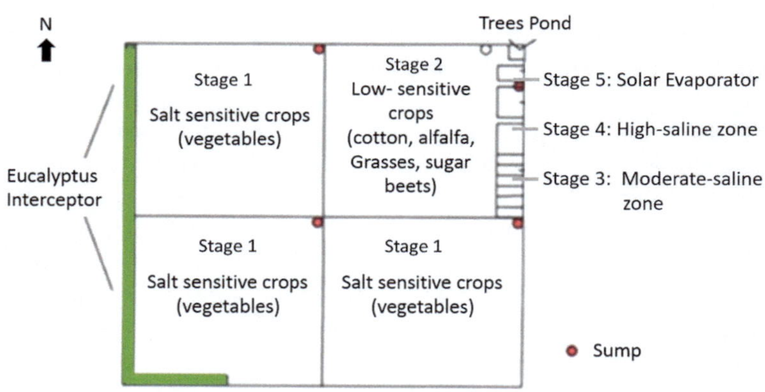

Figure 7 Integrated on-Farm Drainage Management for a sequential reuse of drainage water at farm level: starting with salt-sensitive crops (Stage 1), followed by low-sensitive crops (Stage 2) and finally moderate and high saline crops (Stage 3, etc.). Source: Adapted from: Lin et al. (2002).

- Stage 4: finally, using the SSD water from these trees or grasses to irrigate highly salt-tolerant, halophytic plants; and
- Stage 5: as a final step, the remaining drainage water is disposed via sprinklers into a lined solar evaporator where the mineral salts and associated Se and other trace elements are deposited.

Through this sequential reuse, the volume of drainage water is substantially reduced by crop evapotranspiration.

2.5 Drainage materials and methods

Although SSD practices have evolved from manual to large-scale mechanized installations (Ritzema et al., 2006), including the development of appropriate materials (i.e. pipes and envelopes) (Stuyt et al., 2000), drainpipe performance in irrigated agriculture is often below standard due to clogging, siltation and root growth inside the pipe. To tackle these problems, an innovative pipe-envelope concept, Hydroluis (www.hydroluis.net/en), was tested on a 50-ha pilot area in Harran, Turkey, in 2015 and 2016 (Bahçeci et al., 2018). This novel concept consists of a corrugated inner pipe with three rows of perforations at the top and an unperforated outer pipe that covers about 2/3 of the inner pipe leaving only the unperforated bottom part of the inner pipe in contact with the soil (Fig. 8). The main advantage of this new concept is that it works for a wide range of soil textures and there is better protection against root growth inside the pipe. This new concept was compared against a conventional geotextile envelope, a gravel envelope and a control with no envelope. The Hydroluis and gravel envelopes had a significantly lower entrance resistance compared to the geotextile, the best drain performance and no signs of sedimentation

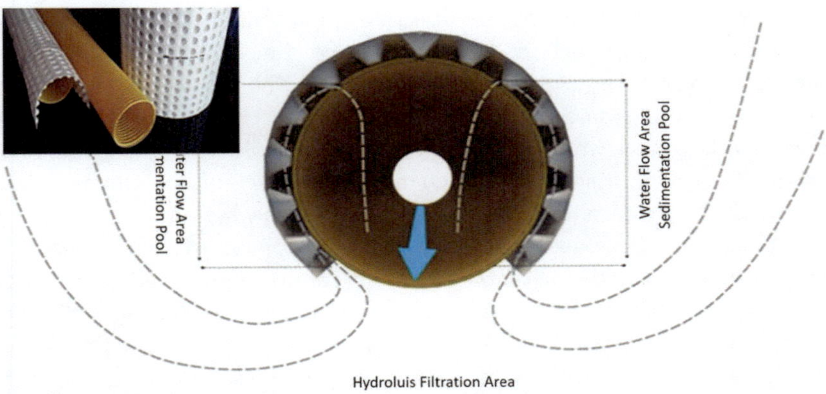

Figure 8 Working principle of the Hydroluis pipe-envelope combination (Bahçeci et al., 2018).

nor of root growth inside the pipe. The production costs of the Hydroluis envelope are comparable to those of pre-wrapped synthetic envelopes and considerably lower than gravel envelopes. To test the performance of this envelope in a saline-sodic soil, which often causes clogging of the normally used synthetic envelopes, the Hydroluis envelope was also tested in a soil tank laboratory model with a saline-sodic problem soil from south-western Iran (Alavi et al., 2021). The results indicate that this envelope is not suitable for this type of saline-sodic soils as the Hydroluis drainpipe gets clogged directly after installation due to invasion of the test soil into the space between the inner and outer pipes. The results show that in stable soils, with very fine sand or coarser soil particles ($D > 0.05$ mm), there are no serious limitations for the Hydroluis drainpipe application, whereas the current design is unsuitable for filtering fine silt particles ($0.002 < D < 0.02$ mm).

2.6 Organization of drainage management

In irrigated areas, especially in developing countries, drainage systems are often designed and implemented by the government, with the users, very often smallholder farmers, having little input or responsibility and with governance structures offering them little opportunity to influence or contribute to existing management approaches. The majority of these farmers are poor and do not have the means to invest in SSD themselves. In the typical 'top-down' standardized approach, the location-specific conditions and farmers' preferences are hardly taken into consideration. Furthermore, the emphasis has traditionally been on the technical aspects (physically engineered infrastructure) of the systems, while the organizational aspects (institutional infrastructure) have been largely neglected. The result has been that the installed systems are not effective, and most beneficiaries do not feel ownership, resulting in poor operation and maintenance (O&M) programmes. A shift to a more service-oriented approach, as promoted since the 1990s, has not gained much of a foothold in practice (Ritzema, 2009). The institutional set-up for drainage is complex and enforcement of rules and regulations is difficult. In contrast to irrigation, where direct benefits to stakeholders are involved, rules and regulations for drainage are much more difficult to enforce. Drainage fees need to be collected and, unlike the irrigation supply system, it is difficult to disconnect uncooperative customers. A case study on participatory research in Vietnam showed that the existing institutional set-up, often based on irrigation system layout, needs to be modified or adapted to improve drainage efficiency (Ritzema and Schultz, 2011). In the Ica Valley of Peru, six public and five private organizations are involved in the governance of local water management. This legal and institutional framework, however, appears to be inadequate to solve the overexploitation of the (ground) water resources (Salmoral et al., 2020; Hepworth et al., 2010). Finding institutional and/or organizational arrangements that are culturally acceptable and functionally effective are a fundamental challenge for achieving the proper role of drainage in integrated water resources management.

3 Case study: modernization of the drainage design and management in Egypt

In this case study, we will discuss the modernization of the drainage design and management in Egypt, a country with one of the largest drainage practices in the world, where almost all the agricultural lands are equipped with a SSD system and with an annual implementation rate of about 25 000 ha/year (NDP3, 2021a).

3.1 Need to modernize the drainage network

Egypt's Nile Valley and Delta, one of the oldest agricultural areas in the world, has been under continuous cultivation for at least 5000 years. Egypt has an arid climate, characterized by high evaporation rates (1500–2400 mm/year) and low rainfall (5–200 mm/year), thus agriculture depends almost entirely on irrigation from the River Nile. From ancient times onwards, irrigated agriculture in the Nile Valley and Delta depended on the annual floods of the River Nile. The receding floods also drained and leached the cultivated areas. The construction of the Aswan High Dam in 1970 ended the annual flooding but made irrigation water available throughout the year. However, the termination of the annual floods reduced the natural drainage capacity and increased the risk of salinization. Since then, two to three crops can be grown each year, resulting in a practically continuous growing season. To avoid salinization, the Egyptian Government started an ambitious programme to drain all of Egypt's agricultural land (approximately 2.5 million ha) (Abdel-Dayem et al., 2007).

The drainage network consists of a composite SSD system of buried field (in Egypt called laterals) and collector drains that discharge by gravity into a large open surface drainage (SD) system from where the drainage water is discharged (pumped or by gravity) into large open outfall drains which eventually discharge into the River Nile or the Mediterranean Sea. Pumping is necessary almost everywhere in the Delta and the Valley, except in some areas in Upper Egypt, where there is enough gradient to dispose of the effluent freely by gravity (Amer and de Ridder, 1989).

The drainage practices in Egypt have evolved from manual to large-scale mechanized installation (Nijland, 2000). The installed drainage networks were very successful in the majority of the old lands of the Nile Valley and Delta (Ali et al., 2001; Abdel-Dayem et al., 2007). There are, however, needs to modernize these networks (Ritzema et al., 2023; NDP3, 2022a):

- The existing gravity SSD systems have no options to control the water table and drain outflow during the growing seasons. Farmers, however, want more control over their drainage practices to optimize their agricultural water management practices, i.e. to match their irrigation and drainage needs. Thus, they illegally block drainage outflow in the manholes of the (sub-) collectors, significantly increasing the maintenance costs and reducing the lifetime of the system.
- The gravity system also results in a rather deep open drainage system; subsequently, the cross-sections of the surface drainage drains are much larger than the hydraulic cross-section needed to remove the excess drainage water (Fig. 9). As a result, the over-dimensioned surface drainage system is misused by many stakeholders to illegally depose of wastewater and solid waste, resulting in high maintenance costs to clean the drains. This also consequently limits the potential to reuse the polluted drainage water for further irrigation.
- Land use and cropping patterns are changing. Currently, one standard design methodology of the SSD system is used all over Egypt. The drainage design criteria are based on cotton, which is considered the most critical crop (Abdel-Dayem and Ritzema, 1990). The implication of this design approach is that if other crops are cultivated, these crops will be over-drained, reducing the irrigation efficiency. Thus, it seems logical that the cropping pattern is taken into consideration in the design of a drainage system. A complicating factor is, however, that it is very hard to predict which combination of crops will be cultivated in a given area, both now and in the future. Therefore, it is logical to not change the existing agricultural design criteria but rather instead to look for options to modify the management practices to reduce over-drainage.
- Water scarcity is increasing because of climate change and the construction of the Grand Ethiopian Renaissance Dam (Ellah, 2020; Omar and Moussa, 2016; Wolters et al., 2016). Climate change will result in less

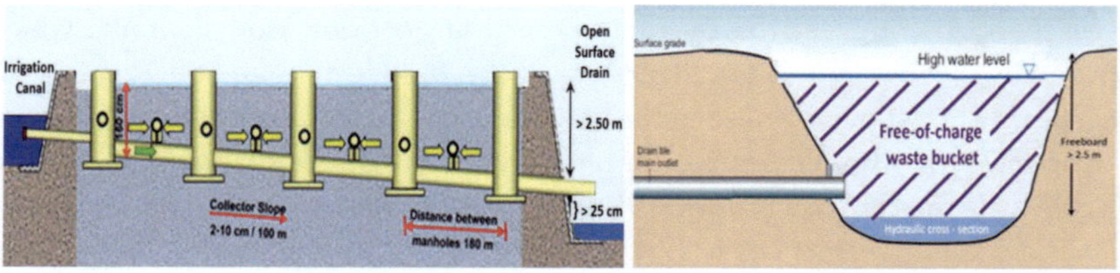

Figure 9 The gravity drainage system results in an over-dimensioned open drainage system.

predictable annual inflow into Lake Nasser, reduce rainfall in northern Egypt and increase crop water requirements (MWRI, 2017). The current water shortage in Egypt is already 13.5 billion cubic metres per year (BCM/year) and is expected to continue to increase. More control in the drainage system to reduce drainage water will be one of the options to reduce water scarcity (Wahba et al., 2005).

- Reuse of drainage water, however, is threatened because the quality of the drainage water has severely deteriorated due to pollution with wastewater from villages and industries and in the north of the Nile Delta the upward seepage of brackish groundwater (El-Agha et al., 2020). Reuse should therefore be applied at field level before the drainage water is polluted in the open drainage system. This requires improved control and pumping from the SSD system.

To address these challenges, the National Drainage Programme Phase 3 (NDP3) aims to improve the efficiency of the drainage system by developing and implementing new design concepts to reduce maintenance costs and extend the lifetime of the drainage system (NDP3, 2022a). The focus is on a decentralized design approach to facilitate a more participatory process and better adjustment to location-specific conditions. The following developments have taken place to address these needs.

3.2 More control in the drainage system

More control in the drainage system is a promising option to avoid misuse of the free-flowing SSD system (Nasralla, 2019; Wahba et al., 2005). Soon after the installation of the SSD systems in the 1970s, it became clear that control in the gravity drainage system was needed to reduce water losses from areas with rice without restricting drainage of other crops (El-Atfy et al., 1991). The abolishment of the land consolidation in the 1990's, however, made controlled drainage more complicated, because the individual farm plots are often quite small (on average 0.2 ha). Thus the fields and (sub) collector drains are serving more than 1 plot and subsequently controlled drainage is only possible at (sub) collector level. In the current design, a (sub) collector unit is based on the existing main infrastructure and does not take into consideration the irrigation system layout and management at *mesqa* level (≈100 feddan). In a *mesqa*, farmers already work together for their irrigation water management; to extend this cooperation to drainage management, sub-collectors should be designed based on the layout of the *mesqa* system. But contrary to irrigation, where each farmer can decide for himself when and how much water he wants to supply to his plot (of course within the rules and regulations of the *mesqas*), field drains will always serve neighbouring fields thus management of the SSD at sub-collector level will always need cooperation between the farmers to agree on a common strategy. A complicating factor is that each crop has its own drainage needs (Muhammada et al., 2021), thus if farmers want to work together on drainage management, they each must agree on the crops they want to grow in the sub-collector. Farmers, however, are reluctant to cooperate at the sub-collector level to agree on which crops they want to cultivate. To convince farmers on the benefits of controlled drainage, more farmer participation in planning, supervision and operation and maintenance is also required (Section 3.6).

In Egypt, the design of the SSD system is based on the most critical crop, cotton, thus water tables are relatively deep when other crops are cultivated, resulting in over-drainage and low irrigation efficiencies. The introduction of controlled drainage can reduce these negative effects. As controlled drainage is only possible at the sub-collector level, farmers must compromise on the crops they want to cultivate and ideally should agree to cultivate crops with similar water table requirements (Table 2).

Controlling the water table at optimum depths can increase irrigation efficiency by 15–30% (DRI, 2013; Nasralla, 2019). Higher irrigation efficiency results in lower drain discharges and thus a lower salt load on the surface drainage network. There is, however, the risk that, because of the higher irrigation efficiencies, the leaching requirements are not satisfied and soil salinity levels will gradually increase. Water and salt balance studies over the whole crop rotation period must be made to check if these higher irrigation efficiencies and the corresponding drain discharges meet the leaching requirements (Ritzema, 2009).

In summary, the biggest challenges for the introduction of controlled drainage in Egypt are as follows:

- Layout of the (sub) collector units must be based on the existing layout of the irrigation/*mesqa* system;
- Farmers in a sub-collector unit have to agree on their cropping pattern during each cropping season, so that they synchronize their (irrigation and) drainage management practices;
- Farmers in a sub-collector unit should be responsible for the operation of the controlled drainage system.

Table 2 Crops with similar water table depth (m) requirements

Crops	Initial	Mid-season	Late season
Rice	<0.2 m	<0.2 m	>1.0 m
Vegetables: potatoes, tomatoes, onions	0.3 m	0.5 - 0.7 m	>1.0 m
Other crops: wheat, maize, sugar beet, cotton, berseem	0.3 m	0.8 m	> 1.0 m

- These operational practices must be developed in cooperation with the farmers.

To address these challenges and to apply the lessons learned in pilot studies (MWRI, 2016; DRI, 2013) at a larger scale, the Egyptian Public Authority for Drainage Project and the Drainage Research Institute are currently implementing controlled drainage in large-scale pilots in Shereshra (23 ha) in the West Delta, Hala (35 ha) in the East Delta and Baha (17 ha) in Upper Egypt.

3.3 From gravity to pumped drainage

Control at the sub-collector level also enables a switch from gravity to pumped drainage. One of the main reasons to install gravity drainage systems was that in the 1970s small pumps were not readily available on the market and were relatively expensive; thus pumping at the sub-collector level was not an option. Note that at that time, most farmers still used the animal-powered *Sakia*'s (Persian waterwheel) to lift the irrigation water to their fields. Nowadays, farmers are using pumps to lift their irrigation water, thus they can also do this to pump their drainage water. The benefits of pumped drainage at the sub-collector level are as follows:

- The drainage discharge can be pumped directly back into the *mesqa* canals, so that farmers can reuse the drainage water directly in their own sub-collector.
- For long collectors, pumping from sub-collector drains can reduce the depth of the downstream sections of the collector.
- A pumped collector outlet will reduce the free board in the open surface drains and thus the cross-section of the open drains, considerably reducing the operation and maintenance cost.

3.4 Reuse of drainage water

Government pumping stations and farmers' small diesel pumps reuse drainage water by lifting water up from drainage ditches and diverting it back into the irrigation canals, increasing the country's available water resources by 20% (Barnes, 2014). In the Delta, especially in the southern part, the drainage water is generally of an acceptable quality. Thus 12 reuse pumping stations were constructed to pump the drainage water back into the irrigation system (El Quosy, 1989). Currently, however, most of these reuse pumping stations have been decommissioned because the drainage water in the surface drainage network has become so heavily polluted with wastewater and is now unsuitable for reuse. In the northern part of the Delta, the salinity of the drainage water in the main drainage system is also negatively affected by the upward seepage of saline groundwater (El-Agha et al., 2020). Salinity levels in these deep open main drains can be up to four times higher than the salinity of the water that comes out of the SSD system. Thus it is beneficial

- to reuse the drainage water directly from the SSD system as the salinity of this water is still acceptable (range 1.2–1.4 dS/m) compared to the salinity of the irrigation water of 0.44 dS/m. This is in contrast to the salinity levels in the main drains (> 6 dS/m); and
- to raise the levels of the open drainage system (by starting to pump from the collectors) to reduce the upward seepage of saline groundwater and subsequently the total salt load in the drainage system.

It should be remembered that the total volume of drainage water available for reuse is small compared to the total crop water requirements. Reuse can only really be used to support supplementary irrigation, as crop water requirements can be as high as 6–8 mm/day (in the summer season) compared to a drainage design rate of around 1 mm/d. The operation of a pumped drainage system to enable the reuse of drainage water should be based on the leaching requirements over the crop rotation period (Ritzema, 2009). During a period of water scarcity, the drainage water can be reused and a slight increase in soil salinity can be accepted, as long as this soil salinity level does not affect crop yield, but these salts still have to leached in the following cropping season(s) to guarantee a long-term equilibrium in the soil salinity level. To avoid the risk of (localized) soil salinity, it is recommended to mix the drainage water with irrigation water by pumping it back into the irrigation system.

3.5 Use of the subsurface drainage system for supplementary sub-irrigation

Controlled drainage makes it possible to store water to support crop growth. But even then, drought stress can occur during prolonged dry periods. Under such conditions, the drainage system can also be used for irrigation. This is known as sub-irrigation or inverse drainage (Bos and Wolters, 2006). Sub-irrigation through the drainage system is an option to reduce the major problem of pollution of the open drainage network. It can be practiced at locations where partially treated wastewater from domestic areas is available (Habib et al., 2015; Hashem and Qi, 2021). Wastewater from domestic areas has a high organic matter content and is low on chemical pollutants, but still contains anthropogenic contaminants that can be harmful when farmers come in direct contact with this wastewater. The supply of the partly treated wastewater to the crop through the soil, from below, via the SSD system reduces the health and operator risks because there will be no direct contact. Another benefit is that the high organic matter can supplement fertilizers.

The direct use of treated wastewater through the SSD system has several advantages. The most important are (Bartholomeus et al., 2016) as follows:

- Better control over the soil moisture regime and therefore better growth conditions during periods of drought;
- Lower irrigation water requirements (water saving) and no direct evaporation losses;
- A reduced load on surface water with fertilizers and anthropogenic contaminants; and
- Less effluent is discharged into the surface drainage network, and/or indirectly because the drainage water only reaches the surface water after it has flown through the soil.

But there are also risks and challenges, which are as follows:

- Groundwater contamination with various micro-contaminants, including pharmaceutical by-products, is an important potential risk; and
- The extent to which and how residues of drugs and pathogens/bacteria/viruses spread during sub-irrigation.

Note however, that if there is no sub-irrigation with wastewater, then direct irrigation with the surface water will also have these problems because, in that case, the wastewater effluent is discharged directly into the surface water system. Around wastewater treatment plants (WWTPs), sub-irrigation with wastewater can be practiced in a controlled drainage system, the only additional construction will be a pipeline to connect the WWTP to the collector drains in neighbouring agricultural areas.

3.6 Improvement of farmer participation

The SSD systems in Egypt are designed and constructed by the government, thus farmers do not feel a sense of ownership. Furthermore, the existing systems do not have options for control, so farmers block the drainage system to save irrigation water or to adapt the system for their needs when they change their cropping pattern. These unofficial practices result in high maintenance costs and a reduced lifetime of the subsurface systems. To create more ownership, farmer participation needs to be increased in the entire project cycle from planning and supervision of works to operation and maintenance of the drainage infrastructure. This is a complicated process as farmers have to foster collaboration and work together with various governmental agencies (Fig. 10).

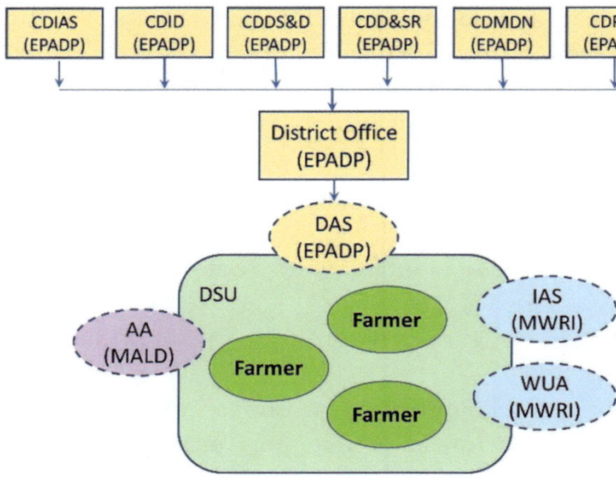

Figure 10 Farmers in Egypt have to cooperate with each other and various governmental agencies in the design, construction and operation and maintenance of their drainage system.

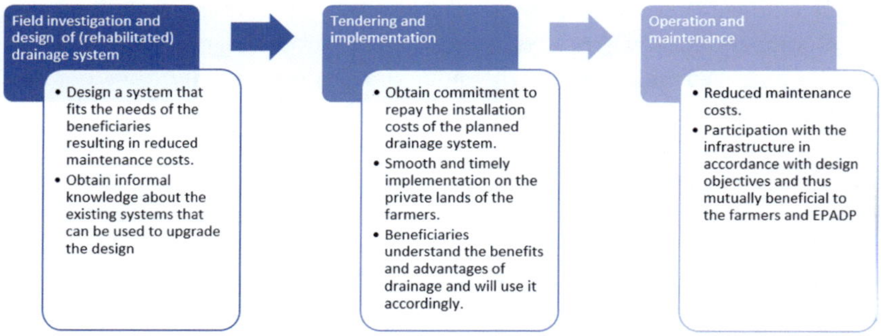

Figure 11 Benefits of farmer participation at different stages for EPADP's project cycle.

The main challenges to achieve improved participation include (i) farmers having to work together in Drainage Collector User Groups (DCUGs) as fields and collector drains always serve more than one farmer's field and (ii) better coordination and cooperation are needed between the governmental agencies, namely the six departments of the Egyptian Public Authority for Drainage Projects (EPADP), two Departments of the Ministry of Water Resources and Irrigation (MWRI) and one Department of the Ministry of Agriculture and Land Reclamation (MALR), that are involved in providing technical extension and advisory services to the farmers. Improved farmer participation is also beneficial for EPAPD in various stages of the lifetime of the drainage infrastructure (Fig. 11). In support of this, a strategy paper and a training manual have recently been prepared to convince both EPADP staff and farmers of the benefits of improved farmer participation (NDP, 2021a,b).

3.7 Drainage materials

3.7.1 Drain pipes

Field drains are made using corrugated and perforated PVC pipes, with collector drains generally being constructed using plain concrete pipes, PVCE or HDPE. Concrete manholes/drop structures are installed at some of the connection points of the laterals and collectors. Currently, the EPADP is transforming its production process to replace the PVC field drains and the concrete manholes with HDPE (NDP3, 2021b). There is absolutely no need to use virgin HDPE granules, as HDPE can be endlessly recycled without loss of quality or quantity. In Egypt, recycled HDPE granules are commercially available at about EGP 12 000/t (US$645). The use of this raw material considerably cuts the costs of the pipes and at the same time creates a market for the abundance of HDPE waste that clogs the irrigation and drainage canals and contributes to landfill and garbage dumps. Recycling therefore offers a win-win opportunity for EPADP. The new Environmental Law 2002–2020 tries to reduce single use plastics and to promote recycling. EPADP as a governmental organization has the opportunity and a moral duty to become a forerunner in this field.

3.7.2 Drain envelopes for calcareous soils

Pre-wrapped synthetic envelopes are used for all soils with a clay content >30% and the synthetic envelope is usually made from voluminous polypropylene fibres with an average thickness of 5 mm and an average pore size index 400 microns (range 200–600 microns) (PP400). The envelopes, however, can get easily clogged in calcareous soils with a calcium carbonate ($CACO_3$) content > 10%. For this type of soil, four drain-envelope combinations of SSD systems were tested in a Laboratory Permeameter Test, followed by an in-field experiment in a problematic calcareous soil in Abu-Masoud, West Nile Delta, Egypt (Sallam, 2018). Combinations of granular and pre-wrapped drain envelopes with drain slope, ranging from 30 to 100 cm/100 m, were tested. The synthetic envelope, made of voluminous polypropylene fibres, had an average thickness of 5 mm and an average pore size index 400 microns (range 200–600 microns) (PP400). The granular drain envelope was determined using the USBR filter design criteria where the particle size distribution of the granular drain envelope is matched with the particle size distribution of the soil. The combination of a granular and pre-wrapped envelope with a normal grade of 10 cm/100 m was found to be the best material regarding their hydraulic and mechanical performance for sub-surface drainage in calcareous soils.

4 Conclusion and future trends in research

In the traditional top-down approach of drainage design and management, standardized designs were generally used with little or no consideration of the location-specific conditions, environmental concerns and farmers' needs and preferences. Furthermore, the emphasis has been primarily on the technical aspects (physical infrastructure), while the organizational aspects (institutional infrastructure) of the drainage systems have been largely neglected (Scheuman, 1997). A shift to a more service-oriented approach, which has been promoted since the 1990s, has not gained much of a foothold in practice. To reverse this trend and allow the benefits of drainage to be more widely realized, a three-step approach is needed (Ritzema, 2016):

- Balancing 'top-down' against 'bottom-up'. This requires a participatory approach focussing on societal choices and decentralized management. Participation must occur throughout the complete cycle: from planning, design and installation to O&M. Recent approaches, such as participatory learning and action research, including participatory modelling, and communities of practice, can enhance this process (Ritzema and Schultz, 2011). Another key part of this is the need to enhance the link between technical aspects (requiring physical solutions) and organizational aspects (requiring institutional changes).
- From standardization to flexibility. Instead of standardized design and implementation practices, a much more flexible approach based on location-specific conditions and farmers' preferences is recommended. Integration between the irrigation and drainage networks needs to be improved. The challenge is to find a balance between the individual need for drainage, which varies from field to field, and the fact that drainage at the farm level is a collective activity. This calls for better operational control (Ritzema and Stuyt, 2015).
- Focus on capacity development. There is ample evidence that capacity building is a prerequisite for success in most drainage projects involving activity by multiple stakeholders (Burns et al., 2015; Ghumman et al., 2012; Van Keulen et al., 2003). A capacity-building process, based on the knowledge-creating process of Nonaka and Takeuchi (1995), is needed to integrate research, education and advisory services to link the local tacit knowledge of the stakeholders with the explicit knowledge of the researchers to create new knowledge (Ritzema et al., 2008). This type of applied research is extremely profitable. Experiences from Egypt, India and Pakistan show that the benefits of applied research easily outweigh the costs (Ritzema et al., 2007b).

5 Where to look for further information

5.1 Handbooks

Drainage principles and applications:

- Ritzema, H. P. (Editor-in-Chief), 2006. *Drainage Principles and Applications*. ILRI Publication 16, Alterra, Wageningen, 1125p.

The book presents the basic principles of land drainage with applications. The book provides coverage of all the various topics useful to those engaged in drainage engineering and includes a glossary.

Drainage systems:

- Skaggs, R. W. and van Schilfgaarde, J. (Ed.), 1999. *Agricultural Drainage*. Number 38 in the series Agronomy, American Society of Agronomy, Madison, USA, 1328p.

This monograph summarizes the many aspects of contemporary agricultural irrigation and drainage systems, placing these systems into the perspective of comprehensive water management. It can serve as the scientific basis for decision-makers in developing management strategies to improve the soil conditions of the field and protect water quality from contamination by cropping practices.

Drainage design:

- Van der Molen, W. H., Martínez Beltran, J. and Ochs, W. J. 2007. Guidelines and computer programs for the planning and design of land drainage systems. FAO Irrigation and Drainage Paper No. 62, Rome, Italy, 228 pp. + CD-ROM.

The paper provides guidelines for the identification of drainage problems, for the planning and design of field drainage systems (surface and subsurface) and the main drainage and disposal systems. The paper considers the integration of technical, socioeconomic and environmental factors and the need for system users' participation in the planning, design and O&M processes.

Drainage materials:

- Stuyt, L. C. P. M., Dierickx, W. and Beltrán, J. M. 2000. Materials for subsurface land drainage systems. Irrigation and Drainage Paper No. 60. FAO, Rome, 183 p.

This paper provides practical information to drainage engineers and contractors for the selection, installation and maintenance of drainage materials as well as specifications and standards for such materials. In addition, the manual also contains practical guidelines for the implementation of laboratory and field investigations to evaluate the performance of drainage materials.

Drainage planning, design and management:

- Smedema, L. K., Vlotman, W. F. and Rycroft, D. W. 2004. *Modern Land Drainage: Planning, Design and Management of Agricultural Drainage Systems*. A.A. Balkema Publishers, Taylor & Francis The Netherlands: Leiden, 450 pp.

The book is based on traditional drainage methods for rainfed agriculture in the humid temperate zone. Significant parts are devoted to drainage for salinity control of irrigated lands in (semi-) arid zones and to drainage of rice land in the humid tropics. Institutional, management and maintenance aspects are covered, as well as the mitigation of adverse impacts of drainage interventions on the environment.

Drainage construction:

- Nijland, H. J., Croon, F. W. and Ritzema, H. P. 2005. *Subsurface Drainage Practices: Guidelines for the Implementation, O&M of Subsurface Pipe Drainage Systems*. ILRI Publication 60, Alterra, Wageningen, p. 608.

The handbook gives practical guidelines for the planning, the implementation and the organization of the implementation process of piped SSD systems with a focus on developing countries. It is composed of three parts: (i) planning of the implementation process of pipe drainage systems; (ii) practical guidelines for the implementation; and (iii) history and present practices of the implementation in a number of countries.

5.2 Professional organizations

Working group on sustainable drainage (http://www.icid.org/wg_sdg.html).

This working group of the International Commission on Irrigation and Drainage has the mandate (a) to promote drainage as part of integrated water resources management and (b) to promote sustainable approaches for drainage and related projects through a balanced integration of (i) environmental, (ii) economic and (iii) social and cultural aspects.

5.3 Key journals

Irrigation and drainage (http://onlinelibrary.wiley.com/journal/10.1002/(ISSN)1531-0361).

Official journal of the International Commission on Irrigation and Drainage (ICID). Published five times a year, the journal is a prestigious peer-reviewed publication and enables ICID to accomplish its objectives of publishing original papers on scientific, engineering, environmental and socio-economic issues associated with irrigation and drainage systems to support sustainable water use and flood management in a clear, concise and uniform style that is easily understood by an international audience.

Journal of Irrigation and Drainage Engineering (http://ascelibrary.org/journal/jidedh).

Official journal of the American Society of Civil Engineers. The journal covers all phases of irrigation, drainage, engineering hydrology, and related water management subjects such as watershed management, weather modification, water quality, groundwater and surface water. The journal emphasizes new developments and results of research, as well as case studies and practical applications of engineering.

Agricultural Water Management (https://www.journals.elsevier.com/agricultural-watermanagement).

The journal publishes papers of international significance relating to science, economics and policy of agricultural water management. In all cases, manuscripts must address implications and provide insight regarding agricultural water management.

5.4 Conferences

International Drainage Workshop (http://www.icid.org/conf_idw.html).

This workshop is organized every 3 years under the auspicious of the ICID. The workshop aims at introducing the challenges, needs and solutions facing the drainage community worldwide and to share experience on the national, regional and international scales in the field of agricultural drainage in relation to the environment.

International Drainage Symposium (https://cceevents.umn.edu/international-drainagesymposium).

This symposium is organized every five years by the American Society of Agricultural and Biological Engineers as an international forum for presenting, discussing and collaborating on recent and ongoing drainage research and implementation throughout the world. The symposium provides an opportunity for drainage research, industry and practitioner communities to interact, share experiences and address emerging concerns. Usually, it includes 2 days of technical presentations, followed by a day of field tours.

5.5 On-line courses on drainage

Drainage in agriculture: controlling water and salt levels in the soil.

Wageningen University offers the MOOC *Drainage in Agriculture: controlling water and salt levels in the soil*. The 5-week course has a study load of 18–20 h per week and was developed for professionals and students from various backgrounds, especially those who are interested in agricultural water management and want to broaden their understanding of drainage planning, design, and management, and drainage-related research and training. Website: https://www.wur.nl/en/show/Drainage-in-Agriculture-controlling-water-and-salt-levels-in-the-soil-2.htm.

Drainage engineering:

- The Indian Institute of Technology Kharagpur, Kharagpur, India offers the course 'Drainage engineering'. The course consists of six modules. Website: http://ecoursesonline.iasri.res.in/course/view.php?id=550.

Agricultural land drainage:

The Agriculture and Food Development Authority, Carlow, Ireland, offers the course 'Agricultural land drainage'. This course seeks to equip anyone involved in the drainage of agricultural land with the necessary skills and knowledge to design and implement agricultural land drainage systems. The course has a study load of 8 days comprising 4 × 2 day training blocks.

6 References

Abdel-Dayem, S. and Ritzema, H. P. 1990. Verification of drainage design criteria in the Nile Delta, Egypt. *Irrigation and Drainage Systems* 4(2), 117–131.

Abdel-Dayem, S., Abdel-Gawad, S. and Fahmy, H. 2007. Drainage in Egypt: a story of determination, continuity, and success. *Irrigation and Drainage* 56(S1), S101–S111.

Alavi, S. A., Naseri, A. A., Bazaz, A., Ritzema, H. and Hellegers, P. 2021. Performance evaluation of the Hydroluis drainpipe-envelope system in a saline-sodic soil. *Agricultural Water Management* 243, 106486. DOI: 10.1016/j.agwat.2020.106486.

Ali, A. M., Van Leeuwen, H. M. and Koopmans, R. K. 2001. Benefits of draining agricultural land in Egypt: results of five years' monitoring of drainage effects and impacts. *International Journal of Water Resources Development* 17(4), 633–646, DOI: 10.1080/07900620120094136.

Anderson, W. and Wall, D. 2006. The Minnesota nutrient reduction strategy. In: The 10th International Drainage Symposium, 6–9 September 2016. The American Society of Agricultural and Biological Engineers and the University of Minnesota Water Resources Center, Minneapolis.

Amer, M. H. and de Ridder, N. A. (Eds.) 1989. *Land Drainage in Egypt*. Drainage Research Institute, Cairo, Egypt and the International Institute for Land Reclamation and Improvement, The Netherlands, 377 p.

Ayars, J. E. and Evans, R. G. 2015. Subsurface drainage–what's next? *Irrigation and Drainage* 64(3), 378–392.

Ayers, R. S. and Westcot, D. W. 1994. *Water Quality for Agriculture*. Food and Agriculture Organization of the United Nations, Rome.

Bahçeci, I., Nacar, A. S., Topalhasan, L., Tari, A. F. and Ritzema, H. P. 2018. A new drain pipe-envelope concept for subsurface drainage systems in irrigated agriculture. *Irrigation and Drainage* 67 (Suppl. 2), 40–50, DOI: 10.1002/ird.2247.

Barnes, J. 2014. Mixing waters: the reuse of agricultural drainage water in Egypt. *Geoforum* 57, 181–191, DOI: 10.1016/j.geoforum.2012.11.019.

Bartholomeus, R. P., van den Eertwegh, G. A. P. H. and Cirkel, D. G. 2016. *Reuse of Treated Waste Water for Sub-irrigation in a Rural Area: Monitoring Sub-irrigation with the WWTP Effluent Haaksbergen*. BTO Report 2016.050. KWR Water Research Institute, Nieuwegein, The Netherlands.

Borin, M., Bonaiti, G. and Giardini, L. 2001. Controlled drainage and wetlands to reduce agricultural pollution: a lysimetric study. *Journal of Environmental Quality* 30(4), 1330–1340.

Bos, M. G. and Wolters, W. 2006. Influences of irrigation on drainage. In: Ritzema, H. P. (Ed.). *Drainage Principles and Applications* (3rd edn.). ILRI Publication 16. Wageningen, pp. 513–531.

Brouwer, C., Prins, K. and Heibloem, M. 1989. *Irrigation Scheduling. Irrigation Water Management Training Manual*. Rome, Italy, p. 36.

Burns, M. J., Wallis, E. and Matic, V. 2015. Building capacity in low-impact drainage management through research collaboration. *Freshwater Science* 34(3), 1176–1185.

Christianson, L., Tyndall, J. and Helmers, M. 2013. Financial comparison of seven nitrate reduction strategies for Midwestern agricultural drainage. *Water Resources and Economics* 2–3, 30–56.

De Fraiture, C., Molden, D. and Wichelns, D. 2010. Investing in water for food, ecosystems, and livelihoods: an overview of the comprehensive assessment of water management in agriculture. *Agricultural Water Management* 97(4), 495–501.

Drainage Research Institute 2013. *Studying the Effect of Controlled Drainage on Water Saving Efforts Final Report. Center Unit Integrated Irrigation Improvement and Management Project*, Drainage Research Institute, 40 p.

Drury, C. F., Tan, C. S., Reynolds, W. D., Welacky, T. W., Oloya, T. O. and Gaynor, J. D. 2009. Managing tile drainage, subirrigation, and nitrogen fertilization to enhance crop yields and reduce nitrate loss. *Journal of Environmental Quality* 38(3), 1193–1204.

El-Agha, D. E., Molle, F., Rap, E., El Bialy, M. and El-Hassan, W. A. 2020. Drainage water salinity and quality across nested scales in the Nile Delta of Egypt. *Environmental Science and Pollution Research International* 27(26), 32239–32250, DOI: 10.1007/s11356-019-07154-y.

El-Atfy, H. E., Abdel-Alim, M. Q. and Ritzema, H. P. 1991. A modified layout of the subsurface drainage system for rice areas in the Nile Delta, Egypt. *Agricultural Water Management* 19(4), 289–302.

Ellah, R. G. A. 2020. Water resources in Egypt and their challenges, Lake Nasser case study. *Egyptian Journal of Aquatic Research* 46(1), 1–12, DOI: 10.1016/j.ejar.2020.03.001.

El Quosy, D. 1989. Drainage water re-use project in the Nile Delta, the past, the present and the future. El-din. In: Amer, M. H. and de Ridder, N. A. (Eds). *Land Drainage in Egypt*. Drainage Research Institute, Cairo, Egypt and the International Institute for Land Reclamation and Improvement, The Netherlands, pp. 163–174.

Faures, J. M., Svendsen, M. and Turral, H. 2007. Reinventing irrigation. In: Molden, D. (Ed.). *Water for Food, Water for Life: A Comprehensive Assessment of Water Management in Agriculture*. International Water Management Institute, Colombo, Sri Lanka, pp. 353–394.

Ghumman, A. R., Ghazaw, Y. M., Hashmi, H. N., Kamal, M. A. and Niazi, M. F. 2012. Environmental and socio-economic impacts of pipe drainage in Pakistan. *Environmental Monitoring and Assessment* 184(3), 1671–1681.

Habib, M. F., Ali, M. E., Awad, M. A. and Abdel Hai, M. M. S. 2015. Some soil properties as influenced by different quality of irrigation water under sub-surface drip irrigation systems. *Journal of Soil Sciences and Agricultural Engineering* 6(1), 81–95, DOI: 10.21608/jssae.2015.41808.

Hashem, M. S. and Qi, X. 2021. Treated wastewater irrigation: a review. *Water* 13(11), 1527; DOI: 10.3390/w13111527.

Hepworth, N., Postigo, J. C. and Güemes Delgado, B. 2010. Drop by drop–understanding the impacts of the UK's water footprint through a case study of Peruvian asparagus. *Progressio*. CEPES and Water Witness International, London.

International Commission on Irrigation and Drainage 2022. *ICID Database*. Available at: https://icid-ciid.org/inner_page/162, accessed March 11 (vol. 6).

Jafari-Talukolaee, M., Ritzema, H., Darzi-Naftchali, A. and Shahnazari, A. 2016. Subsurface drainage to enable the cultivation of winter crops in consolidated paddy fields in Northern Iran. *Sustainability* 8(3), 249, DOI: 10.3390/su8030249.

Jonsson, A. C., Andersson, L., Olsson, J. A. and Johansson, M. 2011. Defining goals in participatory water management: merging local visions and expert judgements. *Journal of Environmental Planning and Management* 54(7), 909–935.

Jouni, H. J., Liaghat, A., Hassanoghli, A. and Ritzema, H. 2018. Managing controlled drainage in farmer's fields: case study in Moghan Plain, Iran. *Agricultural Water Management* 208, 393–405, DOI: 10.1016/j.agwat.2018.06.037.

Kalanaki, M., Karandish, F., Afrasiab, P., Ritzema, H., Khamari, I. and Tabatabai, S. M. 2022. Assessing the influence of integrating soil amendment applications with saline water irrigation on Ajwain's yield and water productivity. *Irrigation Science* 40(1), 71–85. DOI: 10.1007/s00271-021-00759-7.

Lin, Z. Q. and Cervinkab, V., Pickeringc, I. J., Zayeda, A., Terry, N. 2002. Managing selenium-contaminated agricultural drainage water by the integrated on-farm drainage management system: role of selenium volatilization. *Water Research* 36, 3150–3160.

McCready, W. 1987. *Left Bank Outfall Drain in Pakistan*, ICID Bulletin 36. International Commission on Irrigation and Drainage, pp. 15–19.

Mejia, M. N., Madramootoo, C. A. and Broughton, R. S. 2000. Influence of water table management on maize and soybean yields. *Agricultural Water Management* 46(1), 73–89.

Ministry of Water Resources and Irrigation 2017. Water Security for All. National Water Resources Plan 2017–2037, Planning Section, Ministry of Water Resources and Irrigation, Cairo, Egypt, 61 p.

Ministry of Water Resources and Irrigation 2016. *Enhanced Water Resources Management Project – Egypt Controlled Drainage Pilot*. Ministry of Water Resources and Irrigation, Egypt and GEF World Bank, 66 p.

Muhammada, E. S., Ibrahim, M. M. and El-Sayed, A. 2021 Effects of drain depth on crop yields and salinity in subsurface drainage in Nile Delta of Egypt. *Ain Shams Engineering Journal* 12(2), 1595–1606.

Nash, P. R., Nelson, K. A., Motavalli, P. P., Nathan, M. and Dudenhoeffer, C. 2015. Reducing phosphorus loss in tile water with managed drainage in a clay pan soil. *Journal of Environmental Quality* 44(2), 585–593.

Nasralla, M. R. 2019. *Controlled Drainage Strategies for Integrated Water Management in Semi-arid Areas Such as the Nile Delta in Egypt*. National Water Research Centre, Ministry of Water Resources and Irrigation, Cairo, 35 p.

National Drainage Project. Phase 3 2022a. *Training Manual on Design and Construction of Subsurface Drainage Networks*. Egyptian Public Authority for Drainage Projects, Ministry of Water Resources and Irrigation, Cairo, 138 p.

National Drainage Project. Phase 3 2022b. *Awareness Manual on Farmer Participation. Egyptian Public Authority for Drainage Projects*. Ministry of Water Resources and Irrigation, Cairo, 73 p.

National Drainage Project. Phase 3 2021a. *Strategy paper. Improvement of Farmers' Participation in Planning, Supervision of Works and Maintenance of Drainage Infrastructure, Egyptian Public Authority for Drainage Projects*. Ministry of Water Resources and Irrigation, Cairo, 21 p.

National Drainage Project. Phase 3 2021b. *Technical Note on Drainage Pipe Manufacturing - Least Cost Supply Analysis*. Egyptian Public Authority for Drainage Projects, Ministry of Water Resources and Irrigation, Cairo, 97 p.

Ng, H. Y. F., Tan, C. S., Drury, C. F. and Gaynor, J. D. 2002. Controlled drainage and subirrigation influences tile nitrate loss and maize yields in a sandy loam soil in Southwestern Ontario. *Agriculture, Ecosystems and Environment* 90(1), 81–88.

Nijland, H. J. (Ed.) 2000. Drainage along the River Nile. RIZA Nota nr. 2000.052. *Egyptian Public Authority of Drainage Project, Directorate-General of Public Works and Water*. Management, The Netherlands, 323 p.

Nonaka, I. and Takeuchi, H. 1995. *The Knowledge-Creating Company; How Japanese Companies Create the Dynamics of Innovation*. Oxford University Press, New York.

Noory, H. and Liaghat, A. 2009. Water table management to improve drainage water quality in semiarid climatic conditions of Iran. *Journal of Irrigation and Drainage Engineering* 135(5), 665–670.

Omar, M. E. D. M. and Moussa, A. M. A. 2016. Water management in Egypt for facing the future challenges. *Journal of Advanced Research* 7(3), 403–412, DOI: 10.1016/j.jare.2016.02.005.

Oosterbaan, R. J. 1988. Agricultural criteria for subsurface drainage: a system analysis. *Agricultural Water Management* 14(1–4), 79–90.

Poole, C., Burchell, M. and Youssef, M. 2018. Controlled Drainage – an important practices to protect water quality that can enhance crop yields. Available at: https://content.ces.ncsu.edu/controlled-drainage, Assessed March 17, 2022. NC State Extension Publications.

Quereshi, A. S. and Sarwar, A. 2009. Managing salinity in the Indus basin of Pakistan. *International Journal of River Basin Management* 7(2), 111–117.

Renault, D., Facon, T. and Wahaj, B. 2007. Modernizing irrigation management–the MASSCOTE approach: mapping system and services for canal operation techniques, irrigation and drainage. *Food and Agricultural Organization of the United Nations*, Rome, Italy, pp 238, Paper No. 63.

Ritzema, H. P. 2007. *Ecosystem Management in Tidal Areas: Ecological Restoration on Lakes in the Netherlands–Concepts and Dilemmas., INWERF and ICIDWG-SDTA Joint Workshop Sustainable Paddy Farming in Asian Monsoon Region*, July 1–3. Rural Research Institute of KRC, Ansan, Korea, pp. 43–56.

Ritzema, H. P. 2009. Drain for gain – making water management worth its salt. PhD thesis, Wageningen University and UNESCO-IHE Delft. CRC Press/Balkema, 208 p.

Ritzema, H. P. 2016. Drain for Gain: managing salinity in irrigated lands–a review. *Agricultural Water Management* 176, 18–28, DOI: 10.1016/j.agwat.2016.05.014.

Ritzema, H. P. 2018. Drainage systems to support sustainable water use. In: Oweis, T. (Ed.). *Water Management for Sustainable Agriculture*. Burleigh Dodds Science Publishing, Cambridge, UK, 467–493. DOI: 10.19103/AS.2017.0037.20.

Ritzema, H. P. and Braun, H. M. H. 2006. Environmental impact of drainage. In: Alterra, I. L. and Wageningen, R. I. (Eds) *Drainage Principles and Applications* (3rd edn.). ILRI Publication (vol. 16), pp. 1041–1064.

Ritzema, H. P. and Schultz, B. 2011. Optimizing subsurface drainage practices in irrigated agriculture in the semi-arid and arid regions: experiences from Egypt, India and Pakistan. *Irrigation and Drainage* 60(3), 360–369.

Ritzema, H. P. and Stuyt, L. C. P. M. 2015. Land drainage strategies to cope with climate change in the Netherlands, Acta Agriculturae Scandinavica. *Soil & Plant Science* 65 (Suppl.1), 80–92, DOI: 10.1080/09064710.2014.9945572014, Section B.

Ritzema, H. P. and van Loon-Steensma, J. M. 2018. Coping with climate change in a densely populated delta: a paradigm shift in flood and water management in the Netherlands. *Irrigation and Drainage, Special Issue "Sustainable Development of Tidal Areas: Climate change and environmental impacts"* 67, 52–65, DOI: 10.1002/ird.2128.

Ritzema, H. P., Nijland, H. J. and Croon, F. W. 2006. Subsurface drainage practices: from manual installation to large-scale implementation. *Agricultural Water Management* 86(1–2), 60–71, DOI: 10.1016/j.agwat.2006.06.026.

Ritzema, H. P., Wolters, W., Bhutta, M. N., Gupta, S. K. and Abdel-Dayem, S. 2007. The added value of research on drainage in irrigated agriculture. *Irrigation and Drainage* 56(S1), S205–S215.

Ritzema, H. P., Wolters, W. and Terwisscha van Scheltinga, C. T. H. M. 2008. Lessons learned with an integrated approach for capacity development in agricultural land drainage. *Irrigation and Drainage* 57(3), 354–365. DOI: 10.1002/ird.431.

Ritzema, H. P., Kirkpatrick, H., Stibinger, J., Heinhuis, H., Belting, H., Schrijver, R. and Diemont, H. 2016. Water management supporting the delivery of ecosystem services for grassland, heath and moorland. *Sustainability* 8(5), 440; DOI: 10.3390/su8050440.

Ritzema, H. P., Abdel-Dayem, S., El-Atfy, H., Nasralla, M. R. and Shaheen, H. S. 2023. Challenges in modernizing the subsurface drainage systems in Egypt. *Agricultural Water Management* 288, DOI: 10.1016/j.agwat.2023.108484.

Sallam, G. A. H. 2018. Drain envelopes for problematic calcareous soils of irrigated agricultural lands in Egypt. *Journal of Experimental Agriculture International* 20(4), 1–13, Article no.JEAI.39072.

Salmoral, G., Vinarta Carbo, A. V., Zegarra, E., Knox, J. W. and Rey, D. 2020. Reconciling irrigation demands for agricultural expansion with environmental sustainability: A preliminary assessment for the Ica valley, Peru. *Journal of Cleaner Production* 276, 123544. DOI: 10.1016/j.jclepro.2020.123544.

Scheuman, W. 1997. *Managing Salinization - Institutional Analysis of Public Irrigation Systems, Department of Environment and Society*. Technical University of Berlin, Berlin, p. 274.

Schultz, B., Zimmer, D. and Vlotman, W. F. 2007. Drainage under increasing and changing requirements. *Irrigation and Drainage: The journal of the International Commission on Irrigation and Drainage* 56, S3–S22. DOI: 10.1002/ird.372.

Singh, A. 2014. Conjunctive use of water resources for sustainable irrigated agriculture. *Journal of Hydrology* 519, 1688–1697.

Skaggs, R. W., Fausey, N. R. and Evans, R. O. 2012. Drainage water management. *Journal of Soil and Water Conservation* 67(6), 167A–172A.

Skogena, M. D., Eilolab, K., Hansend, J. L. S., Meierb, H. E. M., Molchanove, M. S.and Ryabchenkoe, V. A. 2014. Eutrophication status of the North Sea, Skagerrak, Kattegat and the Baltic Sea in present and future climates: a model study. *Journal of Marine Systems* 132, 174–184.

Speelman, S., D'Haese, M. and D'Haese, L. 2011. Simulation of efficiency impact of drainage water re-use: case of small-scale vegetable growers in North West Province South Africa. *Agrekon* 50(1), 89–101.

Stuyt, L., Dierickx, W. and Beltrán, J. M. 2000. *Materials for Subsurface Land Drainage Systems, FAO Irrigation and Drainage*. Food and Agriculture Organization of the United Nations, Rome, Italy, p. 183, Paper 60 Rev. 1.

Sunohara, M. D., Gottschall, N., Wilkes, G., Craiovan, E., Topp, E., Que, Z., Seidou, O., Frey, S. K. and Lapen, D. R. 2015. Long-term observations of nitrogen and phosphorus export in paired-agricultural watersheds under controlled and conventional tile drainage. *Journal of Environmental Quality* 44(5), 1589–1604.

Tanji, K. K. and Kielen, N. C. 2002. Agricultural drainage water management in arid and semi-arid areas. *FAO Irrigation and Drainage*, Paper 61. Food and Agriculture Organization of The United Nations, Rome.

Thompson, J., Rahardianto, A., Gu, H., Uchymiak, M., Bartman, A., Hedrick, M., Lara, D., Cooper, J., Faria, J. P. D. C., Christofides, P. D. and Cohen, Y. 2013. Rapid field assessment of ROdesalination of brackish agricultural drainage water. *Water Research* 47(8), 2649–2660.

Tolomio, M. and Borin, M. 2019. Controlled drainage and crop production in a long-term experiment in North-Eastern Italy. *Agricultural Water Management* 222, 21–29.

Uléna, B., von Brömssen, C., Johansson, G., Torstensson, G. and Forsberg, L. S. 2012. Trends in nutrient concentrations in drainage water from single fields under ordinary cultivation. *Agriculture, Ecosystems and Environment* 151, 61–69.

Van Hoorn, J. W. and Van Alphen, J. G. 2006. Salinity control. In: Ritzema, H. P. (Ed.). *Drainage Principles and Applications* (3rd edn.). ILRI Publication 16. Alterra-ILRI. Wageningen, pp. 533–600.

Van Keulen, A., Smith, M. and Renault, D. 2003. Capacity building in farmers' water management: the FAO experience. In: 9th International Drainage Workshop, International Institute for Land Reclamation and Improvement, September 10–13, Wageningen, The Netherlands, Utrecht, The Netherlands.

Vlotman, W. F. and Jansen, H. C. 2003. Controlled drainage for integrated water management. 9th International Drainage Workshop, 10–13 September, Utrecht, The Netherlands.

Vlotman, W. F., Wong, T. and Schultz, B. 2007. Integration of drainage, water quality and flood management in rural, urban and lowland areas. *Irrigation and Drainage* 56(S1), S161-S177/ird.1369.

Wahba, M. A. S., Christen, E. W. and Amer, M. H. 2005. Irrigation water saving by management of existing subsurface drainage in Egypt. *Irrigation and Drainage* 54(2), 205–215, DOI: 10.1002/ird.164.

Wolters, W., Smit, R., Nour El-Din, M., Sayed Ahmed, E., Froebrich, J. and Ritzema, H. 2016. Issues and challenges in spatial and temporal water allocation in the Nile Delta. *Sustainability* 8(4), 383; DOI: 10.3390/su8040383.

Chapter 6

Crop physiological responses to irrigation and regulatory mechanisms

Ian C. Dodd, Lancaster University, UK; Katharina Huntenburg, Wageningen University and Research, The Netherlands; and Juan Gabriel Pérez-Pérez , Valencian Institute of Agricultural Research, Spain

1 Introduction

Irrigated croplands produce 40% of food globally on only 22% of arable land, while consuming approximately 90% of all fresh water (Rosa, 2022). Increasing pressure on our agricultural systems to produce more food to nourish a growing global population warrants considerable future investment in irrigation infrastructure. However, heightened political concerns over the wise management of water resources particularly in arid catchments (Varela-Ortega et al., 2011; Colloff and Pittock, 2022) and legislation to curtail agricultural abstractions (Dunn et al., 2003; Zeff et al., 2019) exacerbate uncertainties over future farmer access to irrigation. To cope with limited irrigation supplies, farmers might need to accept 'deficit irrigation': applying less water than that calculated to sustain evapotranspiration at optimal levels. Even with abundant water supplies, rising infrastructure (e.g. pipes, sprinklers) and abstraction (e.g. pumping and storage) costs mean that wise irrigation management can maximise economic benefits while minimising environmental impacts. Understanding

http://dx.doi.org/10.19103/AS.2023.0123.25

plant agronomic and physiological responses to variations in soil water status is thus critical to understanding how irrigation regulates crop yield.

Irrigation boosts crop yields per unit area by attenuating crop water deficits. Deficits occur when transpirational demands (determined by atmospheric conditions and plant phenology) exceed the ability of the soil–plant system to supply water to the transpiring shoots. While temporary imbalances (e.g. during the middle of the day when transpirational demands are greatest) are unlikely to significantly affect biomass accumulation, sustained water deficits diminish many yield-determining processes such as leaf expansion, leaf gas exchange and ultimately fruit and grain development. While some shoot processes are highly sensitive to decreased water status, roots can continue to grow at a water status that inhibits shoot growth (Westgate and Boyer, 1985). Preferential allocation of photosynthate to the roots and fruits sustains their activity, thereby attenuating the economic impacts of water deficits. Understanding the sensitivity of different physiological responses to variations in soil water status therefore offers opportunities to mitigate economic losses, or even to increase profits by implementing targeted deficit irrigation strategies that maximise both fruit yield and quality.

While most irrigation science literature rightly emphasises the importance of the volumes of irrigation required or applied, other factors such as irrigation placement (e.g. surface vs subsurface irrigation), frequency (e.g. 'little and often' vs 'feast and famine') and timing (with respect to crop phenology, especially reproductive development) also affect crop responses. Carefully controlled laboratory and field experiments demonstrate that these factors influence crop responses independently of irrigation volume by affecting soil and/or plant water deficits. Managing these factors might simplify irrigation scheduling decisions by implementing creative irrigation strategies for specific agronomic benefits.

This chapter first analyses the relationships between crop yields and irrigation volume to define irrigation objectives. It then considers how soil and/or plant water deficits regulate plant physiological responses, before reviewing how different irrigation techniques affect these responses in two key crops: citrus and tomato. Understanding variation in the sensitivity of these physiological responses to plant water deficits has helped design appropriate irrigation strategies. Finally, future trends in instrumentation and modelling to facilitate improved farmer irrigation scheduling decisions are briefly considered.

1.1 Crop yield vs irrigation volume

As crop yield integrates multiple physiological processes across the growing season, understanding its relationship to irrigation volume is critical for the practitioner. Except when crops are grown in deep, moisture-retentive soils and/or with frequent rainfall that prevents the build-up of soil water deficits, crops typically respond positively to irrigation (Figure 1a). Initially, crop yields increase linearly with irrigation volume (point W in Figure 1a) as irrigation alleviates the damaging effects of soil and/or plant water deficits. This relationship becomes non-linear as maximal yields are approached, and the practitioner should seek to irrigate to the on-farm economic threshold (point X in Figure 1a), beyond which the additional irrigation costs are not financially compensated by the economic benefits of additional yield. While theoretically attractive, irrigating to maximise yield (point Y in Figure 1a) may not make economic sense and incurs the risk of over-irrigating, such that additional irrigation diminishes yield (point Z in Figure 1a). Saturating the soil decreases oxygen availability to the roots (via waterlogging – Marino et al., 2015; Fiebig and Dodd, 2016) and leaches nutrients beyond the crop rootzone (nutrient deficiency), thereby limiting yields and polluting neighbouring water bodies (Phogat et al., 2014). Grower experience can be valuable in determining irrigation volumes to achieve optimal yields although physiological measurements (Section 2) and instrumentation such as wetting front detectors (Stirzaker et al., 2017; Hasnain and Singh, 2022) can also help achieve this goal.

Crop water use efficiency (WUE) represents the slope of the yield vs the volume of water applied relationship. Although WUE attracts considerable academic interest (Morison et al., 2008; Farooq et al., 2019), the societal pressures mentioned previously suggest it is becoming increasingly important as an irrigation metric. Popularisation of the 'water footprint' concept (Mekonnen and Hoekstra, 2011) that defines the water required to produce a specific commodity such as a piece of fruit or bale of cotton may result in supermarkets benchmarking their suppliers in terms of WUE. As excessive irrigation diminishes the quality of some crops (Liu et al., 2022) but not others (Espinoza-Meza et al., 2023), successful growers are likely to use water efficiently. Nevertheless, regional water scarcity also drives increased efficiency,

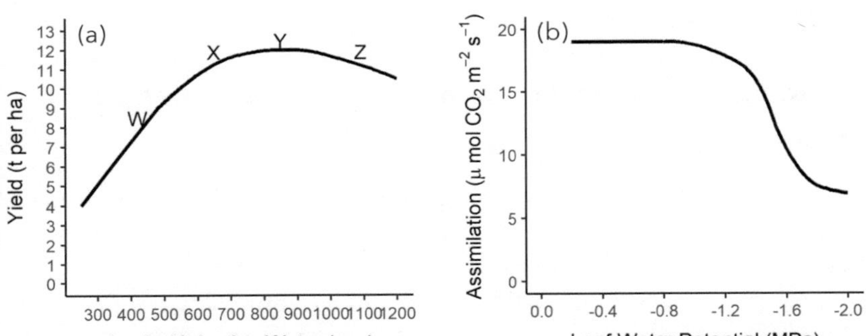

Figure 1 Generalised relationships between crop yield vs water applied (a) and photosynthesis vs leaf water potential (b). Source: Adapted from Fereres and Soriano, 2007 (a) and Boyer, 1970 (b), respectively.

with growers turning to some of the irrigation practices described in Section 3 to enhance their economic productivity, especially as the price of irrigation water increases (Upadhyaya et al., 2022).

Whilst the theoretical relationship in Figure 1 is informative over an entire cropping season, irrigators require a finer temporal resolution to make informed in-season irrigation scheduling decisions. Empirical relationships between instantaneous plant performance (e.g. leaf and fruit expansion, leaf gas exchange including photosynthesis) and soil or plant water status should be considered (Figure 1B). Intuitively, irrigators should maintain soil and plant water status at levels that do not impair plant performance. As different plant processes vary in their sensitivity to water status and their ease of measurement (Section 2), different threshold values can be empirically established (e.g. leaf water potential, Ψ_{leaf} of −0.8 MPa in Figure 1b). As biomass accumulation ultimately determines crop yields, measuring photosynthesis may be especially valuable but the cost of and training required to use the instrumentation is usually beyond the reach of most irrigators. Instead, defining irrigation thresholds with more accessible measurements (Section 2) assumes these values are not inhibiting photosynthesis. End-of-season yields establish whether irrigating to avoid exceeding these threshold values provided sufficient irrigation to prevent loss of crop yield and/or quality.

Fortunately, photosynthesis is less sensitive to decreased plant water status than many other more easily measured variables (Hsiao, 1973). Threshold values of these variables (e.g. stem water potential) can assist in irrigation scheduling by aiming to prevent limitations in the rate of photosynthetic activity. While measuring all or most of the variables in Section 2 might experimentally validate defined thresholds that do not inhibit photosynthesis (e.g. Marino et al., 2018), irrigation practitioners will determine the success of irrigation scheduling by evaluating crop yield. Nevertheless, understanding the mechanisms by which soil drying limits crop physiology can help choose suitable plant-based irrigation scheduling metrics.

2 Crop physiological responses to soil water deficit

2.1 Leaf water relations

Turgor loss (wilting) often indicates plant stress and the need for irrigation to home gardeners, but it is less responsive to withholding water than other physiological measurements (Figure 2). Significant stomatal closure of the youngest fully expanded leaves (Figure 2a) occurred 3 days before Ψ_{leaf} declined (Figure 2b) and 6 days before older leaves visibly wilted. Practitioners may still use plant visible appearance to inform irrigation scheduling when scientific instruments and/or time-consuming physiological measurements are unavailable.

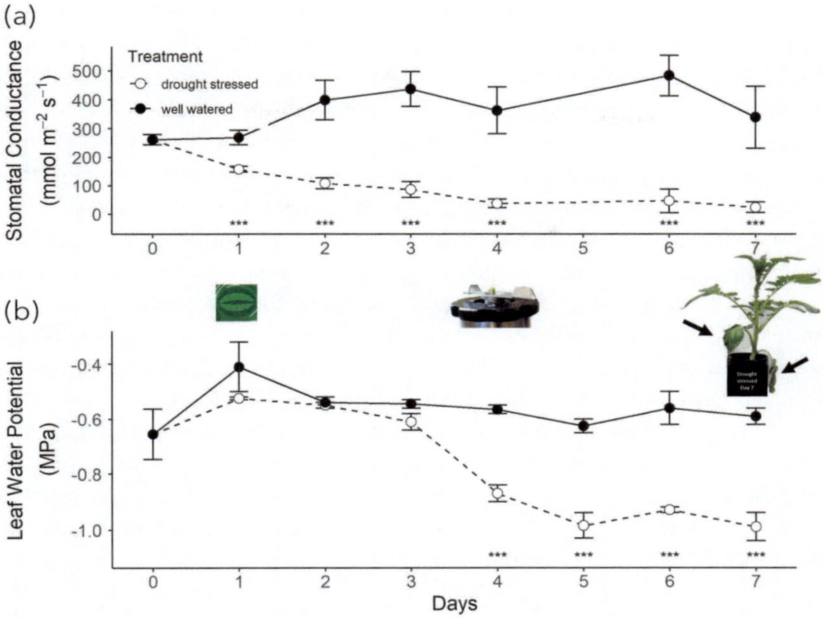

Figure 2 Changes in stomatal conductance (a) and leaf water potential (b) of young fully expanded leaves of well-watered (●) and drought-stressed (○) potato plants, with daily replacement of evapotranspirational losses (●) and water withheld from drought-stressed plants from day 0 (○). Data are means +/− SE of 2–14 replicates, with 14 plants measured on day 1, and two plants removed each subsequent day to measure Ψ_{leaf}, then discarded. Instrument unavailability precluded stomatal measurements on day 5. Images between panels indicate when treatment differences in stomatal closure, Ψ_{leaf} and wilting occurred.

Within plant cells and tissues, turgor (P) represents an outward force against the cell walls generated by water uptake and is responsible for plant structural integrity. It is typically positive (except in wilted cells) and mathematically defined as the difference between tissue water potential (Ψ) and osmotic potential (Ψ_s) thus:

$$P = \Psi - \Psi_s$$

While the cellular pressure probe (Tomos and Leigh, 1999) allows direct P measurements in the laboratory, pressure sensors coupled to magnetic devices allow continuous leaf turgor measurements in irrigated crops (Fernández et al., 2011; Martínez-Gimeno et al., 2017). Measuring the turgor components (Ψ and Ψ_s) allows P to be calculated (e.g. Gupta et al., 2008), although irrigation scheduling has more usually relied on measuring tissue water potentials.

The pressure chamber allows field measurements of leaf water potential, Ψ_{leaf} (Scholander et al., 1965; Powell and Coggins, 1985). However, the utility of such measurements for irrigation scheduling depends to some extent on the crop type. For example, some isohydric species (e.g. maize – Tardieu and Simonneau, 1998; peach – Losciale et al., 2023) maintain Ψ_{leaf} by maintaining or increasing water uptake (e.g. by proliferating roots and/or increasing their hydraulic conductance) and/or restricting their water losses (e.g. through partial stomatal closure and/or decreasing leaf surface area). Stomatal closure of such crop types might restrict photosynthesis and greatly compromise biomass accumulation before Ψ_{leaf} declines when soil moisture is almost depleted. In contrast, Ψ_{leaf} declines with soil drying in anisohydric crops (e.g. sunflower – Tardieu and Simonneau, 1998; apricot – Losciale et al., 2023), thus pressure chamber measurements can be useful for irrigation scheduling.

As Ψ_{leaf} represents a balance between its water uptake and losses, it fluctuates diurnally as atmospheric evaporative demand changes during the day. Fluctuating environmental conditions (e.g. shade from clouds, variation in windspeed) change Ψ_{leaf}. Spatial and temporal variation in the microclimate experienced by each leaf within the canopy causes intra-plant variability in Ψ_{leaf}. Such 'noise' in the Ψ_{leaf} 'signal' diminishes the value of individual Ψ_{leaf} measurements and requires additional replication to minimise variability. Alternatively, stem water potential (Ψ_{stem}) can be measured by enclosing leaves (with aluminium-foil envelopes) to prevent transpiration for the hour preceding measurement. Variability in Ψ_{stem} determinations is minimised as Ψ_{leaf} equilibrates with the water potential of the xylem supplying that leaf. Thus Ψ_{stem} has been widely adopted in irrigation scheduling (McCutchan and Shackel, 1992; Naor, 2000). Portable pressure chambers facilitate field measurements of Ψ_{leaf} or Ψ_{stem} for irrigation scheduling, but removal of leaf tissue, the time taken for measurements and the need for a skilled operator represent operational constraints.

More importantly, the interpretation of Ψ measurements relies on understanding the relationships between Ψ and yield-determining processes such as leaf expansion (in vegetative crops such as sugar cane) and fruit growth (in orchard crops). Each operator must therefore establish empirical relationships between the threshold Ψ values required to trigger irrigation and the irrigation volume required to restore Ψ to values above the threshold. For that reason, directly measuring plant performance may be preferred.

2.2 Tissue expansion

Although one of the most sensitive processes to water deficit (Hsiao, 1973), leaf expansion is rarely measured to assist irrigation scheduling. Using a ruler to measure leaf elongation requires no external power or specialist training

(Howell et al., 2015) and is especially informative in monocotyledons as growth is essentially unidirectional. Growing dicotyledon leaves require two-dimensional measurements, with allometric relationships between leaf length and width and actual leaf area determining relative leaf elongation rate. Researchers usually make these time-consuming measurements, although leaf elongation can be automatically, continuously measured with linearly variable displacement transducers (LVDT) that detect micrometer changes in leaf length (Sharp et al., 1979). As canopy expansion is critical for radiation interception, more integrative measurements than a single leaf seem necessary to inform irrigation scheduling although canopy leaf area index is often measured on a weekly or less frequent basis (e.g. Cetin et al., 2023).

Leaf elongation is especially sensitive to soil water deficit as for expansion to occur, turgor must exert a force on cell walls that can expand, as formalised in an equation, with organ growth (G) expressed as:

$$G = m\,(P - Y)$$

where m is a coefficient that represents the ability of cell walls to yield, P is cellular turgor (as defined earlier) and Y is a threshold turgor above which irreversible cell extension occurs (Lockhart, 1965; Passioura and Fry, 1992). While cell wall properties are important in regulating organ growth, water influxes into growing organs must be maintained to sustain P at levels greater than Y. Thus, leaf expansion declines rapidly with small changes in Ψ or P (Boyer, 1970) at absolute Ψ values that scarcely affected photosynthesis of several monocotyledon and dicotyledon crop species. Elongating maize leaves maintained P of growing cells, while P of exposed (mature) leaves declined considerably (Michelena and Boyer, 1982), indicating that biochemical changes regulated cell wall properties and thus leaf growth. Such non-hydraulic growth regulation is considered later in more detail.

Differences in the relative sensitivity of leaf elongation and leaf gas exchange (including photosynthesis) might inform irrigation scheduling. While maize leaf growth was more sensitive to soil drying than stomatal conductance (Saab and Sharp, 1989), cucumber showed the opposite response (Melkonian and Wolfe, 1995). The variability associated with each measurement (signal to noise ratio) likely not only explains these variable responses but also suggests that irrigation scheduling measurements should be tailored to the individual species, and its relationship to the harvested product. While any decrease in leaf area diminished cucumber yield (Ramirez et al., 1988), maize yields seem more tolerant of defoliation (Battaglia et al.,2018). Identifying the sensitivity of crop yields to the measured physiological variable, as well as the technical feasibility of measuring it, should inform whether plant-based irrigation scheduling is viable.

Using LVDTs to schedule irrigation of orchard crops has been evaluated, with continuous stem or trunk diameter measurements throughout the day. Stem diameter is typically maximal before dawn (coinciding with maximum daily Ψ) and minimal in the early afternoon (coinciding with high evaporative demand and minimum daily Ψ). Late afternoon decreases in evaporative demand coincide with lower transpiration rates as light intensity dwindles and stem tissue rehydrates which continues overnight. Although differences between maximal stem diameter on consecutive days indicate trunk growth, atmospheric evaporative demand and soil drying determine the magnitude of the daily stem contraction or maximum daily shrinkage (MDS), which can be used in irrigation scheduling. Compared to reference well-watered trees grown in the same orchard, trunk MDS can be used to trigger irrigation when threshold values are exceeded (Goldhamer and Fereres, 2001). These sensors have been extensively evaluated in a range of orchard crops, with efficacy varying between species according to phenological and fruit load impacts on daily shrinkage values.

The fact that fruits grow in three dimensions adds complexity to measuring their growth rate using LVDTs, even if they are often the primary economic output. Commercially available sensors do not appear to have been widely applied for irrigation scheduling. For example, in potted apple plants, installing trunk diameter sensors allowed the relative sensitivity of different tissues to be compared, with fruit growth rate declining when an absolute MDS of 200 μm was exceeded (Bonany et al., 2000). In pears, the fruit diameter growth was closely correlated temporally with trunk shrinkage (Maas, 2007). In contrast, daily grape berry shrinkage and recovery occurred much later (hours) compared to trunk shrinkage, with berry recovery times varying with fruit growth stage according to the relative water supply to the berry from xylem and phloem (Otarola et al., 2014). While fruit growth is often measured at harvest to determine the impacts of different irrigation schedules, such dynamic fruit growth measurements are rarely utilised by researchers.

Likewise, soil's opacity and soil particle blockage of LVDTs prevent their use to measure tuber growth in crops such as potatoes. Nevertheless, laboratory studies with non-destructive magnetic resonance imaging (MRI) demonstrated pronounced periodicity of tuber growth in drying soils with detectable night-time tuber growth (when there is minimal competition for water between the tubers and canopy) and limited day-time tuber expansion (Huntenburg et al., 2023). As successful MRI imaging of tubers requires screening out strongly ferromagnetic particles from the soil, this technique is not practical for field use, but other technologies such as ground-penetrating radar (GPR) can image tubers *in vivo*. GPR estimated 41% of the variation in cassava tuber mass when tuber mass varied across an order of magnitude although positive values of mass sometimes occurred when tubers were absent (Agbona et al., 2021).

Improved measurement precision and ease of operation may ultimately facilitate irrigation scheduling based on tuber growth dynamics, but so far GPR has only been used as a research tool to estimate soil water content (Wu et al., 2022), for which other minimally invasive technologies (e.g. neutron probe, time domain reflectometry) are available. Whilst applying geophysical techniques offers the potential to non-destructively measure tuber growth, dynamic shoot measurements of these crops are more likely to inform irrigation scheduling and be (retrospectively) correlated with tuber mass post-harvest.

2.3 Leaf gas exchange

As transpiration represents a loss (to the atmosphere) of water that otherwise might be used by the plant to expand growing tissues, estimating its magnitude either directly or indirectly might assist irrigation scheduling. When transpirational losses exceed root water uptake for a prolonged period, transpiration inevitably decreases even if plant internal water stores might delay this decline. Directly measuring whole plant or crop transpiration is not straightforward, requiring weighing lysimeters (essentially giant balances on which the crop is grown, which are not practical for routine use) or sap flow (SF) gauges fitted to plants with a lignified stem. More typically, crop evapotranspiration is estimated by applying physical equations to micrometeorology-based measurements (e.g. Penman–Monteith equation – Kumar et al., 2012). Infra-red gas analysers (IRGAs) or porometers directly and rapidly (within seconds to minutes) measure photosynthesis and transpiration of individual leaves. Furthermore, transpirational cooling of the canopy or leaf can be readily measured and has attracted considerable attention from the irrigation science community.

All SF sensors apply heat to the plant stem and then measure changes of temperature along it. Applied heat is assumed to be transported in the sap at a rate that is proportional to the whole plant's transpiration rate. Two approaches have been developed to measure SF rate. In the heat-balance method, part of the entire stem circumference is continuously heated electrically, with axial and radial heat-loss measured. Heat dissipation by the ascending sap allows SF rate to be calculated. In the heat-pulse method, temperature sensor probes are mounted inside the stem both upstream and downstream of the point of heat application. SF rates can also be calculated from the time required for heat transport to a specific location. While SF rates are calculated as the mass change in water per unit time, sap flux densities calculate water volumes per unit of sapwood volume and time. These techniques operate most effectively when SF rates are high, but advanced instrumentation (Pearsall et al., 2014; Siqueria and Ferreira, 2020) can detect low, null and or even reverse (sap descent according to Ψ gradients) flows.

Measuring CO_2 uptake in photosynthesis (A) using an IRGA directly links irrigation scheduling with crop economic outputs that depend on carbohydrate use in leaf expansion or grain/fruit filling, but IRGAs are expensive and typically used only in research. Transpiration rate (E) is more likely to be measured with (much cheaper) dynamic and steady-state diffusion porometers. While some models directly quantify transpiration, others measure stomatal conductance (g_s) which depends on the number of stomata per unit leaf area (stomatal density) and how open they are. Most transpiration occurs through the stomata, but some transpirational water losses occur through the leaf cuticle, including at night-time when the stomata are essentially closed. While higher g_s should increase transpiration of well-watered plants, thereby decreasing Ψ_{leaf}, lower g_s limits transpiration and thus further decreases in Ψ_{leaf}. Understanding species variation in stomatal sensitivity to leaf and soil water deficit determines the utility of these measurements in irrigation scheduling.

Transpirational cooling of the leaf allows leaf temperature measurements to inform irrigation scheduling, especially as the costs of thermal cameras have declined from a level commensurate with IRGAs to that more typical of mobile phones (Puértolas et al., 2019). Since the radiation environment determines absolute leaf or canopy temperatures, these are normalised according to prevailing environmental conditions in a Crop Water Stress Index (CWSI) that relies on reference (dry and wet) surfaces. Thus, CWSI can be expressed as

$$CWSI = ((T_{canopy}-T_{air}) - (T_{wet}-T_{air})) / ((T_{dry}-T_{air}) -(T_{wet}-T_{air}))$$

where T_{canopy} and T_{air} are canopy and air temperature respectively, with T_{dry} and T_{wet} indicating the temperatures of the reference surfaces. This equation's numerator reflects the measured degree of transpirational cooling while the denominator indicates the maximal transpirational cooling possible in that environment at the time of measurement. Reference surfaces may comprise pieces of card, or detached leaves, held in the same orientation as the canopy and either sprayed with water (wet reference) or covered in Vaseline (dry reference). Unlike IRGAs or porometers in which thermocouples contact the leaf, remote measurements are possible from a convenient platform such as an outstretched arm or a platform above the crop via a drone or aircraft. Although the camera's field of view determines the cropped area being measured, this technique is especially valuable in diagnosing irrigation hardware failure (e.g. blocked drip lines). Maintaining the crop above a crop-specific CWSI might assist in irrigation scheduling (Gonzalez-Dugo et al., 2013; Alordzinu et al., 2021).

2.4 Root physiology

While irrigation aims to sustain root water uptake to support transpiration, root morphology measurements rarely (if ever) inform irrigation scheduling.

Nevertheless, irrigation (volume, frequency, location) profoundly affects root architecture, that is typically assessed at the end of an irrigation season. Applying precision irrigation often restricts roots to the wetted soil zones (Fernández et al., 1991), which helps define the placement of soil moisture sensors that measure root water uptake (soil moisture depletion). Many different techniques have been utilised to determine root water uptake (reviewed in Whalley et al., 2017) although commercially available sensors typically determine either soil water potential, Ψ_{soil} (e.g. tensiometers) or volumetric soil water content (e.g. TDR probes). While the former determines whether roots can take up water from their environment (Ψ_{root} must be lower than Ψ_{soil}), the latter allows the irrigator to calculate the irrigation volumes required to replenish the soil profile.

Scheduling irrigation according to soil moisture depletion is challenged by understanding where in the soil profile roots are extracting water. Commercially available probes can continuously measure soil moisture to depths of 1 m (e.g. Dodd et al., 2011), yet roots might exist at much greater depths (up to 3 m) even in herbaceous crops (Dardanelli et al., 1997). This problem is exacerbated in well-established perennial tree crops where soil drying outside of the irrigation season can promote root extension. However, surface (and even sub-surface) irrigation can 'hydraulically prune' root systems by ensuring roots are mostly located where irrigation is supplied (Mohamadzade et al., 2021). While this limits vertical heterogeneity in water uptake, considerable horizontal heterogeneity in soil moisture within a field challenges the irrigator to determine where to locate soil moisture sensors. Surveys that characterise spatial heterogeneity in soil moisture helped to characterise distinct irrigation management zones within a field (Ortuani et al., 2019), with statistical techniques informing sensor location (Ohana-Levi et al., 2021). Nevertheless, the ability of the root system to secure water uptake from a heterogeneous soil, with a small percentage of the root system capable of disproportionate water uptake (Sharp and Davies, 1985), suggests that plant-based measurements (discussed above) may remove some of the uncertainty associated with irrigation scheduling based on soil measurements.

2.5 Root-to-shoot signalling

While root water uptake sustains shoot growth and transpiration, isohydric species that tightly regulate g_s maintain Ψ_{leaf} as the soil dries. Field observations of stomatal closure without corresponding decreases in Ψ_{leaf} (e.g. Bates and Hall, 1981) triggered creative controlled environment investigations into the physiological mechanisms regulating stomatal behaviour. Pressurising the root system to maintain leaves close to full turgor (xylem sap flowed from a leaf incision) as the soil dried did not inhibit stomatal closure (Gollan et al., 1986). Furthermore, split-root experiments broke the link between declining Ψ_{soil} and decreased Ψ_{leaf}. When some roots dried part of the soil volume and others were well-irrigated, pronounced stomatal closure (relative to plants with the whole root system in well-watered soil) occurred while Ψ_{leaf} and leaf turgor increased (Blackman and Davies, 1985). Such experiments stimulated a 'root-to-shoot chemical signalling' paradigm, suggesting that the roots sensed drying soil and altered their metabolism to synthesise and export chemical signals to the shoot to regulate stomatal closure (Davies and Zhang, 1991). Much subsequent work has since investigated the nature and physiological effects of these chemical signals (Huntenburg et al., 2022), even if more recent experiments have questioned the direction and efficacy of such signalling (McAdam et al., 2016). Irrespective of these debates, this paradigm has influenced the development of different irrigation techniques.

To understand how different irrigation techniques affect the physiological responses described earlier and their potential utility in irrigation scheduling, two case studies considering contrasting globally important crop types with differing irrigation techniques are described. Variations in the sensitivity of these crops to water deficits, and the adaptive hydraulic and chemical signalling induced by soil drying, have influenced their development. In some cases, utilising some of the physiological measurements described earlier to implement deficit irrigation strategies has increased crop economic profitability and/or improved their environmental sustainability.

3 Regulated deficit irrigation of citrus

Regulated deficit irrigation (RDI) was originally developed to apply water deficits to control vegetative growth and increase fruiting (Chalmers et al., 1981; Mitchell et al., 1989). While only a percentage of crop potential evapotranspiration (ETc) is required during less-sensitive phenological stages (e.g. hard-pit and post-harvest in stone fruit), the most critical fruit growth stages (e.g. during flowering, fruit set and rapid fruit growth) are fully irrigated (100% ETc). This approach offers several significant advantages.

RDI strategies aim to improve WUE by maintaining plant water status within prescribed limits to match crop water demand, thereby conserving water resources. Moderate water stress can stimulate root development and enhance drought tolerance (Chaves et al., 2002; Flexas et al., 2018). However, its success depends on regulating the timing, duration and severity of imposed water stress, requiring a sound understanding of crop physiology. Various strategies have been successfully evaluated under diverse soil and climatic conditions in the Mediterranean on crops such as persimmon (Buesa et al., 2013), pomegranate (Intrigliolo et al., 2013), Japanese plum (Intrigliolo and Castel, 2010), loquat (Ballester et al., 2018) and olive (Martínez-Gimeno et al., 2022). RDI can help balance vegetative and reproductive growth of stone fruit crops, essential to obtain well-formed fruits of optimal quality (de la Rosa et al., 2016). Applying RDI can also mitigate the effects of climate change on fruit quality (Pérez-Pérez et al., 2009). Applying moderate water stress during fruit maturation promoted sugar and organic acid accumulation, improving fruit quality. It enhanced the accumulation of flavour, aroma and nutraceutical compounds, thereby increasing fruit value (Navarro et al., 2010). RDI also benefits vineyard canopy management and grape and wine quality. Applying moderate water stress from budburst to fruit set and controlling the plant water status within the range $-1.2 > \Psi_{stem} > -1.3$ MPa decreased vine vigour, maintained yield and improved grape quality (Romero et al., 2013). Overall, judiciously implementing RDI strategies balances water conservation, crop productivity and fruit quality.

Conversely, as citrus is sensitive to water deficits and has high water requirements, deficit irrigation strategies are less adopted despite promising results (Ballester et al., 2013a, 2014). Growers are concerned that water deficits may compromise fruit size and delay the harvest of early varieties (Pérez-Pérez et al., 2014), which could diminish crop profitability. Nevertheless, applying RDI during the rapid fruit growth period (phase II) enhances final fruit quality (increased acidity and Brix) and prolongs fruit retention on the tree (Pérez-Pérez et al., 2010). These results underscore RDI's potential to add greater value to horticultural production. Enhancing productivity in sensitive crops like citrus requires precisely determined stress thresholds during critical periods and understanding how water deficit affects yield and fruit quality. Standardising these strategies requires ongoing research to develop new RDI approaches in different soil and climatic conditions and rootstock–variety combinations.

Several field studies defined the most sensitive phenological periods in different citrus species. In mandarin, for example, moderate water stress during flowering and fruit-set periods compromised yield by increasing fruit drop (Ginestar and Castel, 2000). Applying RDI during phase I of fruit growth (fruit cell division) decreased grapefruit vegetative growth much more than yield, slightly changing fruit composition (Pérez-Pérez et al., 2014). Applying RDI during phase II (fruit cell elongation) strongly affected plant water relations and decreased final fruit size, yield and water productivity, with severe water stress during phase II delaying external and internal fruit maturation. However, applying RDI during phase III (fruit ripening) improved fruit quality, by increasing total soluble solids (TSSs) and acidity. Long-term application of RDI (50% ETc, during the second part of phase II, when juice accumulation begins) to grapefruit using saline reclaimed water decreased vegetative development and crop load, with 21% less fruit yield (Romero-Trigueros et al., 2017). Thus, the agronomic effects of RDI depend on the timing and severity of water restriction.

Initial studies recommended applying less irrigation than well-watered citrus trees (50% ETc) during phase II (González-Altozano and Castel, 2003). Extrapolating this strategy to different soil and climatic conditions caused variable responses, modulated by rootstock tolerance to water deficit (Pérez-Pérez et al., 2008a; Robles et al., 2017) and irrigation water quality (Pedrero et al., 2014). Monitoring crop water status is crucial to prevent excessive stress, to maintain productivity. Crop management criteria typically impose irrigation reductions based on specific Ψ_{stem} thresholds in mandarin (-1.4 MPa in summer months – Ballester et al., 2011), orange (-2.0 MPa in summer months – Gasque et al., 2016), and grapefruit (-1.5 MPa for fruit growth phase II – Pérez-Pérez et al., 2014), allowing water savings of 10-20% without significantly compromising yield. However, grower adoption of these recommendations has been limited to periods of severe water restriction.

Rootstock genotype affects citrus water relations under low soil moisture availability. The 'Carrizo'-type root system (high root hydraulic conductivity and more fibrous roots restricted in a small soil volume) was suitable for fully irrigated clementine 'Clemenules' (Romero et al., 2006) and sweet orange 'Lane late' trees (Pérez-Pérez et al., 2008b), as this rootstock maintained higher midday Ψ_{stem} and gas exchange, thereby increasing vigour and yield. By ameliorating plant water status and gas exchange during dry periods and achieving better gas exchange when water stress is relieved, the deeper and more densely branched 'Cleopatra'-type root system seems suitable for RDI. Rootstocks of different vigour (*Citrus macrophylla*-CM and 'sour orange'-SO *Citrus aurantium*) modulated water status of field-grown lemon trees (Robles et al., 2017), with Ψ_{stem} of well-irrigated trees varying from ≈ −1.0 MPa (CM) to ≈ −1.12 MPa (SO). Although both rootstocks reached a similar Ψ_{stem} of ≈ −2.0 MPa under soil water deficit, vegetative growth and yield of CM trees decreased to the same extent as the reduction of irrigation water (28%), while SO trees maintained similar yields to well-irrigated trees. Rootstock selection should therefore consider water resources availability, with different Ψ_{stem} thresholds established according to the drought tolerance conferred by the rootstock.

Applying 50% ETc to mandarin trees between mid-July and mid-September (phase II of fruit growth) decreased irrigation requirements by 15% without impairing yield capacity and improving fruit quality, with moderate water stress (−1.3 MPa < Ψ_{stem} < −1.5 MPa) decreasing vegetative growth by 20% (Ballester et al., 2014). However, applying the same RDI strategy to sweet orange *cv.* 'Navelina' induced similar water stress and allowed similar (19%) water savings but decreased final fruit weight by 9%, risking economic returns (Ballester et al., 2013a). In contrast, restricting irrigation (25% ETc) to late lemon *cv.* 'Verna' trees grown on sour orange rootstock, during phases I and III of fruit growth, induced severe water stress (Ψ_{stem} ≈ −2.0 MPa) but achieved water savings of 28% without affecting yield and maintained commercially acceptable fruit size for a longer period (Robles et al., 2017). Thus, the efficacy of RDI strategies in different citrus cultivars depends on various factors such as the specific fruit growth stage, cultivar characteristics and rootstock. Optimising RDI strategies requires a nuanced understanding of crop physiology and water requirements to balance water conservation with economic viability.

Effectively applying RDI in sensitive species such as citrus requires physiological indicators to inform irrigation scheduling. While Ψ_{stem} is widely accepted to indicate overall water status of woody species, other variables indicate soil water availability. Citrus strongly regulates g_s to maintain water status (isohydric behaviour) under soil water deficits (Pérez-Pérez et al., 2012). Field studies of mandarin (Romero et al., 2006) and sweet orange (Pérez-Pérez et al., 2008b) show partial stomatal closure during periods of water restriction. When previously exposed to severe water stress, rewatering did not immediately result in stomatal re-opening, as Ψ_{stem} slowly recovered. However, environmental (e.g. atmospheric evaporative demand) and endogenous (e.g. fruit development) factors might influence leaf gas exchange measurements, affecting their use in irrigation scheduling.

Alternatively, physiological indicators derived from daily SF and trunk diameter fluctuations correlate with changes in plant water status (Fernández et al., 2001; Goldhamer and Fereres, 2001). In potted young lemon trees, the first detectable response to the deficit irrigation was a marked increase in maximum daily trunk diameter shrinkage followed by decreased SF, predawn Ψ_{leaf} and g_s (Ortuño et al., 2004). When trunk growth rate was higher or with more severe water stress ($\Psi_{stem} < -1.3$ MPa), MDS was less sensitive. In adult lemon trees, an MDS threshold was established at $\Psi_{stem} < -1.8$ MPa (Ortuño et al., 2006). Below this threshold of Ψ_{stem}, MDS was more sensitive to detect water stress than Ψ_{stem} and SF. Relative SF values (normalised to well-watered trees) from the compensation heat-pulse method were better correlated with Ψ_{stem} than absolute values in commercial citrus orchards (Ballester et al., 2013c).

Measuring citrus canopy temperatures (T_{canopy}) punctually with hand-operated thermal cameras or continuously with infrared thermometers (IRTs) can assess plant water status (Ballester et al., 2013b). When normalised by air temperature ($T_{canopy} - T_{air}$), T_{canopy} measurements from fixed IRTs were poorly related to g_s, but thermal imaging-based T_{canopy} was strongly correlated with Ψ_{stem} when treatment differences in Ψ_{stem} exceeded 1.0 MPa. Average fruit weight at harvest was most closely correlated with thermal camera measurements of T_{canopy}, then Ψ_{stem} and g_s, suggesting that thermal imaging can reliably predict water deficit impacts on citrus fruit weight under Mediterranean conditions. Canopy temperature and reference leaf temperatures (dry and wet leaves) agreed well when portable, low-cost sensors (e.g. FLIR One Pro) were connected to smartphones, demonstrating their potential utility to assess orange tree water status (Pappalardo et al., 2023). However, CWSI was not correlated with Ψ_{stem}, as citrus is isohydric.

Since strong stomatal regulation of citrus offers several measurement opportunities to inform irrigation scheduling as elaborated previously, the physiological basis of such stomatal closure has captured attention. Stomatal conductance of well-irrigated trees was sensitive to leaf-to-air vapour pressure deficit (LAVPD) and negatively correlated with leaf xylem abscisic acid (ABA) concentration ($[X\text{-}ABA]_{leaf}$) at LAVPDs <2.5 kPa (Pérez-Pérez et al., 2012). In plants grown with RDI at 75% ETc, LAVPD primarily regulated g_s without any correlation with $[X\text{-}ABA]_{leaf}$. Both foliar and xylem ABA concentrations were relatively stable despite irrigation volume and g_s varying (Pérez-Pérez et al., 2012; Romero-Conde et al., 2014; Puglisi et al., 2019), highlighting complex stomatal responses with interacting atmospheric and edaphic water deficits. That chemical treatments can modulate leaf gas exchange of droughted citrus plants by regulating foliar ABA concentrations (Kondo et al., 2009) suggests that ABA contributes to stomatal regulation of citrus, but field-grown plants typically show ABA homeostasis.

4 Partial rootzone drying of tomato

Partial rootzone drying (PRD) was developed from the laboratory split-root experiments described earlier, which aimed to decouple leaf water deficit from soil drying by watering part of the rootzone and irrigating the other (Blackman and Davies, 1985). PRD partially closed the stomata and inhibited leaf growth (Gowing et al., 1990), prompting field-based adaptations. Initial experiments in drip-irrigated vineyards physically split the root system by placing plastic sheeting along the vine row, but in sandy soils with limited lateral water movement distinct wet and dry bulbs occurred on either side of the vine row under the drippers (During et al., 1996; Loveys et al., 2000). Concerns regarding the additional investment costs for irrigation laterals and controllers to run two parallel driplines to implement PRD (Marsal et al., 2008) may have inhibited the uptake and development of this technique. Nevertheless, PRD can easily be applied with surface (furrow) irrigation by irrigating every second furrow at each event. While applying alternate furrow irrigation (AFI) pre-dates the original split-root experiments (Grimes et al., 1968), early AFI experiments with tomatoes occurred (Murray, 1999, Ramalan and Nwokeocha, 2000) roughly when PRD was first applied to pot- and greenhouse-grown tomato (Davies et al., 2000). Before reviewing tomato crop responses to PRD, the aims of PRD irrigation are briefly considered.

PRD aimed to alter root phytohormone accumulation and thereby root-to-shoot signalling, by applying sufficient water to roots in irrigated soil to prevent leaf water deficits, while allowing roots in drying soil to accumulate signals such as ABA and export these to the shoot. Enhanced root-to-shoot signalling partially closed the stomata (limiting transpiration to maintain Ψ_{leaf}, potentially counteracting the effects of decreased irrigation volume), ideally without photosynthetic limitation thereby enhancing leaf WUE. Furthermore, these signals should restrict vegetative growth (again limiting plant transpirational losses) thereby favouring photoassimilate distribution to reproductive tissues at the expense of vegetative growth.

While wet and dry parts of the root system may be maintained throughout the irrigation season (Fixed PRD = PRD-F), usually they are regularly alternated (Alternate PRD = PRD-A). Across different crops, this alternation seems to increase yields compared to Fixed PRD (Dodd et al., 2015). Alternation increases root biomass accumulation (Mingo et al., 2004) allowing root foraging into moister parts of the soil profile to increase water uptake, thereby acting to maintain Ψ_{leaf}. Furthermore, alternation enhances root-to-shoot phytohormone signalling (Dodd et al., 2006; Pérez-Pérez et al., 2020), which seems beneficial as prolonged drying of part of the root system diminished signal export to the shoot (Dodd et al., 2008a). A common question when applying Alternate PRD is when to alternate the wet and dry sides. Measuring some of the variables described in Section 2 should help. For example, maintaining leaf water status above a threshold Ψ_{leaf} during PRD may be important in preventing photosynthetic limitations or preventing stomatal re-opening if signalling is diminished after prolonged soil drying of the dry side. Alternating wet and dry sides at fixed time intervals might explain some of the variation in the agronomic and physiological responses considered below. .

Whether tomato plants maintain Ψ_{leaf} during PRD depends on the irrigation volume supplied. PRD plants receiving 100% of evapotranspiration (100%PRD) maintained their Ψ_{leaf} (Mingo et al., 2004) whereas those receiving half that volume (50%PRD) typically showed decreased Ψ_{leaf} (Dodd, 2007) compared to well-watered plants. Whether well-watered and PRD plants differed in Ψ_{leaf} depended on when Ψ_{leaf} was measured relative to irrigation timing and the time of day. Following morning irrigation, Ψ_{leaf} of 50%PRD plants declined throughout the day (Dodd et al., 2006). In contrast, daily irrigation late in the afternoon with 50%PRD maintained midday Ψ_{leaf} close to well-watered plants, whereas Ψ_{leaf} of deficit irrigated plants declined progressively as the experiment progressed (Tahi et al., 2007). Similarly, greenhouse-grown plants supplied with 50%PRD maintained their Ψ_{leaf} throughout the day and growing season (71–161 days after sowing) when irrigated daily but supplying half the irrigation volume twice daily (with plants receiving the same cumulative irrigation volume) decreased midday Ψ_{leaf} by up to 0.5 MPa when transpiration demands were maximal (Zegbe et al., 2004). In the same experimental system, Ψ_{leaf} of mature fruiting plants decreased more and earlier in the day than plants that had just formed the first truss associated with much greater soil depletion (Zegbe et al., 2006), presumably due to greater vegetative growth and thus transpirational demand. Whether irrigation placement (watering only part, PRD or all the root system, DI) affects Ψ_{leaf} has also been investigated (Figure 3). Applying 70%PRD to greenhouse-grown plants increased Ψ_{leaf} (by up to 0.3

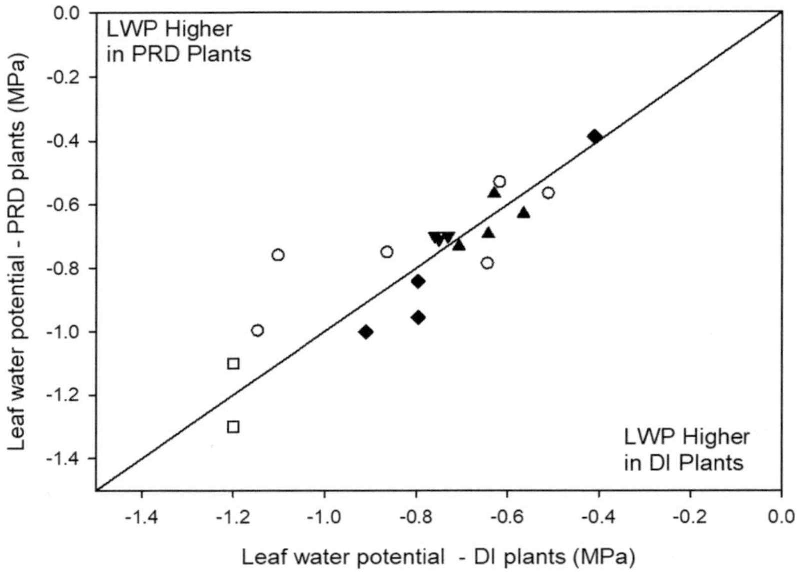

Figure 3 Leaf water potential of PRD plants plotted against leaf water potential of DI plants in the same experiment, when xylem sap was collected by pressurising the root system (hollow symbols) or detached leaves (filled symbols). The 1:1 relationship is indicated, and each symbol represents a separate occasion that measurements were made in the experiments reported in Dodd, 2007 (◆), Pérez-Pérez et al., 2020 (▼), Savic et al., 2009 (○), Topcu et al., 2007 (▲) and Wang et al., 2012 (□).

MPa) relative to DI plants on some measurement occasions while the opposite response occurred on other occasions (Savic et al., 2009), with different tomato cultivars responding differently even if treatment differences in Ψ_{leaf} were <0.1 MPa (Pazzagli et al., 2016). Similarly, Ψ_{leaf} only occasionally differed between DI and PRD plants with no systematic variation (Wang et al., 2012). Applying 50% PRD to field-grown, drip-irrigated plants (irrigated once or twice weekly) maintained Ψ_{leaf} similar to well-watered plants throughout most of the diurnal cycle, with midday Ψ_{leaf} transiently decreasing in both DI and PRD plants (Topcu et al., 2007). Whether Ψ_{leaf} measurements can usefully schedule PRD of tomato likely depends on the % of evapotranspiration applied, as minimal differences were detected compared to well-watered plants when 70% PRD or higher was applied. Nevertheless, Ψ_{leaf} has been used to schedule deficit irrigation of tomato (Lipan et al., 2021), with lower Ψ_{leaf} thresholds (−1.0 MPa) during vegetative growth and from fruit ripening than during fruit expansion (−0.85 MPa). Impacts of leaf (and fruit) water status on tissue expansion during PRD are critical to determine the technique's agronomic success.

Relatively few investigators have dynamically measured fruit expansion in PRD-grown plants. Applying 50% PRD to pot-grown plants in a controlled environment restricted fruit growth rate (as did withholding water from the entire soil profile) even though fruit mesocarp turgor was maintained slightly (0.03 MPa) higher than well-watered plants throughout the experiment (Mingo et al., 2003). Sub-epidermal apoplastic alkalisation correlated with fruit growth rate, suggesting non-hydraulic regulation of fruit growth. Supplying half the irrigation volume to greenhouse-grown plants via Alternate PRD decreased fruit weight and fruit diameter by 15% and 10% respectively, without affecting fruit number (Davies et al., 2000), indicating fruit biomass accumulation and expansive growth were more sensitive than flower initiation and fruit set. Similar decreases in fruit weight (7–20%) and fruit diameter (3–5%) occurred in field-grown plants although higher (7–10%) yields were achieved than with deficit irrigation that watered the entire rootzone (Kirda et al., 2004). Alternate furrow irrigation repeatedly maintained fruit yields of processing tomatoes in different seasons and cultivars with substantial (25%) water savings depending on irrigation season duration (Barrios-Masias and Jackson, 2016). The relative insensitivity of fruit expansion to soil drying compared to leaf expansion likely partially explained PRD's success in maintaining fruit growth.

Leaf area index of 50%PRD plants was substantially less than well-watered plants (Kirda et al., 2004; Topcu et al., 2007), with differences most apparent once fruiting started. PRD limited tomato leaf expansion without Ψ_{leaf} decreasing, suggesting non-hydraulic regulation of fruit growth via increased foliar evolution of the growth inhibitor ethylene (Sobeih et al., 2004). PRD also decreased foliar cytokinin concentrations, which is expected to constrain cell division and expansion (Kudoyarova et al., 2007). Further evidence that non-hydraulic signals limited tomato leaf expansion was provided by drying part of the root system of PRD plants to reach the same soil water content but allowing three different levels of soil drying on the wet side. Although all three treatments had a similar Ψ_{leaf} (no more than 0.10 MPa less than well-watered plants), leaf area declined linearly with overall soil water content (Campos et al., 2009). While mild water deficits affect tomato leaf growth, and LVDTs can continuously monitor tomato leaf elongation (Price et al., 2001), monitoring stem diameter fluctuations with these sensors has been more common.

The suitability of LVDTs to measure tomato stem diameter fluctuations with plant water status depended on crop age and atmospheric evaporative demand (Gallardo et al., 2006). In rapidly growing young plants with high stem growth rates (SGRs), this was the most sensitive variable to soil drying while in older plants MDS was more sensitive than SGR. Indicator suitability for irrigation scheduling depended not only on response time to soil drying but their signal-(amplitude of response)-to-noise (coefficient of variation) ratio (Gallardo et al., 2006). While simultaneously measuring SF and stem diameter diagnosed water deficits in tomatoes (de Swaef and Steppe, 2010) by overcoming fruit load and plant age effects on stem diameter fluctuations, these sensors do not appear to have been widely applied for tomato irrigation scheduling even though they can be more sensitive than pre-dawn Ψ_{leaf} and midday Ψ_{stem} (Wang et al., 2017). Nevertheless, MDS successfully discriminated treatment differences in 70%PRD and 50%PRD plants but only at high evaporative demands in greenhouse-grown crops (Affi et al., 2014). In dwarf (cv. Micro-Tom) pot-grown plants maintained in controlled environment, MDS increased with soil drying with a uniform relationship irrespective of whether all or only part of the rootzone was dried, and irrigation alternation enhanced MDS of PRD plants (Puértolas and Dodd, 2022). While stem diameter fluctuations have been measured in tomato plants including those exposed to PRD, variable responses according to evaporative demand, crop phenology and fruit load suggest that other plant-based variables likely have more utility in irrigation scheduling.

Early experiments with soil- and container-grown tomatoes in the greenhouse demonstrated that supplying half the irrigation volume via Alternate PRD-induced substantial stomatal closure (Davies et al., 2000) although the effects of decreased irrigation volume couldn't be separated from PRD alternation. While prolonged soil drying with 100% PRD caused some stomatal closure (Mingo et al., 2004), alternating the wet and dry parts of the root system of plants receiving 50% PRD induced additional stomatal closure (Dodd et al., 2006; Pérez-Pérez et al., 2020). With 70% PRD in greenhouse-grown plants, PRD and DI plants had a similar g_s during the first drying cycle, but thereafter g_s of PRD plants exceeded that of DI plants (Savic et al., 2009). In contrast, although g_s of DI and PRD plants were similar with 50% less water supplied, an organic N fertiliser greatly increased g_s of DI plants compared to those supplied with mineral N, while g_s of PRD plants was more stable (Wang et al., 2012). More substantial stomatal closure occurred in PRD treatments that also sub-optimally irrigated the wet side of PRD plants at similar dry side soil moisture contents, independent of changes in Ψ_{leaf} (Campos et al., 2009). The sectorial nature of tomato vasculature (Bledsoe and Orians, 2006) may partially explain variable stomatal responses to PRD between different leaves as specific roots (on either the wet and dry side of the plant) supply water to specific leaves. Although the coefficient of variation of g_s did not appreciably vary between DI and PRD plants (Dodd et al., 2006; Wang et al., 2012), measuring whole plant (or canopy) conductance might better integrate plant responses to the soil environment. In dwarf (cv. Micro-Tom) plants, PRD alternation suppressed whole-plant transpiration and canopy conductance while plants exposed to Fixed PRD maintained similar values to well-watered plants (Puértolas and Dodd, 2022). Treatment differences in gas exchange of PRD plants, despite similar overall soil water status, highlight the need to understand stomatal regulation during PRD which might limit the usefulness of leaf gas exchange measurements for irrigation scheduling of PRD.

Canopy temperature measurements, with infra-red thermometers (±0.5°C precision in Ben-Asher et al., 1992) and thermal cameras (±0.1°C precision in Meron et al., 2010), integrate the variable responses of many individual leaves experiencing different microclimates. Despite environmental instability associated with variation in windspeed or cloud cover, CWSI was linearly correlated with Ψ_{leaf} in both studies, suggesting thermal measurements could assist irrigation scheduling. To date, we are not aware of any such use to schedule PRD of tomatoes or in other crops.

While altered root growth and physiology are unlikely to be used for irrigation scheduling, root water uptake determines the above-mentioned variables. In well-watered greenhouse-grown plants, irrigation placement (watering all or half of the rootzone) did not affect overall plant biomass but Alternate PRD increased root biomass by c50% at the expense of shoot tissues (Mingo et al., 2004). More detailed controlled environment measurements indicated that soil drying did not increase root growth (although such effects have been reported - Reid and Renquist, 1997) but that rewatering previously dry roots massively stimulated root growth (Mingo et al., 2004). Greater investment in root growth with PRD may be especially pronounced when plants are grown under elevated CO_2 concentrations (Wei et al., 2023). These changes in root growth dynamics may enhance root system efficiency in taking up fertilisers, thereby minimising nutrient leaching (Wang et al., 2019). While root hydraulic conductance can evaluate root system efficiency, such measurements do not appear to have been made for PRD-grown tomato plants. The other key mechanism by which roots might alter shoot responses is by modulating root-to-shoot chemical signalling, as envisaged when PRD was conceived.

Most studies of such signalling in PRD-grown tomatoes have focused on the phytohormone ABA as it causes stomatal closure. Tomato seems especially suitable for such studies as xylem sap can be collected from individual tomato leaves in the field using a pressure chamber (after measuring Ψ_{leaf}) and from the roots if pots can be inserted into the chamber (Netting et al., 2012). Unravelling ABA dynamics during PRD has proved challenging as responses can be variable (Figure 4) but has spawned innovations such as grafted 'two root, one shoot' tomato plants (Dodd, 2007) to evaluate how different parts of the root system contributed to overall ABA export.

Initial controlled environment experiments determined whether PRD alternation affects ABA dynamics. Supplying half the irrigation volume to PRD plants increased leaf xylem ABA concentration (X-ABA_{leaf}) compared to well-watered plants. With prolonged drying cycles over a week, PRD alternation enhanced X-ABA_{leaf} compared to Fixed PRD plants, attributed to ABA mobilisation from previously dry roots (Dodd et al., 2006). Further studies revealed divergent relationships between PRD and DI plants when plotting X-ABA_{leaf} against Ψ_{leaf} or overall soil water content, with prolonged drying of

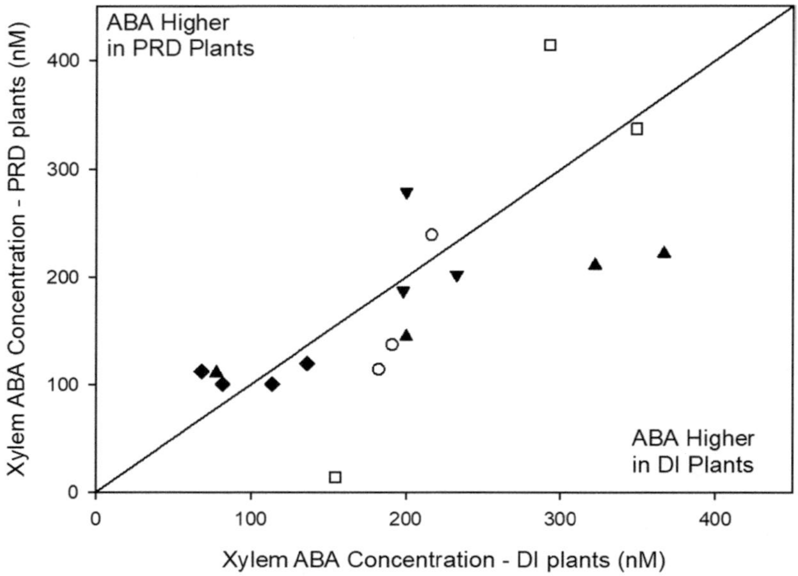

Figure 4 Xylem ABA concentration of PRD plants plotted against xylem ABA concentration of DI plants in the same experiment, when xylem sap was collected by pressurising the root system (hollow symbols) or detached leaves (filled symbols). The 1:1 relationship is indicated, and each symbol represents a separate occasion that sap was collected from the experiments reported in Dodd, 2007 (◆), Pérez-Pérez et al., 2020 (▼), Savic et al., 2009 (○), Topcu et al., 2007 (▲) and Wang et al., 2012 (□). Note that actual values of Wang et al., 2012, have been divided by ten to ensure they fit within the graph axes.

part of the rootzone decreasing X-ABA$_{leaf}$ even as overall soil water content decreased (Dodd, 2007). Collecting root xylem sap from 'two root, one shoot' tomato plants revealed that X-ABA$_{root}$ from the wet side underestimated X-ABA$_{leaf}$ while X-ABA$_{root}$ from the dry side overestimated X-ABA$_{leaf}$ (Dodd, 2007), implying that SF from both soil compartments influenced root ABA export and thus X-ABA$_{leaf}$. Subsequently measuring SF rates and X-ABA$_{root}$ in sunflower (its cylindrical stems facilitated installing SF gauges) helped to develop a model for PRD plants, with X-ABA$_{leaf}$ best explained by adding two terms, each comprising the fractional SF from a (wet or dry) root system multiplied by its X-ABA$_{root}$ (Dodd et al., 2008a) The X-ABA$_{root}$ of sap collected above the graft union was explained by X-ABA$_{root}$ of the constituent roots (Dodd et al., 2008b). Further split-root tomato experiments validated this model, with soil moisture depletion from the different parts of the rootzone replacing SF measurements (Pérez-Pérez and Dodd, 2015). This model helps to interpret other ABA measurements made with PRD tomato plants, with moderate soil drying stimulating X-ABA$_{leaf}$ (maintaining some SF, and thus root

ABA export, from dry roots) but prolonged soil drying diminishing X-ABA_{leaf} (as SF from roots in drying soil ceases and only roots from the wet side export ABA).

When root xylem sap was collected from PRD plants at the end of a drying cycle, X-ABA_{root} was similar to well-watered plants and much lower than DI plants (Savic et al., 2009), apparently validating the model. However, lower Ψ_{leaf} of DI plants (than PRD plants) on one occasion suggests an alternative explanation: enhanced foliar ABA synthesis resulting in greater phloem export of ABA to the roots (Manzi et al., 2015). When X-ABA_{root} was normalised according to Ψ_{leaf} in a similar PRD experiment that also collected root xylem sap at the end of a drying cycle, PRD plants initially had lower ABA concentrations than DI plants but later had higher ABA concentrations than DI plants (Wang et al., 2012). These differential ABA responses suggests the agronomic viability of varying the irrigation technique during the crop's life cycle according to the intended physiological aim. Enhancing ABA signalling early in the crop life's cycle may minimise soil water depletion during root establishment, while attenuating ABA signalling later in the crop life's cycle may restrict stomatal closure to maximise photosynthesis rates.

Despite multiple experiments applying PRD to tomato, this fundamental 'signalling science' has scarcely affected irrigation scheduling. Most field experiments arbitrarily alternate PRD every 1 (Tahi et al., 2007; Topcu et al., 2007; Casa and Rouphael, 2014; Hashem et al., 2019; Mattar et al., 2020) or 2 (Mingo et al. 2004; Casa and Rouphael, 2014) weeks or didn't report the timing (Mahmoud and Fayad, 2022). With less frequent PRD alternation (every second irrigation), yield was 7% less than alternation every irrigation when half the irrigation volume decreased yield by 25% overall (Kirda et al., 2004). Similarly, less frequent PRD alternation (every second week) decreased yield by 4–13% compared to weekly alternation, when half the irrigation volume decreased yield by 40–60% overall (Casa and Rouphael, 2014). In another year, less frequent PRD alternation (every second week) increased yield by 10% compared to weekly alternation when supplying half the irrigation volume decreased yield by 57% overall (Casa and Rouphael, 2014), suggesting that specific environmental conditions modulate effects of PRD alternation. In some cases, alternating PRD after plants were no longer extracting soil moisture from the dry side (Davies et al., 2000) suggested that the irrigated side modulated root-to-shoot signalling at that time. The availability of thermal cameras to rapidly detect stomatal closure within the canopy might allow more dynamic scheduling of PRD in the future according to whether water-saving (partial stomatal closure to enhance crop water use efficiency) or water-spending (maintain stomata open to boost photosynthesis) is desirable, according to the water supply.

5 Conclusion and future trends

Understanding the sensitivity of various physiological responses to changes in soil water status creates opportunities to mitigate economic losses and potentially augment profits while allowing more efficient and sustainable agricultural production practices. Empowering farmers with tools and knowledge to implement innovative and creative irrigation strategies may realise site-specific agronomic benefits to optimise fruit yield and quality and could alter future water management paradigms in agriculture. As global water scarcity intensifies, enhancing efficiency in agricultural water management becomes increasingly imperative. Considering the escalating costs of irrigation and reductions in abstraction reliability, growers are expected to increasingly gravitate towards more precise and optimised irrigation practices rooted in a nuanced understanding of crop physiology and water requirements. Water saving strategies such as RDI and PRD can help address the demands for increased water conservation, crop productivity, whilst maintaining fruit quality. However, achieving greater standardisation and refinement of these strategies necessitates ongoing research endeavours, particularly in diverse soil and climatic contexts. Leveraging advancements in instrumentation, such as thermal cameras for rapidly detecting stomatal closure within the canopy, holds much promise for dynamic irrigation scheduling custom-made to specific agronomic aims.

6 Where to look for further information

PRIMA (Partnership for Research and Innovation in the Mediterranean area) is a strategic programme supporting the development of irrigation technologies and practices research for arid and semi-arid Mediterranean areas (www.prima-med.org).

7 Acknowledgements

Dodd thanks the Global Challenges Research Fund (Project RECIRCULATE, ES/P010857/1) and the European Union (SHui; Project Number 773903) for supporting work on partial rootzone drying of tomato. Huntenburg thanks AHDB (Agricultural and Horticultural Development Board) for supporting work on drought stress responses in potato (PhD Fellowship Ref: 11140035). Pérez-Pérez thanks the European Union and the Spanish Research Agency throughout the 'HANDYWATER' PRIMA project (PCI2021-121940 funded by MCIN/AEI/10.13039/501100011033) for supporting work on water-saving strategies and innovative irrigation technologies and practices.

8 References

Affi, N., El-Fadl, A., El-Otmani, M., Benismail, M. C. and El Mastor, A. (2014). Improvement of tomato water productivity by the use of a partial rootzone drying irrigation strategy. *Acta Horticulturae* 1038: 359–366.

Agbona, A., Teare, B., Ruiz-Guzman, H., Dobreva, I. D., Everett, M. D., Adams, T., Montesinos-Lopez, O. A., Kulakow, P. A. and Hays, D. B. (2021). Prediction of root biomass in cassava based on ground penetrating radar phenomics. *Remote Sensing* 13: 4985.

Alordzinu, E., Li, J., Lan, Y., Appiah, S. A., Aasmi, A. A. and Wang, H. (2021). Rapid estimation of crop water stress index on tomato growth. *Sensors* 21(15): 542.

Ballester, C., Castel, J., Intrigliolo, D. S. and Castel, J. R. (2011). Response of Clementina de Nules trees to summer deficit irrigation. Yield components and fruit composition. *Agricultural Water Management* 98(6): 1027–1032.

Ballester, C., Castel, J., Intrigliolo, D. S. and Castel, J. R. (2013a). Response of Navel Lane late citrus trees to regulated deficit irrigation: yield components and fruit composition. *Irrigation Science* 31(3): 333–341.

Ballester, C., Castel, J., Jiménez-Bello, M. A., Castel, J. R. and Intrigliolo, D. S. (2013b). Thermographic measurement of canopy temperature is a useful tool for predicting water deficit effects on fruit weight in citrus trees. *Agricultural Water Management* 122: 1–6.

Ballester, C., Castel, J., Testi, L., Intrigliolo, D. S. and Castel, J. R. (2013c). Can heat-pulse sap flow measurements be used as continuous water stress indicators of citrus trees? *Irrigation Science* 31(5): 1053–1063.

Ballester, C., Castel, J., Abd El-Mageed, T. A., Castel, J. R. and Intrigliolo, D. S. (2014). Long-term response of 'Clementina de Nules' citrus trees to summer regulated deficit irrigation. *Agricultural Water Management* 138: 78–84.

Ballester, C., Buesa, I., Soler, E., Besada, C., Salvador, A., Bonet, L. and Intrigliolo, D. S. (2018). Postharvest regulated deficit irrigation in early- and intermediate-maturing loquat trees. *Agricultural Water Management* 205: 1–8.

Barrios-Masias, F. H. and Jackson, L. E. (2016). Increasing the effective use of water in processing tomatoes through alternate furrow irrigation without a yield decrease. *Agricultural Water Management* 177: 107–117.

Bates, L. M. and Hall, A. E. (1981). Stomatal closure with soil-moisture depletion not associated with changes in bulk leaf water status. *Oecologia* 50(1): 62–65.

Battaglia, M. L., Lee, C. and Thomason, W. (2018). Corn yield components and yield response to defoliation at different row widths. *Agronomy Journal* 110(1): 210–225.

Ben-Asher, J., Phene, C. J. and Kinarti, A. (1992). Canopy temperature to assess daily evapotranspiration and management of high-frequency drip irrigation systems. *Agricultural Water Management* 22(4): 379–390.

Blackman, P. G. and Davies, W. J. (1985). Root to shoot communication in maize plants of the effects of soil drying. *Journal of Experimental Botany* 36(1): 39–48.

Bledsoe, T. M. and Orians, C. M. (2006). Vascular pathways constrain 13C accumulation in large root sinks of *Lycopsersicon esculentum* (Solanaceae). *American Journal of Botany* 93(6): 884–890.

Bonany, J., Camps, F., Salvia, J. and Cohen, M. (2000). Relationship between trunk diameter fluctuations, stem water potential and fruit growth rate in potted adult apple trees. *Acta Horticulturae* 511: 43–50.

Boyer, J. S. (1970). Leaf enlargement and metabolic rates in corn, soybean and sunflower at various leaf water potentials. *Plant Physiology* 46(2): 233–235.

Buesa, I., Badal, E., Guerra, D., Ballester, C., Bonet, L. and Intrigliolo, D. S. (2013). Regulated deficit irrigation in persimmon trees (*Diospyros kaki*) cv. 'Rojo Brillante'. *Scientia Horticulturae* 159: 134–142.

Campos, H., Trejo, C., Pena-Valdivia, C. B., Ramirez-Ayala, C. and Sanchez-Garcia, P. (2009). Effect of partial rootzone drying on growth, gas exchange, and yield of tomato (*Solanum lycopersicum* L.). *Scientia Horticulturae* 120(4): 493–499.

Casa, R. and Rouphael, Y. (2014). Effects of a partial root-zone drying irrigation on yield, fruit quality, and water-use efficiency in processing tomato. *Journal of Horticultural Science and Biotechnology* 89(4): 389–396.

Cetin, M., Alsenjar, O., Aksu, H., Golpinar, M. S. and Akgul, M. A. (2023). Estimation of crop water stress index and leaf area index based on remote sending data. *Water Supply* 23(3): 1390–1404.

Chalmers, D. J., Mitchell, P. D. and van Heek, L. (1981). Control of peach tree growth and productivity by regulated water supply, tree density, and summer pruning. *Journal of the American Society for Horticultural Science* 106(3): 307–312.

Chaves, M. M., Pereira, J. S., Maroco, J., Rodrigues, M. L., Ricardo, C. P. P., Osório, M. L., Carvalho, I., Faria, T. and Pinheiro, C. (2002). How plants cope with water stress in the field? Photosynthesis and growth. *Annals of Botany* 89(7): 907–916.

Colloff, M. J. and Pittock, J. (2022). Mind the gap! Reconciling environmental water requirements with scarcity in the Murray-Darling Basin, Australia. *Water* 14(2): 208.

Dardanelli, J. L., Bachmeier, O. A., Serano, R. and Gil, R. (1997). Rooting depth and soil water extraction patterns of different crops in a silty loam Haplustoll. *Field Crops Research* 54(1): 29–38.

Davies, W. J. and Zhang, J. (1991). Root signals and the regulation of growth and development of plants in drying soil. *Annual Review of Plant Physiology and Plant Molecular Biology* 42(1): 55–76.

Davies, W. J., Bacon, M. A., Stuart Thompson, D., Sobeih, W. and González Rodríguez, L. (2000). Regulation of leaf and fruit growth in plants growing in drying soil: exploitation of the plants' chemical signalling system and hydraulic architecture to increase the efficiency of water use in agriculture. *Journal of Experimental Botany* 51(350): 1617–1626.

de la Rosa, J. M., Conesa, M. R., Domingo, R., Aguayo, E., Falagán, N. and Pérez-Pastor, A. (2016). Combined effects of deficit irrigation and crop level on early nectarine trees. *Agricultural Water Management* 170: 120–132.

de Swaef, T. and Steppe, K. (2010). Linking stem diameter variations to sap flow, turgor and water potential in tomato. *Functional Plant Biology* 37(5): 429–438.

Dodd, I. C., Theobald, J. C., Bacon, M. A. and Davies, W. J. (2006). Alternation of wet and dry sides during partial rootzone drying irrigation alters root-to-shoot signalling of ABA. *Functional Plant Biology* 33(12): 1081–1089.

Dodd, I. C. (2007). Soil moisture heterogeneity during deficit irrigation alters root-to-shoot signalling of abscisic acid. *Functional Plant Biology* 34(5): 439–448.

Dodd, I. C., Egea, G. and Davies, W. J. (2008a). Abscisic acid signalling when soil moisture is heterogeneous: decreased photoperiod sap flow from drying roots limits ABA export to the shoots. *Plant, Cell and Environment* 31(9): 1263–1274.

Dodd, I. C., Egea, G. and Davies, W. J. (2008b). Accounting for sap flow from different parts of the root system improves the prediction of xylem ABA concentration in plants grown with heterogeneous soil moisture. *Journal of Experimental Botany* 59(15): 4083–4093.

Dodd, I. C., Whalley, W. R., Ober, E. S. and Parry, M. A. J. (2011). Genetic and management approaches to boost UK wheat yields by ameliorating water deficits. *Journal of Experimental Botany* 62(15): 5241–5248.

Dodd, I. C., Puértolas, J., Huber, K., Pérez-Pérez, J. G., Wright, H. R. and Blackwell, M. S. A. (2015). The importance of soil drying and re-wetting in crop phytohormonal and nutritional responses to deficit irrigation. *Journal of Experimental Botany* 66(8): 2239–2252.

Dunn, S. M., Stalham, M., Chalmers, N. and Crabtree, B. (2003). Adjusting irrigation abstraction to minimise the impact on stream flow in the East of Scotland. *Journal of Environmental Management* 68(1): 95–107.

During, H., Loveys, B. R. and Dry, P. R. (1996). Root signals affect water use efficiency and shoot growth. *Acta Horticulturae* 427: 1–13.

Espinoza-Meza, S., Ortega-Farias, S., López-Olivari, R., Araya-Alman, M. and Carrasco-Benavides, M. (2023). Response of fruit yield, fruit quality, and water productivity to different irrigation levels for a microsprinkler-irrigated apple orchard (cv. Fuji) growing under Mediterranean conditions. *European Journal of Agronomy* 145: 126786.

Farooq, M., Hussain, M., Ul-Allah, S. and Siddique, K. H. M. (2019). Physiological and agronomic approaches for improving water-use efficiency in crop plants. *Agricultural Water Management* 219: 95–108.

Fereres, E. and Soriano, M. A. (2007). Deficit irrigation for reducing agricultural water use. *Journal of Experimental Botany* 58(2): 147–159.

Fernández, J. E., Moreno, F., Cabrera, F., Arrúe, J. L. and Martín-Aranda, J. (1991). Drip irrigation, soil characteristics and the root distribution and root activity of olive trees. *Plant and Soil* 133(2): 239–251.

Fernández, J. E., Palomo, M. J., Díaz-Espejo, A., Clothier, B. E., Green, S. R., Girón, I. F. and Moreno, F. (2001). Heat-pulse measurements of sap flow in olives for automating irrigation: tests, root flow and diagnostics of water stress. *Agricultural Water Management* 51(2): 99–123.

Fernández, J. E., Rodriguez-Dominguez, C. M., Perez-Martin, A., Zimmermann, U., Rüger, S., Martín-Palomo, M. J., Torres-Ruiz, J. M., Cuevas, M. V., Sann, C., Ehrenberger, W. and Diaz-Espejo, A. (2011). Online- monitoring of tree water stress in a hedgerow olive orchard using the leaf patch clamp pressure probe. *Agricultural Water Management* 100(1): 25–35.

Fiebig, A. and Dodd, I. C. (2016). Inhibition of tomato shoot growth by over-irrigation is linked to nitrogen deficiency and ethylene. *Physiologia Plantarum* 156(1): 70–83.

Flexas, J., Bota, J., Cifre, J., Escalona, J. M., Galmés, J., Gulías, J., Lefi, E., Martínez-Cañellas, S. F., Moreno, M. T., Ribas-Carbo, M., Riera, D., Sampol, B. and Medrano, H. (2018). Understanding down-regulation of photosynthesis under water stress: future prospects and searching for physiological tools for irrigation management. *Annals of Applied Biology* 173(3): 245–257.

Gallardo, M., Thompson, R. B., Valdez, L. C. and Fernandez, M. D. (2006). Use of stem diameter variations to detect plant water stress in tomato. *Irrigation Science* 24(4): 241–255.

Gasque, M., Martí, P., Granero, B. and González-Altozano, P. (2016). Effects of long-term summer deficit irrigation on 'Navelina' citrus trees. *Agricultural Water Management* 169: 140–147.

Ginestar, C. and Castel, J. R. (1996). Responses of young Clementine citrus trees to water stress during different phenological periods. *Journal of Horticultural Science* 71(4): 551–559.

Goldhamer, D. A. and Fereres, E. (2001). Irrigation scheduling protocols using continuously recorded trunk diameter measurements. *Irrigation Science* 20(3): 115–125.

Gollan, T., Passioura, J. B. and Munns, R. (1986). Soil-water status affects the stomatal conductance of fully turgor wheat and sunflower leaves. *Australian Journal of Plant Physiology* 24: 685–692.

González-Altozano, P. and Castel, J. R. (2003). Riego deficitario en 'Clementina de Nules'. *Spanish Journal of Agricultural Research* 1(2): 81–92.

González-Dugo, P., Zarco-Tejada, Z., Nicolás, E., Nortes, P. A., Alarcón, J. J., Intrigliolo, D. S. and Fereres, E. (2013). Using high resolution UAV imagery to assess the variability in the water status of five fruit trees species within a commercial orchard. *Precision Agriculture* 14: 660–678.

Gowing, D. J. G., Davies, W. J. and Jones, H. G. (1990). A positive root-sourced signal as an indicator of soil drying in apple, *Malus domestica* Borkh. *Journal of Experimental Botany* 41(12): 1535–1540.

Grimes, D. W., Walhood, V. T. and Dickens, W. L. (1968). Alternate-furrow irrigation for San Joaquin valley cotton. *California Agriculture* 22: 4–6.

Gupta, N. K., Gupta, S. and Kumar, A. (2008). Effect of water stress on physiological attributes and their relationship with growth and yield of wheat cultivars ate different stages. *Journal of Agronomy and Crop Science* 186: 55–62.

Hashem, M. S., El-Abedin, T. Z. and Al-Ghobari, H. M. (2019). Rational water use by applying regulated deficit and partial root-zone drying irrigation techniques in tomato under arid conditions. *Chilean Journal of Agricultural Research* 79(1): 75–88.

Hasnain, S. and Singh, A. (2022). Development of an electronic wetting front detector. *Agricultural Water Management* 274: 107980.

Howell, K. R., Shrestha, P. and Dodd, I. C. (2015). Alternate wetting and drying irrigation maintained rice yields despite half the irrigation volume, but is currently unlikely to be adopted by smallholder lowland rice farmers in Nepal. *Food and Energy Security* 4(2): 144–157.

Hsiao, T. C. (1973). Plant responses to water stress. *Annual Review of Plant Physiology* 24(1): 519–570.

Huntenburg, K., Puértolas, J., de Ollas, C. and Dodd, I. C. (2022). Bi-directional, long-distance hormonal signalling of soil water availability. *Physiologia Plantarum* 174(3): e13697.

Huntenburg, K., Pflugfelder, D., Koller, R., Dodd, I. C. and van Dusschoten, D. (2023). Diurnal water fluxes and growth patterns in potato tubers under drought stress. *Plant and Soil* in press. DOI: 10.1007/s11104-023-06108-1

Intrigliolo, D. S. and Castel, J. R. (2010). Response of plum trees to deficit irrigation under two crop levels: tree growth, yield and fruit quality. *Irrigation Science* 28(6): 525–534.

Intrigliolo, D. S., Bonet, L., Nortes, P. A., Puerto, H., Nicolás, E. and Bartual, J. (2013). Pomegranate trees performance under sustained and regulated deficit irrigation. *Irrigation Science* 31(5): 959–970.

Kirda, C., Cetin, M., Dasgan, Y., Topcu, S., Kaman, H., Ekici, B., Derici, M. R. and Ozguven, A. I. (2004). Yield response of greenhouse grown tomato to partial root drying and conventional deficit irrigation. *Agricultural Water Management* 69(3): 191-201.

Kondo, S., Kittikorn, M., Ohara, H., Okawa, K., Suguya, S., Kitihata, N. and Asami, T. (2009). Effect of a cytochrome P450 inhibitor on abscisic acid biosynthesis and stomatal regulation in citrus trees in water-stressed conditions. *Scientia Horticulturae* 120(1): 146–149.

Kudoyarova, G. R., Vysotskaya, L. B., Cherkozyanova, A. and Dodd, I. C. (2007). Effect of partial rootzone drying on the concentration of zeatin-type cytokinins in tomato (*Lycopersicon esculentum* Mill.) xylem sap and leaves. *Journal of Experimental Botany* 58(2): 161–168.

Kumar, R., Jat, M. K. and Shankar, V. (2012). Methods to estimate irrigated reference crop evapotranspiration – a review. *Water Science and Technology* 66(3): 525–535.

Lipan, L., Issa-Issa, H., Moriana, A., Zurita, N. M., Galindo, A., Martin-Palomo, M. J., Andreu, L., Carbonell-Barrachina, Á. A., Hernandez, F. and Corell, M. (2021). Scheduling regulated deficit irrigation with leaf water potential of cherry tomato in greenhouse and its effect on fruit quality. *Agriculture* 11(7): 669.

Liu, Y., Snider, J. L., Bhattarai, A. and Collins, G. (2022). Economic penalties associated with irrigation during high rainfall years in the southeastern United States. *Agricultural Water Management* 272: 107825.

Lockhart, J. A. (1965). An analysis of irreversible plant cell elongation. *Journal of Theoretical Biology* 8(2): 264.

Losciale, P., Gaeta, L., Corsi, M., Galeone, C., Tarricone, L., Leogrande, R. and Stellaci, A. M. (2023). Physiological responses of apricot and peach cultivars under progressive water shortage: different crop signals for anisohydric and isohydric behaviours. *Agricultural Water Management* 286: 107825.

Loveys, B. R., Dry, P. R., Stoll, M. and McCarthy, M. G. (2000). Using plant physiology to improve the water use efficiency of horticultural crops. *Acta Horticulturae* 537: 187–197.

Maas, F. M. (2007). Dynamics of fruit growth in "Conference" pear as affected by root pruning, irrigation and climatic conditions. *Acta Horticulturae* 732: 555–563.

Mahmoud, M. M. A. and Fayad, A. M. (2022). The effect of deficit irrigation, partial root drying and mulching on tomato yield, and water and energy saving. *Irrigation and Drainage* 71(2): 295–309.

Manzi, M., Lodo, J., Rodrigo, M. J., Zacarias, L., Arbona, V. and Gomez-Cadenas, A. (2015). Root ABA accumulation in long-term water-stressed plants is sustained by hormone transport of aerial organs. *Plant and Cell Physiology* 56(12): 2457–2466.

Marino, S., Cocozza, C., Tognetti, R. and Alvino, A. (2015). Use of proximal sensing and vegetation indexes to detect the inefficient spatial allocation of drip irrigation in a spot area of tomato field crop. *Precision Agriculture* 16(6): 613–629.

Marino, G., Caruso, T., Ferguson, L. and Marra, F. P. (2018). Gas exchanges and stem water potential define stress thresholds for efficient irrigation management in olive (*Olea europea* L.). *Water* 10(3): 342.

Marsal, J., Mata, M., del Campo, J., Arbones, A., Vallverdu, X., Girona, J. and Olivo, N. (2008). Evaluation of partial rootzone drying for potential field use as a deficit irrigation technique in commercial vineyards according to two different pipeline layouts. *Irrigation Science* 26: 347–356.

Martínez-Gimeno, M. A., Castiella, M., Rüger, S., Intrigliolo, D. S. and Ballester, C. (2017). Evaluating the usefulness of continuous leaf turgor pressure measurements for the assessment of persimmon tree water status. *Irrigation Science* 35(2): 159–167.

Martínez-Gimeno, M. A., Zahaf, A., Badal, E., Paz, S., Bonet, L. and Pérez-Pérez, J. G. (2022). Effect of progressive irrigation water reductions on super-high-density olive orchards according to different scarcity scenarios. *Agricultural Water Management* 262: 107399.

Mattar, M. A., El-Abedin, T. K. Z., Alazba, A. A. and Al-Ghobari, H. M. (2020). Soil water status and growth of tomato with partial root-zone drying and deficit drip irrigation techniques. *Irrigation Science* 38(2): 163–176.

McAdam, S. A. M., Manzi, M., Ross, J. J., Brodribb, T. M. and Gomez-Cadenas, A. (2016). Uprooting and abscisic paradigm: shoots are the primary sources. *Plant Signaling and Behavior* 11: e1169359.

McCutchan, H. and Shackel, K. A. (1992). Stem-water potential as a sensitive indicator of water stress in prune trees (*Prunus domestica* L. cv. French). *Journal of the American Society for Horticultural Science* 117(4): 607–611.

Mekonnen, M. M. and Hoekstra, A. Y. (2011). The green, blue and grey water footprints of crips and derived crop products. *Hydrology and Earth System Sciences* 5: 1577–1600.

Melkonian, J. and Wolfe, D. W. (1995). Relative sensitivity of leaf elongation and stomatal conductance of cucumber plants to changes in leaf and soil water potentials. *Canadian Journal of Plant Science* 75(4): 909–915.

Meron, M., Tsipris, J., Orlov, V., Alchanatis, V. and Cohen, Y. (2010). Crop water stress mapping for site-specific irrigation by thermal imagery and artificial reference surfaces. *Precision Agriculture* 11(2): 148–162.

Michelena, V. A. and Boyer, J. S. (1982). Complete turgor maintenance at low leaf water potentials I the elongating region of maize leaves. *Plant Physiology* 69(5): 1145–1149.

Mingo, D. M., Bacon, M. A. and Davies, W. J. (2003). Non-hydraulic regulation of fruit growth in tomato plants (*Lycopersicon esculentum* cv. Solairo) growing in drying soil. *Journal of Experimental Botany* 54(385): 1205–1212.

Mingo, D. M., Theobald, J. C., Bacon, M. A., Davies, W. J. and Dodd, I. C. (2004). Biomass allocation in tomato (*Lycopersicon esculentum*) plants grown under partial rootzone drying: enhancement of root growth. *Functional Plant Biology* 31(10): 971–978.

Mitchell, P. D., van den Ende, B., Jerie, P. H. and Chalmers, D. J. (1989). Responses of `Barlett' pear to withholding irrigation, regulated deficit irrigation, and tree spacing. *Journal of the American Society for Horticultural Science* 114(1): 15–19.

Mohamadzade, F., Gheysari., M. and Kiani, M. (2021). Root adaptation of urban trees to a more precise irrigation system: mature olive as a case study. *Urban Forestry and Urban Greening* 60: 127053.

Morison, J. I. L., Baker, N. R., Mullineaux, P. M. and Davies, W. J. (2008). Improving water use in crop production. *Philosophical Transactions of the Royal Society of London. Series B, Biological Sciences* 363(1491): 639–658.

Murray, M. (1999). Lower Sacramento valley processing tomato irrigation field test results. *Acta Horticulturae* 487: 519–524.

Naor, A. (2000). Midday stem water potential as a plant water stress indicator for irrigation scheduling in fruit trees. *Acta Horticulturae* 537: 447–454.

Navarro, J. M., Pérez-Pérez, J. G., Romero, P. and Botía, P. (2010). Analysis of the changes in quality in mandarin fruit, produced by deficit irrigation treatments. *Food Chemistry* 119(4): 1591–1596.

Netting, A. G., Theobald, J. C. and Dodd, I. C. (2012). Xylem sap collection and extraction methodologies to determine in vivo concentrations of ABA and its bound forms. *Plant Methods* 8: 11.

Ohana-Levi, N., Derumigny, A., Peeters, A., Ben-Gal, A., Bahat, I., Katz., I., Netzer, K., Naor, A. and Cohen, Y. (2021). A multifunctional matching algorithm for sample design in agricultural plots. *Computers and Electronics in Agriculture* 187: 106262.

Ortuani, B., Facchi, A., Mayer, A., Bianchi, D., Bianchi, A. and Brancadoro, L. (2019). Assessing the effectiveness of variable-rate drip irrigation on water use efficiency in a vineyard in northern Italy. *Water* 11(10): 1964.

Ortuño, M. F., Alarcón, J. J., Nicolás, E. and Torrecillas, A. (2004). Interpreting trunk diameter changes in young lemon trees under deficit irrigation. *Plant Science* 167(2): 275–280.

Ortuño, M. F., García-Orellana, Y., Conejero, W., Ruíz-Sánchez, M. C., Alarcón, J. J. and Torrecillas, A. (2006). Stem and leaf water potential, gas exchange, sap Flow, and trunk diameter fluctuations for detecting water stress in lemon trees. *Trees* 20(1): 1–8.

Otarola, J., Fereryra, R., Aspigall, C., Zuniga, C., Selles, G., Silva, C. and Silva, H. (2014). Growth relationships between trunk and berry using linear variable differential sensors in *Vitis vinifera* "Crimson" and "Thompson Seedless". *Acta Horticulturae* 1038: 617–624.

Pappalardo, S., Consoli, S., Longo-Minnolo, G., Vanella, D., Longo, D., Guarrera, S., D'Emilio, A. and Ramírez-Cuesta, J. M. (2023). Performance evaluation of a low-cost thermal camera for citrus water status estimation. *Agricultural Water Management* 288: 108489.

Passioura, J. B. and Fry, S. C. (1992). Turgor and cell expansion – beyond the Lockhart equation. *Australian Journal of Plant Physiology* 19(5): 565–576.

Pazzagli, P. T., Weiner, J. and Liu, F. L. (2016). Effects of CO_2 elevation and irrigation regimes on leaf gas exchange, plant water relations, and water use efficiency of two tomato cultivars. *Agricultural Water Management* 169: 26–33.

Pedrero, F., Maestre-Valero, J. F., Mounzer, O., Alarcón, J. J. and Nicolás, E. (2014). Physiological and agronomic mandarin trees performance under saline reclaimed water combined with regulated deficit irrigation. *Agricultural Water Management* 146: 228–237.

Pérez-Pérez, J. G., Romero, P., Navarro, J. M. and Botía, P. (2008a). Response of sweet orange cv 'Lane late' to deficit-irrigation strategy in two rootstocks. II. Flowering, fruit growth, yield and fruit quality. *Irrigation Science* 26(6): 519–529.

Pérez-Pérez, J. G., Romero, P., Navarro, J. M. and Botía, P. (2008b). Response of sweet orange cv 'Lane late' to deficit-irrigation strategy in two rootstocks. I. Water relations, leaf gas exchange and vegetative growth. *Irrigation Science* 26(5): 415–425.

Pérez-Pérez, J. G., Robles, J. M. and Botía, P. (2009). Influence of deficit irrigation in phase III of fruit growth on fruit quality in 'lane late' sweet orange. *Agricultural Water Management* 96(6): 969–974.

Pérez-Pérez, J. G., Robles, J. M., García-Oller, M. I., Quinto, V., Frutos, J. M., Córdoba, F. and Botía, P. (2010). Aplicación de estrategias de riego deficitario controlado durante la fase de crecimiento rápido de fruto en naranjo navel 'Lane late'. X Simposium Hispano-Portugués de Relaciones Hídricas en las Plantas, 6–8 Octubre 2010, Cartagena, pp. 187–190.

Pérez-Pérez, J. G., Dodd, I. C. and Botía, P. (2012). Partial rootzone drying increases water-use efficiency of lemon Fino 49 trees independently of root-to-shoot ABA signalling. *Functional Plant Biology* 39(5): 366–378.

Pérez-Pérez, J. G., Robles, J. M. and Botía, P. (2014). Effects of deficit irrigation in different fruit growth stages on 'Star Ruby' grapefruit trees in semi-arid conditions. *Agricultural Water Management* 133: 44–54.

Pérez-Pérez, J. G. and Dodd, I. C. (2015). Sap fluxes from different parts of the rootzone modulate xylem ABA concentration during partial rootzone drying and re-wetting. *Journal of Experimental Botany* 66(8): 2315–2324.

Pérez-Pérez, J. G., Puértolas, J., Albacete, A. and Dodd, I. C. (2020). Alternation of wet and dry sides during partial rootzone drying irrigation enhances leaf ethylene evolution. *Environmental and Experimental Botany* 176: 104095.

Phogat, V., Skewes, M. A., Cox, J. W., Sanderson, G., Alam, J. and Šimůnek, J. (2014). Seasonal simulation of water, salinity and nitrate dynamics under drip irrigated mandarin (*Citrus reticulata*) and assessing management options for drainage and nitrate leaching. *Journal of Hydrology* 513: 504–516.

Powell, D. B. B. and Coggins, S. J. (1985). A portable Schölander-type pressure chamber for small-leafed cereals. *Agricultural and Forest Meteorology* 34(4): 277–284.

Price, L. E., Bacon, M. A., Young, P. C. and Davies, W. J. (2001). High-resolution analysis of tomato leaf elongation: the application of novel time-series analysis techniques. *Journal of Experimental Botany* 52(362): 1925–1932.

Puértolas, J., Johnson, D., Dodd, I. C. and Rothwell, S. A. (2019). Can we water crops with our phones? Smartphone technology application to infrared thermography for use in irrigation management. *Acta Horticulturae* 1253: 443–448.

Puértolas, J. and Dodd, I. C. (2022). Evaluating soil evaporation and transpiration responses to alternate partial rootzone drying to minimise water losses. *Plant and Soil* 480(1–2): 473–489.

Puglisi, I., Nicolosi, E., Vanella, D., Lo Piero, A. R. L., Stagno, F., Saitta, D., Roccuzzo, G., Consoli, S. and Baglieri, A. (2019). Physiological and biochemical responses of orange trees to different deficit irrigation regimes. *Plants* 8(10): 423.

Ramalan, A. and Nwokeocha, C. U. (2000). Effects of furrow irrigation methods, mulching and soil water suction on the growth, yield and water use efficiency of tomato in the Nigerian Savanna. *Agricultural Water Management* 45(3): 317–330.

Ramirez, D. R., Wehner, T. C. and Miller, C. H. (1988). Source limitation by defoliation and its effect on dry matter production and yield of cucumber. *HortScience* 23(4): 704–706.

Reid, J. B. and Renquist, A. R. (1997). Enhanced root production as a feed-forward response to soil water deficit in field-grown tomatoes. *Australian Journal of Plant Physiology* 24(5): 685–692.

Robles, J. M., Botía, P. and Pérez-Pérez, J. G. (2017). Sour Orange rootstock increases water productivity in deficit irrigated 'Verna' lemon trees compared with *Citrus macrophylla*. *Agricultural Water Management* 186: 98–107.

Romero, P., Navarro, J. M., Pérez-Pérez, J. G., García-Sánchez, F., Gómez-Gómez, A. G., Porras, I., Martínez, V. and Botía, P. (2006). Deficit irrigation and rootstock: their effects on water relations, vegetative development, yield, fruit quality and mineral nutrition of Clemenules mandarin. *Tree Physiology* 26(12): 1537–1548.

Romero, P., Gil-Muñoz, R., Del Amor, F. M., Valdés, E., Fernández-Fernández, J. I. and Martínez-Cutillas, A. (2013). Regulated deficit irrigation based upon optimum water status improves phenolic composition in Monastrell grapes and wines. *Agricultural Water Management* 121: 85–101.

Romero-Conde, A., Kusakabe, A. and Melgar, J. C. (2014). Physiological responses of citrus to partial rootzone drying irrigation. *Scientia Horticulturae* 169: 234–238.

Romero-Trigueros, C., Parra, M., Bayona, J. M., Nortes, P. A., Alarcón, J. J. and Nicolás, E. (2017). Effect of deficit irrigation and reclaimed water on yield and quality of grapefruit at harvest and postharvest. *LWT – Food Science and Technology* 85: 405–411.

Rosa, L. (2022). Adapting agriculture to climate change via sustainable irrigation: biophysical potentials and feedbacks. *Environmental Research Letters* 17(6): 063008.

Saab, I. N. and Sharp, R. E. (1989). Non-hydraulic signals from maize roots in drying soil-inhibition of leaf elongation but not stomatal conductance. *Planta* 179(4): 466–474.

Savic, S., Liu, F., Stikic, R., Jacobson, S. E., Jensen, C. R. and Jovanovic, Z. (2009). Comparative effects of partial rootzone drying and deficit irrigation on growth and physiology of tomato plants. *Archives of Biological Sciences* 61(4): 801–810.

Scholander, P. F., Bradstreet, E. D., Hemmingsen, E. A. and Hammel, H. T. (1965). Sap pressure in vascular plants - negative hydrostatic pressure can be measured in plants. *Science* 148(3668): 339.

Sharp, R. E., Osonubi, O., Wood, W. A. and Davies, W. J. (1979). Simple instrument for measuring leaf extension in grasses, and its application to study the effects of water-stress on maize and sorghum. *Annals of Botany* 44(1): 35–45.

Sharp, R. E. and Davies, W. J. (1985). Root-growth and water-uptake by maize plants in drying soil. *Journal of Experimental Botany* 36(9): 1441–1456.

Siqueria, J. M. and Ferreira, M. I. (2020). Optimisation of CGA technique: an algorithm to estimate low and reverse sap flow in roots. *Agricultural and Forest Meteorology* 294: 108129.

Sobeih, W. Y., Dodd, I. C., Bacon, M. A., Grierson, D. and Davies, W. J. (2004). Long-distance signals regulating stomatal conductance and leaf growth in tomato (*Lycopersicon esculentum*) plants subjected to partial rootzone drying. *Journal of Experimental Botany* 55(407): 2353–2363.

Stirzaker, R. J., Maeko, T. C., Annadale, J. G., Steyn, J. M., Adhanom., G. T. and Mpuisiang, T. (2017). Scheduling irrigation from wetting front depth. *Agricultural Water Management* 179: 306–313.

Tahi, H., Wahbi, S., Wakrim, R., Aganchich, B., Serraj, R. and Centritto, M. (2007). Water relations, photosynthesis, growth and water-use efficiency in tomato plants subjected to partial rootzone drying and regulated deficit irrigation. *Plant Biosystems* 141(2): 265–274.

Tardieu, F. and Simonneau, T. (1998). Variability among species of stomata control under fluctuating water status and evaporative demand: modelling anisohydric and isohydric behaviours. *Journal of Experimental Botany* 49: 419–432.

Tomos, A. D. and Leigh, R. A. (1999). The pressure probe: a versatile tool in plant cell physiology. *Annual Review of Plant Physiology and Plant Molecular Biology* 50: 447–472.

Topcu, S., Kirda, C., Dasgan, Y., Kaman, H., Cetin, M., Yazici, A. and Bacon, M. A. (2007). Yield response and N-fertiliser recovery of tomato grown under deficit irrigation. *European Journal of Agronomy* 26(1): 64–70.

Upadhyaya, A., Jeet, P., Singh, A. K., Kumari, A. and Sundaram, P. K. (2022). Efficacy of influencing factors in the decision-making of irrigation water pricing: a review. *Water Policy* 24(6): 963-979.

Varela-Ortega, C., Blanco-Gutierrez, I., Swartz, C. H. and Downing, T. E. (2011). Balancing groundwater conservation and rural livelihoods under water and climate uncertainties; an integrated hydro-economic modelling framework. *Global Environmental Change* 21(2): 604–619.

Wang, C., Shu, L., Zhou, S., Yu, H. and Zhu, P. (2019). Effects of alternate partial root-zone irrigation on the utilization and movement of nitrates in soil by tomato plants. *Scientia Horticulturae* 243: 41–47.

Wang, X. S., Meng, Z. J., Chang, X., Deng, Z., Li, Y. and Lv, M. (2017). Determination of a suitable indicator of tomato water content based on stem diameter variation. *Scientia Horticulturae* 215: 142–148.

Wang, Y. S., Liu, F. and Jensen, C. R. (2012). Comparative effects of deficit irrigation and alternate partial root-zone drying on xylem pH, ABA and ionic concentrations in tomato. *Journal of Experimental Botany* 63(5): 1907–1917.

Wei, G. Y., Xu, X. N., Cui, B. J., Zhang, M. Y., Liu, J., Wei, Z. H. and Liu, F. L. (2023). Effect of partial root-zone irrigation on plant growth, root morphological traits and leaf elemental stoichiometry of tomato under elevated CO_2. *Agronomy* 13(12): 3069.

Westgate, M. E. and Boyer, J. S. (1985). Osmotic adjustment and the inhibition of leaf, root, stem and silk growth at low water potentials in maize. *Planta* 164(4): 540–549.

Whalley, W. R., Binley, A., Watts, C. W., Shanahan, P., Dodd, I. C., Ober, E. S., Ashton, R. W., Webster, C. P., White, R. P. and Hawkesford, M. J. (2017). Methods to estimate changes in soil water for phenotyping root activity in the field. *Plant and Soil* 415(1): 407–422.

Wu, K., Desesquelles, H., Cockenpot, R., Guyard, L., Cuisiniez, V. and Lambot, S. (2022). Ground-penetrating radar full-wave inversion for soil moisture mapping in trench-hill potato field for precise irrigation. *Remote Sensing* 14(23): 6046.

Zeff, H., Kaczan, D., Characklis, G. W., Jeuland, M., Murray, B. and Locklier, K. (2019). Potential implications of groundwater trading and reformed water rights in Diamond Valley, Nevada. *Journal of Water Resources Planning and Management* 145(6): 05019009.

Zegbe, J. A., Behboudian, M. H. and Clothier, B. E. (2004). Partial rootzone drying is a feasible option for irrigating processing tomatoes. *Agricultural Water Management* 68(3): 195–206.

Zegbe, J. A., Behboudian, M. H. and Clothier, B. E. (2006). Responses of "Petopride" processing tomato to partial rootzone drying at different phenological stages. *Irrigation Science* 24(3): 203-210.

Chapter 7

Managing energy demands in irrigated agriculture

J. A. Rodríguez Díaz, A. Mérida García, R. González Perea and I. Fernández García, University of Córdoba, Spain*

1 Introduction

Irrigated agriculture plays an essential role in securing the food demand of a growing world population. In fact, the current global irrigated area is estimated to be approximately 300 Mha, almost double that of 1960, and uses 70% of the world's available freshwater (Alexandratos and Bruinsma, 2012). However, this area represents only 20% of the total global cultivated area, although it produces 40% of total food production. However, the areas irrigated vary significantly between countries and regions depending on local agroclimate conditions and the importance of agriculture to rural economies. For example, the Mediterranean basin is a clear example of the importance of irrigation. While in Europe, on average, agriculture uses 33% of water resources, for selected countries in Mediterranean Europe, this percentage rises to 80% (Collins, 2012), due to climatic conditions with virtually no rainfall in the central months of the year coupled with high evapotranspiration (ET) rates.

* Corresponding author: jarodriguez@uco.es.

http://dx.doi.org/10.19103/AS.2023.0123.10

Given the importance of irrigation for food production, but at the same time the high dependence on dwindling water resources, in countries such as Spain, major modernisation programmes have been undertaken in recent years to improve the efficiency of water use. These modernisations have primarily consisted of replacing open-channel gravity-based networks with piped pressurised networks. Thus, surface irrigation has been progressively replaced by pressurised systems such as drip or sprinkler irrigation. Spain is the clear example where approximately 1 Mha of the total 3.5 Mha irrigated area in the country has been modernised in recent years (MAGRAMA, 2010). As a result, the area under gravity irrigation has decreased to 30%, while pressurised irrigation systems have risen from 41% 20 years ago to 70% today. According to Tarjuelo et al. (2015), the major advantages of modernised systems include

- reduction of water applied to crops, which also implies a reduction of drainage water, which may carry a significant amount of dissolved agrochemicals;
- increased productivity per unit of water and sometimes per unit of irrigated area;
- possibility of changing to new crop rotations, of higher economic value;
- automation of the system. Increased possibilities to use remote control and remote telemetry systems that facilitate improved irrigation management; and
- improved quality of life for farmers and greater attraction to incorporate the new generation of young people into the sector. This is helping to avoid land abandonment and desertification in rural areas.

However, this commitment to modernisation has led to a notable increase in the efficiency of water use, but, on the other hand, it has led to a much greater dependence on energy for irrigation. In fact, Corominas (2010) estimated that while water use per unit of irrigated area in Spain had fallen by 21% at the national level between 1950 and 2007, energy demand had increased by 657%. Thus, irrigation today is very water-efficient, but it is also highly energy-dependent.

2 Water and energy use in irrigated agriculture: modernisation impacts

The transformation processes involved in converting from open channels to pressurised networks have significantly impacted on water and energy use in irrigated agriculture and several studies have analysed the consequences. For example, Fernández García et al. (2014c) evaluated the effects of the modernisation process on water and energy use in five irrigation districts from southern Spain (Bembézar Margen Izquierda; Bembézar Margen Derecha; Sector BXII; Genil Margen Derecha and Guadalmellato) in which upgrades of the irrigation system had been implemented. The total irrigated area of these irrigation districts was 38 285 ha, which represented 11% of the total area being modernised in Andalusia (southern Spain) at that time (Lopez-Gunn et al., 2012). In their study, indicators relating to water use and energy requirements, among others, for the period both prior to modernisation (1996–2002) and for a period after implementing pressurised irrigation networks (2010–12) were determined and analysed. They observed a large reduction in annual irrigation water supply of 1693 m^3ha^{-1} (23%) on average due to the adoption of more efficient water distribution and irrigation systems. Conversely, the energy costs increased considerably, accounting for up to 50% of the total water costs (Table 1). The significant increase in energy cost was due to the implementation of pressurised irrigation networks, in which energy input by pumps is required in contrast to open-channel distribution systems which rely on gravity. In addition, the rise in electricity costs also explained the high energy cost for these irrigation networks. Over the period studied, the total irrigation costs, consisting of water tariffs, maintenance and energy, have risen from 22% to 189% in these irrigation districts. This has significantly impacted on the viability of production and challenged the economic profitability of individual agricultural holdings.

Although pressurised irrigation networks often result in higher water use efficiencies compared to open-channel systems, their energy requirements and the continued rise in electricity cost have triggered interest and demand for new measures focused on reducing the amount of energy used for irrigation in these systems. One consequence is that farmers often choose to apply irrigation rates that are much lower than the theoretical needs for their crops. This is due to the fact that, given the low profitability of certain extensive crops, the farmers cannot even cover the energy costs and therefore decide to apply strict deficit irrigation strategies (Rodríguez Díaz et al., 2011). Thus, the implementation of strategies that can lead to joint optimisation of both water and energy use is critical to ensure both the economic and environmental sustainability of agriculture dependent on these modernised irrigation systems.

3 Carbon footprint of irrigated agriculture

The increase in water use efficiency has come at the expense of increased energy consumption, which highlights how there is a close relationship between the use of both resources. It is therefore necessary not to consider the two resources independently but rather to assess the water–energy nexus, which implies the optimisation of their simultaneous use (Rodríguez-Díaz et al., 2011). So far, the impacts of irrigation modernisation have largely been assessed according to

Table 1 Water costs in five irrigation districts (BMI, BMD, BXII, GMD and GU) during pre-modernisation process, characterised by open-channel distribution systems (1996–2002), and during post-modernisation process, with pressurised irrigation networks (2010–12)

Irrigation district	BMI		BMD		BXII		GMD		GU	
Indicator	Pre	Post	Pre	Post	Pre	Post	Pre	Post	Pre	Post
C_c, €ha^{-1}	49.9	70.0	50.2	66.9	87.5	89.0	64.5	66.2	88.3	66.9
C_M, €ha^{-1}	43.9	77.0	70.5	76.5	69.0	95.6	42.9	96.1	80.1	106.4
C_E, €ha^{-1}	0.0	124.3	2.9	140	115.4	147.6	*	*	14.1	48.9
C_T, €ha^{-1}	93.9	271.3	123.6	283.5	271.9	332.2	107.4	162.3	182.4	222.2
C_E/C_T, %	0	46	2	49	42	45			8	22

*Not available.BMD, Bembézar Margen Derecha; BMI, Bembézar Margen Izquierda; BXII, Sector BXII; CC, water agency tariff cost; CE, energy cost; CM, maintenance cost; CT, total water cost; GMD, Genil Margen Derecha; GU, Guadalmellato.
Source: Fernández García et al. (2014c).

their water savings. However, it is also necessary to consider the increased demand for energy in terms of both cost, which directly affects farmers, and the environmental impact of higher CO_2 emissions. In fact, it is estimated that the whole agricultural sector, not only irrigation, is responsible for one-fifth of all CO_2 emissions, and this proportion has increased in recent years. Between 2001 and 2011, annual emissions from agriculture alone increased by 14% from 4684 to 5335 $MtCO_2e$ per year (Tubiello et al., 2014). This is particularly important, given that agriculture will be one of the sectors most affected by climate change, with higher temperatures and more frequent extreme events such as droughts and floods. In the particular case of irrigation, higher ET rates, together with changes in rainfall distribution, could lead to significant increases in irrigation demand. Rodríguez Díaz et al. (2007) estimated for typical crops grown in southern Spain that irrigation requirements would increase by 20% by 2050, which would lead to higher water resources and energy needs to meet irrigation demand.

Daccache et al. (2014) estimated, using Food and Agriculture Organization of the United Nations (FAO) data, the CO_2 emissions from irrigation pumping for all countries in the Mediterranean basin. The research showed that Spain, which has the largest area of pressurised irrigation, was by far the country with the highest emissions, despite not having the largest irrigated area. It also showed how in the hypothetical case that if the total irrigated area in the basin were to be modernised to pressurised systems, emissions would increase considerably (+135%). Table 2 shows the emissions for the seven countries with the highest emissions for both scenarios, together with the equivalent values for the basin as a whole.

As a consequence, there has been significant interest in switching to renewable energies in the irrigation sector as it could provide the opportunity

Table 2 Estimated CO_2 emissions (t CO_2e) due to water pumping for irrigation in seven selected Mediterranean countries, for the current situation and for a hypothetical scenario in which all irrigation would be pressurised

	Current situation		Hypothetical scenario where all irrigation is pressurised	
Country	t CO_2e	t CO_2e/mm^3	t CO_2e	t CO_2e/mm^3
Spain	321 584	41	429 843	55
Turkey	236 805	21	627 080	56
Syria	220 096	20	766 469	69
Libya	174 095	125	288 800	208
France	156 223	71	135 148	61
Morocco	140 437	33	432 266	102
Italy	92 276	31	188 205	64
Total	1 780 170	29	4 199 412	68

Source: Daccache et al. (2014).

to reduce greenhouse gas (GHG) emissions. In this way, the partial or even total supply of the energy demand for water pumping, previously supplied from the electricity grid or from diesel generators, could be supplied by other alternative energy sources with low GHG emissions. This is the case for photovoltaic (PV) energy, for example, with virtually zero environmental impact during the operating period since its installation (Mérida García et al., 2019). However, it is also necessary to analyse the GHG emissions linked to the manufacturing stage for the equipment, as well as to its end-of-life disposal as waste. The total impact of renewable energy supply compared to the use of traditional energy sources has been quantified in numerous life cycle assessment (LCA) studies (Mérida García et al., 2021a), where an inventory analysis of all materials and flows involved in the production and operation of the equipment has been carried out, allowing for the impact estimation for a wide variety of categories, such as global warming potential, among others, quantified as g CO_2 equivalent emitted. However, LCA studies applied to energy supply in agricultural irrigation settings are still quite limited and usually applied to small-scale case studies.

4 Factors determining energy demand

For both economic reasons, which affect the farmer's income, and for environmental reasons, associated with emissions and their contribution to climate change, it is necessary to reduce the energy demand for irrigation. This energy demand can be estimated using the following equation:

$$E = \frac{\gamma V H}{\eta} \tag{1}$$

where γ is the water-specific weight, V is the volume of water supplied for irrigation, H is the pressure head and η is the pumping efficiency.

Therefore, if energy demand is to be reduced, it would be necessary to act on all the variables that constitute Equation 1, either by decreasing the numerator or by increasing the denominator. Each of these is briefly considered later:

- Reduction in the volume used for irrigation (V). Precision irrigation systems, with high efficiency and low losses, imply there can be reductions in the volume of water pumped. However, appropriate irrigation scheduling is also necessary, in which the right amount of water is applied to the plant at the right time when needed.
- Reduction of the manometric height (H). Manometric head depends on factors that are difficult to influence, such as the difference in elevation

between the source and the point of water consumption, but also on other factors such as the pressure requirements of the irrigation system. It is generally possible to reduce this head by changes in the management of the distribution network and migration to lower-pressure irrigation systems.

- Pumping performance (η). This can be improved by installing variable speed and high-efficiency pumps.

However, these interventions represent relatively simple and basic approaches to the fundamental problem. Generally, in irrigation districts, the irrigated cropping covers large areas, often several thousand hectares, and water is supplied through large distribution networks with hundreds of kilometres of pipelines and large pumping stations (Fig. 1), which allow water to be supplied to all users delivering optimal water supply conditions of both flow rate and pressure. In addition, these networks are often organised based on a system termed 'on demand', where water is continuously available to all farmers, which generates great variability in flow rates and pressure fluctuations. All these conditions understandably complicate the overall management of these systems, making it difficult to decide on what measures might be most relevant to reduce energy consumption, and therefore requiring detailed analyses that are developed specifically for each irrigation district.

Figure 1 Pumping station in the Valle Inferior del Guadalquivir irrigation district (southern Spain).

5 Measures to reduce energy dependency in irrigation distribution networks

Given the difficulty of optimising energy consumption in such complex systems, several authors have been working on the development of measures to reduce energy dependence in these types of pressurised networks. Among the various measures being evaluated, network sectoring, critical point detection and the optimum pump station design and management have shown promising results. However, none of these measures implemented in isolation ensure a significant reduction in energy demand, with optimal results only generally being obtained through a combination of interventions or measures. A brief description of each of these options is provided later.

5.1 Sectoring

Network sectoring consists of the establishment of irrigation turns named 'sectors'. Thus, instead of providing a water supply whenever the farmer decides to irrigate, the hydrants in an irrigation district are grouped into sectors according to their energy requirements, with each hydrant then scheduled to irrigate at a specific time. When irrigation is organised on-demand and farmers are free to irrigate whenever they want, they tend to irrigate at the same time, leading to peaks in water demand and hence, to high energy consumption. Conversely, irrigation management by sectors enables the system to adapt the energy demand to each sector, thus leading to significant energy savings (Rocamora et al., 2013). However, the determination of sectors in a network is not trivial since a number of factors such as the topography and the topology of the network, the energy requirements of the irrigation system and the crop water needs must all be considered (Carrillo Cobo et al., 2011). Thus, several methodologies focused on the determination of sectors in branched networks with one or more water supply points have been proposed (Rodríguez Díaz et al., 2009; Jiménez-Bello et al., 2010; Moreno et al., 2010; Carrillo Cobo et al., 2011; Navarro Navajas et al., 2012; Mérida García et al., 2018). The analysis of network sectoring in these studies has shown significant energy reductions, in some cases accounting for more than 30% savings. This strategy does imply a reduction in the time available for farmers to irrigate their fields. However, the energy reductions achieved by rearranging the irrigation management justify the loss of this freedom to irrigate. It should be mentioned that this measure becomes especially advantageous in irrigated areas with centralised irrigation management and monoculture cropping systems, such as olive trees orchards in southern Spain.

5.2 Critical point detection

Critical points are specific hydrants with much higher energy requirements than the remaining hydrants making up the irrigation network. The factors which contribute to the elevated energy demand for these hydrants are the differences in elevation and the distance between these hydrants and the source point, the existence of undersized pipes in the distribution network and/or the pressure requirements for in-field irrigation systems. Rodríguez Díaz et al. (2009) analysed the energy performance of hydrants in an irrigation network located in southern Spain. They detected five critical hydrants (out of 85 hydrants) in the network for which their activation involved an additional pressure head of 10 m at the pumping station. Thus, when all hydrants were open across this irrigation network, the pressure head at the pumping station to ensure the service pressure at all hydrants was 80 m.

To improve the performance of critical hydrants measures such as the installation of booster pumps or the replacement of undersized pipes can be implemented. However, optimisation models are also required to determine the number of critical hydrants in a network and to propose the most effective measure to reduce the energy consumption at these points. Accordingly, the Combined Optimisation and Performance Analysis Model (COPAM) enables the evaluation of individual hydrant performance considering different demand conditions obtained by Monte Carlo simulations (Lamaddalena and Sagardoy, 2000). Rodríguez Díaz et al. (2012) developed the Water and Energy optimisation by Critical Points control model which focused on determining the critical points in a network by dimensionless coordinates and the proposal of measures to improve the performance of critical hydrants. An extended version of this model for irrigation networks with several water supply points was proposed by Fernández García et al. (2014a). The estimated energy savings by applying these models accounted for more than 30% in the studied irrigation networks. Using this approach, Fernández García et al. (2016) developed a model for network rehabilitation focused on the optimal redesign of the network considering the replacement of undersized pipes and the optimal operation of the pumping station to reduce operational costs. In contrast to network operation by sectors, critical points control allows the 'on-demand' operation of the irrigation network to be maintained.

5.3 Optimum pumping station design and management

The optimum design and management of pumping station plays an important role in ensuring efficient energy use. In a study related to water and energy use in ten irrigation networks in southern Spain, Rodríguez Díaz et al. (2011) detected that the efficiency of the pumping station was 58% on average, but as low as 30% in some cases. The reason for this low efficiency was explained by the mismatch between the pumping station, which was usually designed to operate at the peak demand condition, and the fluctuating demand conditions across the district, which were especially frequent at the beginning and at the end of the irrigation season.

To improve the efficiency of the pumping station across a wide range of required flow rates, Moreno et al. (2009) proposed the installation of variable speed drives and the installation of low-power pumps to work under reduced flow conditions. Since measures such as network sectoring and critical points control involve changes in both flow and pressure head conditions, efficient management of the pumping station is also required. Hence, Fernández García et al. (2014b) evaluated the joint combination of a sectoring strategy and the optimal management of the pumping station through the inclusion of variable speed drives. The analysis carried out in an irrigation network in southern Spain showed potential energy savings of +26% considering the network operation in two sectors and including two variable speed drives. They also highlighted that in this case study example, the network management in two sectors without optimising the operation of the pumping station would only have led to energy saving of 2%.

6 Renewable energy sources

The price of electricity and fossil fuels has followed a marked upward trend in recent years and is expected to continue to increase, which has led to a situation in which a significant proportion of the cost of water is now devoted to meeting energy consumption. At the same time, economic issues are compounded by growing environmental concerns linked to the impacts of agriculture on atmospheric pollution and global warming. This has led to an urgent demand for the search for alternative and sustainable energy sources that not only guarantee the supply and profitability of farms, whilst also maintaining current production levels. In this context, the use of solar PV energy is presented as an option with great potential for the Mediterranean basin, where irradiance levels throughout the year are high. Renewable energies are thus becoming a new alternative to achieve low-cost, low-carbon water supply and energy availability in areas not reached by the electricity grid. In this context, solar PV energy and energy recovery through microturbines, which allow the generation of electricity from localised excess pressure in water distribution networks, are considered the best options.

6.1 Photovoltaic energy

Solar PV energy has great advantages over other renewable energies for use in irrigation water supply. On the one hand, solar radiation and crop ET follow parallel curves. In other words, solar is an ideal energy source for irrigation because it will be available just when the plant needs water. On the other hand, the significant reduction in the cost of solar panels in recent decades has made their cost very competitive compared with other energy sources. In addition, PV irrigation solutions provide electricity access even to off-grid areas, replacing the use of diesel generators and thus, avoiding GHG emissions linked to fossil fuel combustion (Mérida García et al., 2019). However, the main drawback of this energy source lies in its high dependence on climatic variables, since energy production depends on varying irradiance levels (López-Luque et al., 2015). This not only generates a great variability in the availability of power throughout the different months of the year but also entails great difficulty in managing daily irrigation scheduling. Thus, on a clear day, the daily irradiance follows a normal curve, reaching maximum values at solar midday. However, on a day with cloudy intervals, the irradiance curve over the course of the day fluctuates considerably (Fig. 2). This discontinuous nature of the power generated makes it difficult to manage irrigation installations, so new techniques and methodologies are required to cope with the intermittency of supply and PV management.

Several authors have worked on the integration of solar PV energy as a source of energy supply in irrigation networks. Hamidat et al. (2003) used PV energy as a supply source in small plots in the Sahara region. Carrillo-Cobo

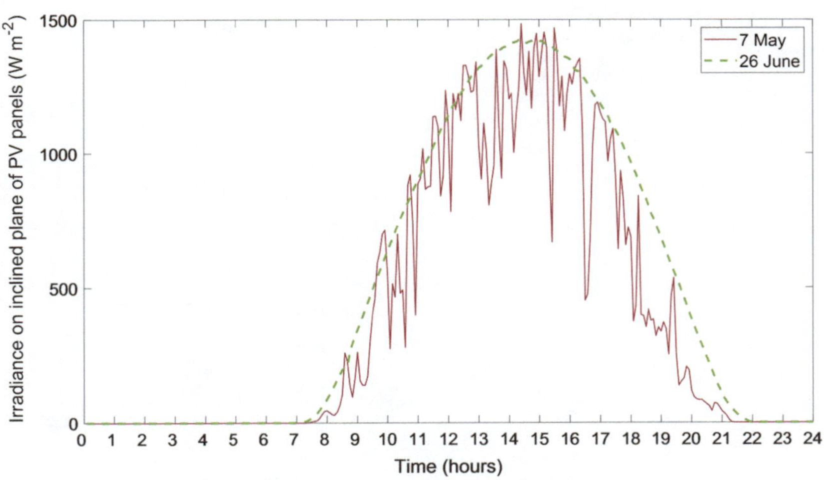

Figure 2 Example of variation in irradiance levels during the day 'with' (07 May) and 'without' (26 June) cloudy intervals in southern Spain (Córdoba).

et al. (2014) combined, in a theoretical approach, the sectoring of irrigation water distribution networks with solar energy as a partial source of supply, which showed an estimated 72% potential savings in energy costs. Hybrid or combined energy supply systems for irrigation based on renewable energies have also been an important focus of research, with the objective of compensating for the variability of energy production due to its dependence on climatic variables. For example, Vick and Almas (2011) combined PV energy with wind turbines to increase the irrigation system's autonomy. Rodríguez-Gallegos et al. (2017) proposed a hybrid solution including PV technology, batteries and the use of a diesel generator, looking for reduced total costs and the fulfilment of the load demand. Yahyaoui et al. (2017) developed a model for energy supply management using PVs supported by batteries to fill a reservoir that fed the irrigation network. In other cases, the smart management of the PV-powered irrigation network has been explored, through the development of specific models, already tested especially for permanent crops grown on heavy soils, with soil water retention capacity playing an important back-up system role. This was demonstrated by Mérida García et al. (2018), who developed a smart PV irrigation system based on a real-time synchronisation between the PV power production and the water and energy demands within the irrigation network. An example of this synchronisation is shown in Fig. 3, in which the PV power production during the day with cloudy intervals, as well as the activation sequence of three sectors that constitute the irrigation network, is represented.

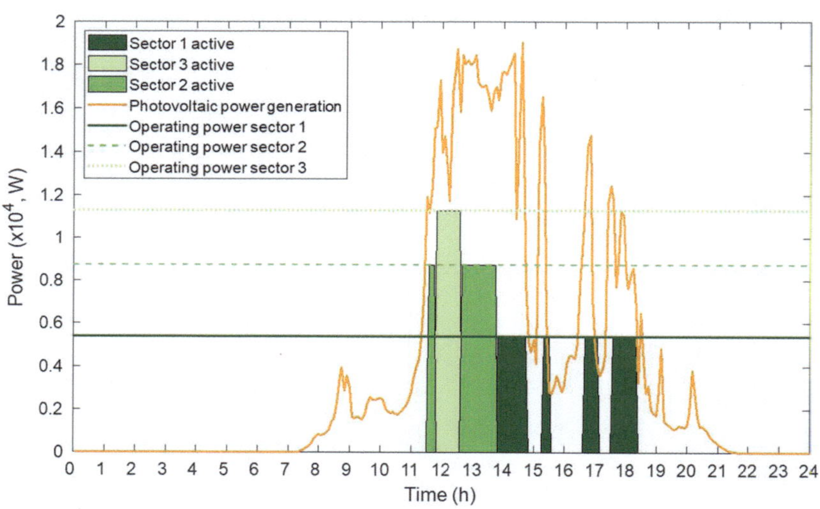

Figure 3 Example of the irrigation activation sequence in a three-sector irrigation network powered by solar photovoltaic (PV) energy during the day with the presence of clouds in the experimental PV Irrigation System of the University of Cordoba (southern Spain).

Figure 4 Photovoltaic (PV) plant, irrigation pond and irrigated olive orchard as part of the smart PV irrigation system on the experimental farm at the University of Cordoba (2018, southern Spain).

Thus, this approach can be used to model and evaluate the instantaneous PV power production and then decide which sector should start irrigating and for how long it should remain active, prioritising the activation of sectors with the highest power demand, with the intervals with the highest PV production. The model also considered soil moisture content, as a support system for days with low irradiance. The system was tested in a PV drip irrigation system of a 13.4 ha-intensive olive orchard in Southern Spain, with a 15-kW PV plant (Fig. 4), being able to supply the full water allocation. In this case, the water was pumped from the reservoir to the irrigation field (1 km away) using PV energy, with the network organised in three independent sectors, and activation controlled by independent electro-valves. The simulation of an irrigation season (with approximately 600 ha irrigated area) with actual climate data proved the energy autonomy of the system, which satisfied the irrigation requirements of the crop during the whole season by using PV energy as the only energy source. In this case, the substitution of the use of the electricity grid by PV energy avoided the emission of 1.2 t CO_2 equivalent each year, considering an average ratio of kg of CO_2 emissions per kWh of energy of 0.264.

However, to date, most previous work has still been based on theoretical studies or, in some cases, only with small- to medium-scale applications. Therefore, there is still extensive unexplored potential for smart management of large irrigation networks organised on demand and powered by PV energy as the main energy source. The design of a solar PV irrigation system should include (1) the sectorisation and hydraulic analysis of the network, (2) the estimation of the crop irrigation requirements, (3) the characterisation of soil and (4) the irradiance availability.

The irrigation network sectorisation and its subsequent organisation must consider the total daily hours with solar irradiance levels above 1000 W/m^2, that is the time during which the solar PV plant will produce at least its nominal power (function of the nominal power of the PV panel and the number of panels). These total available hours, together with the irrigation time required by the crop, will limit the maximum number of irrigation sectors (or group of sectors) that could be irrigated in a single day. The hydraulic analysis of the network will provide the head pressure and flow requirements (power demand) to ensure uniform and quality irrigation downstream of the pumping system. Pressure head and flow requirements will depend, mainly, on topography, the covered area, and the specific characteristics of the irrigation network, such as pipe diameters and roughness, type of irrigation and emitters and the number of emitters. On the other hand, the irrigation time will be defined by crop characteristics and the ambient meteorological conditions (ET and precipitation), although it should be also recommended to consider the soil water balance for an optimal irrigation scheduling.

Once the pressure head and flow demands are known, the power demand for the different irrigation sectors or groups of sectors can be estimated, according to which the PV plant is then sized, under the most restrictive conditions (considering the most demanding sector and the peak irrigation water demand). Finally, the irradiation levels will determine the real instant energy production of the system throughout the irrigation season.

The PV system design will also need to define the size and space requirements for its installation which, depending on the total required PV surface area, may result in the loss of arable land. This area will depend not only on the required nominal PV power but also on the performance of the selected PV panels. In small to medium installations, it is possible to use the roof of farmhouses or warehouses to fix the panels by means of light metal structures. Another option is to install the panels on a floating structure on the irrigation reservoirs, where these exist. This last option also reduces evaporation losses from the water-free surface in the reservoir. However, when the required surface area is larger, the options are reduced to installing the panels on the ground. In that way, so-called agrivoltaic production sustains the design of PV installations with a dual objective to not only achieve self-energetic autonomy of the farm but also to produce excess energy for sale, as a complementary additional source of income (Reca-Cardeña and López-Luque, 2018). Thus, the agrivoltaic concept can be defined as the amalgamation between food and energy production in the same field. In these cases, it is important to study how the PV installation affects the productivity of the crop, mainly due to partial shading (Casares de la Torre et al., 2022).

In the initial applications, experiences with solar irrigation focused predominantly on small farms with very low power used to drive submersible pumps in boreholes. The high power required to supply water in large networks, in addition to the high variability in flow rates for such 'on demand' systems was a major problem that was difficult to address. However, in recent years, there are areas where solar plants have been installed to reduce their energy

Figure 5 Solar plant installed in the Valle Inferior del Guadalquivir irrigation district (southern Spain).

dependency. These include, for example, the 'Valle Inferior del Guadalquivir' and 'Bembézar Margen Derecha' which were installed in 2019 and 2020 with solar plants of 6 MWp and 5 MWp, respectively (Fig. 5).

One of the main successes of these solar plants is that the irrigation districts have not been disconnected from the national grid. In this way, conventional energy always remains as a backup supply, for times when there is insufficient solar radiation or night-time irrigation. In addition, being connected to the grid allows any surplus solar energy to be sold when it is not needed for irrigation. This net balance means that the amortisation or return on investment for this type of installation investment can be very short, usually no more than 6 years.

Today, electricity consumption from the national electricity grid has been reduced by approximately 50% in both irrigation districts. This is very impressive given that most networks are still organised on demand and many users continue to prefer to irrigate at night. This approach is quite reasonable given that before the PV investment, farmers were recommended to irrigate at night to avoid evaporation and to take advantage of the significantly cheaper conventional electricity night tariff. However, the irrigation districts are now looking at how to incentivise farmers to irrigate mainly during the day to maximise the use of the solar plant.

6.2 Energy recovery with microturbines

Energy recovery through micro-hydropower (MHP) is an alternative opportunity to generate green energy from excess pressure in water distribution networks. Water distribution systems are optimally sized to meet the pressure requirements at different water supply points. However, excess pressure locations usually appear in specific areas of the network, which are then normally addressed by the installation of pressure-reducing valves, to avoid pipe bursts and leakage (Gupta et al., 2017). The installation of micro-turbines or pumps as turbines (PAT) can contribute to mitigating the excess pressure generating at the same time green energy, that can be used on-site. The use of PATs is based on operating a pump in reverse, which has been shown to be a cost-effective solution for energy recovery in water distribution systems (Crespo Chacón et al., 2019). For small capacities, PATs are much cheaper than conventional turbines and their payback periods are very short, making them very viable investments. On the other hand, they have the limitation that their best efficiency point is produced for a rather narrow flow range, so special care must be taken in their design to ensure they always operate under optimal conditions from flow and pressure perspectives.

The potential of this technology has been examined within the water-intensive sector, such as pressurised water distribution networks for drinking water supply (McNabola et al., 2011; Fernández García and Mc Nabola, 2020), in wastewater systems (Power et al., 2014), the processing industry (Mérida García et al., 2021b) and for irrigation (Crespo Chacón et al., 2020). For irrigated agriculture, flow rates are commonly variable over time, and sometimes zero during periods of the year when irrigation is not required. This was addressed in different studies that focused on the development of methodologies based on probabilities and flow distribution in real networks (Pérez-Sánchez et al., 2016; Crespo Chacón et al., 2019). The main advantage is that the energy production of the turbine or PAT coincides with the operation period for the irrigation network so that energy generated can then be used on-site during the irrigation season.

However, since irrigation networks cover large areas, it is very common that the optimal points for energy recovery are located in areas far from the power grid. This makes it difficult to feed this energy into the grid and sell it. On the other hand, pico-scale energy recovery installations (<5 kW) could be used to generate sufficient energy to monitor operation parameters throughout the network, feeding energy demands of a sensor network (i.e. pressure sensors for monitoring and control of the correct operation of the hydraulic installation). Also, it is very common for farms to require small amounts of energy to supply different equipment such as filters, fertigation equipment and irrigation programmers. This energy is usually supplied by diesel generators, which are expensive and dependent on fossil fuels. Thus, PATs represent an alternative to these generators, providing clean and low-cost energy at the hydrant level.

Such an example of this approach is the pilot MHP plant installed as part of the Redawn Project (www.redawn.eu) in Palma del Río (South of Spain), in which a diesel generator (6 kVA) was substituted by a 4 kW-PAT, supplemented by a small solar PV installation (660 W) and a battery bank (10.6 kW). The PV energy and batteries supplied the energy demands outside the irrigation season when no power can be recovered by the PAT. The PAT was installed in a by-pass at the hydrant level (Fig. 6) and operates with the head pressure of the incoming water, to ensure the continuous operation of the irrigation network downstream

Figure 6 Pumps as turbines installed (left) and render of the energy recovery installation in the pilot plant in Palma del Rio (Southern Spain) as part of the Redawn Project.

during PAT repairs or maintenance. The hybrid system was installed to supply the energy demands coming from the off-grid fertigation system in an irrigation network of a 170-ha walnut farm. This system was able to fully satisfy the energy demands of the fertigation system, replacing the use of the diesel generator, with economic and environmental savings equating to €2258 and 8.4 t CO_2 equivalent per year (based on 2019 irrigation season data), respectively, for an approximate total demand of 2400 irrigation hours between May and September. However, the advantages were not only in the form of savings in diesel consumption but also in a more reliable, stable and less noisy system. In addition, the farm business was very positive about the impacts of reducing its carbon footprint as part of its marketing strategy for its agricultural produce.

At the network level, previous studies related to the irrigation sector have identified significant potential for energy recovery with turbines and PATs based on analysis from several case studies. For example, García Morillo et al. (2018) identified an annual potential energy recovery of 270 MW in one pressurised network in a 400 ha irrigation district in southern Spain, with a potential saving of 108 t CO_2 equivalent. At the regional scale, Crespo Chacón et al. (2020) predicted the energy recovery potential for MHP installed in on-demand pressurised irrigation networks, with 21 GWh energy recovery potential identified in 164000 ha in southern Spain. These studies highlight the scope for significant energy consumption reductions and the economic and environmental benefits associated with MHP energy recovery within the agricultural sector.

7 Future trends and the role of artificial intelligence

As described earlier, irrigation networks are subject to a large variability in the flows demanded, especially when they are organised to be operated 'on demand'. This creates problems for the scheme managers, as they cannot predict the network situation and its pattern of demand in advance, which would be important for aspects such as power contracting, pumping station management and maintenance works. The uncertainty associated with irrigation water demands and the water–energy costs have been raised in many irrigation districts (Fernández García et al., 2014c) causing serious challenges for daily management. However, the demand for irrigation is highly variable as farmers are not only guided by the needs of the crop but also by other cultural and market aspects. For these reasons, the demand during the irrigation season and during the day can be very variable, with moments of high and low demand concentration. This large variability contrasts with other sectors such as urban water supply, where water consumption is much more predictable and stable. To solve these problems, new trends are moving towards the use of artificial intelligence (AI) techniques to try and predict energy–water demand and thus facilitate improved decision-making by irrigation network managers.

In Spain, during the months of June and July (peak irrigation season), most of the irrigation occurs during periods with expensive tariffs. Recently, to minimise energy costs, irrigation district managers have had the option to hire electric energy through different modalities with the aim of adjusting the electrical power contracted to the actual power required by the water distribution network. One of these modalities allows a central purchasing body to have its own strategy of purchasing power (buying futures market, pass through and intraday markets) and users (such as water user associations, WUA) must pay in advance their estimated electricity consumption. Therefore, to improve the daily management of the water and energy use in a WUA, managers need tools to estimate accurately, and as soon as possible (previous hours, days, weeks), the real daily water demand and energy consumption of the entire irrigation network. Better control of the irrigation process as well as better management of the pressurised irrigation network is essential to convert irrigation to be a more precise activity where efficiency of all the resources is maximised. This highlights the need to improve efficiency in the water–energy nexus, which is essential for the economic, social and environmental development of the sector.

Traditionally the decision-making processes in the agricultural sector were based on the farmer's experience that typically does not represent the optimum alternative in terms of water and energy use. However, the recent revolution of Information and Communication Technologies, as well as the development of new low-cost sensors capable of automatically collecting thousands of robust data observations across the water–soil–plant–atmosphere system, has significantly impacted on the technological environment within irrigated agriculture. These technologies offer new possibilities for the optimisation of the water supply and irrigation processes that may lead to improved water and energy use efficiencies, making agricultural production more sustainable, from environmental, social and economic aspects. In addition, cloud computing development and virtual storage have reduced the storage cost of Big Data (Bin and Xin, 2013). Thus, there are options to develop new tools to improve water–energy use based on AI techniques such as artificial neural networks (ANN), decision trees (DT), fuzzy logic (FL) or genetic algorithms (GA). The integration of these techniques within geographical information systems visualisation and hydraulic simulation frameworks allows soft computing to incorporate human knowledge effectively, deal with imprecisions and uncertainty, and design adaptation strategies for better performance of WUA and irrigated agriculture in general.

The AI techniques have been applied to solve different problems of water resources management and planning such as modelling monthly, daily and hourly rainfall–runoff process (Anctil and Rat, 2005; Agarwal et al., 2006), real-time river level and lake stage forecasting (Ondimu and Murase, 2007) or the optimisation of the multi-crop pattern plans using FL and GA (Rezaei et al., 2017). There are also applications of the Big Data approach in urban distribution networks such as the prediction of flow rates in water supply systems with the aim to improve both energy efficiency and water quality (Bakker et al., 2011) or several fault detection models on distribution networks based on data mining (Wang et al., 2013), distributed Principal Component Analysis (PCA) (Sanchez-Fernandez et al., 2015) and spectral clustering (Candelieri et al., 2014). Deep learning was also applied by Wu et al. (2015) to improve the operational efficiency on urban water networks. Koo et al. (2015) highlighted the relevance of the sustainable water supply developing, in a theoretical way, a schematic of Internet of Things application for Big Data collection through a myriad of water clients. However, there remains little experience in their application within the irrigated agriculture sector, which is surprising given its importance since irrigation is the main freshwater user globally.

Irrigation water demand forecasting at irrigation district scale was addressed by Pulido-Calvo and Gutiérrez-Estrada (2009). In that study, the performance of a hybrid methodology combining feed-forward ANNs, FL and GA to forecast 1-day ahead daily water demands at irrigation districts were analysed. The forecasts for each of the individual models were then corrected via an FL approach whose parameters were adjusted using GAs to improve the prediction accuracy. The methodology was applied to a real irrigation district located in southern Spain. A major limitation in this model was the determination of the ANN architecture by trial and error.

González Perea (2017) developed the approach to build a real smart water network model considering the different infrastructure levels that constitute a water distribution network. It was the first attempt to predict irrigation demand down to the farm level because previous studies only focused on the pumping station in the water distribution network. Thus, initially, González Perea et al. (2015) developed a short-term forecasting model of daily irrigation water

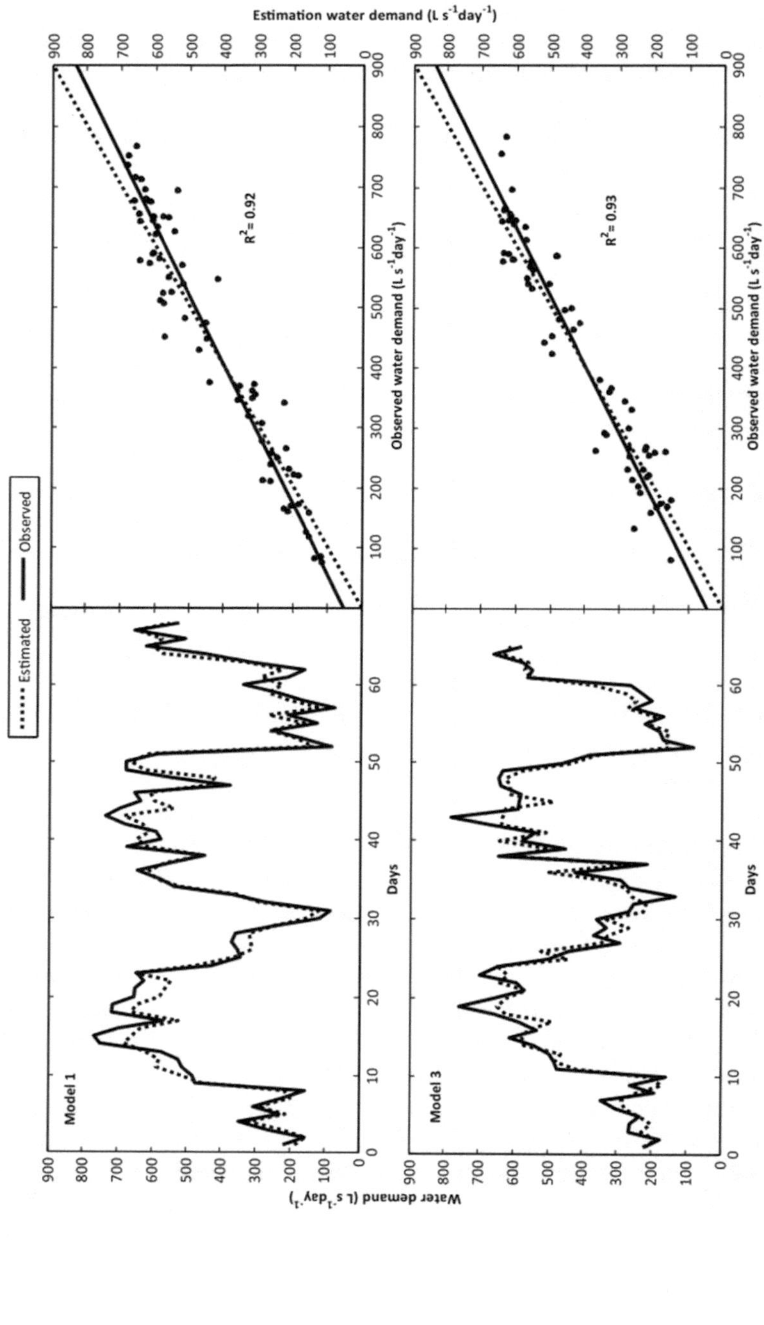

Figure 7 Daily water demand forecasting versus observed values in a real irrigation district in southern Spain.

demand at the district level. It predicted the daily water diverted to the whole irrigation district and consequently the energy consumption. The study also addressed water use at the farm level using a farmer's behaviour model. Individual behavioural models are extremely complex since the behaviour of one farmer is even much more complex to predict than a group, in the irrigation district. Regarding the irrigation network, Gonzalez Perea et al. (2015) tackled this drawback improving the results obtained in previous research. This novel model was applied to a real water distribution network with forecasting error close to 12% explaining more than 90% of the total variance observed. Figure 7 shows the water demand forecasting and observed water demand at a daily scale (validation period) of the best two models (a and b) obtained after implementation of the ANN architecture optimisation process.

These systems are based on data collected from various sensors located at the pumping station or hydrant and, from there, are used to predict the behaviour of the entire system under future scenarios or management situations. Therefore, the availability of data with adequate temporal and spatial resolution is essential for such a predictive modelling system. However, the availability of data is currently one of the main constraints in irrigation distribution networks when designing and optimising a forecasting system. It is therefore necessary to establish new models that work optimally with as little data as possible. For example, González Perea et al. (2019a) developed a model which improved the prediction accuracy by between 3% and 11% with respect to previous studies. The best ANN model had a standard error prediction (SEP) and a determination coefficient (R^2) of 8.7% and 96%, respectively.

At the farm level, irrigation scheduling tries to answer two important questions: when to irrigate (frequency including day and intraday scale) and how much to irrigate (application depth). According to this, three models were developed to answer both questions (González Perea et al., 2018a, 2019, 2021). Thus, the occurrence of the irrigation event was firstly approached as a classification problem combining DTs (Figs 8 and 9) and GAs (González Perea

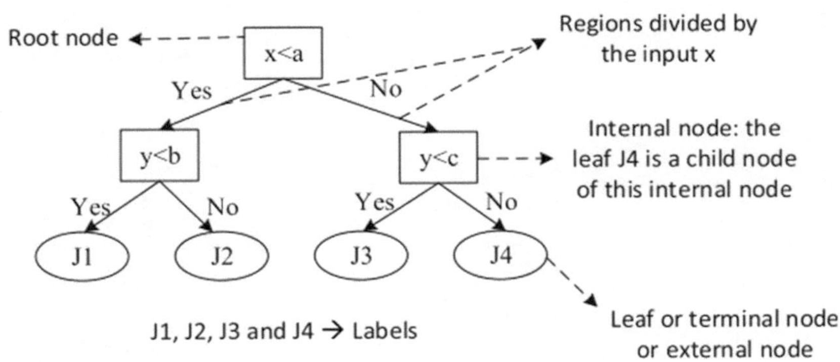

Figure 8 Classical structure of a decision tree.

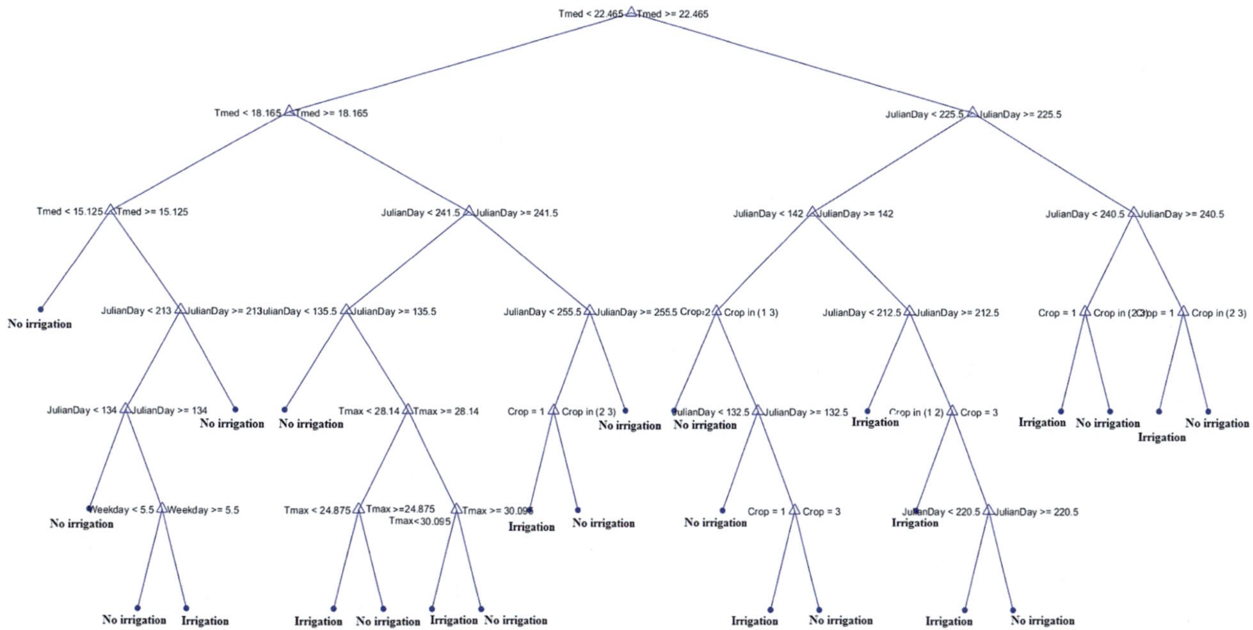

Figure 9 Schemes of the classification tree optimised by the genetic algorithms applied to an irrigation district. Source: González Perea et al. (2019).

et al., 2019). This approach found the optimal DT to model farmer's behaviour, predicting the occurrence of the required individual irrigation events. The model was applied to an irrigation district and managed to forecast between 93% and 100% of all irrigation events. A brief detail of one of the best DTs found after the optimisation process is shown in Fig. 9.

When to irrigate question was addressed by a hybrid methodology developed by González Perea et al. (2018). This methodology combined ANNs, FL and GAs to model farmer behaviour and forecast the daily amount of water used by each end-user within a water distribution network. This model was again applied to an irrigation district to estimate the representativity and accuracy of the developed model which was close to 90% and higher than 90%, respectively. Figure 10 shows these results for the main crops of the tested WUA.

Finally, a new model combining multiple input–multiple output, fuzzy logic, ANN and multi-objective GA was developed to model farmer behaviour and for short-term forecasting of the distribution of the irrigation depth applied

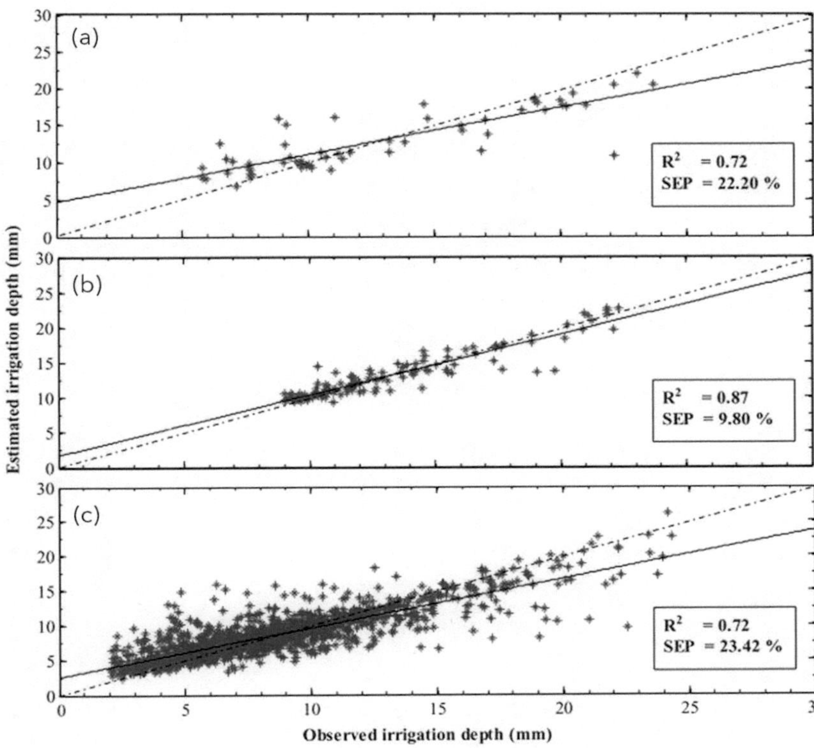

Figure 10 Observed versus estimated irrigation depths (mm) for the testing period for the three main crops grown in the irrigation district. Source: González Perea et al. (2018b).

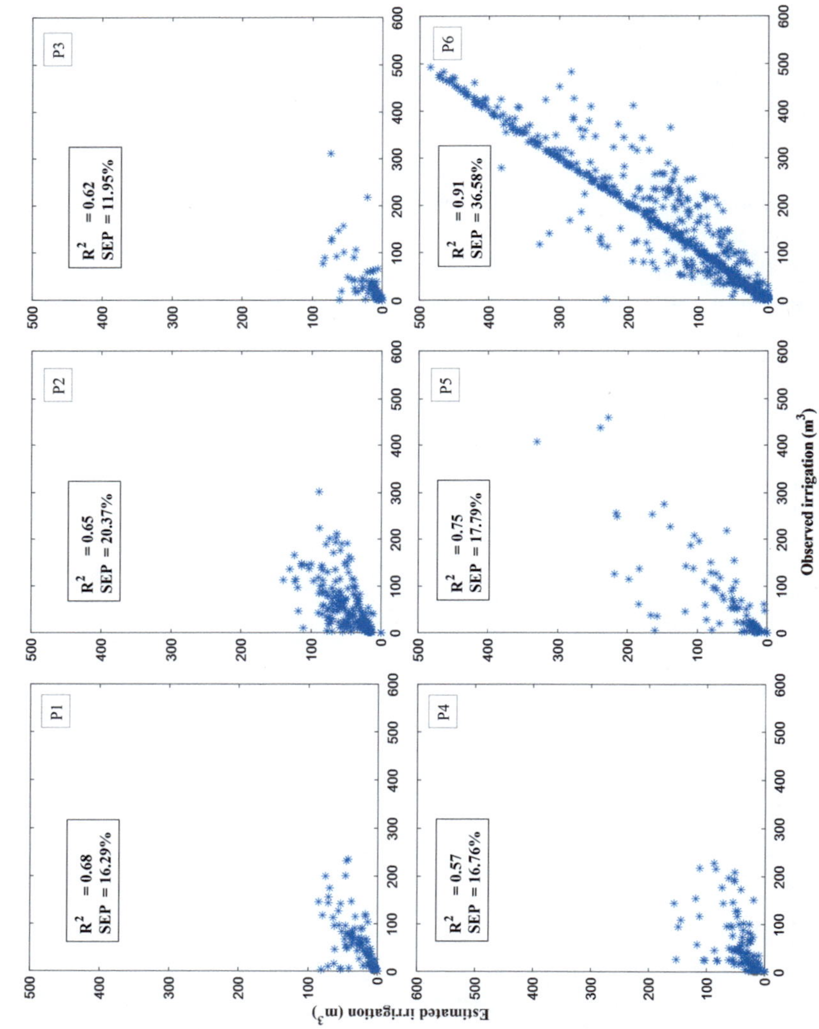

Figure 11 Observed versus estimated irrigation volumes (test set) applied in each tariff period.

at farm level by tariff period (six tariff periods). In Spain, electricity tariffs are organised in hourly periods whose prices vary during the day and within the irrigation season. This model enabled the energy to be more efficiently managed in the pressurised water distribution network. Three optimal models for the main water uses in a real irrigation district were obtained. The average representativity (R^2) and accuracy of the forecasts (SEP) for all tariff periods ranged from 0.70 to 0.85 and from 19.5% to 22%, respectively (Fig. 11).

New research priorities are now focusing on predicting demand further in advance. If the early models that were developed have been able to predict at least 1 day in advance, future models should strive to predict at least 1 week in advance, which will allow decisions to be taken even earlier. In relation to this, and focusing on middle (several days) to long-term (several weeks) water demand forecasting, Gonzalez Perea et al. (2021a) developed a new methodology and decision support tool for managers based on water demand forecasting at the field scale for 1 week. This tool, termed WatergyForecaster, combined AI techniques, satellite remote sensing (Sentinel 2) and open-source climate data to automatically build a water forecasting model at farm scale for a week in advance with an accuracy greater than 80%.

Although AI is being incorporated into the management of the water–energy nexus in irrigation distribution networks, there is still a long way to go to achieve integrated intelligent management of these resources. In this sense, new fields of knowledge and tools are emerging to help make more efficient decisions. Thus, the implementation of what are known as digital twins, a natural evolution of traditional decision support systems, or the incorporation of 5G technology, artificial vision and augmented reality in the field of managing the water–energy nexus in pressurised water distribution networks for irrigation will be fundamental in the next decade for the sustainable economic and environmental development of irrigated agriculture.

8 Conclusion

In today's increasingly technically orientated irrigated agriculture sector which relies on an increasing predominance of pressurised systems, energy efficiency plays as important a role as good water use. Energy generally represents the largest component of water costs, which has a direct impact on farmer's incomes. Furthermore, it is necessary to consider the environmental impact of CO_2 emissions, especially in the context of climate change. It is therefore necessary to jointly optimise the water–energy nexus, using the amount of water necessary for adequate crop growth, but with the lowest possible energy consumption.

There are various strategies to reduce energy dependency in irrigated areas, which on average can lead to savings of 20–30% in energy consumption.

Some of these measures involve changes in network management and others may lead to investments in facility upgrades. Some of these measures include sectorisation, critical points control and optimum pumping station design and management. However, the optimal application of these measures is not easy and requires the use of optimisation methods that take account of the complexity and large number of variables involved in the process.

Renewable energies represent a real alternative that will provide clean energy at low cost. The PV solar energy has undergone a great evolution in recent years. If the first applications were focused on its use in small farms, today they are beginning to be implemented in large irrigable areas with high-power solar plants. In the second case, the possibility of selling excess energy produced to the grid makes investments much more viable and also turns irrigation districts into renewable energy providers.

In water distribution networks, the use of energy from excess pressure via PATs is also an alternative that needs to be considered. However, their application is possibly more interesting at the hydrant level, to provide energy on farms when there is no alternative energy supply, replacing diesel generators that are currently most widely in use.

In the coming years, new techniques based on AI will become increasingly important in water management. Predicting the water demand, increasingly in advance, will make it possible to better manage the network and optimise the management and energy contracting. Today, the main limitation lies in the amount of data required for calibration, which must be taken in real-time by sensors in the field. The lower cost of these sensors will increasingly encourage more irrigation districts to implement this technology.

9 Where to look for further information

The strategies to minimise water and energy use in irrigated agriculture described in this chapter are in accordance with the 'farm to fork strategy' included in the European Green Deal. This strategy focused on the adoption of a sustainable food system, which considers:

- 'a neutral or positive environmental impact'; and
- 'mitigate climate change and adapt to its impacts'.

A detailed information about 'farm to fork strategy' is available at: https://food.ec.europa.eu/horizontal-topics/farm-fork-strategy_en.

In relation to the optimisation of energy consumption in irrigation water distribution networks, the Spanish Institute for Energy Diversification and Saving offers freely distributed publications:

- In 'Ahorro y eficiencia energética en las Comunidades de Regantes', they discuss the different alternatives to reduce energy consumption in irrigation water distribution networks (https://www.idae.es/publicaciones/ahorro-y-eficiencia-energetica-en-las-comunidades-de-regantes).
- They also develop a protocol for carrying out energy audits in irrigation water distribution systems, which is available at: https://www.idae.es/publicaciones/protocolo-de-auditoria-energetica-en-comunidades-de-regantes.

The Interreg Atlantic Area Project REDAWN (Reducing the Energy Dependency in Atlantic Area Water Networks) aims to foster the adoption of hydropower energy recovery technology in built water networks, mainly using pumps working as turbines (PATs). In their website (https://www.redawn.eu/), there are examples of demonstration sites for different water users (agrotourism and family home in Portugal, drinking water networks in France and irrigation in Spain). They also provide links to the main research outputs of the project.

The Food and Agriculture Organization of the United Nations (FAO) published, *The use of solar energy in irrigated agriculture. A sourcebook for irrigation water management with alternative energy solutions* (ISBN 978-92-5-135699-9)' (https://www.fao.org/documents/card/en/c/cb8459en). This sourcebook provides information on the design, operation, inspection, troubleshooting and maintenance of solar PV pumping systems.

The Toolbox SPIS (Solar Powered Irrigation Systems) was developed by GIZ (Deutsche Gesellschaft für Internationale Zusammenarbeit GmbH) and the FAO, from a global initiative of Powering Agriculture: An Energy Grand Challenge for Development, supported by United States Agency for International development (USAID), the Swedish International Development Cooperation Agency (SIDA), the German Federal Ministry for Economic Cooperation and Development (BMZ), Duke Energy and the Overseas Private Investment Cooperation (OPIC).

The Toolbox SPIS was designed to provide an easy tool for solar irrigation which combines informative modules supplemented by user-friendly software tools. SPIS includes a series of guides, checklists and spreadsheets to determine the crop irrigation needs, sizing and economic aspects of solar PV irrigation systems. This toolbox is free available for download (https://energypedia.info/wiki/Toolbox_on_SPIS).

The World Bank Group provides a Global Solar Atlas (https://globalsolaratlas.info/map) in addition to a series of global, regional and country GIS data layers to support scale-up solar power. This tool provides quick and easy access to solar resources and photovoltaic power potential data globally.

ISO 14040:2006 describes the principles and framework for life cycle assessment (LCA) with definition of the goal and scope of the LCA, life cycle inventory analysis (LCI) phase, life cycle impact assessment (LCIA), interpretation, reporting and critical review of the LCA.

Ecoinvent is a not-for-profit association based in Zurich, Switzerland, dedicated to promoting and supporting the use of environmental data worldwide. The main activity is the publication of Ecoinvent database, used worldwide as background database for LCA and other environmental studies. The access to the database is licensed.

10 References

Agarwal, A., Mishra, S. K., Ram, S. and Singh, J. K. (2006). Simulation of runoff and sediment yield using artificial neural networks. *Biosyst. Eng.* 94(4):597–613. doi: 10.1016/j.biosystemseng.2006.02.014.

Alexandratos, N. and Bruinsma, J. (2012). *World Agriculture towards 2030/2050: ESA Working Paper No. 12-03*. Food and Agriculture Organization of the United Nations.

Anctil, F. and Rat, A. (2005). Evaluation of neural network streamflow forecasting on 47 watersheds. *J. Hydrol. Eng.* 10(1): 85–88. doi: 10.1061/(ASCE)1084-0699(2005)10:1(85).

Bakker, M., Verberk, J. Q. J. C., Palmen, L. J., Sperber, V. and Bakker, G. (2011). *Higher Energy Efficiency and Better Water Quality by Using Model Predictive Flow Control at Water Supply Systems*. IWW symposium Water & Innovation - Watertechnology, Amsterdam, Nederland.

Bin, H. X. and Xin, Z. H. (2013). Build enterprise competitive intelligence system model based on Big Data. *J. Intell.* 3: 37–43.

Candelieri, A., Contib, D. and Archetti, F. (2014). A graph based analysis of leak localization in urban water networks. *Procedia Eng.* 70: 228–237. doi: 10.1016/j.proeng.2014.02.026.

Carrillo Cobo, M. T., Rodriguez Díaz, J. A., Montesinos, P., López Luque, R. and Camacho Poyato, E. (2011). Low energy consumption seasonal calendar for sectoring operation in pressurized irrigation networks. *Irrig. Sci.* 29(2): 157–169 doi: 10.1007/s00271-010-0228-2.

Carrillo-Cobo, M. T., Camacho-Poyato, E., Montesinos, P. and Rodriguez-Diaz, J. A. (2014). Assessing the potential of solar energy in pressurized irrigation networks. The case of Bembézar MI irrigation district (Spain). *Span. J. Agric. Res.* 12(3) doi: 10.5424/sjar/2014123-5327.

Casares de la Torre, F. J., Varo, M., López-Luque, R., Ramírez-Faz, J. and Fernández-Ahumada, L. M. (2022). Design and analysis of a tracking/backtracking strategy for PV plants with horizontal trackers after their conversion to agrivoltaic plants. *Renew. Energy* 187: 537–550. doi: 10.1016/j.renene.2022.01.081.

Collins, B. (2012). Towards Efficient Use of Water Resources in Europe (EEA Report No 1/2012). *European Environment Agency Report*.

Corominas, J. (2010). Agua y energía en el riego en la época de la sostenibilidad. *Ing. Agua* 17(3): 219–233.

Crespo Chacón, M., Rodríguez Díaz, J., García Morillo, J. and McNabola, A. (2019). Pump-as-turbine selection methodology for energy recovery in irrigation networks: minimising the payback period. *Water* 11(1): 1–20. doi: 10.3390/w11010149.

Crespo Chacón, M., Rodríguez Díaz, J. A., García Morillo, J. and McNabola, A. (2020). Estimating regional potential for micro-hydropower energy recovery in irrigation networks on a large geographical scale. *Renew. Energy* 155: 396–406. doi: 10.1016/j.renene.2020.03.143.

Daccache, A., Ciurana, J. S., Rodriguez Diaz, J. A. and Knox, J. W. (2014). Water and energy footprint of irrigated agriculture in the Mediterranean region. *Environ. Res. Lett.* 9(12) doi: 10.1088/1748-9326/9/12/124014.

Fernández García, I. and McNabola, A. (2020). Maximizing hydropower generation in gravity water distribution networks: determining the optimal location and number of pumps as turbines. *J. Water Resour. Plann. Manage.* 146(1): 04019066. doi: 10.1061/(ASCE)WR.1943-5452.0001152.

Fernández García, I., Creaco, E., Rodríguez Díaz, J. A., Montesinos, P., Camacho Poyato, E. and Savic, D. (2016). Rehabilitating pressurized irrigation networks for an increased energy efficiency. *Agric. Water Manag.* 164: 212–222. doi: 10.1016/j.agwat.2015.10.032.

Fernández García, I., Moreno, M. A. and Rodríguez Díaz, J. A. (2014a). Optimum pumping station management for irrigation networks sectoring: case of Bembezar MI (Spain). *Agric. Water Manag.* 144: 150–158 doi: 10.1016/j.agwat.2014.06.006.

Fernández García, I., Rodríguez Díaz, J. A., Camacho Poyato, E., Montesinos, P. and Berbel, J. (2014b). Effects of modernization and medium term perspectives on water and energy use in irrigation districts. *Agric. Syst.* 131: 56–63. doi: 10.1016/j.agsy.2014.08.002.

García, I. F., Montesinos, P., Poyato, E. C. and Díaz, J. A. R. (2014). Methodology for detecting critical points in pressurized irrigation networks with multiple water supply points. *Water Resour. Manag.* 28(4): 1095–1109 doi: 10.1007/s11269-014-0538-x.

García Morillo, J., McNabola, A., Camacho, E., Montesinos, P. and Rodríguez Díaz, J. A. (2018). Hydro-power energy recovery in pressurized irrigation networks: A case study of an Irrigation District in the South of Spain. *Agric. Water Manag.* 204:17–27 doi: 10.1016/j.agwat.2018.03.035.

González Perea, R. (2017). *Optimum Management of Pressurized Irrigation Networks at Different Scales Using Artificial Intelligent Techniques*. University of Cordoba.

Gonzalez Perea, R., Poyato, E. C., Montesinos, P. and Díaz, J. A. R. (2015). Irrigation demand forecasting using artificial neuro-genetic networks. *Water Resour. Manag.* 29(15): 5551–5567 doi: 10.1007/s11269-015-1134-4.

González Perea, R., Camacho Poyato, E., Montesinos, P. and Rodríguez Díaz, J. A. (2018a). Prediction of applied irrigation depths at farm level using artificial intelligence techniques. *Agric. Water Manag.* 206: 229–240 doi: 10.1016/j.agwat.2018.05.019.

González Perea, R., Camacho Poyato, E., Montesinos, P. and Rodríguez Díaz, J. A. (2018b). Prediction of applied irrigation depths at farm level using arti fi cial intelligence techniques. *Agric. Water Manag.* 206: 229–240. doi: 10.1016/j.agwat.2018.05.019.

González Perea, R., Camacho Poyato, E., Montesinos, P. and Rodríguez Díaz, J. A. (2019). Prediction of irrigation event occurrence at farm level using optimal decision trees. *Comput. Electron. Agric.* 157: 173–180 doi: 10.1016/j.compag.2018.12.043.

Gonzalez Perea, R., Ballesteros, R., Ortega, J. F. and Moreno, M. Á. (2021a). Water and energy demand forecasting in large-scale water distribution networks for irrigation using open data and machine learning algorithms. *Comput. Electron. Agric.* 188: 1–13 doi: 10.1016/j.compag.2021.106327.

González Perea, R., Poyato Camacho, E. and Rodríguez Díaz, J. A. (2021b). Forecasting of applied irrigation depths at farm level for energy tariff periods using Coactive neuro-genetic fuzzy system. *Agric. Water Manag.* 256: 1–12 doi: 10.1016/j.agwat.2021.107068.

Gupta, A., Bokde, N., Marathe, D. and Kulat, K. (2017). Leakage reduction in water distribution systems with efficient placement and control of pressure reducing valves using soft computing techniques. *Eng. Technol. Appl. Sci. Res.* 7(2): 1528–1534. doi: 10.48084/etasr.1032.

Hamidat, A., Benyoucef, B. and Hartani, T. (2003). Small-scale irrigation with photovoltaic water pumping system in Sahara regions. *Renew. Energy* 28(7):1081–1096. doi: 10.1016/S0960-1481(02)00058-7.

Jiménez-Bello, M. A., Martínez Alzamora, F., Bou Soler, V. and Ayala, H. J. B. (2010). Methodology for grouping intakes of pressurised irrigation networks into sectors to minimise energy consumption. *Biosyst. Eng.* 105(4): 429–438. doi: 10.1016/j.biosystemseng.2009.12.014.

Koo, D., Piratla, K. and Matthews, C. J. (2015). Towards sustainable water supply: schematic development of big data collection using Internet of things (IoT). *Procedia Eng.* 118: 489–497. doi: 10.1016/j.proeng.2015.08.465.

Lamaddalena, N. and Sagardoy, J. A. (2000). *Performance Analysis of On-demand Pressurized Irrigation Systems*. Food and Agriculture Organization. Irrigation and Drainage Paper No. 59.

Lopez-Gunn, E., Zorrilla, P., Prieto, F. and Llamas, M. R. (2012). Lost in translation? Water efficiency in Spanish agriculture. *Agric. Water Manag.* 108: 83–95. doi: 10.1016/j.agwat.2012.01.005.

López-Luque, R., Reca, J. and Martínez, J. (2015). Optimal design of a standalone direct pumping photovoltaic system for deficit irrigation of olive orchards. *Appl. Energy* 149: 13–23. doi: 10.1016/j.apenergy.2015.03.107.

MAGRAMA (2010). *Dirección General Del Agua Estrategia Nacional Para la Modernización Sostenible de los Regadíos H2015 Informe de Sostenibilidad Ambiental*. Ministerio de Medio Ambiente y Medio Rural y Marino.

McNabola, A., Coughlan, P. and Williams, A. P. (2011). The technical & economic feasibility of energy recovery in water supply networks. *ReNew Energy Power Qual. J.* 1: 1123–1127. doi: 10.24084/repqj09.569.

Mérida García, A., Fernández García, I., Camacho Poyato, E., Montesinos Barrios, P. and Rodríguez Díaz, J. A. (2018). Coupling irrigation scheduling with solar energy production in a smart irrigation management system. *J. Clean. Prod.* 175: 670–682. doi: 10.1016/j.jclepro.2017.12.093.

Mérida García, A., Gallagher, J., McNabola, A., Camacho Poyato, E., Montesinos Barrios, P. and Rodríguez Díaz, J. A. (2019). Comparing the environmental and economic impacts of on- or off-grid solar photovoltaics with traditional energy sources for rural irrigation systems. *Renew. Energy* 140: 895–904. doi: 10.1016/j.renene.2019.03.122.

Mérida García, A., Gallagher, J., Crespo Chacón, M. and McNabola, A. (2021a). The Environmental and Economic Benefits of a Hybrid Hydropower Energy Recovery and Solar Energy System (PAT-PV), under varying energy demands in the Agricultural Sector. *J. Clean. Prod.* 303: 127078. doi: 10.1016/j.jclepro.2021.127078.

Mérida García, A., Rodríguez Díaz, J. A., García Morillo, J. and McNabola, A. (2021b). Energy recovery potential in industrial and municipal wastewater networks using micro-hydropower in Spain. *Water* 13(5). doi: 10.3390/w13050691.

Moreno, M. A., Planells, P., Córcoles, J. I., Tarjuelo, J. M. and Carrión, P. A. (2009). Development of a new methodology to obtain the characteristic pump curves that minimize the total cost at pumping stations. *Biosyst. Eng.* 102(1): 95–105. doi: 10.1016/j.biosystemseng.2008.09.024.

Moreno, M. A., Córcoles, J. I., Tarjuelo, J. M. and Ortega, J. F. (2010). Energy efficiency of pressurised irrigation networks managed on-demand and under a rotation schedule. *Biosyst. Eng.* 107(4): 349–363. doi: 10.1016/j.biosystemseng.2010.09.009.

Navarro Navajas, J. M., Montesinos, P., Poyato, E. C. and Rodríguez Díaz, J. A. (2012). Impacts of irrigation network sectoring as an energy saving measure on olive grove production. *J. Environ. Manage.* 111: 1–9 doi: 10.1016/j.jenvman.2012.06.034.

Ondimu, S. and Murase, H. (2007). Reservoir level forecasting using neural networks: Lake Naivasha. *Biosyst. Eng.* 96(1): 135–138. doi: 10.1016/j.biosystemseng.2006.09.003.

Pérez-Sánchez, M., Sánchez-Romero, F. J., Ramos, H. M. and López-Jiménez, P. A. (2016). Modeling irrigation networks for the quantification of potential energy recovering: A case study. *Water* 8(6). doi: 10.3390/w8060234.

Power, C., McNabola, A. and Coughlan, P. (2014). Development of an evaluation method for hydropower energy recovery in wastewater treatment plants: case studies in Ireland and the UK. *Sustain. Energy Technol. Assess.* 7: 166–177. doi: 10.1016/j.seta.2014.06.001.

Pulido-Calvo, I. and Gutiérrez-Estrada, J. C. (2009). Improved irrigation water demand forecasting using a soft-computing hybrid model. *Biosyst. Eng.* 102(2): 202–218. doi: 10.1016/j.biosystemseng.2008.09.032.

Reca-Cardeña, J. and López-Luque, R. (2018). *Design Principles of Photovoltaic Irrigation Systems*.

Rezaei, F., Safavi, H. R. and Zekri, M. (2017). A hybrid fuzzy-based multi-objective PSO algorithm for conjunctive water use and optimal multi-crop pattern planning. *Water Resour. Manag.* 31(4): 1139–1155 doi: 10.1007/s11269-016-1567-4.

Rocamora, C., Vera, J. and Abadía, R. (2013). Strategy for Efficient Energy Management to solve energy problems in modernized irrigation: analysis of the Spanish case. *Irrig. Sci.* 31(5): 1139–1158. doi: 10.1007/s00271-012-0394-5.

Rodríguez Díaz, J. A., Weatherhead, E. K., Knox, J. W. and Camacho, E. (2007). Climate change impacts on irrigation water requirements in the Guadalquivir River basin in Spain. *Reg. Environ. Change* 7(3): 149–159. doi: 10.1007/s10113-007-0035-3.

Rodríguez Díaz, J. A., Luque, R. L., Cobo, M. T. C., Montesinos, P. and Poyato, E. C. (2009). Exploring energy saving scenarios for on-demand pressurised irrigation networks. *Biosyst. Eng.* 104(4): 552–561. doi: 10.1016/j.biosystemseng.2009.09.001.

Rodríguez-Díaz, J. A., Pérez-Urrestarazu, L., Camacho-Poyato, E. and Montesinos, P. (2011a). The paradox of irrigation scheme modernization: more efficient water use linked to higher energy demand. *Span. J. Agric. Res.* 9(4) doi: 10.5424/sjar/20110904-492-10.

Rodríguez Díaz, J. A., Camacho Poyato, E. and Blanco Pérez, M. (2011b). Evaluation of water and energy use in pressurized irrigation networks in Southern Spain. *J. Irrig. Drain. Eng.* 137(10): 644–650. doi: 10.1061/(ASCE)IR.1943-4774.0000338.

Rodríguez Díaz, J. A., Montesinos, P. and Poyato, E. C. (2012). Detecting critical points in on-demand irrigation pressurized networks - A new methodology. *Water Resour. Manag.* 26(6): 1693–1713 doi: 10.1007/s11269-012-9981-8.

Rodríguez-Gallegos, C. D., Gandhi, O., Yang, D., Álvarez-Alvarado, M., Zhang, W., Reindl, T. and Panda, S. K. (2017). A siting and sizing optimization approach for PV-battery-diesel hybrid systems. *IEEE Trans. Ind. Appl.* 9994: 1–1. doi: 10.1109/TIA.2017.2787680.

Sanchez-Fernandez, A., Fuente, M. J. and Sainz-Palmero, G. I. (2015). Fault detection with Distributed PCA methods in water Distribution Networks. In 23th Mediterr. Conf Control. Autom. 156–161 (vol. 2015). doi: 10.1109/MED.2015.7158744.

Tarjuelo, J. M., Rodriguez-Diaz, J. A., Abadía, R., Camacho, E., Rocamora, C. and Moreno, M. A. (2015). Efficient water and energy use in irrigation modernization: lessons from Spanish case studies. *Agric. Water Manag.* 162: 67–77 doi: 10.1016/j.agwat.2015.08.009.

Tubiello, F. N., Salvatore, M., Cóndor Golec, R. D. Ferrara, A., Rossi, S., Biancalani, R., Federici, S., Jacobs, H. and Flammini, A. (2014). *Agriculture , Forestry and Other Land Use Emissions by Sources and Removals by Sinks*. Food and Agriculture Organization, Climate, Energy and Tenure Division.

Vick, B. D. and Almas, L. K. (2011). Developing a hybrid solar/wind powered irrigation system for crops in the Great Plains. *Appl. Eng. Agric.* 27: 235–245.

Wang, R., Dong, W., Wang, Y., Tang, K. and Yao, X. (2013). Pipe failure prediction: A data mining method 29th Int Conf Data Eng. (vol. 2013). IEEE Publications 1208–1218. doi: 10.1109/ICDE.2013.6544910.

Wu, Z. Y., El-Maghraby, M. and Pathak, S. (2015). Applications of deep learning for smart water networks. *Procedia Eng.* 119: 479–485. doi: 10.1016/j.proeng.2015.08.870.

Yahyaoui, I., Tadeo, F. and Segatto, M. V. (2017). Energy and water management for drip-irrigation of tomatoes in a semi- arid district. *Agric. Water Manag.* 183: 4–15. doi: 10.1016/j.agwat.2016.08.003.

Chapter 8

Solar-powered irrigation systems: recent developments and future trends

Muhammad Arif Watto, CSIRO Environment Business Unit, Australia; Maria Shabir, University of Foggia, Italy; Ali Sher, School of Management, Guangzhou University, China; and Muhammad Awais Ali Khan, Independent Researcher, UK*

1 Introduction

Globally, irrigated agriculture heavily depends on both surface and groundwater resources. Even though irrigated agriculture covers less than 20% of the world's cropland, it contributes to over 40% of total food production (Döll and Siebert, 2002). Over the last few decades, significant advancements in water pumping and distribution technologies have led to a substantial increase in groundwater use for irrigation purposes (Scanlon et al., 2012). Overall, the agricultural sector accounts for ~70% of worldwide water usage (Siebert et al., 2010) with ~42% share from groundwater resources. Globally, the irrigated cropped area has increased twofold over the past half-century, now occupying nearly one-third of the global cropped area. It is projected that the global population will reach 10 billion by 2050 (Bwambale et al., 2022), significantly increasing demand for food, with consequent impacts for irrigation requirements (Bwambale et al., 2022). In addition to a growing population, climate change is expected to add further impetus to rely more extensively on groundwater abstraction for irrigation. Given this challenge, sustainable agriculture supported by climate-resilient irrigation strategies are essential to meet future food security challenges. In this

* Corresponding author: arif.watto@csiro.au.

http://dx.doi.org/10.19103/AS.2023.0123.11

regard, modern irrigation techniques and technologies can help revolutionize the agricultural sector, allowing farmers to increase crop production, thereby positively impacting food security in many parts of the world (Balasubramanya and Stifel, 2020). It is an *a priori* expectation that reliable irrigation water supplies increase agricultural production, minimize variability of farm production, and hence variability of farm income (Watto et al., 2018). More broadly, irrigation plays an important role in ensuring food security, generating employment opportunities, and driving rural development goals, ultimately improving overall socioeconomic well-being of rural communities. However, agricultural production is being challenged by various factors such as climate change and water scarcity, which hamper the productivity and profitability potential of agricultural production (United Nations, 2015). Roughly, 570 million farmers across the world belong to the small farmers' category (Grant et al., 2022). Smallholder farmers are key to sustainable agricultural intensification, but they are resource constrained. Without access to resources, they are unable to invest in sustainable practices, such as in adopting modern irrigation systems and thus taking advantage of new technologies. In terms of achieving sustainable agricultural production, reliable irrigation is one of the most crucial factors as it helps farmers to hedge against the uncertainties of climate (rainfall) variability and consequently increase crop yields (Watto et al., 2018). On a larger scale, irrigation is important for food security, creating employment opportunities and realizing rural development agendas. However, the reliability of irrigation supplies is linked with reliable access to energy supplies (Agrawal and Jain, 2019). In this context, solar-powered irrigation systems (SPIS) offer enormous potential to ensure reliable access to energy, transforming agricultural production and improving the socioeconomic well-being and livelihoods for rural and remote communities.

1.1 Solar-powered irrigation systems: a technological transformation

Due to its high dependence on groundwater, irrigated agriculture faces significant challenges due to the economic and environmental impacts associated with using fossil fuel-powered pumping technologies. Besides the rising costs, the environmental sustainability of these technologies remains a serious concern. The water–energy–food (WEF) nexus is a framework that interconnects water, energy, and food systems, and which affects each component in one or another way. Different factors such as a growing population, economic development, and social change have increased competition among all actors and aspects of the WEF nexus, e.g. agricultural, domestic, industrial, hydropower, and environmental uses. These challenges have led to a need to identify methods and synergies that minimize losses and impacts and increase the efficiency of all aspects in the WEF nexus. In the agricultural sector, the use of groundwater for irrigation requires using fossil fuel-based pumping technologies which are increasingly expensive but more importantly are environmentally unsustainable. These technologies contribute to greenhouse gases (GHGs) and increase carbon emissions primarily carbon dioxide (CO_2) and methane (CH_4), thereby increasing carbon footprints (Foster and Cota, 2014). In this context, using alternative solar energy sources can help farmers curtail the environmental impacts and reduce financial burdens associated with the use of conventional energy-based water pumping technologies (Verma et al., 2021).

It is against this backdrop that solar powered irrigation systems (SPISs) have emerged as an affordable, reliable, and sustainable source of energy compared to conventional energy-based irrigation systems. In addition, SPISs also complement and support the United Nation's Sustainable Development Goal 7 (SDG 7) which focuses on ensuring sustainable, reliable, and affordable access to modern energy for all in developing countries. Solar-powered irrigation potentially offers a win–win solution, contributing to SDGs – including poverty alleviation and hunger elimination (Bizimana et al., 2021), while concurrently providing clean energy through climate-smart irrigation technologies (Shukla et al., 2017). As the costs associated with SPIS gradually decline, with projections indicating a population of 10 billion people by 2050 (Bwambale et al., 2022), solar energy becomes increasingly attractive to both large-and small-scale farmers. SPISs offer a flexible and eco-friendly energy alternative in rural areas where diesel fuel is economically unfeasible or where reliable access to the electricity grid is lacking, thereby avoiding negative environmental externalities (Xie et al., 2015). Moreover, SPIS can be integrated into both centralized, large-scale irrigation operations and decentralized, for small-scale irrigation setups (Grant et al., 2022).

Different countries are now widely promoting the adoption of SPIS as a climate change combating technology because the use of solar pumps does not generate GHG emissions. Based on the entire life cycle, i.e. right from manufacturing to disposal stages, assessments indicate that SPIS unequivocally curtail GHG emissions per unit of energy used for water pumping (CO_2-eq/kWh) by c95–97% compared to pumps relying on fossil fuel or grid-based electricity (global average energy mix) and by an impressive 97–98% compared to conventional diesel pumps (GIZ, 2016). The per watt cost of solar photovoltaic (PV) has also decreased dramatically over the past few decades. For instance, in the early 1980s, the solar PV cost was around $3.50 per watt, but which has decreased to $1 per watt by 2020. The International Renewable Energy Agency (IRENA) estimates that the unit cost of solar PV will drop further to $0.50 per watt by 2030, which if it happens will be a major milestone for the agricultural sector, marking a milestone and accentuating the cost-effectiveness of solar energy (Moosavian et al., 2013). Globally, the paradigm shift towards renewable energy resources has encouraged numerous development agencies to further explore the potential of solar PV technology. The reduced costs, improved reliability, and escalating demand within the private sector, coupled with much reduced environmental impacts, have collectively created considerable interest among irrigators in solar-powered irrigation water pumping technologies in many parts of the world (Khoodaruth et al., 2017).

2 Historical developments

The efficiency of PV cells have increased dramatically over time. Previously, the efficiency of PV cells were reported to be around 4%, which was later improved to around 11%. With the help of this development, the use of solar energy started to increase, and it became a more viable and cost-effective alternative to other sources of electricity for many purposes, not just for irrigation pumping. Since then, the efficiency of solar cells has gradually increased (Mohan et al., 2023). The introduction of solar-powered irrigation pumps/systems in the late 1970s revolutionized PV water pumping systems. Early systems relied on centrifugal pumps powered by direct current (DC) or variable frequency alternative current (AC) motors, renowned for their reliability and hydraulic efficiency in the range of 25–35%. As time progressed, alternative pumping mechanisms such as positive displacement pumps, progressive cavity pumps, and diaphragm pumps were introduced. These systems catered to lower water volumes, requiring less input power and capital investment while delivering higher hydraulic efficiencies (Chandel, 2015). Extensive literature highlights that replacing inefficient irrigation systems with solar-powered systems could significantly reduce water wastage during irrigation (by 20–76%) and increase crop productivity by up to 30% (Aliyu et al., 2018; Jägermeyr et al., 2015). Figure 1 provides a schematic summary of the workings of a typical solar irrigation system.

2.1 Advances in solar-powered irrigation systems functionality and application

Over time, SPIS have undergone significant advancements, resulting in improved functionality and application for a range of purposes and benefits for farmers, such as water and energy cost savings and increased crop yields (Hussain et al., 2023). Recent technological advancements have made SPIS an attractive option for farmers in both developed and developing countries. One of the most recent developments has been the emergence of 'smart' irrigation systems powered by solar energy, which can be monitored and operated remotely and in real-time. This technology allows farmers to make more informed decisions about their crop's water requirements, such as when and how much irrigation water is required. Smart systems also enable farmers to optimize water use by improving irrigation efficiency (Wazed et al., 2018). The

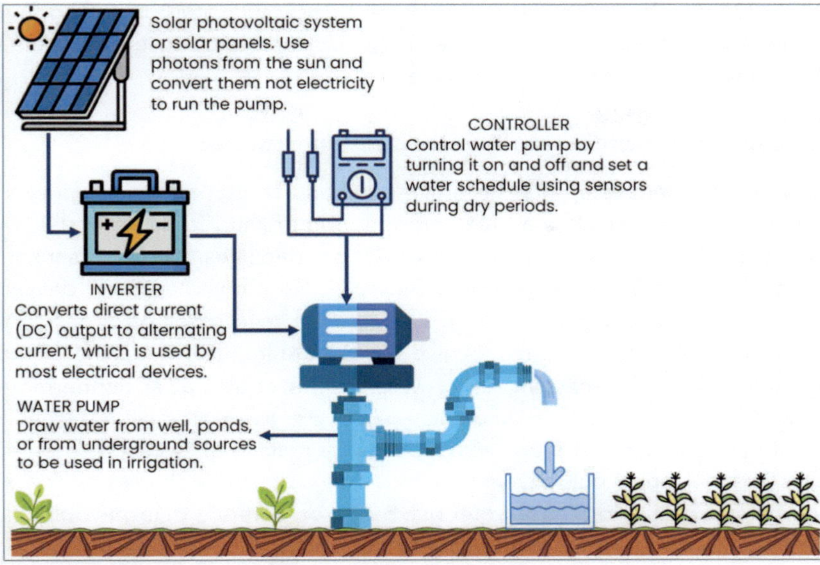

Figure 1 Components of a typical solar irrigation system.

recent advancements in harnessing solar energy have enabled SPIS to become a viable and reliable source of energy, making it an economically, technically, and environment friendly option (Siyal et al., 2021; Aliyu et al., 2018; Kishore et al., 2014). In the following section, we explore the latest developments in solar energy technology and its impact on SPIS, in terms of enhanced efficiency and applicability within diverse agricultural settings. We review into specific innovations and a case study to highlight the potential and role of such advancements in the context of sustainable agriculture and water resource management.

2.2 Improved solar cell efficiency for irrigation systems

Recent developments in materials technology has significantly improved the efficiency of solar cells, which can capture and convert solar energy more efficiently (Mohan et al., 2023). The integration of nano-structured materials and quantum dots, and cutting-edge designs have significantly improved solar cell conversion efficiencies, durability, and performance even in low-light level conditions. These transformative developments have made SPIS more cost-effective and reliable and environmentally compatible (Li et al., 2023). Notably, Zhao et al. (2021) introduced a groundbreaking approach using semi-transparent organic solar cells that can capture and allow the sunlight of desired and selected wavelengths to maximize output in energy production. Given the improved solar cell efficiency, SPIS are being advocated as a promising technology for improving water use efficiency and reducing environmental footprints of irrigation by providing clean energy.

2.3 Advances in automated water delivery systems

One of the latest developments is the integration of automated water delivery systems with SPIS, which is a breakthrough in improving irrigation water use efficiencies. These systems include networking capabilities, advanced sensors, and intelligent computations (Angelin Blessy, 2021). Automated water delivery systems can precisely measure and distribute water to crops based on their needs. This helps ensure crops receive the right amount of water they need, thus minimizing the excessive use of water (Sunny et al., 2023). Furthermore, automation can reduce labour costs, which are associated with manually operated irrigation systems and also reduces the maintenance costs compared to the costs incurred on conventional irrigation systems (Mohan et al., 2023). Automated SPIS can optimize water use based on different datasets obtained from *in situ* soil moisture monitoring and from weather prediction, using GSM technology. It can automatically control the motor pump by taking feedback information from *in situ* soil moisture sensors in the field. After providing the demanded water based on pre-determined data, the sensor will turn off the motor. Such a system entails all the features which help make an ideal irrigation system a smarter, eco-friendly, and economically feasible (Shufian et al., 2019). Therefore, automated water delivery systems powered by solar energy are expecting to revolutionize traditional irrigation systems, methods, and practices.

2.4 Development of low-cost solar irrigation controllers

Traditional methods of irrigation have long relied on qualitative irrigation techniques that lack an objective approach to accurately estimating crop and environmental conditions. However, various modern techniques involve a quantitative approach with instruments and sensors for soils, climate, and crop measurement (Alkhalidi et al., 2023). A key component of these is the use of smart controllers, which enable farmers to monitor and adjust irrigation schedules based on real-time data regarding soil moisture levels, rainfall, and weather conditions. By automating on-off cycles, these controllers play a pivotal role in conserving precious water resources while also promoting energy efficiency (Ramachandran et al., 2022).

Solar irrigation controllers integrate wireless technology into irrigation systems, ensuring precise and conveniently managed irrigation watering schedules (Matekaire, 2021). Similarly, more efficient solar irrigation controllers help optimize energy storage and utilization (Kocaman et al., 2020), with some models incorporating specialized sensors capable of measuring soil moisture levels and weather conditions, thereby facilitating optimized irrigation schedules and water conservation efforts (Gu et al., 2020). These controllers collect climatic data and enable real-time monitoring from a nearby automatic wireless system (AWS). However, obtaining real-time data is a challenge in scheduling an irrigation controller and validating and re-constructing the missing data (Hema and Kant, 2019). For example, Poyen et al. (2020) developed a fuzzy rule-based irrigation controller prototype to control water use for optimal irrigating management. The system consists of a wireless sensor unit (WSU) that measures weather and soil conditions and calculates actual water loss due to evapotranspiration. The wireless information processing unit (WIPU) then uses this data to take the necessary control actions, ensuring that sufficient water flows into the field.

2.5 Solar photovoltaic water irrigation system with wireless control

A solar-powered water irrigation system with wireless control is a system that uses solar energy to power an irrigation system and wirelessly controls the system remotely (Sudharshan et al., 2020). According to Waleed et al. (2019) irrigating field crops using solar PV is the best technological solution. PV-powered pumps usually comprise solar panels and DC/AC water pumping systems. Technological advances have enabled the control of such irrigation systems through novel control features, such as sensors. Global systems for mobiles and humidity control sensors are also being installed for wireless control and automation. Such automation reduces water and energy waste and helps reduce labour use. Hence, automatic irrigation systems with wireless controls have made solar PV-led irrigation more cost-effective and sustainable.

3 Future trends

In many regions of the world, SPISs have become popular due to recent advancements in the science and technology development of solar energy generation, storage, and use. They reduce reliance on fossil fuels and minimize the associated costs such as environmental costs. In the future, it is expected that the use of SPIS will continue to increase, driven by a growing awareness of the need to conserve energy and reduce carbon emissions. Similarly, advancements in the solar technology are expected to improve the efficiency and reliability of SPIS, which will make them more viable and an attractive option for irrigators. Consequently, it is likely that SPIS will become a standard technology for many farming operations in the near future. The integration of artificial intelligence (AI) and machine learning (ML) algorithms may further improve the solar technology and introduce even further low-cost solar irrigation controllers, allowing for improved automation and precision in water management. There is thus enormous potential for SPIS technology to revolutionize the agricultural industry and is an area of active research and development (Patro et al., 2023). Recently, Yoo et al. (2021), developed metal halide perovskite solar cells (PSCs) that could revolutionize the PV technology market. It has a huge potential because it has been manufactured after several developments in fabrication protocols, stabilization phases, and chemical processes. It is made of cost-effective processable PV technologies and offers the most efficient output for agricultural uses. Thus, such an important innovation would help support widespread SPIS uptake globally.

3.1 Solar-powered drip irrigation techniques

Solar-powered drip irrigation is an efficient technique that utilizes solar energy to deliver water precisely to the roots of plants, promoting controlled and optimal irrigation practices. This approach significantly reduces water wastage, enhances water usage efficiency, and offers several advantages for agricultural practices (Dahnill et al., 2021). By minimizing evaporation and ensuring the delivery of the right amount of water at the right time, solar-powered drip irrigation contributes to water conservation. Furthermore, this method has led to increase in crop yields while simultaneously reducing labour costs and improving water quality by minimizing runoff and contamination, resulting in healthier crops and a more sustainable agricultural process. Research by Hussan et al. (2023) highlights the effectiveness and sustainability of solar-powered drip irrigation systems. Their study demonstrates that these systems not only improve yields but also reduce water contamination caused by chemicals and fertilizers. Hussan et al. (2023) also assessed the economic and environmental sustainability of solar-powered drip irrigation systems in maize cultivation and showed that solar-powered drip irrigation systems have a higher benefit-cost ratio compared to diesel-powered systems. Solar-powered drip irrigation systems also have lower CO_2 emissions compared to equivalent diesel-powered systems.

3.2 Introduction of smart irrigation technologies

The introduction of smart irrigation technologies represents a significant milestone in agriculture. By integrating sensory systems and IoT (Internet of Things) technologies, farmers can gain a deeper understanding of their crops' needs and development cycles. This knowledge allows for more precise and resource-efficient farming techniques, minimizing environmental impacts while conserving valuable resources including water (Obaideen et al., 2022). Smart irrigation technologies, driven by sensors and advanced algorithms, are set to revolutionize the way water is managed in agriculture. These systems continuously monitor soil moisture levels, weather conditions, and crop requirements (see Fig. 2). The data collected is then processed by algorithms that determine optimal irrigation schedules. This data-driven approach leads to substantial water savings, improved crop yields, and a reduced ecological footprint (Ali et al., 2023). The widespread adoption of these technologies holds the potential to reshape the future of water management, crop production, and rural development.

3.3 Widespread adoption of solar-powered irrigation systems

The latest SPIS technologies are relatively easy to install and maintain and can be used in large-scale agricultural operations (Powell et al., 2021). The adoption of SPIS will be especially useful in areas where access to electricity is limited. The rising cost of traditional energy sources and the increasing concern over water scarcity are also likely to drive the further adoption of SPIS, helping reduce water waste and energy costs and providing a valuable source of additional income for farmers. This could benefit communities and the environment by promoting sustainable agricultural practices. Reducing water consumption will also help ensure the sustainability of agricultural production, particularly in water scarce areas (Qazi et al., 2022).

3.4 Potential use of artificial intelligence and machine learning in solar irrigation systems

Artificial intelligence (AI) and machine learning (ML) in solar irrigation systems has being developed in recent years. These can be used to optimize the operation of solar irrigation systems, helping to maximize the energy generated from solar energy. AI and ML can also be used to monitor and control irrigation systems, providing insights into water usage, irrigation schedules, and water conservation. For instance, Ayoub Shaikh et al. (2022) evaluated the role of ML and AL in the agricultural sector and concluded that they have enabled systems to become more efficient and accurate in controlling water usage, by using algorithms to predict weather, and soil conditions. Similarly, Obaideen et al. (2022) investigated that smart irrigation system enhances crop production and improves environmental quality. Radoglou-Grammatikis et al. (2020) reported that AI- and ML-based systems can help farmers to better manage and monitor their irrigation systems, allowing them to identify and troubleshoot any issues quickly. The increased use of AI and ML in solar irrigation systems is critical to realizing their full potential. AI and ML algorithms can be used to optimize the functioning of SPIS, e.g., by enabling the approaches to adjust the amount of water used in response to weather conditions, resulting in reduced water consumption. Additionally, AI and ML can monitor irrigation systems, providing farmers with real-time data, and enabling them to better manage their water resources. Ünal (2020) predicated that deep learning has an ability to automate

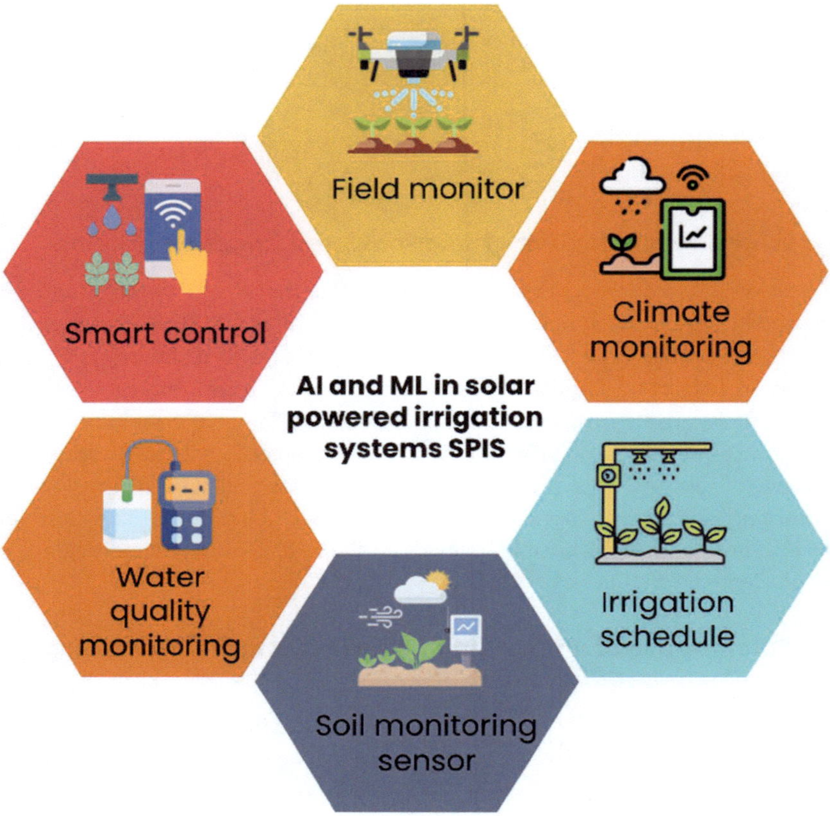

Figure 2 Smart irrigation and use of artificial intelligence in the agricultural sector.

decision-making and provide more accurate predictions to farmers for their irrigation decisions.

SPISs are also becoming increasingly popular in the agricultural sector due to their potential to reduce irrigation energy costs. Solar irrigation systems offer several advantages, including cost savings, energy efficiency, environmental benefits, and convenience. However, there are also some potential disadvantages to consider, as summarized in Table 1.

4 A case study of solar-powered irrigation systems from Pakistan

4.1 Groundwater-irrigated agriculture in Pakistan

Pakistan has an estimated 18.3 million hectares (ha) of cultivated land, of which 8.2 million ha is irrigated using automated decision-making to provide

Table 1 Summary of main advantages and disadvantages of solar-powered irrigation systems

Benefits	Drawbacks
Relies on solar energy, lowering the operational costs significantly compared to conventional fossil energy-based equivalents	Although the technology is becoming economically attractive, the installation costs still remain high
Does not rely on unreliable or costly energy/fuel supply chains	Many systems require greater water storage capacity compared to diesel or conventional energy grid systems
Some systems such as non-moving solar panels and inverters require minimal routine maintenance	In some areas, the theft of solar panels is a potential challenge
Reduction in noise and pollution	Performance of the system is contingent upon solar radiation levels
High-quality solar panels are typically covered by a 25-year warranty, while inverters usually carry a 6–8-year warranty	Lack of skilled SPIS technicians available in remote areas for installation and maintenance
Helps reduce production costs, increasing farmers' profitability	Generally requires more land to be economically effective
Helps improve rural livelihoods by creating jobs and providing access to water in remote areas	Weather conditions can be challenging to SPIS operations
Reduced carbon footprint, resulting in lower greenhouse gases and improvements to the environment	Geographical limitations can be an issue as SPIS can be installed in areas with abundant sunlight
Requires less manual labour and uses free energy	Generally, requires a large initial investment to purchase and install the necessary components and infrastructure

more accurate predictions to farmers for their canal and groundwater sources (Pakistan Bureau of Statistics, 2014–15). With 94% of groundwater being used for irrigation, the remaining 6% is allocated for domestic and industrial purposes (Briscoe et al., 2005). In the last few decades, groundwater has overweighed the role of surface water and become increasingly important source in meeting the country's irrigation requirements, currently accounting for around 60% of total irrigation requirements (Qureshi and Ashraf, 2019). This is primarily due to the development of tube well technology and the unrestricted growth of privately installed tube wells across the country (Watto et al., 2018). There are now over 1.4 million tube-wells in Pakistan (automate decision-making and provide more accurate predictions to farmers for their (Fig. 3), collectively abstracting 62 billion m^3 groundwater, which is used to irrigate 11.96 million ha, either alone or in conjunction with surface water (Young et al., 2019; Pakistan Bureau of Statistics, 2014–15). Nonetheless, the reliance on groundwater for irrigation differs significantly across the country, due to the considerable disparities in surface water availability, groundwater quality, and accessibility. In areas where

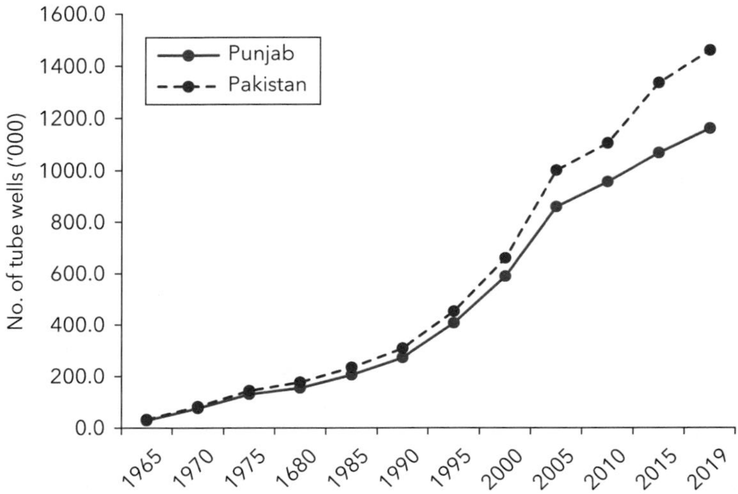

Figure 3 Reported increase in the number of tube wells in Pakistan and in Punjab province between 1965 and 2019.

surface water is either seasonal or not available, groundwater meets up to 80% and even the entire irrigation requirements (Nabeel, 2021), especially in the Punjab province. Given that the Punjab is the largest groundwater user and agricultural producing province in Pakistan, here we focus on SPIS adoption in the Punjab province only.

Currently, more than half the overall irrigation water requirements are being met through groundwater abstractions in the Punjab province, with over three-quarters of the irrigated area (78%) depending on groundwater, either exclusively through tube wells, wells, or in combination with water from canals (Fig. 4). Over 1.15 million tube wells in the Punjab are being used to collectively abstract 55.6 Km^3 (90% of the overall groundwater abstraction in Pakistan) for irrigation annually (ACE, AGC, and SMEC, 2011). Of these, 70% of tube-wells are located in canal command areas and utilized as a supplemental source of irrigation, while the remaining 30% are used solely for groundwater-based irrigation, primarily in the tail-end areas of canals where farmers do not receive adequate canal water supply (Jehangir et al., 2003; Qureshi, 2015).

4.2 Solar-powered irrigation systems' potential in the Punjab, Pakistan

Located between latitude 24N and 37N and longitude 62E and 75.5E, Pakistan receives abundant sunshine on an annual basis. It is estimated that Pakistan has an overall potential of 2.9 million megawatts of solar energy through PV and solar thermal systems. Considering the potential for solar energy and c11.76

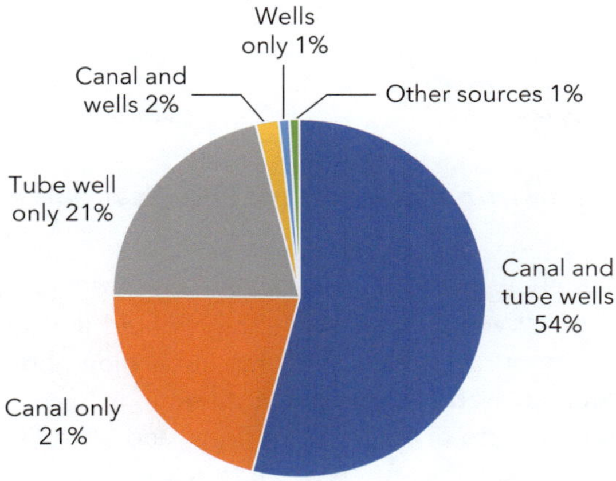

Figure 4 Reported area under cultivation dependent on different water sources in the Punjab province. Source: Adapted from: Punjab Bureau of Statistics (2014–15).

million ha dependent on groundwater and 1.15 million electric and diesel tube wells, there is an enormous potential to unleash the energy–groundwater nexus and limit 1 million metric tonnes of CO_2 emissions by converting diesel- and electric-powered tube wells to solar. Punjab has the highest concentration of tube wells in the country, with around 90% of diesel tube wells and 83% of electric tube wells (Punjab Development Statistics, 2020).

In Punjab, there are more than 1.1 million tube wells, representing a density of 77 tube wells per ha of irrigated land (Punjab Development Statistics, 2020). Across Pakistan, tube wells consume 51 billion kWh of energy for groundwater abstraction, with almost 46 billion kWh used in Punjab alone (Qureshi, 2014). At a national level, 1.18 million tube wells are powered by diesel, while 197 000 are electric. Qureshi (2014) estimated that electric tube wells require 6 billion kWh of electricity and diesel-powered tube wells consume 3.5 billion litres of diesel, resulting in a collective consumption of 41 billion kWh of energy to abstract 50 Km^3 of groundwater. Based on Qureshi (2014) calculations, 0.820 kWh of energy is needed to extract 1 m^3 of groundwater. Given that groundwater abstraction has reached 62 Km^3, and the number of tube wells has increased, the extrapolated 0.820 kWh energy consumption per cubic metre suggests that approximately 51 billion kWh of energy is used for groundwater pumping for irrigation in Pakistan. As such, the total energy use in Punjab is calculated to be around 45.5 billion kWh.

A recent study by Lytton et al. (2021) revealed that the groundwater–energy nexus within the region is set to become more intense. This is due to the declining groundwater levels, along with increasing waterlogging and salinity, leading to increased energy use for groundwater abstraction, and thus increased pumping costs. In light of this, solar-powered tube wells can offer a cost-effective renewable energy alternative to conventional tube wells, allowing for groundwater pumping while avoiding the burning of scarce fossil fuels and their associated GHG emissions.

4.3 Solar-powered irrigation systems: initiatives and efforts

Various feasibility studies and reports have been conducted to assess the use of solar energy in the agricultural sector in Pakistan and Punjab province. For example, the National Water Policy (2018) highlighted the potential of solar energy, calling for the provinces to guide the installation of different types of tube wells depending on specific geographical settings. In addition, incentives such as duty-free imports of solar PV modules and attractive financing for borrowers and banks/institutions have been made available. In 2015, the Ministry of National Food Security and Research, the State Bank of Pakistan, and the Ministry of Finance introduced an interest-free lending scheme to support small farmers (landholdings <12.5 acres) with the installation or conversion of 30 000 solar-powered tube wells over 3 years. The Punjab government, in collaboration with the World Bank, allocated common reporting standard 4 billion (US$25 million) to provide an 80% subsidy to farmers for the installation of solar-powered tube wells with high efficiency irrigation systems, such as drip/sprinkler systems (Hassan et al., 2019).

Punjab is at the threshold of a revolution in the use of solar-powered tube wells. Over the past few years, more affluent farmers have been installing these without government support, leading to a steady growth in the number of installations. However, the exact number of solar-powered tube wells in Punjab is unknown. The Punjab Irrigation Department recently declared that it will soon begin data collection on tube wells, which will be available within the next 1–2 years. According to Ali Shah and Akbar (2021), the National Agricultural Research Council (NARC) reported that various private companies and authorized dealers have installed 374 solar-powered tube wells at various locations in the Soan Basin. So far, the largest investment in Punjab was made under the Punjab Irrigated Agriculture Productivity Improvement Project which plans to install high efficiency irrigation systems (HEISs) such as drip coupled with solar-powered irrigation pumps on 48 000 acres by 2021 which was completed on 8000 acres by 2016–17. Under this scheme, adopters of the complete system of HEIS including solar-powered pumps were given 60% subsidies on total system costs by the Punjab government. In addition, other organisations such as the Agency for Barani Areas Development (ABAD) are promoting profitable agriculture involving high-value crops and reducing groundwater pumping costs through provision of alternative sources of energy.

4.4 Concerns related to solar-powered irrigation systems use in Pakistan

The groundwater–energy nexus is a highly intricate phenomenon. This complexity makes groundwater governance even more challenging, not just in the case of traditional energy systems but also for solar-powered groundwater pumping systems. To effectively manage groundwater resources, a holistic approach is required, which should include consistent monitoring, education, and awareness. Nevertheless, evidence suggests that solar-powered tube wells can bear a relatively higher risk of encouraging groundwater over-exploitation as there is no recurring input cost involved rather there is an incentive to intensify and expand the scale of operation through increased groundwater abstraction. This is particularly concerning in Pakistan and the Punjab province, where over-drafting of groundwater is already occurring to support one of the world's least water productive systems. Pakistan's average rice water productivity is 0.45 kg/m^3 – this is 55% lower than the average water productivity in Asian countries (1.0 kg/m^3) (Water Watch, 2003). Pakistan also exports one of the most 'water-expensive' rice at competitive prices in the global market. Approximately 11% of the world's depleted groundwater is estimated to be embedded in transboundary food trade. Pakistan is the leading contributor with nearly 29% of the global non-renewable groundwater exports embedded in food exports, with 82% of this exported groundwater embedded in rice exports alone (Dalin et al., 2017). Without effective measures and incentives to reduce groundwater use, installing solar-powered tube wells and transitioning to more water-intensive crops and higher cropping intensities could further exacerbate the groundwater problem. This is due to the low running costs of solar-powered tube wells. Therefore, it is important to make sure that the potential economic and environmental costs associated with rapid groundwater depletion should not outweigh the benefits of reduced energy consumption and carbon emissions.

4.5 Pathways to ensure long-term sustainability of solar-powered irrigation systems

Over the last one decade, unreliable grid supplies and the vagaries of energy costs have forced farmers to seek alternative more reliable and cost-efficient alternatives for ensuring timely irrigation water supplies. Fortunately, recent improvements in efficiency and notable reductions in cost have made solar-powered tube wells the best alternative. Solar-powered tube wells have the potential to unleash the typical energy–groundwater nexus and limit significant CO_2 emissions as they use clean, cheap, and easily accessible energy. However, open and easy access to water pumping can lead to over-exploitation on the other hand unless the long-term sustainability of groundwater resources is ensured. In this regard, a multidimensional approach with actions involving institutions, policies, tools, and technologies for groundwater monitoring, education, and awareness raising is required for up-scaling solar-powered tube wells. The following recommendations focus on facilitating the long-term sustainability of solar tube wells and groundwater resources.

4.5.1 Groundwater management actions

To ensure more sustainable use of groundwater, it is important to enforce measures including groundwater licensing and establishing a water commission. In addition, it will be necessary to adopt a comprehensive approach by implementing a set of complementary policies for monitoring and allocating groundwater, particularly in relation to solar-powered tube wells. A few examples are as follows:

1 To ensure groundwater abstractions do not exceed recharge rates, all solar-powered tube wells (whether subsidized or private, new or replaced) should be equipped with smart meters linked to an advanced metering infrastructure that tracks and reports abstracted volumes. This data should then be used to inform the licensing process for groundwater abstraction.
2 The expansion of solar-powered tube wells should be tied to mandatory measures that maximize irrigation water efficiency by implementing suitable agricultural practices. In addition, agricultural subsidies and other incentives related to solar-powered tube wells could be linked to regular monitoring and reporting of groundwater usage.
3 Mapping and analysing groundwater levels and quality should be conducted in order to identify areas with and without groundwater potential, serving as a guide for selecting areas for the expansion of solar-powered tube wells.
4 Different cropping zones should be mapped based on groundwater availability to identify the optimal cropping pattern for each zone. For instance, encouraging olive cultivation in the dry Pothohar region in Punjab and discouraging the cultivation of rice and sugarcane in south-eastern Punjab in favour of high-value, lower-water demanding horticultural crops could be considered.
5 A quota-based groundwater allocation system should be implemented to cap groundwater abstraction, along with the mandatory use of solar-powered tube wells and irrigation efficient methods and technologies. Since solar-powered tube wells do not provide any economic incentive to limit groundwater abstraction, alternative economic incentives should be explored. This could include incentives to farmers who comply with recommended irrigation water usage and application practices and

imposing sanctions, such as restrictions on groundwater allocations, on those who do not comply.

6 Exploring the optimal minimum distance between tube wells in a specific area is essential to prevent excessive tube well density, since the extraction of groundwater from one tube well can diminish the pumping rate of a neighboring one, potentially leading to derogation of water rights and social conflicts.

7 Rather than using energy pricing to control groundwater abstractions, as done historically, solar-powered tube wells should be seen not only as a way to reduce energy consumption but also as an opportunity to promote more sustainable energy and groundwater use.

4.5.2 Addressing the social equity challenge

Despite significant improvements in efficiency and reductions in the cost of solar panels, the high installation and capital investment costs still limit scaling-up of solar-powered tube wells, especially for subsistence and marginalized farmers as subsidies often fail to reach those who need them most. In Punjab, around 70% of farmers are smallholders with <5 ha of cultivated land. Smallholder farmers generally have limited access to resources and opportunities and insufficient funds to invest in solar-powered tube wells, even when subsidized. The following recommendations relate to ways of linking solar-powered tube well to social equity benefits.

1 Gathering and disseminating data to demonstrate how policy-based lending can effectively address smallholders' issues in the energy and irrigation sectors should be a priority. To maximize social equity benefits, policy-based lending approaches should be formulated and implemented for solar-powered tube wells.

2 The provision of smaller capacity solar-powered tube wells to poor and marginalized farmers should be investigated as an option to ensure that a wide range of farmers have access to solar-powered tube wells. In some areas, subsistent farmers might benefit from smaller capacity units suitable for vegetable gardening and livestock farming.

3 Subsidies for conventional pumps should be withdrawn and redirected towards converting or growing solar-powered tube wells to encourage using more sustainable energy sources.

4.5.3 General requirements to facilitate solar-powered irrigation systems

The implementation of solar-powered tube wells in Pakistan is still in its infancy, thus requiring thoughtful policy and planning in order to maximize the advantages of their proliferation. A number of challenges must be considered to ensure the successful long-term adoption of this technology.

1. The current lack of regulation for private sector involvement in the solar business has become an increasing issue with regard to solar-powered tube wells. To ensure uniform regulations, a regulatory body should be established to oversee the implementation of clear institutional and administrative procedures. This body should provide an effective enforcement mechanism, a system for resolving complaints and disputes, and ensure administrative transparency.
2. Given the land constraints, farmers should be encouraged to install solar panels over water bodies/ponds (floatovoltaics). Moreover, land fragmentation can be a challenge for smallholders. Mobile solar panel units should be developed to address this, similar to mobile diesel engines.
3. Exploring options to reduce the cost of installing solar batteries, particularly in wintery, cloudy, or foggy conditions, is highly recommended.
4. In order to ensure effective institutional coordination for the successful planning, development, and implementation of solar-powered irrigation projects, it is essential to clearly define the processes of who, what, when, where, and why. This will create sustainable SPISs by establishing clear institutional arrangements, roles, and responsibilities.
5. A thorough evaluation of the irrigation requirements of farmers in different geographical locations, with varying cropping patterns, soil types, groundwater table depths, and other relevant factors, should be conducted in order to customize the solar-powered tube well technology to suit the preferences of farmers and encourage their willingness to accept and pay for the technology.
6. The creation of PV waste from used and discarded solar panels is inevitable, so it is essential to plan appropriately for their future handling, disposal, and recycling.
7. The Punjab region holds a vast potential for businesses and employment opportunities with its estimated 1.15 million irrigation tube wells currently dependent on traditional energy sources. In order to capitalize on this market opportunity, short-term courses, training, and workshops should be organized to inform and educate stakeholders, particularly youth and women, with the necessary skills and knowledge to seize the opportunities presented by the solar energy market.

5 Conclusion

This chapter explored the recent developments in harnessing solar energy that have revolutionized solar-powered or PV water pumping technologies as technically, financially, and environmentally more viable and sustainable than in the past. In the future, as solar energy science and technology progress, SPIS will become more efficient, cost-effective, and more accessible to farmers worldwide. Moreover, further advancements in the technology such as more advanced automated water delivery systems, wireless control systems, and the use of AI and ML can lead to widespread adoption of SPISs in the future. As the climate continues to change, these systems will inevitably become even more critical in helping farmers to have reliable access to irrigation water supplies while minimizing the environmental footprint of irrigated agriculture.

In Pakistan, there is an enormous potential to unleash the typical energy-irrigation (groundwater) nexus and limit million metric tonnes of CO_2 emissions by replacing the existing diesel and electric tube wells with SPIS. Conventional tube wells consume about 51 billion kWh energy for pumping groundwater. Evidences suggest that the groundwater–energy nexus in the region will be further intensified. Given this context, SPIS can offer a potentially lower cost renewable energy alternative to conventional tube wells and enable groundwater pumping while minimizing consumption of scarce fossil fuels and decreasing their environmental footprints through reduced CO_2 emissions in Pakistan and in many other countries with similar geographic and climatic settings. However, there is a challenge that without effective measures to ensure sustainable groundwater use, free solar energy can encourage groundwater exploitation and transition to more water-intensive crops and increase cropping intensities. It is, therefore, imperative to envision that the economic and environmental costs incurred as a result of groundwater depletion should not exceed the benefit of reduced consumption of electricity and diesel and reduced carbon emissions.

6 References

Agrawal, S. and Jain, A. 2019. Sustainable deployment of solar irrigation pumps: key determinants and strategies. *WIREs Energy and Environment* 8(2): 1–14.

Ali, A., Hussain, T., Tantashutikun, N., Hussain, N. and Cocetta, G. 2023. Application of smart techniques, internet of things and data mining for resource use efficient and sustainable crop production. *Agriculture* 13: 397. https://doi.org/10.3390/agriculture13020397.

Ali Shah, M. A. and Akbar, M. Z. B. 2021. *Solar Irrigation in Pakistan: A Situation Analysis Report*. International Water Management Institute (IWMI), Colombo, 35p. https://doi.org/10.5337/2021.219.

Aliyu, M., Hassan, G., Said, S. A., Siddiqui, M. U., Alawami, A. T. and Elamin, I. M. 2018. A review of solar-powered water pumping systems. *Renewable and Sustainable Energy Reviews* 87: 61–76. https://doi.org/10.1016/j.rser.2018.02.010.

Alkhalidi, A., Assaf, M. N., Alkaylani, H., Halaweh, G. and Salcedo, F. P. 2023. Integrated innovative technique to assess and priorities risks associated with drought: impacts, measures/strategies, and actions, global study. *International Journal of Disaster Risk Reduction* 94: 103800.

Angelin Blessy, J. 2021. Smart irrigation system techniques using artificial intelligence and IoT. *Third International Conference on Intelligent Communication Technologies and Virtual Mobile Networks (ICICV)* : 1355–1359.

Ayoub Shaikh, T., Rasool, T. and Rasheed Lone, F. 2022. Towards leveraging the role of machine learning and artificial intelligence in precision agriculture and smart farming. *Computers and Electronics in Agriculture* 198: 107119. https://doi.org/10.1016/j.compag.2022.107119.

Balasubramanya, S. and Stifel, D. 2020. Viewpoint: water, agriculture & poverty in an era of climate change: why do we know so little? *Food Policy* 93: 101905.

Bizimana, J.-C., Bryant, H. and Richardson, J. W. 2021. Feed the future innovation lab for small scale irrigation preliminary economic impacts assessment of tariff reduction on water lifting technologies in Ethiopia .

Briscoe, J., Qamar, U., Contijoch, M., Amir, P. and Blackmore, D. 2005. *Pakistan's Water Economy: Running Dry*. World Bank, Washington, DC.

Bureau of Statistics Punjab. 2020. *Punjab Development Statistics 2020*. Bureau of Statistics, Planning and Development Department, Government of Punjab. Available at: http://bos.gop.pk/developmentstat.

Bwambale, E., Abagale, F. K. and Anornu, G. K. 2022. Smart irrigation monitoring and control strategies for improving water use efficiency in precision agriculture: a review. *Agricultural Water Management* 260: 107324. https://doi.org/10.1016/j.agwat.2021.107324.

Dahnill, D. P., Hood, Z., Adam, A., Ab Razak, M. Z. A. and Ismail, A. G. 2021. Drip irrigation detection for power outage-prone areas with internet-of-things smart fertigation management system. *International Journal of Advanced Computer Science and Applications* 12(7): 745-755.

Dalin, C., Wada, Y., Kastner, T. and Puma, M. J. 2017. Groundwater depletion embedded in international food trade. *Nature* 543(7647): 700–704.

Döll, P. and Siebert, S. 2002. Global modeling of irrigation water requirements. *Water Resources Research* 38(4): 8.

Foster, R. and Cota, A. 2014. Solar water pumping advances and comparative economics. *Energy Procedia* 57: 1431–1436. http://doi.org/10.1016/j.egypro.2014.10.134.

Grant, F., Sheline, C., Sokol, J., Amrose, S., Brownell, E., Nangia, V. and Winter, A. G. 2022. Creating a solar-powered drip irrigation optimal performance model (SDrOP) to lower the cost of drip irrigation systems for smallholder farmers. *Applied Energy* 323: 119563. https://doi.org/10.1016/j.apenergy.2022.119563.

Gu, Z., Qi, Z., Burghate, R., Yuan, S., Jiao, X. and Xu, J. 2020. Irrigation scheduling approaches and applications: a review. *Journal of Irrigation and Drainage Engineering* 146(6): 04020007.

Hassan, D., Rais, M. N., Ahmed, W., Bano, R., Burian, S. J., Ijaz, M. W. and Bhatti, F. A. 2019. Future water demand modeling using water evaluation and planning: a case study of the Indus Basin in Pakistan. *Sustainable Water Resources Management* 5(4): 1903–1915. https://doi.org/10.1007/s40899-019-00343-0.

Hema, N. and Kant, K. 2019. Cost-effective smart irrigation controller using automatic weather stations. *International Journal of Hydrology Science and Technology* 9(1): 1–27.

Hussain, F., Maeng, S. J., Cheema, M. J. M., Anjum, M. N., Afzal, A., Azam, M., Wu, R., Noor, R. S., Umair, M. and Iqbal, T. 2023. Solar irrigation potential, key issues and challenges in Pakistan. *Water* 15(9): 1727.

Hussan, I. U., Nadeem, M., Yamin, M., Ali, S., Omar, M. M., Ahmad, S., Zulfiqar, M. and Mahmood, T. 2023. Socioeconomic and environmental impact assessment of different power-sourced drip irrigation systems in Punjab, Pakistan. *Agric. Eng*. 5(1): 236–256.

Jägermeyr, J., Gerten, D. and Heinke, J. 2015. Water savings potentials of irrigation systems: dynamic global simulation. *Hydrol. Earth Syst. Sci.* 19: 3073–3091.

Jehangir, W. A., Turral, H. and Khan, S. 2003. Net gains from conjunctive use of surface and ground water. *Pakistan Development Review* 42(4II): 823–838. https://doi.org/10.30541/v42i4IIpp.823-38.

Khoodaruth, A., Oree, V., Elahee, M. K. and Clark, W. W. 2017. Exploring options for a 100% renewable energy system in Mauritius by 2050. *Utilities Policy* 44: 38–49.

Kocaman, A. S., Ozyoruk, E., Taneja, S. and Modi, V. 2020. A stochastic framework to evaluate the impact of agricultural load flexibility on the sizing of renewable energy systems. *Renewable Energy* 152: 1067.

Li, B., Zhao, J., Zhang, Y. and Bai, X. 2023. Optimal configuration and economic operation of wind – solar storage complementary system for agricultural irrigation in mountainous areas. *Journal of Electrical Engineering and Technology*. https://doi.org/10.1007/s42835-022-01373-z.

Lytton, L., Ali, A., Garthwaite, B., Punthakey, J. F. and Saeed, B. 2021. *Groundwater in Pakistan's Indus Basin: Present and Future Prospects*. World Bank, Washington, DC.

Matekaire, K. 2021. Smart farming irrigation systems using internet of things. Doctoral dissertation, Faculty of Engineering and the Built Environment, University of the Witwatersrand, Johannesburg.

Mohan, Ch., Singh, S., Gupta, M. K., Nimdeo, Y. M., Raushan, R., Deorankar, A. V., Ananda Kumar, T. M., Rout, P. K., Chanotiya, C. S., Pakhale, V. D. and Nannaware, A. D. 2023. Solar energy: a promising renewable source for meeting energy demand in Indian agriculture applications. *Sustainable Energy Technologies and Assessments* 55: 102905. https://doi.org/10.1016/j.seta.2022.102905.

Moosavian, S. M., Rahim, N. A., Selvaraj, J. and Solangi, K. H. 2013. Energy policy to promote photovoltaic generation. *Renewable and Sustainable Energy Reviews* 25: 44–58. http://doi.org/10.1016/j.rser.2013.03.030.

Obaideen, K., Yousef, B. A. A., AlMallahi, M. N., Tan, Y. C., Mahmoud, M., Jaber, H. and Ramadan, M. 2022. An overview of smart irrigation systems using IoT. *Energy Nexus* 7: 100124. https://doi.org/10.1016/j.nexus.2022.100124.

Pakistan Bureau of Statistics (2014–15). *Pakistan Statistical Year Book 2014-15 (Provisional)*. Pakistan Bureau of Statistics, Ministry of Planning Development and Special Initiatives, Government of Pakistan. Available at: http://www.pbs.gov.pk/content/pakistan-statistical-year-book-2014-provisionalPatro.

Powell, J. W., Welsh, J. M., Pannell, D. and Kingwell, R. 2021. Factors influencing Australian sugarcane irrigators ' adoption of solar photovoltaic systems for water pumping. *Cleaner Engineering and Technology* 4: 100248. https://doi.org/10.1016/j.clet.2021.100248.

Poyen, F. B., Ghosh, A., Kundu, P., Hazra, S. and Sengupta, N. 2020. Prototype model design of automatic irrigation controller. *IEEE Transactions on Instrumentation and Measurement* 70: 9456(c).

Qazi, S., Khawaja, B. A. and Farooq, Q. U. 2022. IoT-equipped and AI-enabled next generation smart agriculture: a critical review, current challenges and future trends. *IEEE Access* 10: 21219–2123535.

Qureshi, A. S. 2014. Reducing carbon emissions through improved irrigation management: a case study from Pakistan. *Irrig. and Drain* 63: : 132–138.

Qureshi, R. and Ashraf, M. 2019. *Water Security Issues of Agriculture in Pakistan*. Pakistan Academy of Sciences (PAS), Islamabad. Available at: https://www.paspk.org/wp-content/uploads/2019/06/PAS-Water-Security-Issues.pdf.

Radoglou-Grammatikis, P., Sarigiannidis, P. and Lagkas, T. 2020. A compilation of UAV applications for precision agriculture. *Computer Networks* 172: 107148. https://doi.org/10.1016/j.comnet.2020.107148.

Ramachandran, V., Ramalakshmi, R., Kavin, B. P., Hussain, I., Almaliki, A. H., Almaliki, A. A., Elnaggar, A. Y. and Hussein, E. E. 2022. Exploiting IoT and its enabled technologies for irrigation needs in agriculture. *Water* 14(5): 719.

Scanlon, B. R., Faunt, C. C., Longuevergne, L., Reedy, R. C., Alley, W. M., McGuire, V. L. and McMahon, P. B. 2012. Groundwater depletion and sustainability of irrigation in the US High Plains and Central Valley. *Proceedings of the National Academy of Sciences of the United States of America* 109(24): 9320–9325.

Shufian, A., Mominur Rahman, Md., Islam, R. and Dey, S. K. 2019. Smart Irrigation System with Solar Power and GSM Technology. In: 2019 5th International Conference on Advances in Electrical Engineering (ICAEE). IEEE, 301–305.

Shukla, A. K., Sudhakar, K. and Baredar, P. 2017. Renewable energy resources in South Asian countries: challenges, policy and recommendations ☆ *Resource-Efficient Technologies* 3: 342–346.

Siebert, S., Burke, J., Faures, J. M., Frenken, K., Hoogeveen, J., Döll, P. and Portmann, F. T. 2010. Groundwater use for irrigation – a global inventory. *Hydrology and Earth System Sciences* 14(10): 1863–1880.

Siyal, A. W., Gerbens-leenes, P. W. and Nonhebel, S. 2021. Energy and carbon footprints for irrigation water in the lower Indus Basin in Pakistan, comparing water supply by gravity fed canal networks and groundwater pumping. *Journal of Cleaner Production* 286: 125489. https://doi.org/10.1016/j.jclepro.2020.125489.

Sudharshan, N., Karthik, A. K., Kiran, J. S. and Geetha, S. 2020. Renewable energy based smart irrigation system renewable energy based smart irrigation system. *Procedia Computer Science* 165: 615–623. https://doi.org/10.1016/j.procs.2020.01.055.

Sunny, F. A., Islam, M. A., Karimanzira, T. T. P., Lan, J., Rahman, Md. S. and Zuhui, H. 2023. Adoption impact of solar based irrigation facility by water-scarce northwestern areas farmers in Bangladesh: evidence from panel data analysis . *Front. Energy Res.* 10.

Ünal, Z. 2020. Smart farming becomes even smarter with deep learning – a bibliographical analysis.*IEEE Access* 8: 105587–105609.

Verma, S., Mishra, S., Chowdhury, S., Gaur, A., Mohapatra, S., Soni, A. and Verma, P. 2021. Solar PV powered water pumping system – a review. *Materials Today: Proceedings* 46: 5601–5606. https://doi.org/10.1016/j.matpr.2020.09.434.

Waleed, A., Riaz, M. T., Muneer, M. F., Ahmad, M. A., Mughal, A., Zafar, M. A. and Shakoor, M. M. 2019. Solar (PV) water irrigation system with wireless control. In: 2019 International Symposium on Recent Advances in Electrical Engineering (RAEE). IEEE, pp. 1–4.

WaterWatch. 2003. Crop and water productivity of the Pakistan wheat and rice systems. In: Technical Report.Wageningen, The Netherlands, pp. 1–6.

Watto, M. A., Mugera, A. W., Kingwell, R. and Saqab, M. M. 2018. Re-thinking the unimpeded tube-well growth under the depleting groundwater resources in the Punjab, Pakistan. *Hydrogeology Journal* 26(7): 2411–2425.

Wazed, S. M., Hughes, B. R., O'Connor, D. and Kaiser Calautit, J. 2018. A review of sustainable solar irrigation systems for sub-Saharan Africa. *Renewable and Sustainable Energy Reviews* 81: 1206–1225. http://doi.org/10.1016/j.rser.2017.08.039.

Xie, H., Ringler, C., Zhu, Tingju and Waqas, A. 2015. Droughts in Pakistan: a spatiotemporal variability analysis using the standardized precipitation index. *Water International* 38: 37–41.

Yoo, J. J., Seo, G., Chua, M. R., Park, T. G., Lu, Y., Rotermund, F., Kim, Y. K., Moon, C. S., Jeon, N. J., Correa-Baena, J. P., Bulović, V., Shin, S. S., Bawendi, M. G. and Seo, J. 2021. Efficient perovskite solar cells via improved carrier management. *Nature* 590(7847): 587–593.

Young, W., Anwar, A., Bhatti, T., Borgomeo, E., Davies, S. W., Garthwaite III, W. R., Gilmont, E. M., Leb, C., Lytton, L., Makin, I. and Saeed, B. 2019. *Pakistan: Getting More from Water*. World Bank, Washington. Available at: https://openknowledge.worldbank.org/handle/10986/31160.

Zhao, Y., Zhu, Y., Cheng, H., Zheng, R., Meng, D. and Yang, Y. 2021. A review on semitransparent solar cells for agricultural application. *Materials Today Energy* 22: 100852.

Part 3

Improving water resources allocation and management

Chapter 9

Advances in farmer-led irrigation development in Africa

Phil Woodhouse, University of Manchester, UK

1 Introduction

The centrality of farmers' participation for successful irrigation has been a recurring theme within irrigation management literature for the past four decades (Uphoff, 1986; Bottrall, 1985; Chambers, 1989; Wade, 1988) but has assumed a particular significance in sub-Saharan Africa (SSA) in the twenty-first century. This is because the region's irrigation development and food production are widely perceived as lagging compared to other less-developed regions such as Asia and Latin America (AFDB et al., 2008; World Bank, 2011; FAO, 2020). A representative study presenting this argument (Svendsen et al., 2009) states that SSA's area equipped with irrigation infrastructure accounts for only 3.5% of total cultivated area, compared to almost 35% in Asia (Table 1). Moreover, in this regard SSA contrasts with arid North Africa and the Indian Ocean islands (mainly Madagascar and Mauritius) which are similar to the Asian profile of irrigation development. In these two regions irrigation development has approached its maximum potential, at 88% and 71% in North Africa and the Indian Ocean Islands respectively, compared to less than 20% across sub-Saharan Africa.

A number of studies in the 1980s (Moris and Thom, 1985; Diemer and Vincent, 1992; Adams, 1992) identified the reasons for poor irrigation performance in SSA as arising partly from flawed technical design, but particularly from a lack of understanding of how irrigation was to fit within the

http://dx.doi.org/10.19103/AS.2023.0123.14

Table 1 Indicators of irrigation performance in Africa (after Svendsen et al., 2009 Tables 3 and 11)

African agro-ecological region	'equipped' irrigation area/ total area cultivated	% use of 'equipped' irrigation area	Total area of water management/total area cultivated	'Equipped' irrigation as % of potential
Northern	28.1	80.4	28.1	88
Sudano-Sahelian	6.9	63.3	9.2	50
Gulf of Guinea	1.5	73.5	3.3	8
Central	0.7	47.5	2.8	1
Eastern	2.6	24.0	1.8	11
Southern	4.2	80.7	4.8	36
Indian Ocean Islands	30.4	99.4	30.7	71
Average Sub-Saharan Africa	3.5	71	4.5	18
Average Asia	33.6	66.9	34.3	

broader farming system and rural economy. This resulted in unforeseen labour shortages, failure of crop revenues to cover production costs, insufficient infrastructure maintenance and, consequently, low rates of use of irrigable areas. This kind of analysis supported a decade-long moratorium on new irrigation schemes in the region, until the new millennium witnessed continent-wide initiatives to increase African agricultural productivity. These included: the development of the Comprehensive Africa Agricultural Development Programme (CAADP) under the African Union's New Partnership for Africa's Development (NEPAD). The CAADP is organised through four pillars of which the first is 'extending the area under sustainable land management and reliable water control systems' (NEPAD, 2003). An increase in global food prices in 2008 highlighted failures to increase productivity of sub-Saharan African agriculture during the previous two decades and an alliance of five influential international organisations called for new large-scale investments into irrigation in the region (AFDB et al., 2008). Nonetheless, concerns about the high cost and low productivity of irrigation schemes in Africa remain (Inocencio et al., 2007). A recent study (Higginbottom et al., 2021) of 79 irrigation schemes in SSA shows the problem of poor productivity of irrigation persists, with a median of only 16% of the designed irrigable area in production and only a quarter of schemes attaining more than 80% of their designed production area while a fifth were completely inactive. The authors concluded that productivity had not increased over the previous six decades.

One response to this situation is to focus on rainfed production, exemplified by a World Bank assessment of potential for expanding commercial agriculture in 'Guinea savannah' zones in Africa, based entirely upon the premise of rainfed systems of production (World Bank, 2009). While there are undoubtedly gains achievable from improving rainfed agriculture, notably by improved soil management (Reij et al., 1996), there are reasons to argue that irrigation will be crucial for the future development of sub-Saharan agriculture. First, while less apparent than in arid North Africa, where irrigation is an obvious requirement, for the two thirds of sub-Saharan Africa that lie outside the equatorial humid zone, water is the key constraint to agricultural production. Annual rainfall may vary from as little as 400 mm in Sahelian zones to 1200 mm in 'Guinea savanna', but in all cases is strongly seasonal, being restricted to 4–5 months in a year. In addition to the long dry season, lack of moisture for crops arises from significant deficits ('meteorological' droughts) in total annual rainfall once or twice a decade, and high probability (in two out of three years) of dry spells at critical crop growth stages during the rainy season. This means that significant risk attaches to all other investments in agriculture if crops are dependent on rainfall alone (Rockström et al., 2003; Rockström et al., 2010). Studies on maize in semi-arid areas of Kenya and Tanzania suggests the crop may be expected to be exposed to dry spells of 10 days or longer in 74–80% of rainy seasons (Barron et al., 2003).

Historical and contemporary African agriculture includes many different 'indigenous soil and water conservation' strategies to reduce risks associated with low and unreliable rainfall. These include: digging pits and laying stone lines to increase interception and retention of rain (Reij et al., 1996); traditions of stream diversion for 'hill furrow' crop irrigation in the East African highlands (Adams and Anderson, 1988; Tagseth, 2008); cultivation of floodplains following a receding flood (e.g. on the Senegal and Niger rivers); and construction of terraces on hillsides and cultivation of raised beds in wetlands in Zimbabwe (Soper, 2006). More generally, crop planting is often split to occupy a variety of topographical positions and thus mitigate rainfall hazards: floods in lower lying sites in wet years, drought on higher, better-drained sites in drier years (Richards, 1985).

Under colonial administration in the early 20th century these aspects of water management in African agriculture were largely ignored. Instead, investment focused on large-scale infrastructure on major river floodplains, such as the Gezira scheme on the Nile in Sudan (Barnett, 1977; Ertsen, 2016) and the Office du Niger in Mali (Aw and Diemer, 2005), both built with the aim of settling small-scale tenant farmers within government-managed large schemes to produce cotton. Later, these schemes were used as the development model for rural resettlement programmes, such as at Mwea and Tana river in Kenya (Kenya Government n.d.). Irrigation investment increased particularly during the decade after independence of many African countries, accelerated by instances of famine due to extended drought periods in the 1970s. While researchers documented the poor performance of many of these investments in the 1980s, irrigation inventories (Hocombe et al., 1986; Underhill, 1984) started to reveal significant 'informal' irrigation practised by small-scale farmers using a variety of technologies, both modern (e.g. motor pumps and boreholes) and traditional (e.g. stream diversions using weirs and canals; managed flooding and drainage of valley bottoms). While official irrigation statistics have struggled to measure accurately the extent of this irrigation activity spread across many small plots (Venot et al., 2021), research over the past decade has generated increasing evidence that small-scale farmers across SSA are investing both capital and labour in a diverse range of irrigation methods. We may classify these loosely under four broad headings: (1) 'hill furrows', (2) management of seasonally flooded valleys, (3) use of small motor pumps and (4) re-use of urban wastewater.

Hill furrow irrigation, historically established in mountainous and Rift valley terrain of East Africa (Adams and Anderson, 1988; Grove, 1993; Adams et al., 1994; Tagseth, 2008) has been observed more recently in hilly terrain of central and northern Mozambique (Nkoka et al., 2014; Bolding et al., 2010). Research in the upper catchment of the Revué River in Mozambique (Bolding et al., 2010; Beekman et al., 2014) shows that typically several furrow irrigation systems take water from one stream, sometimes additionally capturing water from side streams, springs or neighbouring catchments. Beekman et al. (2014) suggest that these systems have recently expanded significantly, such that in the Mozambican border area of Zimbabwe there are currently more than 100,000 ha irrigated in this manner.

Low-lying land and valley bottoms are cultivated throughout SSA. Such areas have a variety of names in different countries including *bolis* in Sierra Leone, *fadama* in Nigeria, *bas fonds* in Niger, Mali and Burkina Faso, the Swahili term *mbuga* in East Africa and *vlei* in Zimbabwe and South Africa. In Southern African countries, they are also referred to as *dambos* (Bell et al., 1987). Characteristically high in fertility, they may be cultivated by digging channels and bunds to manage seasonal floodwaters or maintain high water tables to grow rice, by ridging and draining to grow other crops and by lifting water from the shallow water table to irrigate during the dry season. World Bank funding for pumps and boreholes has supported small-scale irrigation in the extensive fadamas of northern Nigeria since the 1980s (Carter et al., 1983; Carter, 1989), and by 2004 fadama irrigation was estimated at 114 000 ha (Vermillion, 2004), equivalent to more than half of Nigeria's official total 220 000 ha of irrigation. More generally, the wetland areas of SSA have been sites of agricultural intensification by small-scale farmers as cultivation has replaced dry season grazing (Woodhouse et al., 2000) wherever markets for high-value crops are accessible. They are now highly valued and sites of struggles for ownership of land (Woodhouse, 2003; Peters and Kambewa, 2007). Where such small-scale systems are very extensive, as in the upper catchment of the Great Ruaha River in Tanzania, they may significantly affect downstream water use and attract political opposition and state intervention (Lankford, 2004; Lankford et al., 2004; Lankford and Beale, 2007).

The re-use of urban wastewater for irrigation is the most ubiquitous form of small-scale irrigation development by farmers. Rising demand for fresh fruit and vegetables has driven an informal sector of small-scale horticultural producers in and around almost all African urban centres (Drechsel and Keraita, 2014). The use of wastewater, diluted or raw, creates health risks for both irrigators and consumers of the crops. Despite this, and often precarious access to land, small-scale producers have developed substantial irrigated areas in many African urban areas. In Ghana, where urban agriculture is officially recognised as a means of meeting food demands, Drechsel and Keraita (2014, 3) estimate that irrigated production of lettuce and cabbage for salad consumption involves 'up to 2000 urban vegetable farmers, 5300 street food sellers, and 800 000 daily consumers within the major cities plus an unknown number of traders'. Plot sizes may be very small, between 0.01 and 0.02 ha per farmer in Accra (Danso et al., 2014), and watering cans are the most common technology, although motorised pumps may be used to bring water from a more distant source to a reservoir nearer the fields (Keraita and Cofie, 2014).

The reduction in capital cost of imported small gasoline or diesel water pumps has had a major impact on irrigation by small-scale farmers in many different locations, often supplementing other irrigation technologies but also opening up new opportunities. Documented examples include: the fadama irrigation in Northern Nigeria referred to above, irrigation from Lake Victoria in Kenya (Hebinck et al., 2019); use of groundwater to irrigate in an area of failed surface irrigation in Kahe, Tanzania (de Bont et al., 2019a); and exploitation of land upstream of small reservoirs in Burkina Faso (Venot et al., 2012; de Fraiture et al., 2014). Namara et al. (2014) report that official data showed that in Ghana alone over 65 000 pumps and accessories were imported between 2003 and 2010, worth more than USD 8 million.

As observations of farmers' small-scale irrigation initiatives have proliferated, a number of terms have been used to characterise them, such as 'distributed irrigation' (Burney et al., 2013) and 'small private irrigation' (de Fraiture and Giordano, 2014; Giordano et al., 2012) as well as 'farmer-led irrigation development' (Nkoka et al., 2014). It is the latter term that has come to be widely used in irrigation policy (Woodhouse et al., 2017; Veldwisch et al., 2019; Wiggins and Lankford, 2019; IWMI, 2019; Izzi et al., 2021). It has been defined as:

> a process where farmers assume a driving role in improving their water-use for agriculture by bringing about changes in knowledge production, technology use, investment patterns and market linkages, and the governance of land and water. In the process, farmers exhibit entrepreneurial and risk-taking behaviour and interact with a range of other actors. (Woodhouse et al., 2017: 216)

This definition highlights the role of small-scale farmers in the *process* of irrigation development, rather than any particular technique of irrigation. This focus on process further emphasises that, while farmers are active in determining the purpose, location and methodology of irrigation, they do not act in a vacuum but engage with a variety of different government, non-government and commercial development agencies. While many commentators may regard farmers' irrigation initiatives as 'informal', the above definition makes clear that, from a farmer's perspective, they are neither 'spontaneous' nor 'unplanned', but rather constitute intentional development that requires work and investment.

2 Key issues and challenges of farmer-led irrigation development

Farmer-led irrigation development is a phenomenon that, once named, has come to be recognised widely across SSA and has become incorporated in policy frameworks (e.g. African Union, 2020) and investment programmes (e.g. World Bank. See: Izzi et al., 2021). We can identify five key areas in which research is needed to clarify the trajectory and impact of farmer-led irrigation development:

1. How significant is it quantitatively, in terms of land area irrigated?
2. What conditions drive farmer-led irrigation?
3. How is technological change achieved?
4. What outcomes are evident, in terms of productivity, sustainability and equity effects?
5. What policy challenges arise?

Each of these key knowledge gaps are briefly considered in the following sections.

2.1 What is the quantitative significance of farmer-led irrigation?

There are no definitive data for the overall extent of farmer-led irrigation development in SSA. Data for imports of pumping equipment (Namara et al., 2014) or the extent of hand-dug canals for 'hill-furrow' irrigation (Beekman et al., 2014) suggest that, in aggregate, such small-scale irrigation activities cover areas at least as large as formal, engineer-designed large-scale irrigation schemes. However, official statistics are often incomplete since data collected by government agencies tend to emphasise areas 'fully-equipped' with permanent irrigation infrastructure while ignoring small-scale irrigation using more improvised (e.g. temporary stream diversions) or portable (e.g. pump) technology.

Yet assessments of the potential benefits from irrigation in Africa highlight the greater returns from small-scale irrigation. In particular, modelling techniques integrating different datasets on crop yield potential, rainfall and hydrology (You et al., 2011; Xie et al., 2014) have assessed irrigation investment profitability for a number of economic scenarios in different agroecological zones. These indicated that, for internal rates of return (IRR) of greater than 12%, irrigation is expected to be profitable on 1.3 million ha of large-scale (dam-based) schemes but about 3.7 million hectares small-scale irrigation in five regions of sub-Saharan Africa (i.e. excluding North Africa and Indian Ocean Islands). This type of exercise is based on many assumptions one would want to question in specific contexts, and the authors of the work have underlined the high sensitivity of the analysis to cost assumptions. Moreover, Wiggins and Lankford (2019) argue such studies are likely to over-estimate potential irrigation because irrigable areas depend on water required. This means potential is greater in the rainy season (when irrigation is supplemental to rainfall) or where significant rainfall storage is available for dry season usage. More work is needed to evaluate the impact of such factors on irrigation in specific contexts.

Nonetheless, recent advances in the availability and increased resolution of satellite imagery has enabled remote sensing studies to detect smaller areas of irrigation. Such studies show that "broadly, in Africa areas receiving improved water management practices including irrigation are about 2 to 3 times greater than was previously thought. About half of these areas have some level of active water management which is beyond traditional definition of 'rainfed' but not yet 'irrigated'" (IWMI, 2016). Studies in Tanzania (Venot et al., 2021) using radar remote sensing have also shown that areas of irrigated rice may be very much larger than official statistics record. These authors highlight that the discrepancy between official statistics and such remote sensing measurements derive from problems of official definition of 'irrigation' exclusively in terms of engineering infrastructure (see also Harrison, 2018). This makes it difficult for irrigation statistics to include much of the farmer-led irrigation development that does not have permanent infrastructure.

2.2 What conditions drive farmer-led irrigation?

Farmer-led irrigation development appears oriented primarily towards the production of crops for local sale, rather than for consumption within irrigators' households. Irrigators focus on growing high-value vegetables and fruit for sale in urban markets or rice, which is a preferred staple in urban areas in many parts of Africa. A primary driver of farmer-led irrigation development is therefore urbanisation, estimated at about 50% across SSA but varying between about 33% and 70% in individual countries (OECD/SWAC, 2020). Moreover, the projected annual population growth rate of 2.45%, the highest in the global South, is primarily in urban centres. Access to these growing centres of food consumption, and hence the condition of road infrastructure, is therefore a major factor in stimulating small-scale farmers' irrigation initiatives.

Other factors that may catalyse farmer-led irrigation development are supply-chain development, particularly through reduction in barriers to importation of motor pumps (Namara et al., 2014; Dessalegn and Merrey, 2015) and provision of seasonal credit by crop purchasers (Veldwisch and Woodhouse, 2022), and immigration that may bring irrigation expertise as well as additional labour (Bolding et al., 2010). These observations mark a quite different perspective from that based on 'agro-ecological potential' that typically informs conventional irrigation planning. From a farmer-led perspective, irrigation development is primarily a matter of economic opportunities and the availability of means with which to develop land and water resources to exploit them.

More broadly, farmers' initiatives to develop irrigation cannot be viewed in isolation from their social, economic, ecological or political context. Woodhouse et al. (2017) document instances where farmers have lobbied government and non-government agencies to provide funding, material or technical support to upgrade their irrigation. In contrast, governments may see farmers' irrigation initiatives as contrary to their visions for agricultural development (De Bont et al., 2019b), or as having negative downstream impacts on other water users (Lankford et al., 2004; Walsh, 2012; Mdee and Harrison, 2017). To analyse and understand the contemporary dynamics of farmer-led irrigation development a dichotomy between 'external' and 'internal' (farmers' own) investment needs to be replaced by an understanding of irrigation development as a range of interactions between government, donor and non-government agencies, markets and the rural economy, and farmers.

2.3 How is technological change achieved?

Following from the discussion above, we can understand technological change in farmer-led irrigation as occurring within broad socio-technical networks that involve landholders, tenant farmers, intermediaries such as pump-owners, traders, masons and mechanics, and agents from governmental, non-governmental and international organisations. Evidence of farmers' initiatives in irrigation shows they copy water management technologies (the knowledge of which is being greatly enhanced by migration) and adapt them to local circumstances. Early examples of use of water pumps for irrigation in the Senegal River Valley during the 1970s trace the introduction of pumps to migrants returning from work in France (Adams, 1981; Diemer and van der Laan, 1987). Another example arose when Japanese technicians constructed the Lower Moshi irrigation scheme in Tanzania, small-scale irrigators copied the techniques of rice cultivation (using transplanting) and set up their own system of irrigation upstream (de Bont et al., 2019a).

Farmers may also build new relationships in order to tap into engineering advice and support, where available. For instance, in several West African countries petrol pump irrigation developed in a dynamic triangle of relations between pump owners, land owners and cultivators (De Fraiture et al., 2014). Namara et al. (2014, 197) describe in Ghana a network, functioning on market principles, in which 'pumps can be rented for a day, for a season, for a year or even on an hourly basis'. Hebinck et al. (2019) describe the role of mechanics and agro-dealers in supporting the effective spread of irrigation using petrol pumps in Western Kenya.

Farmer-led perspectives on irrigation may inform different technological choices from those of irrigation planners (Beekman et al., 2014). Farmers are likely to give priority to the ease of development, maintenance and operation, either individually or within a small group. This may lead, for instance, to a preference for developing irrigation using small streams on land with relatively steep slopes (because this combination makes it easier to move water around), or around stable open water sources or in wetland areas instead of looking for large flat areas and abundantly flowing rivers favoured by irrigation planners and development projects. The importance of understanding farmers' constraints and priorities in technological change is fundamental to the idea of farmer-led irrigation development as a process, rather than as a 'package' of technologies to be 'adopted' by small-scale farmers (Veldwisch et al., 2019). This needs to be emphasised particularly because the history of international funding support to irrigation development in Africa is replete with technologies, such as treadle pumps and drip irrigation that have commonly failed to achieve the impacts often claimed for them (Lankford, 2009; Venot, 2016; Merrey and Sally, 2017).

2.4 What outcomes: productivity, sustainability and equity effects?

Few systematic assessments of the impacts of farmer-led irrigation are yet available. However, those that have been undertaken show dramatic increases in income from crop production for irrigators compared to non-irrigating households (de Bont et al., 2019b). These are associated with ownership of assets and lower likelihood of food insecurity. These effects, set out in more detail below, raise the prospect of a growing differentiation between those who irrigate and those who do not. This impact may be partially mitigated by increasing demand for labour, as households that irrigate are much more likely to employ paid labour, but nonetheless points to processes of social change within communities where farmer-led irrigation development is taking place. These processes may include gendered intra-household re-distribution of burdens and benefits observed in earlier irrigation developments (Carney, 1998).

A recurring observation is that farmer-led irrigation development is associated with immigration from other rural areas (Woodhouse et al., 2017). Migrants may be seeking work as paid labour on irrigated plots or may be seeking land on which to start their own irrigated production. Farmer-led irrigation development is therefore associated with development of markets for labour and land. With regard to the latter, it is important to note that formal land tenure is not a prerequisite for irrigation development. To the extent that farmers' irrigation initiatives take place outside schemes formally demarcated by government, access to land will be governed by customary tenure (Woodhouse et al., 2017). Incomers wishing to irrigate may purchase or rent land from local customary landholders. A number of studies report land rental arranged for a growing season, sometimes involving partnerships between farmers/farm managers, landowners, and pump-owners (e.g. de Fraiture et al., 2014; de Bont et al., 2019a; Karimba et al., 2022).

While such arrangements are widely used, they seldom involve formal contracts and rely on social relationships for compliance. Particularly where such transactions are between local landholders and immigrants, issues of legitimacy and legality may arise if they are contested by others that feel they have claims to the land or water (Woodhouse et al., 2017; Chimhowu and Woodhouse, 2006; Peters and Kambewa, 2007; Chauveau et al., 2006). Such contestation may pose questions about the longer-term stability, sustainability, and equity dimensions of farmer-led irrigation development. More fundamentally, farmer-led irrigation development requires new forms of organisation to manage shared water resources (Komakech and van der Zaag, 2011; Komakech et al., 2012a,b), whether at local, regional or national levels.

2.5 Challenges for policy

Farmer-led irrigation development is a phenomenon that presents development agencies with both opportunities to raise productivity and food security and challenges to make this process socially equitable, environmentally sustainable and economically broad-based. A meeting of policy makers and researchers in 2018 noted that some countries have adjusted policy and budgets to support existing farmer-led irrigation activity. They cited the example of Ghana, where the government recognised that farmers' own irrigation initiatives covered about ten times the area of formal irrigation schemes and policy is focused on removing 'binding constraints', such as shortage of electricity supply (SAFI, 2018). In the years since that meeting, 'farmer-led irrigation' has been widely promoted by international agencies such as the World Bank, particularly in association with solar-powered water pumps (Lefore et al., 2021). However, not all governments are persuaded that small-scale irrigation initiatives are a productive use of water. De Bont et al. (2019b) document cases where government policy has defined farmers' irrigation initiatives as inefficient and emphasises the regulation and formalisation of small-scale farmers' irrigation initiatives in order to improve productivity and protect water resources. The 2018 SAFI meeting concluded that interventions by government and non-government agencies to address farmer-led irrigation development should address four main areas:

1. Framing farmer-led irrigation development as part of economic and social security strategies by:
    a. seeking ways to reduce vulnerability and spread benefits of irrigation among different social groups in rural communities;
    b. facilitating access to reliable markets for inputs (including technology) and agricultural produce;
    c. identifying and removing systemic constraints, such as transport infrastructure, electricity supply and taxation of key inputs.
2. Learning from existing cases of irrigation development by farmers by:
    a. analysing the dynamics and constraints of irrigation development in specific contexts;
    b. exercising caution over expectations for replicability of experience from one site to another;
    c. encouraging opportunities for farmer-to-farmer learning.
3. Informing policy through more accurate irrigation data, by:
    a. evaluating alternative, and possibly complementary, methods of mapping and measuring irrigation beyond formal 'schemes';
    b. revising irrigation statistics to enable recognition of location and extent of farmer-led irrigation development;
    c. identifying the status and support needs of farmer-led irrigation development.
4. Developing a supportive and accessible regulatory framework for small-scale irrigators by:

a. recognising small-scale irrigators as productive water users;
b. avoiding registration requirements that are costly and onerous for small-scale irrigators;
c. identifying 'good-enough' regulation that does not stifle local initiatives;
d. reviewing legislative and regulatory frameworks for water and agriculture to ensure they take account of farmers' irrigation initiatives and identify where state agencies confront capacity limitations to effective intervention;
e. exploring investment and technical strategies for intensification vs expansion of irrigation.

Wiggins and Lankford (2019) argue that interventions by government need to be tailored to the stage of development of farmers' irrigation activity. They outline a three-stage model in which irrigation is developed slowly in an initial phase when technology and/or market access are poorly established, through a second stage of rapid proliferation and adoption of irrigated production and a third stage in which irrigation has a significant impact on water resource availability. The development of regulatory and administrative capacity to address farmer-led irrigation development remains a critical challenge in many African countries. Responsibility for irrigation is often split, or even contested, between different ministries and capacity seldom exists to regulate effectively water use by multiple irrigators in river basins (Mdee and Harrison, 2019).

3 Case studies of farmer-led irrigation development in Tanzania and Mozambique

A survey of 2732 irrigating and non-irrigating households was undertaken at 18 study sites (Fig. 1) in Mozambique and Tanzania to obtain information on farmers' use of irrigation and its socio-economic significance for rural households. Sites were selected where farmer initiative was evident in determining the purpose, design, and management of irrigation – even though some sort of input or external assistance may have occurred in the past or following farmers' own initiatives (e.g. 'upgrading' by government agencies). The selected sites included instances where farmers have: developed irrigation during colonial times (Makanya, Parta); rehabilitated and extended irrigation abandoned by colonial settlers (Vanduzi, Messica); copied technology from neighbouring government irrigation schemes (Mandaka Mnono, Mijongweni); or purchased small motor pumps to introduce irrigation in new areas (Kahe [de Bont et al., 2019a], Tica, Zembe, Macate). The sites surveyed provided examples of a range of technologies: stream diversion for basin, furrow, or sprinkler irrigation; wetland management; and small motor pumping from surface or groundwater.

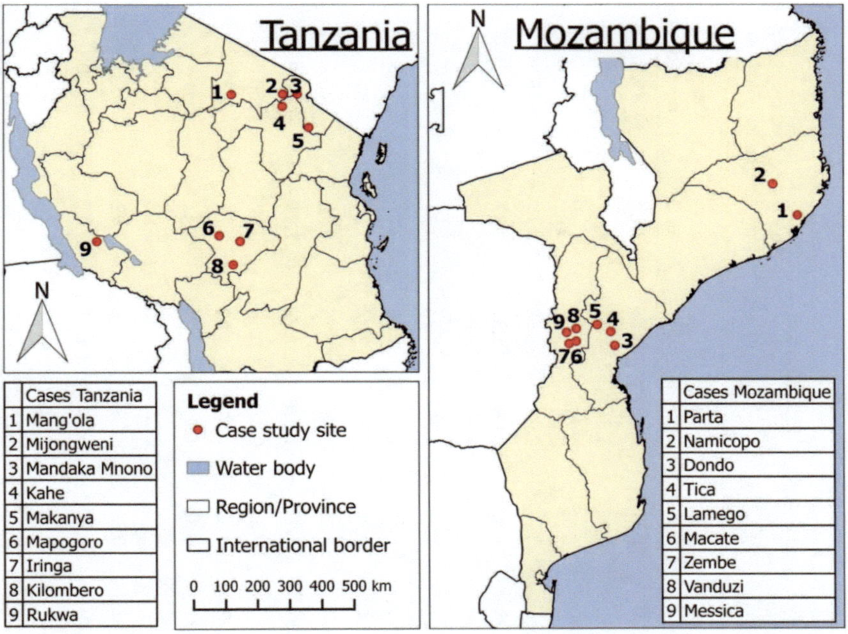

Figure 1 Overview of case study sites in Tanzania and Mozambique (from de Bont et al., 2019b).

Irrigators produced a variety of crops, including paddy rice, maize, and high-value horticultural crops for local and regional markets (tomatoes, cabbages and onions) or for export to Europe (green beans and baby corn).

Some of these areas have been recognised by the government and have received support to upgrade infrastructure (Vanduzi, Iringa, Mandaka Mnono, Mijongweni, Mapogoro), but many have not. One site has witnessed efforts by government to close down irrigation (Rukwa). Table 2 shows that areas irrigated average one to two hectares per irrigating household, although some individuals (for example, a customary chief in Messica and a businessman with rental property in a local town in Rukwa) may cultivate areas 10 times larger. The great majority of irrigating households in Mozambique have less than two hectares under irrigation, while this is less than one hectare in Tanzania. However, aggregate areas covered by such irrigation frequently reach hundreds or even thousands of hectares within an administrative district (Beekman et al., 2014). A study in Tanzania (Venot et al., 2021) found that areas of irrigated rice estimated from remote sensing data acquired by satellite were 36 600 ± 11 800 ha in Rukwa, or 2.6 to 5.5 times greater than recorded in official statistics. In the Shinyanga region, the study estimated 267 000 ± 37 000 ha of irrigated rice

Table 2 Summary of total and irrigated areas of crops grown by irrigating and non-irrigating households at 18 sites in Mozambique and Tanzania (de Bont et al., 2019b)

Case study site	Sample size (N)			Total cropped area per non-irrigating household (ha)		Total cropped area per irrigating household (ha)		Area of irrigated crops per irrigating household (ha)	
	Total	Non-irrigating	Irrigating	Mean	SD	Mean	SD	Mean area	Mean % irrigated
Mozambique	***1372***	***574***	***798***	***2.53***	***2.86***	***3.68***	***3.69***	***1.81***	***49.3***
Dondo	120	72	48	1.35	1.11	1.86	2.12	1.60	86.1
Lamego	192	61	131	2.79	2.05	2.74	1.47	1.07	39.0
Macate	197	86	111	2.86	3.72	4.42	4.62	2.15	48.5
Messica	245	94	151	3.98	4.67	5.69	5.37	2.92	51.3
Namicopo	43	21	22	1.48	1.80	1.32	0.55	0.82	62.0
Parta	100	39	61	1.21	1.26	1.73	1.26	0.71	41.0
Tica	136	73	63	2.10	1.18	2.87	1.96	1.44	50.3
Vanduzi	159	53	106	2.74	1.80	3.22	2.04	1.84	57.2
Zembe	180	75	105	2.54	2.15	4.61	3.75	1.95	42.4
Tanzania	***1361***	***445***	***916***	***0.69***	***1.10***	***1.35***	***1.78***	***1.19***	***87.9***
Iringa	151	54	97	0.76	0.65	0.82	0.84	0.82	100.0
Kahe	150	62	88	0.78	0.67	1.52	1.81	1.31	86.3
Kilombero	152	75	77	0.71	1.04	1.08	1.38	0.84	77.5
Makanya	149	25	124	0.47	1.06	2.06	1.36	2.02	98.0
Mandaka Mnono	152	43	109	0.32	0.36	1.26	1.46	0.92	72.7
Mang'ola	150	30	120	0.00	0.00	0.71	1.11	0.71	100.0
Mapogoro	153	34	119	0.54	0.86	1.62	2.10	1.26	77.9
Mijongweni	153	48	105	0.20	0.40	1.21	1.11	1.15	94.9
Rukwa	151	74	77	1.52	1.87	1.87	3.57	1.60	85.6

Published by Burleigh Dodds Science Publishing Limited, 2024.

Table 3 Comparison of hiring of agricultural labour, and gross value of crop sales, by irrigating and non-irrigating households, and extent of contribution of sale of irrigated crops to total income of irrigating households (de Bont et al., 2019b)

Households	Gross value of crop sales (USD/household/year)				Percentage of irrigating households for whom at least half of total income is derived from sale of irrigated crops	Percentage of households employing agricultural labour	
	Non-irrigating		Irrigating			Non-irrigating	Irrigating
	Mean	SD	Mean	SD			
Mozambique	**51.9**	**178.4**	**703.8**	**2091.2**	**78.7**	**23.5**	**48.1**
Dondo	37.4	92.3	249.5	693.0	87.5	22.2	50.0
Lamego	18.0	57.7	99.1	471.5	72.5	27.9	41.2
Macate	114.8	345.1	676.9	2032.1	72.1	34.9	64.0
Messica	48.1	104.9	1240.0	2790.0	84.8	23.4	41.7
Namicopo	1.7	5.4	187.2	455.3	72.7	0.0	18.2
Parta	53.2	170.1	186.0	260.6	67.2	10.3	29.5
Tica	94.1	233.9	1254.0	4250.0	92.1	31.5	54.0
Vanduzi	21.7	58.0	1049.5	1944.8	88.7	28.3	60.4
Zembe	19.5	65.2	653.3	1123.2	70.5	10.7	49.5
Tanzania	**161.2**	**651.9**	**884.0**	**2458.4**	**90.8**	**29.9**	**72.6**
Iringa	33.9	113.7	634.1	1153.1	96.9	9.3	50.5
Kahe	36.3	117.2	635.3	2250.5	83.0	43.5	69.3
Kilombero	350.6	1119.3	937.9	2384.7	92.2	36.0	68.8
Makanya	9.9	30.5	370.5	402.9	87.9	8.0	77.4
Mandaka Mnono	18.4	55.4	739.9	977.6	85.3	39.5	85.3
Mang'ola	0.0	0.0	1180.2	3303.0	92.5	3.3	70.0
Mapogoro	9.4	33.0	1237.0	1211.4	97.5	26.5	82.4
Mijongweni	0.7	4.5	541.4	1092.6	84.8	12.5	77.1
Rukwa	539.9	1012.7	1920.1	5897.8	98.7	52.7	64.9

Table 4 Use of 'modern' inputs in irrigated and non-irrigated agriculture at 18 sites in Mozambique and Tanzania (de Bont et al., 2019b)

	Survey sample (N)		Applied manure from own livestock		Applied purchased fertiliser		Used improved seeds		Average intensification index[a]	
	Non-irrigated crops	Irrigated crops	Non-irrigated crops (%)	Irrigated crops (%)	Non-irrigated crops (%)	Irrigated crops (%)	Non-irrigated crops (%)	Irrigated crops (%)	Non-irrigating households	Irrigating households
Mozambique	**2323**	**1950**	**1.3**	**7.2**	**2.2**	**41.6**	**8.8**	**51.6**	**0.11**	**0.93**
Dondo	135	79	0	1.3	2.2	34.2	5.9	36.7	0.09	0.52
Lamego	358	204	0.3	6.4	0.6	10.8	6.1	22.5	0.09	0.23
Macate	357	306	0.6	5.6	0.3	23.9	9.5	33.7	0.11	0.73
Messica	353	356	2.5	3.9	3.7	63.8	14.2	49.4	0.23	1.27
Namicopo	91	60	8.8	35.0	1.1	5.0	2.2	46.7	0.08	0.81
Parta	205	81	0	0.0	0.0	0.0	0.0	0.0	0.00	0.00
Tica	280	167	0.7	1.2	5.4	33.5	11.1	56.3	0.13	1.12
Vanduzi	256	420	3.5	8.6	3.5	63.6	17.6	83.1	0.12	1.95
Zembe	288	277	0	13.4	2.4	49.5	4.2	65.7	0.06	1.17
Tanzania	**808**	**1786**	**6.1**	**3.6**	**15.2**	**49.1**	**22.3**	**38.2**	**0.75**	**1.51**
Iringa	60	101	0	1.0	1.7	9.9	3.3	3.0	0.06	0.81
Kahe	123	174	0.8	1.7	11.4	36.8	14.6	29.9	0.63	1.15
Kilombero	127	132	0	1.5	38.6	83.3	27.6	64.4	1.39	2.32
Makanya	44	377	0	4.5	2.3	3.2	18.2	30.8	0.60	0.83
Mandaka Mnono	116	215	7.8	0.9	16.4	88.8	42.2	48.8	1.19	2.20
Mang'ola	29	213	0	1.9	0.0	69.5	0.0	50.7	0.0	2.13
Mapogoro	90	141	1.1	0.7	2.2	70.2	15.6	5.0	0.35	1.17
Mijongweni	82	328	0	2.4	43.9	73.2	40.2	57.0	1.49	2.30
Rukwa	137	105	28	25.7	0.7	2.9	15.3	19.0	0.52	0.75

[a]Intensification index is calculated for each household as: for each crop, score 1 for 'buy fertiliser' or 'buy manure' + 1 for 'use improved seeds' + 1 for 'use pesticides'. Average score across all crops grown by the household = input intensification index for each household.

– up to 10 times higher than in official data. Almost all of this area is cultivated by small-scale farmers using hand-dug ditches and bunds to manage water levels to create and maintain flooded paddy fields.

The survey of irrigating and non-irrigating households used a random sample of about 150 households at each site, based on household lists obtained from administrative authorities and adjusted to ensure the sample contained a minimum of 50 of either irrigating or non-irrigating households at every site. Table 3 shows gross values of crop sales by irrigating households (uncorrected for production costs or amount of crop consumed by the household) are on average higher than those of non-irrigating households by factors of 5 (Tanzania) and 13 (Mozambique). These sales from growing irrigated crops account for at least half of household monetary income for the vast majority of those engaging in irrigation.

The data clearly show the commercial nature of farmer-led irrigation development. Most irrigated crops are grown for the market, and irrigating households are much more likely to use improved seeds and fertilisers (Table 4), and to hire agricultural labour (Table 3), compared to their non-irrigating neighbours. Inputs may be financed by credit from traders and corporate buyers (for local, regional, or export markets) in return for an undertaking from farmers to sell them the crop (Veldwisch and Woodhouse, 2022). Some sites show lower rates of input use by irrigating households, particularly where poor roads mean weak supply chains (e.g. Parta in northern Mozambique) or where water availability is less reliable (as at the spate irrigation site at Makanya, in Tanzania). Elsewhere, the increase in use of modern inputs (agrochemicals and improved seeds) is markedly and consistently higher for irrigated crops, and irrigating households are more consistent purchasers of these inputs (Table 4).

4 Conclusion

Evidence accumulating over the past two decades suggests that the poor performance of many official irrigation schemes in sub-Saharan Africa has distracted attention from the widespread adoption of irrigation by small-scale farmers in the region. This farmer-led irrigation development involves use of a variety of techniques. Some, such as 'hill-furrow' irrigation using streams in hilly areas, pre-date colonial administration while others, such as the use of small motor pumps, have been widely available for the past three or four decades. Farmers' innovation and investment in adopting irrigation technology draws on a variety of sources of information, including equipment dealers, government and non-government technical services, and other farmers. Migrants are also important sources of information regarding irrigation techniques and agricultural innovations used elsewhere.

In aggregate, these irrigation initiatives by small-scale farmers appear likely to support areas of irrigated crops that are much larger than those irrigated by formal engineering schemes but are under-represented in official statistics that focus on areas served by permanent infrastructure. Farmer-led irrigation development is driven by growing local and regional food markets arising from urbanisation and rapid population growth in sub-Saharan Africa. Crop sales by households using irrigation developed by farmers' initiatives appear to be five to ten-fold higher than crop sales by non-irrigating households and therefore offers major scope and opportunity to transform agricultural productivity and food supply. A number of international agencies have recognised this, and 'farmer-led irrigation' is the subject of considerable funding efforts, such as the promotion of financial support for acquisition of solar water pumping equipment. Such funding initiatives emphasise the key role of improved agricultural water management in enabling people in Africa to adapt to changing constraints arising from climate change.

As small-scale irrigation initiatives by farmers multiply, a number of challenges arise over increasing differentiation both within and between households (those irrigating and not irrigating, for example) and also over increasing competition for limited water supplies among irrigators and between irrigators and other water users, such as urban domestic users, hydro-power schemes and ecological requirements. This presents new policy and regulatory challenges that will be very context-specific but for which African government agencies will require significant re-focusing of goals and development of new capacity.

5 Future trends in research

Recognition by development agencies of 'farmer-led irrigation development' as a discrete process dates back less than a decade. Research on technical and socio-economic aspects of irrigation in sub-Saharan has previously focused mostly on formal state-run schemes. However, research is now urgently needed to understand both the processes of technological change pursued by small-scale farmers and the scope for improved productivity, and avenues through which these can be supported by engineers and other science-based professions.

Research will also need to explore challenges to the sustainability of the current proliferation of small-scale farmers' irrigation initiatives. We can identify three aspects. First, what are the medium-term consequences of farmer-led irrigation development in terms of socio-economic change? Who are the 'winners' and 'losers' and what can be done to ensure that the raised incomes from irrigated crop production result in a broad-based benefit to rural communities? Second, how can institutions meet the evolving challenge of organisation and regulation of small-scale irrigation? While studies have identified farmers' capacity for organisation in using shared water resources, few have documented the interface of such farmers' organisations with local, regional or national government and non-government agencies. Third, is there a need to develop water storage infrastructure with a view to serving farmers' irrigation initiatives? In the past, infrastructure such as dams have been constructed to provide hydropower or irrigate large areas into which rural communities have been inserted as labourers or tenants. The proliferation of irrigation developed by farmers' initiatives raises the fundamental question of whether this logic may be reversed and infrastructure design re-thought as supporting such initiatives. These three aspects remain major gaps with important implications for the future sustainability of much farmer-led irrigation development.

6 Where to look for further information

- African Union Framework for Irrigation and Agricultural Water Management: https://au.int/en/documents/20200601/framework-irrigation-development-and-agricultural-water-management-africa.
- Agrilinks: https://agrilinks.org/post/building-better-solar-irrigation-market-ghana.
- DEGRP Synthesis Report. London: Foreign, Commonwealth and Development Office. https://odi.org/en/publications/farmer-led-irrigation-in-sub-saharan-africa-synthesis-of-current-understandings/.
- Food Policy Research Institute: https://www.ifpri.org/project/ilssi.
- International Water Management Institute (IWMI): https://www.iwmi.cgiar.org/what-we-do/farmer-led-irrigation/.
- The SAFI project. Studying African Farmer-led Irrigation: http://www.safi-research.org/.
- World Bank: Farmer-led irrigation: the what, why, and how-to guide https://blogs.worldbank.org/water/farmer-led-irrigation-what-why-and-how-guide.

7 References

Adams, A. 1981. The Senegal river valley. In: Heyer, J., Roberts, P. and Williams, G. (Eds.), *Rural Development in Tropical Africa*. London: Macmillan.

Adams, W. M. 1992. *Wasting the Rain: Rivers, People and Planning in Africa*. Minneapolis, MN: University of Minnesota Press.

Adams, W. M. and Anderson, D. M. 1988. Irrigation before development: Indigenous and induced change in agricultural water management in East Africa. *African Affairs* 87(349): 519–535.

Published by Burleigh Dodds Science Publishing Limited, 2024.

Adams, W. M., Potkanski, T. and Sutton, J. E. G. 1994. Indigenous farmer-managed irrigation in Sonjo, Tanzania. *Geographical Journal* 160(1): 17–32.

AFDB. 2008. Investment in agricultural water for poverty reduction and economic growth: A collaborative program of AFDB, FAO, IFAD, IWMI and the World Bank. Synthesis Report. Available at: http://www.fanrpan.org/documents/d00508/1-agric_water_investments_World_Bank.pdf.

African Union. 2020. Framework for irrigation development and agricultural water management in Africa. Available at: https://au.int/sites/default/files/documents/38632-doc-framework_for_irrigation_development_english.pdf (accessed 27 January 2021).

Aw, D. and Diemer, G. 2005. *Making a Large Irrigation Scheme Work: A Case Study From Mali. Directions in Development*. Washington, DC: The World Bank.

Barnett, T. 1977. *The Gezira Scheme: An Illusion of Development*. Abingdon: Routledge.

Barron, J., Rockström, J., Gichuki, F. and Hatibu, N. 2003. Dry spell analysis and maize yields for two semi-arid locations in east Africa. *Agricultural and Forest Meteorology* 117(1–2): 23–37.

Beekman, W., Veldwisch, G. J. and Bolding, A. 2014. Identifying the potential for irrigation development in Mozambique: Capitalizing on the drivers behind farmer-led irrigation expansion. *Physics and Chemistry of the Earth, Parts A/B/C* 76–78: 54–63.

Bell, M., Faulkner, R., Hotchkiss, P., Lambert, R., Roberts, N. and Windram, A. 1987. *The Use of Dambos in Rural Development With Reference to Zimbabwe*. Loughborough: Loughborough University/University of Zimbabwe.

Bolding, A., Post Uiterweer, N. C. and Schippers, J. 2010. The fluid nature of hydraulic property: A case study of Mukudu, Maira and Penha Longa irrigation furrows in the upper Revue River, Manica District, Mozambique. In: Van der Zaag, P. (Ed.), *What Role of Law in Promoting and Protecting the Productive Uses of Water by Smallholder Farmers in Mozambique?* Delft: UNESCO, pp. 105–136. Challenge Program Project No. 66-IHE Institute for Water Education.

Bottrall, A. 1985. Managing large irrigation systems: A problem of political economy. AAU Occasional Paper No. 5. London: Overseas Development Institute.

Burney, J. A., Naylor, R. L. and Postel, S. L. 2013. The case for distributed irrigation as a development priority in sub-Saharan Africa. *Proceedings of the National Academy of Sciences of the United States of America* 110(31): 12513–12517.

Carney, J. A. 1998. Women's land rights in Gambian irrigated rice schemes: Constraints and opportunities. *Agriculture and Human Values* 15(4): 325–336.

Carter, R. C. 1989. The development of small-scale irrigation in sub-Saharan Africa. *Public Administration and Development* 9(5): 543–555.

Carter, R. C., Carr, M. K. V. and Kay, M. G. 1983. Policies and prospects in Nigerian irrigation. *Outlook on Agriculture* 12(2): 73–76.

Chambers, R. 1989. Managing canal irrigation: Practical analysis from South Asia. In: *Wye Studies in Agricultural and Rural Development*. Cambridge: Cambridge University Press.

Chauveau, J.-P., Colin, J.-P., Jacob, J.-P., Lavigne Delville, P. and Le Meur, P.-Y. 2006. *Changes in Land Access and Governance in West Africa: Markets, Social Mediations and Public Policies*. London: IIED/DFID.

Chimhowu, A. and Woodhouse, P. 2006. Customary vs private property rights? Dynamics and trajectories of vernacular land markets in Sub-Saharan Africa. *Journal of Agrarian Change* 6(3): 346–371.

Danso, G., Drechsel, P., Obuobie, E., Forkuor, G. and Kranjac-Berisavljevic, G. 2014. Urban vegetable farming sites, crops and cropping practices. In: *Drechsel and Keraita*.

de Bont, C., Komakech, H. C. and Veldwisch, G. J. 2019a. Neither modern nor traditional: Farmer-led irrigation development in Kilimanjaro Region, Tanzania. *World Development* 116: 15–27. https://doi.org/10.1016/j.worlddev.2018.11.018.

de Bont, C., Liebrand, J., Veldwisch, G.-J. and Woodhouse, P. 2019b. Modernisation and African farmer-led irrigation development: Ideology, policies and practices. *Water Alternatives* 12(1): 107–128.

De Fraiture, C. and Giordano, M. 2014. Small private irrigation: A thriving but overlooked sector. *Agricultural Water Management* 131: 167–174.

De Fraiture, C., Kouali, G. N., Sally, H. and Kabre, P. 2014. Pirates or pioneers? Unplanned irrigation around small reservoirs in Burkina Faso. *Agricultural Water Management* 131: 212–220.

Dessalegn, M. and Merrey, D. J. 2015. Motor pump revolution in Ethiopia: Promises at a Crossroads. *Water Alternatives* 8(2): 237–257.

Diemer , G. and van der Laan, E. 1987. *l'Irrigation au Sahel*. Paris: Karthala.

Diemer, G. and Vincent, L. 1992. Irrigation in Africa: The failure of collective memory and collective understanding. *Development Policy Review* 10(2): 131–154.

Drechsel, P. and Keraita, B. 2014. *Irrigated Urban Vegetable Production in Ghana Characteristics, Benefits and Risk Mitigation* (2nd edn.). Colombo: International Water Management Institute.

Ertsen, M. 2016. A matter of relationships: Actor-networks of colonial rule in the Gezira irrigation system, Sudan. *Water Alternatives* 9(2): 203–221.

FAO. 2020. *The State of Food and Agriculture: Overcoming Water Challenges in Agriculture*. Rome: The Food and Agriculture Organization of the United Nations. Available at: https://www.fao.org/3/cb1447en/cb1447en.pdf (accessed 7 March 2022).

Giordano, M., de Fraiture, C., Weight, E. and van der Bliek, J. 2012. Water for wealth and food security: Supporting farmer-driven investments in agricultural water management. Synthesis Report of the AgWater Solutions Project. Colombo: International Water Management Institute.

Grove, A. 1993. Water use by the Chagga on Kilimanjaro. *African Affairs* 92(368): 431–448.

Harrison, E. 2018. Engineering change? The idea of the 'scheme' in African irrigation. *World Development* 111: 246–255.

Hebinck, P., Bosma, L. and Veldwisch, G.-J. 2019. Petrol pumps and the making of modernity along the shores of Lake Victoria, Kenya. *Water Alternatives* 12(1): 13–29.

Higginbottom, T. P., Adhikari, R., Dimova, R., Redicker, S. and Foster, T. 2021. Performance of large-scale irrigation projects in sub-Saharan Africa. *Nature Sustainability* 4(6): 501–508. https://doi.org/10.1038/s41893-020-00670-7.

Hocombe, S., Kidane, P., Jazayeri, A. and Gadelle, M. 1986. *Irrigation in Africa South of the Sahara*. Technical Paper No 5. Rome, Italy: Food and Agriculture Organization Investment Centre, Food and Agricultural Organisation of the United Nations.

Inocencio, A., Kikuchi, M., Tonosaki, M., Maruyama, A., Merrey, D., Sally, H. and De Jong, I. 2007. Costs and performance of irrigation projects: A comparison of Sub-Saharan Africa and other developing regions. *IWMI Research Report*. Colombo, Sri Lanka: International Water Management Institute 109, p. 81.

IWMI (International Water Management Institute). 2016. *Irrigated Africa and Asia*. Colombo, Sri Lanka: IWMI. Available at: www.iwmi.cgiar.org/2016/02/irrigated-africa-and-asia/ (accessed 28 February 2022).

IWMI. 2019. Operationalizing farmer-led irrigation: Implementers dialogue. Available at: https://www.iwmi.cgiar.org/events/farmer-led-irrigation-development-fli-webinar-series/ (accessed 27 January 2021).

Izzi, G., Denison, J. and Veldwisch, G.-J. (Eds.). 2021. *The Farmer-Led Irrigation Development Guide: A What, Why and How-to for Intervention Design*. Washington, DC: World Bank.

Karimba, B. M., Duker, A., Prasad, P., Karimi, P., de Fraiture, C. and van der Zaag, P. 2022. Irrigation on the move: How transient farming partnerships facilitate the expansion of smallholder irrigation along ephemeral rivers in dryland areas of Kenya. *Agricultural Water Management* 265: 107526. https://doi.org/10.1016/j.agwat.2022.107526.

Kenya Government. n.d.a. Mwea irrigation scheme. Available at: https://www.irrigation.go.ke/projects/mwea-irrigation-scheme/ (accessed 22 January 2022).

Kenya Government. n.d.a. Tana irrigation scheme. Available at: https://www.irrigation.go.ke/projects/tana-irrigation-scheme/ (accessed 22 January 2022).

Keraita, B. and Cofie, O. O. 2014. Irrigation and soil fertility management practices. In: *Drechsel and Keraita*.

Komakech, H. C. and Van der Zaag, P. 2011. Understanding the emergence and functioning of river committees in a catchment of the Pangani basin, Tanzania. *Water Alternatives* 4(2): 197–222.

Komakech, H. C., Van der Zaag, P., Mul, M. L., Mwakalukwa, T. A. and Kemerink, J. S. 2012. Formalisation of water allocation systems and impacts on local practices in the Hingilili catchment, Tanzania. *International Journal of River Basin Management* 10(3): 213–227.

Komakech, H. C., Van der Zaag, P. and Van Koppen, B. 2012. Dynamics between water asymmetry, inequality and heterogeneity sustain canal institutions in Makanya catchment, Tanzania. *Water Policy* 14(5): 800–820.

Lankford, B. 2004. Irrigation improvement projects in Tanzania: Scale impacts and policy implications. *Water Policy* 6(2): 89–102.

Lankford, B. 2009. Viewpoint–The right irrigation? Policy directions for agricultural water management in sub-Saharan Africa. *Water Alternatives* 2(3): 476–480.

Lankford, B., van Koppen, B., Franks, T. and Mahoo, H. 2004. Entrenched views or insufficient science? *Agricultural Water Management* 69(2): 135–153.

Lankford, B. A. and Beale, T. 2007. Equilibrium and non-equilibrium theories of sustainable water resources management: Dynamic river basin and irrigation behaviour in Tanzania. *Global Environmental Change* 17(2): 168–180.

Lefore, N., Closas, A. and Schmitter, P. 2021. Solar for all: A framework to deliver inclusive and environmentally sustainable solar irrigation for smallholder agriculture. *Energy Policy* 154: 112313. https://doi.org/10.1016/j.enpol.2021.112313.

Mdee, A. and Harrison, E. 2017. Successful small-scale irrigation or environmental destruction? The political ecology of competing claims on water in the Uluguru Mountains, Tanzania. *Journal of Political Ecology* 24(1): 406–424. https://doi.org/10.2458/v24i1.20881.

Mdee, A. and Harrison, E. 2019. Critical governance problems for farmer-led irrigation: State capacity and institutional capability. *Water Alternatives* 12(1): 30–45.

Merrey, D. and Sally, H. 2017. Viewpoint: Another well-intentioned bad investment in irrigation: The Millennium Challenge Corporation's 'compact' with the Republic of Niger. *Water Alternatives* 10(1): 195–203.

Moris, J. R. and Thom, D. J. 1985. African irrigation overview: Summary. In: *Water Management Synthesis II Project*. Department of Agricultural and Irrigation Engineering, Utah State University.

Namara, R. E., Hope, L., Sarpong, E. O., De Fraiture, C. and Owusu, D. 2014. Adoption patterns and constraints pertaining to small-scale water lifting technologies in Ghana. *Agricultural Water Management* 131: 194–203.

NEPAD (New Partnership for Africa's Development). 2003. *NEPAD*. Pretoria, South Africa: Comprehensive Africa Agriculture Development Program. Available at: http://www.nepad.org/foodsecurity/agriculture/land.

Nkoka, F., Veldwisch, G. J. and Bolding, A. 2014. Organisational modalities of farmer-led irrigation development in Tsangano District, Mozambique. *Water Alternatives* 7(2): 414–433.

OECD and SWAC. 2020. *Africa's Urbanisation Dynamics 2020: Africapolis, Mapping a New Urban Geography, West African Studies*. Paris: OECD Publishing. https://doi.org/10.1787/b6bccb81-en.

Peters, P. E. and Kambewa, D. 2007. Social conflicts over access and transfer of land in Malawi and their implications for new land policy. *Journal of Modern African Studies* 45(3): 447–472.

Reij, C., Scoones, I. and Toulmin, C. (Eds.). 1996. *Sustaining the Soil: Indigenous Soil and Water Conservation in Africa*. London, UK: Earthscan Publications.

Richards, P. 1985. *Indigenous Agricultural Revolution: Ecology and Food Production in West Africa*. London, UK: Hutchinson.

Rockström, J., Barron, J. and Fox, P. 2003. Water productivity in rain-fed agriculture: Challenges and opportunities for smallholder farmers in drought-prone tropical agroecosystems. In: Kijne, J., Barker, R. and Molden, D. (Eds.), *Water Productivity in Agriculture: Limits and Opportunities for Improvement*. Wallingford, UK: CAB International, pp. 145–162.

Rockström, J., Karlberg, L., Wani, S. P., Barron, J., Hatibu, N., Oweis, T., Bruggeman, A., Farahani, J. and Qiang, Z. 2010. Managing water in rainfed agriculture – The need for a paradigm shift. *Agricultural Water Management* 97(4): 543–550.

SAFI. 2018. Policy brief: Farmer-led irrigation development and investment strategies for food security, growth and employment in Africa. Available at: http://www.safi-research.org/publications/farmer-led-irrigation-development-and-investment-strategies-for-food-security-growth-and-employment-in-africa/.

Soper, R. 2006. *The Terrace Builders of Nyanga*. Harare, Zimbabwe: Weaver Press.

Svendsen, S., Ewing, M. and Msangi, S. 2009. *Measuring Irrigation Performance in Africa*. IFPRI Discussion Paper 894. Washington, DC: International Food Policy Research Institute.

Tagseth, M. 2008. The expansion of traditional irrigation in Kilimanjaro, Tanzania. *The International Journal of African Historical Studies* 41(3): 461–490.

Underhill, H. 1984. *Small-Scale Irrigation in Africa in the Context of Rural Development*. Rome: Food and Agriculture Organization of the United Nations.

Uphoff, N. 1986. *Improving International Irrigation Management With Farmer Participation: Getting the Process Right*. Boulder, CO: Westview Press.

Veldwisch, G.-J., Venot, J.-P., Woodhouse, P., Komakech, H. and Brockington, D. 2019. Re-introducing politics in African farmer-led irrigation development: Introduction to a special issue. *Water Alternatives* 12(1): 1–12.

Veldwisch, G. J. and Woodhouse, P. 2022. Formal and informal contract farming in Mozambique: Socially embedded relations of agricultural intensification. *Journal of Agrarian Change* 22(1): 162–178.

Venot, J.-P. 2016. A success of some sort: Social enterprises and drip irrigation in the developing world. *World Development* 79: 69–81.

Venot, J.-P., Bowers, S., Brockington, D., Komakech, H., Ryan, C., Veldwisch, G. J. and Woodhouse, P. 2021. Below the radar: Data, narratives and the politics of irrigation in sub-Saharan Africa. *Water Alternatives* 14(2): 546–572.

Venot, J. P., de Fraiture, C. and Nti-Acheampong, E. 2012. Revisiting dominant notions: A review of costs, performance and institutions of small reservoirs in sub-Saharan Africa. *IWMI Research Report*. Colombo, Sri Lanka: IWMI, p. 144.

Vermillion, D. 2004. *Developing Water Users Associations for Small Dams and Fadama Areas in Nigeria*. Report to the Nigeria Country Programme of the FAO.

Wade, R. 1988. The management of irrigation systems: How to evoke trust and avoid prisoner's dilemma. *World Development* 16(4): 489–500.

Walsh, M. 2012. The not-so-Great Ruaha and hidden histories of an environmental panic in Tanzania. *Journal of Eastern African Studies* 6(2): 303–335.

Wiggins, S. and Lankford, B. 2019. Farmer-led irrigation in Sub-Saharan Africa: Synthesis of current understandings: DEGRP synthesis report. London: Foreign, Commonwealth and Development Office. Available at: https://odi.org/en/publications/farmer-led-irrigation-in-sub-saharan-africa-synthesis-of-current-understandings/.

Woodhouse, P. 2003. African enclosures: A default mode of development. *World Development* 31(10): 1705–1720.

Woodhouse, P., Bernstein, H. and Hulme, D. 2000. *African Enclosures? The Social Dynamics of Wetlands in Drylands*. Oxford: James Currey Publishers Ltd.

Woodhouse, P., Veldwisch, G. J., Venot, J.-P., Brockington, D., Komakech, H. and Manjichi, Â. 2017. African farmer-led irrigation development: Reframing agricultural policy and investment? *Journal of Peasant Studies* 44(1): 213–233. https://doi.org/10.1080/03066150.2016.1219719.

World Bank. 2009. *Awakening Africa's Sleeping Giant: Prospects for Commercial Agriculture in the Guinea Savannah Zone and Beyond*. Washington DC: The World Bank.

World Bank. 2011. *Africa's Future and the World Bank Strategy to Support It*. Washington, DC: The World Bank.

World Bank. 2021. Farmer-led irrigation: The what, why, and how-to guide. Available at: https://blogs.worldbank.org/water/farmer-led-irrigation-what-why-and-how-guide.

Xie, H., You, L., Wielgosz, B. and Ringler, C. 2014. Estimating the potential for expanding smallholder irrigation in sub-Saharan Africa. *Agricultural Water Management* 131: 183–193 https://doi.org/10.1016/j.agwat.2013.08.011.

You, L., Ringler, C., Wood-Sichra, U., Robertson, R., Wood, S., Zhu, T., Nelson, G., Guo, Z. and Sun, Y. 2011. What is the irrigation potential for Africa? A combined biophysical and socioeconomic approach. *Food Policy* 36(6): 770–782. https://doi.org/10.1016/j.foodpol.2011.09.001.

Published by Burleigh Dodds Science Publishing Limited, 2024.

Chapter 10

Managing the irrigation efficiency paradox to "free" water for the environment

Bruce Lankford[1], University of East Anglia, UK; and Matthew McCartney, International Water Management Institute, Sri Lanka

1 Introduction

Reducing water consumption in irrigation is widely perceived as a solution to restore flows in rivers if that 'saving' can be reallocated to the environment (Jägermeyr et al., 2017; Siderius et al., 2022; Richter et al., 2017). However, this task is not straightforward, given the many scales, motives, actors, policies and hydrological practices and processes involved (Lankford et al., 2020; Uhlenbrook et al., 2022; van Oel et al., 2019). Our interpretation of this complicated science-and-policy environment is that a more structured layered and comprehensive approach is required. Without this, attempts to save water will be costly, misguided or conflicting (Molle and Tanouti, 2017), or will tend to result in no impact, or inadvertently reallocate water between catchment actors (Lankford and Scott, 2023) or lead to an increase in water depletion (Wheeler

[1] Corresponding author: B.Lankford@uea.ac.uk.

http://dx.doi.org/10.19103/AS.2023.0123.16

et al., 2020) or run the risk of reducing crop production (Pérez-Blanco et al., 2020).

This chapter addresses this problem. It specifically focuses on how to manage water use in irrigated agriculture to reduce its impact on the environment (Batchelor et al., 2014; McCartney et al., 2019) whilst simultaneously controlling for any potential impact on crop production (Pérez-Blanco et al., 2020). It acknowledges the 'paradox' that, if not managed, improvements in irrigation efficiency lead to a rise in water consumption (Ward and Pulido-Velázquez, 2008; Grafton et al., 2018). We propose a six-part 'irrigation savings allocation framework' (ISAF) to address this challenge. The framework provides a systematic approach to managing water flows and apportionment to reduce consumption in irrigated agriculture and, importantly, reallocate these 'savings' to the environment. Our solution is to encourage reductions in water applied at the lower levels of irrigation systems and protect these from being reused in these lower levels or elsewhere in order to redistribute these savings upwards to the basin scale for allocation to the environment and society in the form of enhanced ecosystem services.

It is worth highlighting the 'consumption versus withdrawals' debate that is relevant to ISAF. ISAF arrives at reductions in water consumption by working with extant water flows and withdrawals. This is because real-world farmers on real-world irrigation systems have to work with real-world water. In other words, people manage, on a daily basis, water flows, volumes, withdrawals and their derived ratios (e.g. flow rate to command area). But the problem is that more carefully and efficiently managed real water flows do not result in 'real savings', meaning reductions in consumption. This lack of equivalence is because irrigation systems are complex, multi-scalar, nested and change over time. Via the reuse of drainage water and the redeployment of saved water flows elsewhere (or into another season), water consumption can paradoxically remain high or even increase.

However, it is vitally important to understand what irrigation systems and the people involved in them are actually dealing with. They are working with real water flows that contain within them different fractions of final destinations of 'beneficially consumed', 'non-beneficially consumed', 'non-recovered' and 'recovered' water (Perry, 2011). In contrast to the beliefs of Van Opstal et al. (2021), water flows cannot be easily parsed out into these fractions as water passes through different components (e.g. canals, gates and fields) of the irrigation system to the crop. Furthermore, ISAF's focus on actual water flows and their apportionment contrasts with calls to work directly on water consumption as proposed by Uhlenbrook et al. (2022) who stated '...water consumption caps (instead of limits on withdrawal rates) need to be set, monitored, and enforced'. Nonetheless, ISAF and Uhlenbrook et al. (2022) seek the same outcome, which is reduction in the water consumed in irrigation (provided this does not harm crop production). But what we question about the Uhlenbrook et al. (2022) perspective is that it prioritises consumption caps 'instead' of withdrawal rates, rather than consumption sitting 'alongside' withdrawals. Their research does not explain how consumption caps can be practically operationalised by farmers and by irrigation and basin managers.

2 The irrigation savings allocation framework

The Irrigation Savings Allocation Framework (ISAF) describes how consumption of water may be reduced within an irrigated agricultural setting, or 'freed up', in order to allocate those savings to the environment. The framework seeks to reduce water requirements, applications, withdrawals and consumption while simultaneously controlling for a loss in food production. The conceptual framework is schematically illustrated in Fig. 1. The main terms and acronyms are given in Table 1.

Following Fig. 1, this framework has six parts, each of which can also be expressed as an objective. These are (1) to protect and if possible boost crop production via water productivity gains; (2) to manage and distribute water in order to make water savings at the crop, field, intra-system, irrigation system and catchment levels; (3) to protect and materially track those savings to the environment; (4) to put in place cross-cutting dimensions to monitor those savings and control for rebounds (i.e. changes in water use triggered by changes in farmer behaviour that partly or fully offset the water savings and in cases lead to further depletion through increases in total water consumption); (5) to account for 'before and after' allocations, basin-level consumption and total crop production; and (6) to decide whether this staged process best operates as either a 'bottom up' or 'top down' approach. Without these objectives working in an integrated and controlled way, loopholes and paradoxes can arise. For example, although farmers may reduce water applied to fields, their actions can reduce crop production, or may not translate to water savings at the basin level or be realised as savings transferred to nature. Other important ISAF precepts include the following:

- The allocation of water saved from irrigation to other sectors can only occur via a reduction in basin-level consumption of water withdrawn into irrigation systems, in other words 'wet or real savings' (Seckler, 1996). Recognising the distinction between real (wet) and paper (dry) savings, here we only refer to 'savings' to signify reductions in irrigation water required, applied and withdrawn, which translate to 'real' reductions in basin-level consumption through the control of rebound and/or reuse of drainage flows.
- Note ISAF links savings at the crop and field levels to reductions in water withdrawn into irrigation systems rather than to all water sourced by farmers

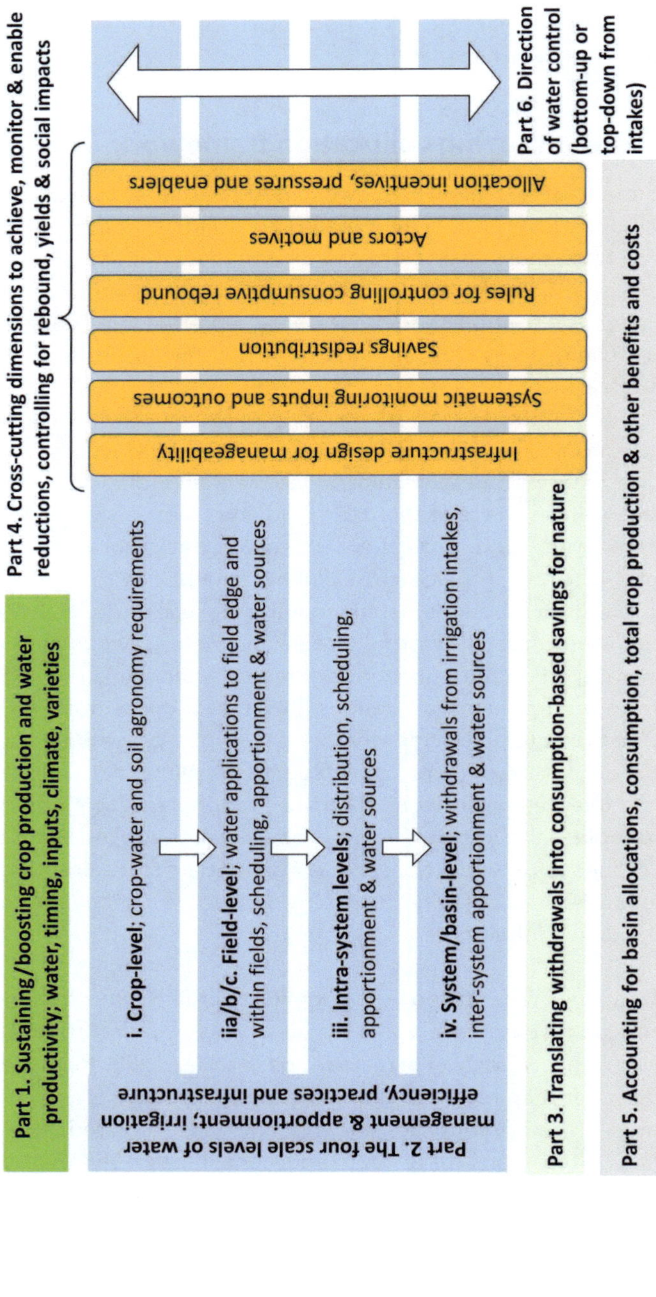

Figure 1 Schematic of the irrigation savings allocation framework.

Table 1 Terms and key definitions

CT Crop type CC Climate change DI Deficit irrigation ETc Evapotranspiration (crop) FEM Field evaporation modification (e.g. shade cloth) IE Irrigation efficiency ISA Irrigated Systems Accounting ISAF Irrigation Savings Allocation Framework WUA Water user association	'Irrigation system(s)' is used in this chapter because this chapter focuses on the system properties and processes that shape water flows and consumption. 'Irrigation scheme' is normally reserved for named and funded projects that have a designed command area.

for irrigation. Other sources of water include increasing effective use of rainfall and capillary rise of water from shallow groundwater and making greater use of wastewater sources (Mendoza-Espinosa et al., 2019).

- The framework is based on progressive stages but an effective savings-based allocation of water out of irrigation results from the simultaneous application of all stages.
- ISAF is conceptualised from the perspective of those seeking to allocate water out of irrigation to nature whilst not harming crop productivity and production. Its main audience are catchment regulators and representatives of water-using sectors including irrigators whose practices and skills are central to reallocation.
- The ISAF framework builds on but differs from the irrigation efficiency matrix (IEM) presented by Lankford et al. (2020) by being much more normative or instructive in how it guides irrigation actors to manage the irrigation efficiency paradox. In other words, IEM is 'descriptive' in terms of unpacking the scale- and motive dimensions of irrigation efficiency, but ISAF is more analytical and directive in its purpose.
- ISAF complements the water accounting analysis that can be used to resolve the irrigation efficiency paradox using a comprehensive accounting method called 'Irrigated Systems Accounting' (Lankford, 2023a).

3 Sustaining/boosting crop production and water productivity (Part 1)

Although the main objective of ISAF is to translate water savings from irrigation systems to nature, the argument is that this should occur without incurring agronomic or economic losses to irrigated crop production. Put another way, cutting 'non-beneficial waste' out of irrigation systems to reallocate to the environment should not reduce the water that enables crop transpiration, growth and production. Pérez-Blanco et al. (2020) term this a 'Pareto efficient solution' or 'Pareto improvement' whereby 'no one is left worse off and at least one user is left better off'. In our case, crop production is not left worse off by water reallocation, while the environment (and other users in the basin) gain from increased water availability.

For this reason, the first objective of ISAF prioritises crop production and water productivity. The crop and field levels are where farmers can implement water conservation in ways that do not harm crop growth and total farm production. This requires farmers to boost crop productivity (as the ratio of crop biomass per water unit consumed (Giordano et al., 2017)). Practices include better crop husbandry via more timely and controlled use of seeding and transplanting, fertilizer, weeding, land preparation, pest control and harvesting. Additional soil and rainfall management practices that improve crop-water productivity are summarised in Section 4.2.

4 Managing and apportioning water at different scales (Part 2)

4.1 Water apportionment, allocation and efficiency

Water apportionment constitutes the distribution, division or allocation of water to different parts and areas of irrigated systems at all levels and scales (Lankford et al., 2020). The accuracy and efficacy of the control of water apportionment over time and space both define well-managed irrigation in a given time period and are central to apportioning 'freed up' water to new users, places and uses. The significance of water apportionment is explained here.

When river water or aquifer water becomes available to an irrigation system, the aim of irrigation managers in managing water flowing through a bifurcating network is five-fold: (1) to withdraw that water (withdrawal); (2) to convey withdrawn bulk water to where it is needed at the head of sub-branches in the network via pipes or channels (conveyance); (3) to convey, divide, apportion and distribute this water within the sub-branches to individual farms and to fields within farms in a timely manner (distribution and scheduling); (4) to apply it to field edges and soils within fields (field application); and (5) to drain off any excess (drainage). All five change in tandem with fluxes in water supply and demand, varying both temporally and spatially. At no point are irrigation systems static and in equilibrium; they face changing seasons, climate variability affecting both evapotranspiration and rainfall, soil-type distribution and behaviours, and farmers' choices for cropping and technologies in turn driven by matters such as input prices, crop markets and policy interventions such as water law reform.

Between bulk water withdrawal and crop water evapotranspiration (ETc), the third and fourth tasks of water distribution, namely, scheduling and field application, are the most complicated. It is this water apportionment which is the key hydrological feature of irrigation systems and management. This is because farmers, carefully trying to match crop evapotranspiration, are constantly dividing conveyed bulk water into smaller and smaller flows that supply smaller and smaller units of land before eventually providing the water to an individual soil parcel and plant in a timely scheduled way (Lankford, 2012). This bifurcating apportionment has to be efficient, equitable and timely so that parts of the irrigation system, soils and crops are not over-, under- or late-watered. In addition, in idealised irrigation networks, this apportionment must take place at least seven times from the basin scale down to the individual crop; river to irrigation system intake, main canal to secondary canal, then to tertiary canals, then to farms, then to fields, sub-fields and then across fields to soils and individual plants. Our framework recognises how these water divisions occur at different levels; ISAF utilises four key levels (i to iv) called 'crop water requirements', 'field applications', 'intra-system distribution' and 'main system/basin-level withdrawals'. These four levels and their role within water control and division are presented in Fig. 1 and Fig. 2. This choice of four levels is to simplify the framework rather than to downplay the intra-system levels each with their own task of accurate water apportionment.

Water apportionment strongly steers and is steered by irrigation efficiency, practices and infrastructure. With this observation, we introduce two points. First irrigation efficiency is as much about suboptimal water apportionment in the irrigation network as it is about 'leakiness and losses', although the two are intimately connected. For example, take 'losses' from a canal. Many irrigation commentators envisage these kinds of losses as canals leaking through cracks in lining materials or through seepage into soil lining (Van Opstal et al., 2021). Not only is this overly leak-fixated but it also mistakenly prioritises the conveying part of water management.

To move beyond conveyance leakage, it is vital to adopt a water apportionment point of view regarding 'efficiency' (Lankford, 1992). A canal or pipe can be completely 'leak-free', but if this canal has overly apportioned water for the command area it serves, that canal will create losses elsewhere either as tail-losses or as excess applications into fields, both reducing irrigation efficiency and crop productivity. Thus, poor water apportionment means a given canal need not show visible seepage or over-topping, but instead, it plays a part in contributing to both over- and under-irrigation.

Second, irrigation practices and infrastructure affect how irrigators manage water apportionment. Irrigation practices are those ways of managing crops, water, soil and land that have a bearing on the amounts and timing of irrigation water required, applied and withdrawn. Examples of practices include changing crop mixes and/or crop rotations, adding shade cloth, harvesting more rainfall and applying deficit irrigation strategies. Irrigation

#	Name of scale level	Scale (ha)	Managing water technology	Monitoring water/crop/soil outcomes	Savings allocation options	Rules for capping rebound	Actors	Incentives Pressures Enablers	Direction of travel
i	Reducing crop/soil/ sub-field water req.	0.1-1.0	CT, DI, FEM,	Leaf/proxy sensors Crop yields	Within crop season To different crop To irrig. duration	Crop type & duration	Crop buyers Farmers	Water pricing/ markets	From crop and field upwards to irrigation intake
iia	Reduce field applications (Application Ea)	1.0-10	Gradients Flatness Channels Field length Field edge	Human Soil sensors Crop profile Areas Time / flow			Farmers	Tariffs/cost of other Dist'n rules Drought	
iib	Reduce farm withdrawals (Unit distribution Ed)	10-100	Channels Pipes Gates	Human Areas Withdrawals	(Direct to drainage) Expand area in level To other area in level To other use in level To storage = reuse Up a level	Area in level Area all levels Time duration Withdrawals No-use drain'ge X-seasonal use No. of offtakes	Farmers & WUAs	New tech & varieties Whose knowledge	
iic	Reduce withdrawals to tertiary & secondary units (Conveyance Ec)	100-1000	Dams Drains Placement Density Capacity	Time duration Time/flow/area (l/sec/ha) Scheduling comparisons			Engineers & other experts	CC & rising demand Reform of water law	From the main intake down to the farmers and fields
iii	Reduce withdrawals to irrigation system (IE = Ea*Ed*Ec)	1000-10^4	Operation Design for management	Total yields				Water allocation plans	
iv	Reduce basin withdrawals	10^4 -10^5		Areas over time Withdrawals	Paracommons; proprietor, neighbour nature, society	Area, time duration, new schemes	Basin officers External agencies	National plans for economy	

Figure 2 The four levels (i to iv) within Part 2 of ISAF (see text for explanation).

infrastructure covers aspects such as operational and storage dams, and network design such as the density of tertiary canals and the size of tertiary units. Each of these determines how easy, difficult or 'automatic' it is to divide water from bulk withdrawal to 'per plant' water application and to do so in an accurate and timely way. For example, capturing more rainfall in-situ reduces the irrigation need at the plant and field level. This ostensibly results in a simple reduction in depth applied and field-edge water demand, but recall that (1) rainfall does not fall or is captured uniformly across irrigation systems; (2) rainfall moistens top soil, which improves the uniformity of water advance and infiltration across a field surface and (3) rainfall, both its incidence and absence, disrupts the rotational schedule of sharing water between farmers. The practice of capturing more rainfall thus alters how farmers individually and collectively manage their irrigation water apportionment. Some farmers make good use of combined rainfall and irrigation water whilst others do not. Those who do not successfully co-manage their soils, land, rainfall and irrigation typically witness either longer drier periods between irrigation doses or increased field runoff during rainfall and irrigation (Lankford and Orr, 2022).

Lastly, it is worth noting that apportionment is not only about the splitting of flows within irrigation systems in real time but it also covers the allocation of water at the basin scale between irrigation and other sectors. ISAF perceives allocation at this scale in two ways: as a 'commons' of available water (Libecap, 2009) and as a 'paracommons' of surplus wastes and wastages found within irrigation systems (Lankford, 2013; Lankford, 2023a). The latter is explained in Section 5.

4.2 Crop-level water requirement

With 'saving' water as the over-arching objective, this stage focuses on how farmers decide to reduce crop-level water requirements without harming crop production, including sometimes through a change in crop type. The aim here is to reduce atmospheric evapotranspiration demand governed by radiative and aerodynamic components rather than reducing non-beneficial evaporation. These include practices such as mild deficit irrigation, changing crop types, using shorter-season varieties, improving seed homogeneity, and applying mulches, greenhouses, and shade cloth (Saitta et al., 2021; Gil et al., 2018; Lankford, 2023a). The control of rebound at this level is covered in Section 6.4.

4.3 Field-level water applications, scheduling and sources

Here, farmers aim to optimise water applications at the field level while not reducing crop productivity, either through technical interventions or via improved irrigation scheduling. Both approaches can reduce the risks of over-watering of crops and thereby save water. There is good evidence for reducing field-level applications at this scale (Jovanovic et al., 2020), with three main approaches recommended:

1. Farmers can apply less water to match the reduced crop water requirements created by the approaches described in Section 4.2. This also means that water applications need to be timed better; otherwise, fields run the risk of being too wet or too dry prior to irrigation, which, in turn, makes it difficult to manage field-level applications efficiently. Therefore, practices, such as irrigation scheduling and deficit irrigation, by default, apply to this level.
2. To enhance field-edge and within-field water control via the many technologies available. Examples include levelling fields; shortening or lengthening fields as relevant, adding bunded plots; changing the shape of furrows and ridges; applying alternate row irrigation; using surge irrigation; reducing runoff at the end of the furrows or fields; zoning fields more accurately to not irrigate low or high-spots or infertile soil; and more drastically, switching to sprinkler or drip irrigation.
3. To use more water from sources that are not withdrawn directly from rivers or aquifers. This happens commonly when farmers capture more rainfall or make use of shallow groundwater. Regarding rainfall capture, farmers experiment with practices, such as ploughing, land-levelling and shaping, and digging soil pits or making bunds. Note that mulches can act both to reduce evapotranspiration and to improve infiltration of rainfall into soil.

The use of sensors and meters to monitor field-level applications and the means to control rebound at the field-level are covered in Sections 6.2 and 6.4, respectively.

4.4 Intra-system water distribution, scheduling and sources

This stage relates to the cascading of reduced field applications explored in the previous sub-section, up the irrigation system to higher levels (farm, tertiary and secondary units) through to the main irrigation intake (a specific level covered below). The following considerations apply:

1. Within irrigation systems, seepage and operational losses can be reduced by, for example, improved operation of sluices, pumps, canal-gates and pipe-valves; the repair of conduits; and removal of bottlenecks

(whereby a flow surcharges a canal so that operators throttle back the flow to stop the over-topping which then adversely affects crop water provisioning).

2. Smaller field applications can reduce the amount of water supplied to farms. This might seem easy as it would simply involve farmers closing down their farm irrigation gate valve or cutting the time duration of farm withdrawals. However, the question of equity comes to the fore. Farmers are naturally cautious of unfairly shouldering cuts. This means improved gate designs or meters, or it might mean changes to the water conveyance and distribution design to improve the manageability and visibility of water apportionment (see Section 6.1).
3. Intra-system distribution of water is not only about splitting flows and volumes but also about managing the timely delivery of water in irrigation schedules that match soil-water and crop parameters.
4. It is possible to improve intra-system water control by adding operational dams and ponds. These tend to be small-scale, providing the capacity to meet only a few hours or days demand. However, they add flexibility by, for example, allowing farmers to work during the day, or by capturing rainfall, or by storing water that has been ordered from the main (i.e. bulk) water provider but is no longer needed due to rainfall.
5. If the total volume of stored water via many small dams represents a substantial proportion of total season demand, then small operational dams also allow farmers to reduce the volumes of water being withdrawn during critical low flow periods.
6. Institutional and organisational measures assist the above, for example, water user associations organise meetings and collect farmer payments to improve operation and maintenance.

4.5 System and basin-level sources and withdrawals

Building on previous steps, if fields, farms and unit areas within irrigation systems withdraw less water, then the amounts of water withdrawn into the irrigation system can be reduced. This completes the bottom-up cascade of 'savings' at the field upwards to the main irrigation system intake; meaning inflows through the intake are reduced because less water is being applied by farmers. Reductions can occur by regulating downwards or switching off sluice gates and pumps and reducing the hours or days of withdrawal. Water stored in large dams can be used to supplement low flow periods and facilitate withdrawals in low flow periods without drying up rivers (McCartney and Smakhtin, 2010; Lankford, 2023a).

This level also expresses and manages basin-wide withdrawals into irrigation. Therefore, it is important to confirm not only that the single irrigation system reduces its withdrawal of water but that this task occurs across all irrigation systems drawing water from the basin or aquifer. Otherwise, one irrigation system withdraws less water, but its downstream neighbour abstracts the residual surplus flowing down to it. Without this basin-wide systematic approach, the risk of rebound negates the delivery of freed-up water to serve downstream ecological needs.

5 Translating withdrawals into consumption-based savings for nature (Part 3)

The next stage is to materially translate any irrigation and basin-level reductions in water withdrawals to nature whilst guarding against rebounds in both withdrawals and consumption. There are four key dimensions to this water transfer for it to be effective and consumption based. First, in terms of theory, this stage can be framed as a 'commons of conserved resources' where competition is tight and occurs between the farmers making the savings, immediate neighbours, other users in the economy and finally nature. Termed a 'paracommons', this predicts nature/the environment will be last in line for water savings in allocations (Lankford, 2013; Lankford, 2023a).

Second, recognising the highly competitive nature of the paracommons, the directness of the pathway of water transfer is important. When irrigation withdrawals and distributions are reduced in one location, care must be taken otherwise water savings may be reused and re-consumed by irrigators elsewhere or by other users, such as towns and cities. Translating water to the environment can occur directly or indirectly. A direct example is when irrigators are immediately upstream of another use of water. Here, cutting withdrawals sends water immediately downstream to other users/uses. It is direct when no other inflows or water abstractions sit between the irrigation systems and downstream users, or where abstraction points are few in number. For example, upstream irrigated fruit producers in Limpopo Province in South Africa manage their withdrawals to deliver environmental flows to Kruger National Park, thereby meeting the ecological reserve enshrined in the 1998 South Africa Water Act (Pollard and du Toit, 2011). Under such circumstances, relatively little additional scrutiny is needed to ensure that releases of water caused by lower irrigation withdrawals flow downstream for ecological benefit. An indirect example is when irrigators withdraw and consume less water from an aquifer so that the water in the aquifer recovers for use elsewhere. This is indirect because the effect of smaller irrigation withdrawals translates to the aquifer rather than directly to another user. This indirect translation of saved water to nature requires a lot more scrutiny because the aquifer is likely to be a common property subject to many withdrawals that require monitoring; hidden abstractions might exist; water levels need to be recorded; the aquifer might be large and not immediately recover; and additional factors might be at play, such as groundwater movement and rainfall (or lack of).

Third, it is important to confirm that reductions in withdrawals into irrigation systems equate to reductions in consumption. This is because environmental allocations secured across a large river basin changing over time can only be derived from verified reductions in consumption. Several dimensions apply:

- The classical irrigation efficiency ratio 'controls' for only three components; withdrawals (or gross inflows), net beneficial consumption and the residual between withdrawals and beneficial consumption (often called losses). Provided that the reduction in 'losses' in each of the ISAF levels, achieved via better or more efficient water management, is translated upwards through ISAF's levels to reductions in withdrawals, then mathematically, rebounds in consumption do not occur. This control over where the losses 'end up' counters the argument that losses are always converted into increased beneficial consumption (Ward and Pulido-Velázquez, 2008), which occurs when losses are subsequently translated to new farmers, uses, areas, durations and seasons (Lankford, 2023a).
- Accounting for water at the system and basin scales is vital to ascertain withdrawals and their consumptive versus non-consumptive destinations (see Section 7).
- The directness of the pathway discussed earlier will be important here. Direct pathways will minimise the burden of accounting for consumption versus withdrawal. Indirect pathways will need more monitoring as outlined previously.

Fourth, as discussed in Section 6.6, other allocation mechanisms can enable and facilitate this material transfer, for example, through legal rights of the environment to water (e.g. environmental reserves). Alternatively, water allocation policy may exclude environmental flows or not give them sufficient priority. Furthermore, although water allocation might prioritise environmental flows, there may be weaknesses in governance to ensure adequate monitoring and enforcement of the physical tracking of water to nature, especially during drought events.

6 Cross-cutting governance architecture (Part 4)

This considers six different dimensions that make up a governance architecture for freeing up water savings in irrigation to restore environmental flows. These include hydrological, infrastructural and institutional considerations.

6.1 Infrastructure design for manageability of apportionment

The subject of how to improve the manageability of water apportionment within an irrigation system is an aspect that is relatively ignored in the literature (Plusquellec et al., 1994; Plusquellec, 2002; Horst, 1999; Lankford and Gowing, 1996). Manageability refers to making it easier via design to carefully and 'automatically' (or with minimal manual adjustment) apportion water to crops. There are two levels of design when improving irrigation infrastructure for better manageability.

The first is the operation of the irrigation equipment to more 'easily' place water where and when it is most needed. Physical infrastructure of canals and gates mediate how farmers receive water and apply water in their fields; physical design also affects whether they are able to reduce their applications. For example, a paddy rice farmer in the middle of a 300-hectare rice system with few canals receiving their water from neighbour's fields has little control over the timing, rates, location and depths of flow. Conversely, a farmer in an irrigation system of 2-hectare farms each fed by a channel and gate, has greater control; e.g. they can switch off their water when not needed. Thus, physical irrigation infrastructure needs to assist new apportionments rather than obstruct them or, worse, pull in the opposite direction. For example, main irrigation intakes might need changes to the dimensions of orifices and weirs to help reduce the volumes withdrawn. Large-scale storage might help store wet season runoff to enable allocation during drought and dry seasons.

The second level of design is more abstract in aim but equally material, requiring real changes to physical components. It is about 'designing in' the ratio of supply (litres per second, l/sec) to area served (hectares, ha) so that the supply water duty (or hydromodule) in l/sec/ha accurately matches the crop and system water requirements in l/sec/ha (Lankford, 1992; 2023b). Design changes should ideally involve adjustments to flow capacities (of gates, valves and conduits) and to command area served; select dimensions of gates and valves so they cannot exceed their desired flow; apply during peak-demand periods; and design in scarcity so that surpluses (the difference between crop water needs and the supply to the field, farm or tertiary unit) are transferred up the system rather than lost in excess evaporation or tail-end sinks. However, applying appropriate infrastructural revisions to irrigation systems has a poor track-record seen through attempts to modernise irrigation (McCartney et al., 2019). For example, government or public large-scale canal systems often have limited scope or willingness to adapt. This sector has been notoriously difficult to reform and witnesses low cost-effectiveness (Lankford et al., 2016).

6.2 Systematic monitoring of inputs and outcomes for crops and water

A significant challenge within irrigation management lies with the concurrent monitoring of 'what is going on' with water and other inputs (such as energy or labour) in ways that make immediate sense for irrigation service managers and farmers alike. Notwithstanding this objective, technically, water measurement and performance assessment in irrigation systems is difficult. This means indicators of 'good water management', such as water productivity, efficiency, per-hectare consumption, total system consumption, irrigation equity and adequacy, and their 'cost' of other inputs, such as energy consumed, are not easy to derive. Underlying these problems are common gaps and weaknesses:

- A lack of measurement of river and aquifer flows, supplies and stores of water;
- A lack of or inaccurate flow measurement of withdrawals into irrigation systems;
- Flow measurement within irrigation systems is often missing or inaccurate;
- Flow measurement of water exiting/draining from irrigation systems is nearly always missing;
- Difficulties in using satellite data to ascertain irrigated fields versus partially or non-irrigated fields;
- Difficulties in accurately assessing evapotranspiration and rainfall as point and areal data;
- Gaps in measuring other sources of water inputs to irrigated soils such as contributions from local springs, small streams and capillary rise from shallow water tables;
- Difficulties in quantifying other water depletion occurring via evaporation and non-recovered flows;
- Difficulties in measuring the above at the right time and scale to help farmers adjust their agricultural water management; and
- Poor data collection of other relevant inputs, such as labour, energy, agrochemicals.

The answers to the challenge of system monitoring do not sit with single solutions, such as satellite area observations, main canal flow metering or soil moisture sensors. The problem with 'single solutions' and especially high-technology solutions is that they (1) typically only measure one part of a wide complex system; (2) are spatially inadequate in the face of high heterogeneity; (3) require additional analysis and synthesis; (4) are undermined by institutional weaknesses, such as their ownership, repair and maintenance and (5) often only feed through information to one type of stakeholder (e.g. experts external to or remote from the irrigation system).

Instead, more carefully designed systematic or optimal approaches to irrigation monitoring are desirable, but this is by no means easy. On commercial irrigation systems growing high-value crops, such as citrus or avocado using piped water, system-wide data can be monitored by a dedicated agronomy office, pulling in information from rainfall gauges, flow meters, evapotranspiration measures, soil and plant sensors and so on, many of them 'smart' and interconnected. While this model might indicate an ideal to be copied, it works because it fits that particular irrigation and farming system. In small scale, smallholder or public irrigation systems, a different approach is needed. This has to be built from a number of principles such as 'fit to scale', 'robustness', 'easy to use', 'immediate' and 'farmer-centred'. Recognizing these gaps and principles will help design appropriate solutions, such as the following:

- Asking local farmers what indicators they use to assess good water management (e.g. total yield, rainfall management, rotational scheduling, rate of irrigation progress, and visual clues);
- Discerning what land and water measurements are already being made, where and by whom and building on this knowledge;
- Identifying willing farmers, engineers and other local experts who might develop services to record basic information;
- Determining priority (must have) data that are easier to collect and analyse (e.g. rainfall, cropping patterns, areas irrigated and times irrigated);
- Determining additional data needs and technologies that might require more substantial support and funding (examples include inflow and outflow measurement, satellite data and soil moisture probes);
- Developing smartphone apps and mechanisms to integrate information (especially informal feedback from individual farmers) into system-wide and systematic information;
- Developing 'relative' measures that minimize the need for detailed information. Examples include the hectares of irrigation completed per day or week of irrigation (ha/day). Over time, this can be used to build trend lines or compare between farmers based at the top end of irrigation systems vs those located at the tail end;
- Building 'no-measure' irrigation systems (Lankford, 1992) whereby water flow distribution is extremely closely tied to both the area served and 'tight' irrigation schedules that fit crop water requirements with little spare capacity; and
- Imposing strict irrigation schedules that enforce deficit irrigation to dictate the dosage and interval of irrigation.

6.3 Redistribution of savings enabled by apportionment

Introducing water apportionment in Section 4.1 as a key hydrological feature of irrigation systems and their management allows the corollary argument to be made; that changes in water apportionment under actions to 'save water' automatically result in a surplus being 'freed up' to be potentially diverted elsewhere. To illustrate if 120 l/sec is ordinarily withdrawn into a farm but later only 100 l/sec is withdrawn, a 'residual surplus' of 20 l/sec is 'freed up' and is now seeking a new destination. However, to be more accurate, because the time duration of withdrawal is also a variable, a volume of water in litres or cubic metres is now seeking a new destination. Therefore, to ensure accurate water apportionment under these changing conditions across the whole irrigation system requires us to manage the new reduced supply not just as a flow or volume but as a hydromodule by expressing its 'specific supply' expressed as a ratio in l/sec/ha (Lankford, 2023b). To be even more accurate, this residual surplus exists only if the new smaller withdrawal is less than the original withdrawal, which runs the risk of being drawn at the original flow rate or volume. Furthermore, if a drought cuts the supply to farmers, a drop in withdrawal by farmers does not create a surplus.

As well as managing reduced flows, volumes and specific flows over time and space, it is also necessary to identify the likely destinations where savings may end up. Bearing in mind that irrigation systems are hydrologically and physically complex, the following destinations for surpluses are possible:

- Within a cropping season, the time duration of withdrawal is extended.
- Within a farm, the surplus is directed to a more consumptive crop.
- A farmer expands their field area and therefore the total farmed area.
- The surplus passes to other farmers in the same tertiary canal unit.
- The surplus is used elsewhere in the irrigation system by farmers by being passed via the main and secondary canal networks.
- The surplus passes through the farm runoff or groundwater drainage for use by other farmers.
- The surplus is banked in storage for reuse in another area or a second season.
- The surplus is forestalled and, therefore, passes up the irrigation to the main intake to remain in the river or aquifer.
- At the basin level, the forestalled surplus is seen via a paracommons and is competed over by the proprietor, neighbour, nature or society.

In summary, the accurate apportionment of flows, volumes and specific flows and their surpluses throughout the irrigation system over time is central to the success of ISAF. Excellence in water management helps to 'control for' irrigation recovery and rebound.

6.4 Controlling consumptive rebound

The ever-present risk is that improvements in water management, leading to reductions in water applied and withdrawn, do not translate to real water savings over time at the farm, system or basin level. This is because, as outlined previously, despite reducing the volume of water applied or withdrawn, farmers find ways of using their own savings elsewhere or because other farmers pick up and recover irrigation losses. ISAF argues that controlling for rebound should occur at the relevant scale level of the irrigation system, rather than *post hoc* being applied at the basin scale. The following points explain this:

- At the crop water requirement level, farmers can extend water withdrawals by selecting more consumptive crops (e.g. bananas use more water than citrus), or crops that have a longer irrigating season or that irrigate for two seasons.
- Within the lower scales of an irrigation system (field and farm), different means for controlling rebound can be applied. The principal control is that of area. Thus, a farmer might expand her field or farm over time having made efficiency gains. Another important control is the duration of irrigation, for example rebound might occur by switching from 20 days of watering over a 120-day season to 25 days of watering over a 140-day season.
- Recognising this, time-duration caps can be applied to water withdrawals in specific parts of the seasonal calendar (for example, no abstraction allowed during the dry season) or to parts of the month (i.e. release water downstream every 24th day).
- However, it is also necessary to control for area and time rebounds within a scale but at geographically distant locations. Thus, while one farmer might not expand the area of their field, the irrigation system inadvertently transfers their unused water to another farmer in the same tertiary unit who subsequently expands their irrigated area. Water transfer might also happen to farmers in other tertiary units supplied by a common secondary canal network.
- Thus, controlling for rebound at the secondary canal, main canal and basin levels can also be achieved by ensuring that new gates, intakes, branches and networks are not established and that within existing infrastructure, water flow rates are not increased or extended over time.
- It is also possible to regulate the reuse of drainage water in order to stop or reduce irrigated areas that depend on drainage from upstream. While, on paper, this appears to be distinguishing between paper and real savings, in reality, peripheral irrigated areas are extremely difficult to

monitor and enforce. This is because tailwater from irrigation cannot easily be separated from water arriving from natural runoff and groundwater base flows. The priority here is to control the total area under irrigation rather than to rule on the provenance of drainage water.

- Controlling for rebound also tracks cross-seasonal shifts in water. Thus, a farmer 'appears' to apply less water during the main cropping season but has banked this in groundwater or an on-farm operational dam/pond, only to use this water in the off-season to irrigate a small area of vegetables.
- To reiterate; as well as their flowrate and volumetric capacity, it is necessary to control for an upwards creep in the number and size of intakes/offtakes that are installed at all levels of irrigation systems but especially at the basin level. With regards to the latter, the physical presence of an intake on a river presages irrigation withdrawals, and, moreover, additional or larger intakes can be visually monitored by concerned downstream farmers and other water users.

6.5 Actors and motives

Each scale level is represented by different actors and their motives and interests (Lankford et al., 2020; Uhlenbrook et al., 2022; van Oel et al., 2019). For example, at the lower levels, farmers are commonly interested in maximising crop production rather than in saving water. This can change under pressure, for example, drought or new incentives and enablers such as drip irrigation (see Section 6.6). Higher up in the irrigation system are other actors, such as water user associations (WUAs), engineers and other experts, and at the basin level, there are river basin officers, regulators and external agencies. These different actors and their individual motives influence debates on saving water, irrigation efficiency and water allocation, regardless of how well- or ill-informed they are on the topic (Molden et al., 2010).

6.6 Economic incentives, pressures and enablers for water allocation

The above technical mechanisms are underpinned by a range of economic, technological and social incentives, enablers and penalties, ideally tailored for specific circumstances and for the scale that they are best suited for. The following dimensions outline this broad topic, divided into three categories:

Pricing, markets

Regarding incentives to reduce field-level applications, there are some very effective simple price proxies for, and influences on, water applied. For example, farmers who pump water using their own fuel or paid-for-electricity start to economize on fuel and water. Irrigators in Southern Tanzania who agreed a small levy on land area to be irrigated also started to economize on water. Farmers also respond to intra-farmer competition for water in their irrigation system, for example, accepting that in rotational systems, they should take their 'turn' and pass on water to the next in turn within a strict timetable.

Regarding basin and system levels, volumetric caps can be effective in limiting water withdrawals. When combined with the capacity to transfer water rights between users (e.g. water markets), it may also stimulate more sustainable development and efficient use of available water. However, successful water markets require significant institutional arrangements and are not viable in all contexts (Wheeler et al., 2017). They are especially difficult in transboundary basins comprising nations with divergent political systems and economies.

In some instances, markets have successfully provided mechanisms to improve the allocation of water to the environment. For example in the USA, water has been strategically purchased to enable salmon passage in California (Lund, 2017). Similarly, in the Murray Darling Basin in Australia, the federal government has spent millions of dollars in water markets to 'buy back' water from irrigators to conserve downstream wetlands, although the effectiveness of the environmental flows generated has been questioned (Settre and Wheeler, 2016; Chen et al., 2021). Despite the theoretical benefits, there remain substantial barriers to the use of markets to serve the environment, especially in developing country contexts (Rosegrant et al., 2014).

There are few examples of economic instruments (pricing) being used successfully to manage irrigation water demand (Molle and Berkoff, 2007) although water pricing can shift farmers' thinking on the economics of their production and lead to changes in irrigation technology and cropping.

Water pricing to bring down net demand is constrained by a number of problems: (1) metering of water (or energy use to pump water) by individual farmers is a prerequisite for success; (2) at low prices, the elasticity of water use (or power used for pumping) is very low; (3) raising tariffs to a level high enough to influence farmer behaviour is often politically unacceptable; and (4) higher prices bring equity impacts, such as costs borne by poorer farmers.

Water allocation policies

Water-literate and numerate allocation policies are needed; recognizing factors such as seasonality (dry versus wet season), the general lack of water measurement, the difference between in-stream and abstracted water needs and rights, and the controlling influences on water from water law and physical infrastructure.

National water allocation policies, while not always successful, nonetheless benefit from being clearly expressed and tied to country and sector priorities. For example, it is generally agreed that the high priority given to sustaining ecological flows in South Africa (with parity to domestic water), while not always strictly enforced, is a strong statement on the allocation of water to the environment.

The legal terms and stipulations within irrigation licenses will need revisiting. Examples include how licences are written to allow for incremental caps on withdrawals over time, effectively applying an annual de-escalator of withdrawals. Historical legacy water rights and the legal right to consume one's own water savings also need revisiting (Norris, 2011).

Policies and those advising on policy design should be 'politics- and society-aware'. For example, raising tariffs to a level high enough to influence farmer behaviour is often not politically acceptable; allocations out of irrigation may result in social unrest and equity problems (with female and poor farmers hit hardest); and penalties and enforcements, despite being important, are difficult to impose.

Fit to national priorities: Water reallocation is a complex area, politically, technically, economically and socially. One way of highlighting this difficulty is by arguing that the most effective way to 'save water' for nature is by not allowing or developing irrigation and/or abstractions in the first place (Lankford and Grasham, 2021). In this way, water is kept in the river/environment. However, decisions such as these require Ministerial-level national plans to be clarified (e.g. for food security, economic growth, power generation, social well-being and freshwater ecologies).

Water reserves legally establish environmental water as a prior right to consumptive water use. They are an attempt to address the issue of legally who owns conserved resources (also asked by the paracommons). This is needed because within volumetric caps and quotas, farmers under given legal frameworks, such as the US's Prior Appropriation Law, may legally continue to make and then consume their own savings, whereas the latter should be subject to alternative distributions including reallocation to nature. The ecological reserve of South Africa enshrined in the National Water Act (1998) is a well-known example but there are others (e.g. Mexico). Reserve requirements are typically determined through environmental flow assessments (EFA) using methods to estimate the quantity and timing of flows required to sustain identified aquatic values in riverine environments. However, there is little evidence that reserves/EFAs are effective in practice. Implementation is hindered by the complexity of translating reserve requirements into operational plans, and compliance is typically low. In recent years, there has been criticism of EFAs as being too technocratic and 'top down'.

All allocation mechanisms require specific legal or regulatory tools to either impose conditions on water users or create specific legal rights for the environment. Often, it is a lack of capacity to monitor compliance and enforce these legal instruments that results in them being ineffective. With these and other weaknesses, allocation mechanisms provide little environmental water during times of scarcity, and once water has been fully allocated (or indeed overallocated) to consumptive water users, it is difficult and costly to reduce these rights and return water to the environment at a future date.

Knowledge, services and social learning

Drought teaches: Drought brings in new technical practices, institutions and agreements that can be carried forward once the drought ends. Examples are deficit irrigation, alternate row irrigation, reducing irrigated areas and so on. 'Drought learning' here is not only about how and what individual farmers change but also how to get group agreements acting across large areas to endure for the long term.

Whose knowledge counts? The best is that farm-scale, local, farmer-originated solutions combine with additional or outside expertise to work together to reduce water applications. This is preferable to externally imposed blue-prints or 'magic bullets' attempting the same. Nonetheless, a rapidly changing complex legal, institutional, physical and technological policy space brings the risk that it should be interpreted and delivered by experts, such as water lawyers and irrigation engineers. This risk needs to be managed by ensuring that the farmers are supported by experts and moreover that these teams collectively work on real-world water problems rather than those manufactured in the minds of either farmers or experts.

7 Accounting for basin-level outcomes (Part 5)

ISAF is based on the premise that water allocation out of irrigation takes place by coordinating reductions in crop water requirements, field applications, intra-system distribution and system withdrawals. Ultimately, however, the allocation of water to nature can only come from reducing water consumption at the basin level. Various mechanisms covered previously, deployed at different levels of irrigation systems, assist in reducing consumption, for example, tighter design for management; controlling for reuse and rebound in consumption; better tracking of water to nature; and enhanced monitoring of these inputs and outcomes. Accordingly, to deliver these objectives, comprehensive water accounting at multiple scales is required. As argued by Lankford (2023) and exemplified by 'Irrigated Systems Accounting' (ISA), comprehensive accounting goes beyond current water accounting (WA) as the following explain:

- ISA covers total crop production at the basin scale in order to control for Pareto impairment whereby allocation of water out of irrigation inadvertently reduces crop production.
- ISA is sensitive to both short-term and long-term 'creep' of irrigation areas across all cropping seasons as possible evidence of rebound.

- ISA tracks the role of small-scale (operational) and large-scale (seasonal) storage in facilitating areal expansion and water consumption.
- ISA compares irrigated systems at three important scales (field, farm/system and catchment) and 'before and after' in order to decide how management inputs result in tangible changes over time for both water-donating irrigation systems and water-receiving sectors such as ecological flows in rivers.
- Comprehensive accounting is also about recording the social impacts of changes in water withdrawal, distribution and consumption on farmers.
- Accounting is dependent on good quality data; although satellite data are improving all the time, on-the-ground verification of crops, areas and water flows remains essential.
- Accounting also covers other inputs such as energy use, and their related performance outcomes (e.g. crop yield per unit of carbon dioxide emitted).

8 Direction of cascading 'savings' between scale-levels (Part 6)

ISAF is constructed as a 'bottom-up' conceptual framework, meaning that reducing withdrawals at the main intake has its origins at the crop and field scale. ISAF's aim is to cascade 'savings' at field and farm levels upwards to the main irrigation system intake because less water is being applied to fields by farmers. In this sense, ISAF recognises the agency that farmers have in managing their water, land and inputs in the face of a constantly changing environment.

However, ISAF also recognises that farmer behaviours can be 'influenced' by reducing withdrawals at the main intake and cascading these 'top-down'. The aim here is to reduce inflows to the irrigation system and then use this within-system water scarcity to encourage farmers to use less water in their farms and on their fields. Reduced flows into systems that have become accustomed to more generous water provisions raise the question of equity; otherwise, tailenders are likely to suffer whilst upstream farmers continue to consume the same water. For this reason, institutional and physical factors described in this chapter such as strengthening WUAs, monitoring social effects, and designing the manageability of irrigation equity and efficiency, are critical.

An alternative version of a top-down 'savings' approach could result in caps being applied to irrigated area instead of water withdrawals because (1) area is a major driver of withdrawals and consumption; (2) area tracks both rebound and recovery of irrigation water; (3) area allows 'per-farm' cuts to be equitably applied; and (4) area is accurately estimated by farmers and can be measured using satellites. An example de-escalator might cut irrigated area by 1% per year. However, in reality, hybridised bottom-up and top-down policies combined with prioritising areas and/or withdrawals are likely to be most effective.

9 Conclusion

Theory and practice regarding water conservation and reallocation continue to be a contested domain, and our understanding is still being developed; for example, the emphasis given to 'pricing to conserve water' in recent decades is being added to, and debates regarding, the utility of the concept of irrigation efficiency are on-going (Lankford et al., 2020). Recent and forthcoming developments include those discussed in this chapter such as water accounting, scale-relevant solutions, and the need to update water law and legal frameworks. However, the evidence for irrigation system- and basin-level savings transferred to nature is currently poor or mixed and is subject to different accounting methods, theoretical contestation, and a general lack of monitoring.

However, the future will be very different as pressure mounts to reallocate water away from irrigated agriculture and into other sectors (Garrick et al., 2019). What ISAF brings to this debate is a structure to water governance and conservation policies that hitherto have been either piecemeal, overly reliant on one tool, do not grapple with the risk of increased consumption, or do not accommodate most or all scales of water management. It is likely that clearer mechanisms for reallocation will have to advance on a number of co-ordinated fronts: working more closely with farmers; regulating the cropped area under irrigation at field, farm, intra- and inter-system scales; tightening up design for manageability of scarce water; building in pricing signals where feasible; reducing the number or volumes of water withdrawals/quotas; revising out-of-date water laws; adding real-time soil, land and water monitoring; and ensuring overall accounting of land, water, crop production, other resources such as energy, and other outputs such as carbon dioxide emitted.

10 Where to look for further information

Two key books in the field of irrigation were written some time ago but are notable for their depth and breadth of treatment of the topic. They came out as twin monographs: (1) Jensen, M. E. (Ed). 1983. Design and operation of farm irrigation systems. Rev. print. ed. St. Joseph, Mich.: American Society of Agricultural Engineers. (2) Hoffman, G. J.; Howell, T. A. and Solomon, K. H. 1990. Management of Farm Irrigation Systems. St. Joseph, Mich.: American Society of Agricultural Engineers.

Two recent broad overviews of irrigation are given by:

- McCarthy, N., Ringler, C., Agbonlahor, M. U., Pandya, A. B., Iyob, B. and Perez, N. (2023). Is irrigation fit for purpose? A review of the relationships between scheme size and performance of irrigation systems. https://www.ifpri.org/publication/irrigation-fit-purpose-review-relationships-between-scheme-size-and-performance.

- Wiggins, S. and Lankford, B. 2019. Farmer-led irrigation in sub-saharan Africa: synthesis of current understandings. Synthesis Report of the DFID-ESRC Growth Research Programme. Overseas Development Institute, UK. https://odi.org/en/publications/farmer-led-irrigation-in-sub-saharan-africa-synthesis-of-current-understandings/.

Globally, the three main organisations that are working on irrigation are the International Water Management Institute (https://www.iwmi.cgiar.org/), the Food and Agriculture Organisation (http://www.fao.org) and the Daugherty Water for Food Global Institute at the University of Nebraska (https://waterforfood.nebraska.edu/). Other important organisations researching irrigation policy include International Food Policy Research Institute (www.ifpri.org/), ODI (https://odi.org/), and various universities. As well as Nebraska University, these include IHE Delft (https://www.un-ihe.org/) and Wageningen University and Research (https://www.wur.nl). Irrigation research is also located across other University departments such as Cranfield University (UK), University of East Anglia (UK), Arizona University (USA), Penn State (USA) and Sokoine University of Agriculture (Tanzania).

11 References

Batchelor, C., Reddy, V. R., Linstead, C., Dhar, M., Roy, S. and May, R. 2014. Do water-saving technologies improve environmental flows? *Journal of Hydrology* 518, 140–149.

Chen, Y., Colloff, M. J., Lukasiewicz, A. and Pittock, J. 2021. A trickle, not a flood: environmental watering in the Murray-Darling Basin, Australia. *Marine and Freshwater Research* 72(5), 601–619.

Garrick, D., De Stefano, L., Yu, W., Jorgensen, I., O'donnell, E., Turley, L., Aguilar-Barajas, I., Dai, X., De Souza Leão, R., Punjabi, B., Schreiner, B., Svensson, J. and Wight, C. 2019. Rural water for thirsty cities: a systematic review of water reallocation from rural to urban regions. *Environmental Research Letters* 14(4), 043003.

Gil, P. M., Lobos, P., Durán, K., Olguín, J., Cea, D. and Schaffer, B. 2018. Partial root-zone drying irrigation, shading, or mulching effects on water savings, productivity and quality of 'Syrah' grapevines. *Scientia Horticulturae* 240, 478–483.

Giordano, M., Turral, H., Scheierling, S., Tréguer, D. and Mccornick, P. 2017. *Beyond "More Crop per Drop": Evolving Thinking on Agricultural Water Productivity*. Colombo, Sri Lanka: Washington, DC: IWMI Research Report 169.

Grafton, R. Q., Williams, J., Perry, C. J., Molle, F., Ringler, C., Steduto, P., Udall, B., Wheeler, S. A., Wang, Y., Garrick, D. and Allen, R. G. 2018. The paradox of irrigation efficiency. *Science* 361(6404), 748–750.

Horst, L. 1999. The failure of adjustable irrigation technology, the options for change and the consequences for research. *Agricultural Water Management* 40(1), 101–105.

Jägermeyr, J., Pastor, A., Biemans, H. and Gerten, D. 2017. Reconciling irrigated food production with environmental flows for Sustainable Development Goals implementation. *Nature Communications* 8, 15900.

Jovanovic, N., Pereira, L. S., Paredes, P., Pôças, I., Cantore, V. and Todorovic, M. 2020. A review of strategies, methods and technologies to reduce non-beneficial consumptive water use on farms considering the FAO56 methods. *Agricultural Water Management* 239, 106267.

Lankford, B. A. 1992. The use of measured water flows in furrow irrigation management: a case study in Swaziland. *Irrigation and Drainage Systems* 6(2), 113–128.

Lankford, B. A. 2012. Fictions, fractions, factorials and fractures; on the framing of irrigation efficiency. *Agricultural Water Management* 108, 27–38.

Lankford, B. A. 2013. *Resource Efficiency Complexity and the Commons: The Paracommons and Paradoxes of Natural Resource Losses, Wastes and Wastages*. Abingdon: Routledge.

Lankford, B. A., Closas, A., Dalton, J., López Gunn, E., Hess, T., Knox, J. W., Van Der Kooij, S., Lautze, J., Molden, D., Orr, S., Pittock, J., Richter, B., Riddell, P. J., Scott, C. A., Venot, J.-P., Vos, J. and Zwarteveen, M. 2020. A scale-based framework to understand the promises, pitfalls and paradoxes of irrigation efficiency to meet major water challenges. *Global Environmental Change* 65, 102182.

Lankford, B. A., Makin, I., Matthews, N., McCornick, P. G., Noble, A. and Shah, T. 2016. A compact to revitalise large-scale irrigation systems using a leadership-partnership-ownership'theory of change. *Water Alternatives* 9, 1–32.

Lankford, B. A. and Orr, S. 2022. Exploring the critical role of water in regenerative agriculture; building promises and avoiding pitfalls. *Frontiers in Sustainable Food Systems* 6, 1-17.

Lankford, B. A. 2023a. Resolving the paradoxes of irrigation efficiency: irrigated systems accounting analyses depletion-based water conservation for reallocation. *Agricultural Water Management* 287, 108437.

Lankford B. A. 2023b. Hydromodule numeracy unlocks the puzzles of irrigation. Posted 28 Sept 2023. https://brucelankford.org.uk/2023/09/28/hydromodule-numeracy-unlocks-the-puzzles-of-irrigation/.

Lankford, B. A. and Gowing, J. 1996. Understanding water supply control in canal irrigation systems. In: Howsan, P. and Carter, R. (Eds). *Water Policy: Allocation and Management in Practice*. London: E & FN Spon.

Lankford, B. A. and Grasham, C. F. 2021. Agri-vector water: boosting rainfed agriculture with urban water allocation to support urban–rural linkages. *Water International* 46(3), 432–450.

Lankford, B. A. and Scott, C. A. 2023. The paracommons of competition for resource savings; irrigation water conservation redistributes water between irrigation, nature, and society. *Journal of Resources, Conservation and Recycling* 198, 107195.

Libecap, G. D. 2009. The tragedy of the commons: property rights and markets as solutions to resource and environmental problems. *Australian Journal of Agricultural and Resource Economics* 53(1), 129–144.

Lund, J. 2017. Why the environment needs water markets. *The New Humanitarian*. Available at: https://deeply.thenewhumanitarian.org/water/community/2017/01/17/why-the-environment-needs-water-markets.

McCartney, M. P. and Smakhtin, V. 2010. Water storage in an era of climate change: addressing the challenge of increasing rainfall variability. *Blue Paper*. Colombo, Sri Lanka: International Water Management Institute.

McCartney, M. P., Whiting, L., Makin, I., Lankford, B. A. and Ringler, C. 2019. Rethinking irrigation modernisation: realising multiple objectives through the integration of fisheries. *Marine and Freshwater Research* 70(9), 1201–1210.

Mendoza-Espinosa, L. G., Burgess, J. E., Daesslé, L. and Villada-Canela, M. 2019. Reclaimed water for the irrigation of vineyards: Mexico and South Africa as case studies. *Sustainable Cities and Society* 51, 101769.

Molden, D., Oweis, T., Steduto, P., Bindraban, P., Hanjra, M. A. and Kijne, J. 2010. Improving agricultural water productivity: between optimism and caution. *Agricultural Water Management* 97(4), 528–535.

Molle, F. and Berkoff, D. 2007. *Irrigation Water Pricing: The Gap Between Theory and Practice*. Wallingford: CABI Publishing.

Molle, F. and Tanouti, O. 2017. Squaring the circle: agricultural intensification vs. water conservation in Morocco. *Agricultural Water Management* 192, 170–179.

Norris, J. 2011. Montana v. Wyoming: is water conservation drowning the Yellowstone River Compact. *U. Denv. Water L. Rev.* 15, 189.

Pérez-Blanco, C. D., Hrast-Essenfelder, A. and Perry, C. 2020. Irrigation technology and water conservation: a review of the theory and evidence. *Review of Environmental Economics and Policy* 14(2), 216–239.

Perry, C. 2011. Accounting for water use: terminology and implications for saving water and increasing production. *Agricultural Water Management* 98(12), 1840–1846.

Plusquellec, H. 2002. *How Design, Management and Policy Affect the Performance of Irrigation Projects*. Bangkok, Thailand. Bangkok, Thailand: Food and Agriculture Organization Regional Office for Asia and the Pacific. Food and Agriculture Organisation.

Plusquellec, H., Burt, C. and Wolter, H. W. 1994. *Modern Water Control in Irrigation: Concepts, Issues, and Applications*. Washington DC: World Bank.

Pollard, S. and Du Toit, D. 2011. Towards adaptive integrated water resources management in Southern Africa: the role of self-organisation and multi-scale feedbacks for learning and responsiveness in the Letaba and crocodile catchments. *Water Resources Management* 25(15), 4019–4035.

Richter, B. D., Brown, J. D., Dibenedetto, R., Gorsky, A., Keenan, E., Madray, C., Morris, M., Rowell, D. and Ryu, S. 2017. Opportunities for saving and reallocating agricultural water to alleviate water scarcity. *Water Policy* 19(5), 886–907.

Rosegrant, M. W., Ringler, C. and Zhu, T. 2014. Water markets as an adaptive response to climate change. In: Easter, K. W. and Huang, Q. (Eds). *Water Markets for the 21st Century: What Have We Learned?* Dordrecht: Springer Netherlands, 35–55.

Saitta, D., Consoli, S., Ferlito, F., Torrisi, B., Allegra, M., Longo-Minnolo, G., Ramírez-Cuesta, J. M. and Vanella, D. 2021. Adaptation of citrus orchards to deficit irrigation strategies. *Agricultural Water Management* 247, 106734.

Seckler, D. W. 1996. The new era of water resources management. In: *"Dry" to "Wet" Water Savings*. Colombo, Sri Lanka: International Irrigation Management Institute.

Settre, C. and Wheeler, S. A. 2016. Chapter 4. Environmental water governance in the Murray-Darling Basin of Australia: the movement from regulation and engineering to economic-based instruments. In: Ramiah, V. and Gregoriou, G. N. (Eds). *Handbook of Environmental and Sustainable Finance*. San Diego: Academic Press.

Siderius, C., Biemans, H., Kashaigili, J. and Conway, D. 2022. Water conservation can reduce future water-energy-food-environment trade-offs in a medium-sized African river basin. *Agricultural Water Management* 266, 107548.

Uhlenbrook, S., Yu, W., Schmitter, P. and Smith, D. M. 2022. Optimising the water we eat—rethinking policy to enhance productive and sustainable use of water in agri-food systems across scales. *The Lancet. Planetary Health* 6(1), e59–e65.

Van Oel, P., Chukalla, A., Vos, J. and Hellegers, P. 2019. Using indicators to inform the sustainable governance of water-for-food systems. *Current Opinion in Environmental Sustainability* 40, 55–62.

Van Opstal, J., Droogers, P., Kaune, A., Steduto, P. and Perry, C. 2021. *Guidance on Realizing Real Water Savings with Crop Water Productivity Interventions*. Wageningen: FAO and FutureWater.

Ward, F. A. and Pulido-Velázquez, M. 2008. Water conservation in irrigation can increase water use. *Proceedings of the National Academy of Sciences of the United States of America* 105(47), 18215–18220.

Wheeler, S. A., Carmody, E., Grafton, R. Q., Kingsford, R. T. and Zuo, A. 2020. The rebound effect on water extraction from subsidising irrigation infrastructure in Australia. *Resources, Conservation and Recycling* 159, 104755.

Wheeler, S. A., Loch, A., Crase, L., Young, M. and Grafton, R. Q. 2017. Developing a water market readiness assessment framework. *Journal of Hydrology* 552, 807–820.

Chapter 11

Developments in water sharing and water trading

Juliane Haensch, University of Göttingen, Germany; and Sarah Ann Wheeler, University of Adelaide, Australia*

1 Introduction

Water use is increasing globally at a rate of about 1% per year resulting from population growth, socio-economic development and shifting consumption patterns (Boretti and Rosa, 2019; UNESCO and WWAP, 2020). Climate change, specifically uncertain rainfall and water availability, further exacerbates water scarcity. Climate change adaptation and mitigation through water management is vital to meet international agreements, such as the Paris Agreement on Climate Change or the UN Sustainable Development Goals (SDGs) (UNESCO and WWAP, 2020), and to raise the capacity of irrigated agriculture to improve its climate change resilience (Ward, 2022).

Water is characterised by several special aspects that can make it difficult to manage and which are likely to require government intervention in water management and allocation mechanisms (Hanemann, 2006). Different water management instruments in agriculture are implemented depending on the specific national and local political, legal, socio-economic and environmental characteristics; for example, water supply augmenting (e.g. targeting storage or distribution systems, drilling wells or treating wastewater) or water demand management measures (e.g. metering, quotas, water conservation education,

* Corresponding author: sarah.wheeler@adelaide.edu.au.

http://dx.doi.org/10.19103/AS.2023.0123.17

water trading/leasing, water pricing, taxes or subsidies) (Wheeler, 2023). Traditionally, the focus is first on improving water supply while water demand management measures are often gradually adopted thereafter (e.g. Settre and Wheeler, 2016; Wheeler and Garrick, 2020). Water demand management also includes a diverse set of agronomic, economic and technical instruments with the general objective of reducing irrigation water demand, whilst trying to maintain or increase yields and crop quality and incomes per unit of water used. Water supply and demand management are interdependent and thus should be used jointly for optimal water security outcomes, but often this is not the case (e.g. Barbier, 2019; Griffin, 2006). Increasingly, it is well known that improving water supply is often a less cost-effective solution (e.g. Grafton et al., 2018; Pérez-Blanco et al., 2021) (as discussed elsewhere in this book), which has led to many countries focusing more on developing and implementing water demand management strategies in the last years. It is expected that this trend will continue to manage water allocation issues given the projected increases in future water demand (Grafton et al., 2016).

Water sharing and water trading schemes are water allocation regimes and are generally regarded as effective water demand management tools to respond to water allocation issues by many countries. Water allocation regimes typically aim to comply with economic efficiency and social equity considerations (Dinar et al., 1997). To achieve optimal resource allocation, water allocation schemes can be evaluated according to several criteria: flexibility, security, real opportunity cost, predictability, equity, political and public acceptability (Howe et al., 1986), as well as efficacy and administrative feasibility/sustainability (Winpenny, 1994).

Water-sharing concepts are vital parts of integrated water resource management (IWRM) and principally involve the control and allocation of water along specific water resources (Davis, 2007). Water allocation conflicts in one political jurisdiction can be alleviated by one political/legal entity. However, developing water-sharing concepts and alleviating water allocation conflicts in transboundary water resources can be a highly complex and contentious process. Globally, there are 268 transboundary river basins, 250 major rivers shared between two or more countries and over 50 major rivers shared between three or more countries. In light of increasing water competition and scarcity, it is important to succeed in establishing and negotiating mutual water-sharing agreements between those countries; otherwise, serious economic and political consequences may occur (Draper, 2007).

Water-trading schemes are special water sharing and water (re-)allocation mechanisms where water resources are (re-)allocated through market forces (i.e. economic water demand management instruments). Economic instruments in water generally focus on measuring and effectively managing the trade-offs across water's competing (non-)uses. Measuring trade-offs implies estimating costs, benefits and risks associated with alternative water uses (e.g. environmental, agricultural or urban uses) (Gómez et al., 2018; Grafton, 2014). Generally, water markets aim to provide a flexible, voluntary and efficient allocation of a scarce resource (e.g. Howe et al., 1986; Randall, 1981) and have been widely evolved and advocated globally for many decades (e.g. Chong and Sunding, 2006; Gardner and Fullerton, 1968; Gómez et al., 2018; Howe, 2000). Many studies illustrate the public gains derived from the reallocation of water resources through water markets (e.g. Dridi and Khanna, 2005; Easter et al., 1999; Grafton et al., 2016; Knapp et al., 2003; Levers et al., 2019; Vaux and Howitt, 1984). Wheeler (2021) provides a recent and comprehensive overview of the extent of water markets worldwide in 20 countries. Other seminal water market books include, for example, Easter et al. (1998), Easter and Huang (2014) and Maestu (2013).

The majority of the literature has focused on water quantity markets. In addition to water quantity markets, water quality trading has received more attention recently. Similar to carbon emissions trading, water quality markets aim to regulate discharges of point sources (e.g. from industries) and allow polluters to trade credits. Financial gains from credit trading are often used to support practices that decrease discharges of nutrients from non-point sources (e.g. agriculture) (Leonard et al., 2019). Another form of water trading that is sometimes discussed is in the trading of agricultural products, i.e. virtual water trading, manifested on agricultural/food trading markets. Virtual water trading has been defined as 'the amount of water, either green (soil moisture) or blue (renewable and non-renewable), that is consumed in the production of agricultural goods that are then traded' (Graham et al., 2020, p. 2). Virtual water trading is not true water trading in markets in the true sense of the water market concept; hence, this chapter, focuses mainly on water-quantity sharing and trading.

2 Developments in water-sharing concepts

Water-sharing plans provide a framework for the conditions and rules associated with the water rights of relevant stakeholders (e.g. farmers, urban water suppliers, industries, environment). Water sharing concepts are essential components of integrated water resource management (IWRM) which aims to facilitate coordination of water sharing while minimising conflict across involved stakeholders and legislative mandates. Overall, it is the aim to provide access to a specific water resource within specified/modelled long-term (sustainable) limits while also sustaining socio-economic and environmental outcomes. Water-sharing concepts can differ substantially depending on the region and may encompass the following issues (Davis, 2007):

- Water extraction and restitution;
- Water impoundments and release schedules;
- Groundwater management and interaction with river flows;
- Environmental allocation;
- Quantity and timing of the surface and groundwater flows; and
- Conflict management (e.g. mediation methods and court assignment).

Water-sharing concepts in one political jurisdiction are built upon the respective water rights system in that country. In addition to allocating the available water resources to different uses and regions, water-sharing concepts may also define different conditions per water right during an emerging drought situation, involve environmental watering or groundwater management issues as well as water storage and reservoir management (Davis, 2007). A major challenge in water sharing and allocation mechanisms is a change in water demand, for example, through increasing urban or industrial water demand competing with other demands from agriculture, environment and recreation.

Generally, studies highlight the importance of participatory approaches for the development stages of water-sharing mechanisms to increase the stakeholder acceptance and better manage conflicts at different scales (e.g. van Huynh et al., 2019). For example, in France, water user groups were encouraged to develop collective water allocation and water-sharing rules which improved the awareness amongst irrigators of their water resources and increased collective action (i.e. 'co-management' between the government and water users) (Rouillard and Rinaudo, 2020). In more humid regions, collaborative water sharing/co-management approaches to water allocations have received increasing attention as an alternative to the formal processes of water trading (Whaley and Weatherhead, 2014). In the UK, for example, it was found that socio-economic benefits from water sharing increased with the spatial scale of water-sharing agreements but, on the other hand, have led to increased levels of water abstraction (Chengot et al., 2021). This study highlights the importance of regulatory and drought management restrictions within such water-sharing agreements.

Water-sharing issues tend to be highly complex when transboundary water resources involve more than one country (e.g. Browder and Ortolano, 2000; Jalilov et al., 2015; Miner et al., 2009; Swain, 2011). Target 6.5 of the SDG of the 2030 Agenda for Sustainable Development calls for implementing IWRM at all levels, including through transboundary water cooperation as appropriate, with its specific indicator *6.5.2: Monitoring the proportion of transboundary basin area with an operational arrangement for water cooperation*. The identification of 'operational arrangements' depends on whether there is a joint (water management) body, members of the countries meet and exchange information at least once per year and joint/coordinated management plans/objectives for the transboundary water body were established (UN and UNESCO, 2021). A recent report finds that an increasing number of countries (129 out of 153 countries sharing transboundary water bodies) submitted reports to the UN improving the data quality and the commitment to transboundary water cooperation globally. On the other hand, only 32 countries have most of their transboundary basin area covered by operational arrangements (only 24 countries covered all their basin area) (UN and UNESCO, 2021). In order to meet the SDG Goal 6.5, the report highlights several areas for action, such as improving data gaps; capacity development to pave the way for negotiation and implementation of operational arrangements; mobilising political will; and coordination with national improvements on IWRM.

The scientific literature has long stressed the need for adequate cooperation over transboundary water resources to minimise the threat of disputes over water (e.g. Dinar, 2004; Wolf, 2007). Furthermore, it is predicted that the nature of water disputes is altering with more varied reasons for conflicts (e.g. local poverty/stability issues), increased water quality issues and use of alternate water sources (e.g. desalination, re-use water, deep fossil aquifers or inter-basin water transfers) (Wolf, 2007). A number of studies compare and analyse different water-sharing agreements according to different criteria, for example, the vulnerability to reduced flows and socio-economic needs (e.g. Ambec et al., 2013; Avarideh et al., 2017; Brumbelow and Georgakakos, 2007). As Ostrom (2010) highlighted, there is no 'one-size-fits all' approach to water-sharing concepts. However, Ostrom (2010) identified major principles that can be used to assess the robustness of water-sharing arrangements. These included clearly defined user and resource boundaries; congruence with local conditions; participatory/collective choice making; user and resource monitoring; sanctions for violations; mechanisms of conflict resolution; recognised self-determination of the community; and hierarchies of relevant organisations according to the size of the common pool resource.

3 Fundamentals of and developments in water trading

Water trading typically takes place in the forms of (i) temporary water transfers; (ii) medium-term water leasing (i.e. securing access to water for a specified period of time); or (iii) permanent water transfers (as a permanent water property right with access to a proportional or fixed water quantity at a given resource, or as a water delivery right) (Wheeler and Garrick, 2020; Wheeler, 2021).

Water markets involve water users, typically irrigators, located in a specific irrigation community/region sharing a water resource. Other participants may enter the water market, depending on how advanced water markets are. For example, in the case of Australia, other water market participants (non-landholders) comprise urban users, environmental water holders and financial investors, such as pure traders and superannuation (pension investment) companies (Seidl et al., 2020b).

Water can be traded in formal or informal water markets. Formal water trading is typically following major water law reforms that allow the separation of land and water rights and the creation of water use rights that are transferable, divisible, privately managed and can be partly or fully traded (Griffin, 2006). Formal water markets can be slow to develop in some regions due to a number of reasons, such as local political circumstances and the interrelated nature of water use (e.g. return flows) (Vaux and Howitt, 1984; Young, 1986). Informal water trading can be a preliminary stage for future formal water markets. Informal water trading typically takes place in a specific region or at a shared water resource (e.g. between neighbours) and can be found in various forms, such as private tube-well groundwater trade in India (Lountain et al., 2021) or informal water swapping in France (de Bonviller and de Bonviller, 2021).

Water trading increases the overall water allocation efficiency, as water is moved to higher value and more efficient uses, by establishing a price signal for the value of that water, and as it provides incentives to increase irrigators' water-use efficiency. Thus, water markets are associated with many different forms of economic efficiency: (i) allocative efficiency (improving short-term water resource allocation through temporary water trade under different seasonal factors (e.g. weather, price or crop choice changes)); (ii) dynamic efficiency (improving long-term water resource decision-making through permanent water trade under different investment opportunities, regulatory (e.g. extraction limits) or personal changes (e.g. retirement); or (iii) productive efficiency (improving water use efficiency as an investment or production input incentivised through water prices) (Wheeler, 2021). These efficiency forms result from several key advantages of water markets over other water allocation schemes, including (i) flexible reallocation of water over time in response to economic, demographic and social-value changes; (ii) involving only willing sellers and buyers; (iii) willing seller/buyers provide security of tenure of property rights; (iv) by providing the value of water, water users are confronted with the real opportunity cost of their water; and (v) measures can be put in place to keep transaction costs low (Howe, 2000). It is essential to keep transaction costs of water trade as low as possible (Loch et al., 2018).

Optimal water allocation is a complex issue, as, for example, it requires the assessment of the value of water in various uses. The total economic value of water comprises direct (i.e. benefits for individuals or agricultural businesses from using the water) and indirect (i.e. non-use water values, such as aesthetic values) use values (e.g. Grafton and Wheeler, 2015; Rolfe, 2008). Water is characterised by several other special conditions, and thus, there are many differences to other traded commodities, and some studies argue that water is not suitable to be treated as an economic good and traded on a market (e.g. Barlow and Clarke, 2003). Water's other special features are, for example, water's private or public good conditions, its mobility, costs (e.g. regarding the transport/storage of water), price, essentialness for all life as well as its heterogeneity (i.e. regarding quantity, location, timing, quality and variability/ uncertainty) (Hanemann, 2006). However, water's public good characteristic may encourage 'free-riders' and complicate the estimation of the correct value of the water (e.g. Grafton and Wheeler, 2015; Young and Haveman, 1985). Furthermore, water resources inherit a diverse set of public values, including social, cultural, environmental, recreational and ecosystem values. Public values are often unnoticed, e.g. in a water market which mainly addresses the private values of buyers/sellers. Other effects, such as unrecognised opportunity costs (i.e. downstream benefits) and positive or negative externalities of water markets further complicate the issue (Howe, 2000). Needless to say, there are ways of dealing with such externalities in markets (Wheeler, 2022). Dai et al. (2017) investigated compensation mechanisms for water transfers in China, Japan, America and Australia which aims to reduce the potential negative effects of water transfers.

Despite the fact that economic modelling (see Settre et al., 2017, for an Australian review) has long shown evidence of how water markets can efficiently reallocate water, water markets have long been subject to criticism because of potential social, economical (e.g. EBC et al., 2011) and environmental (e.g. Bjornlund and McKay, 2002; NWC, 2012) impacts. Grafton et al. (2016) provide a comprehensive discussion of common critical perceptions of water markets and relate them to practical experiences from the water market in Australia's Murray–Darling Basin (MDB). In summary, MDB's water market has (i) improved environmental outcomes; (ii) assisted irrigators' adaptation to climatic changes; (iii) increased the gross value added of farming; and (iv) been regulated aiming to meet social goals (Grafton et al., 2016). Wheeler (2022) also debunks much of the criticisms of MDB water markets. Connor et al. (2013) illustrate how water trading can be used to improve environmental outcomes. Hence, Grafton et al. (2016) and others (e.g. Wheeler et al., 2017; Wheeler, 2022) conclude that if careful consideration is given to establishing fair and effective water governance and water property rights structures, then there is great potential for water markets to deliver positive outcomes. This means governments need to be involved in the governance of water markets, for example, by creating effective institutions for water management and by minimising transaction costs (Easter et al., 1999). Thus, there are a variety of requirements that need to be addressed to establish a successful water market and to prevent water market failure. As a result, implementing water markets often requires a substantial amount of time as well as political commitment and financial resources.

A key criterion for establishing (formal) water markets are well defined, enforced and transferable water use rights (Grafton et al., 2004). Different water rights systems are found in different countries. Water rights are often defined at regional or local scales as water is treated differently in each system and society (e.g. under private ownership or as a common resource). Two common and traditional water rights regimes are riparian (i.e. land adjacent to rivers and water bodies) and appropriative rights (i.e. 'first in time, first in right') (Burness and Quirk, 1979). As Ostrom et al. (2003) pointed out, there is no one best water rights system. A water rights design must be a dynamic, sensitive and ongoing process that needs careful consideration of specific local aspects and a focus on water governance and adaptive management aspects involving all relevant stakeholders (Ostrom et al., 2003).

Grafton et al. (2011) provide an integrated framework to help assess water markets regarding institutional foundations, economic efficiency and environmental sustainability. Using this framework, water markets can be evaluated on how well they contribute to integrated water resource management and in which areas they may require further development. Furthermore, Wheeler et al. (2017) developed a water market readiness assessment framework (WMRA) to facilitate an evaluation of the benefits and institutional foundations for the implementation and sustainability of water markets in different regions. The study identified three fundamental and institutional steps as prerequisites for establishing water markets: (i) enabling institutions (including defining the total water resource pool; assessing institutional, legislative, planning and regulatory capacity to enable water trading); (ii) facilitating gains from water trading (including implementing water trading rules and evaluating benefits and costs of market-based reallocation of water); and (iii) monitoring and enforcement ensuring compliance (including constant development of water trade-enabling instruments; reducing transaction costs, monitoring externalities; developing new water markets products [e.g. options contracts]). The institutional factors in step ii will be arranged differently in each region depending on the size of the water market (enabling water trading for a small number of individuals as opposed to broader water reform legislation which involves, for example, water registers, clear trading rules and public information sources). Wheeler (2021) built on this WMRA framework and compiled different case studies spanning over 28 regions and 20 countries to test the framework globally. Applying the WRMA framework to each case study proved its usefulness in understanding the limitations and gaps before formal water markets can be established. Overall, Wheeler (2021) found that formal water markets have been slow to develop in most regions – with only areas in the US and Australia rated as being at a highly developed market stage. However, formal water markets exist in many countries in some shape or form in a number of other countries in the world, for example, in Spain, China, England, New Zealand and Chile (e.g. Grafton et al., 2011; Hearne and Easter, 1997; Howitt, 1994; Palomo-Hierro et al., 2015; Wheeler, 2021). Regarding informal water markets, Wheeler (2021) found that informal water trading (where water users share a resource and can include arrangements between neighbours or other forms) was found to be more common in a variety of countries (e.g. Africa, Pakistan and India) and indeed could be greater utilised in future.

Finally, Wheeler and Yu (2021) present a comprehensive water market literature review and analysis comprising 465 studies (between 1970 and 2019). The majority of the water market-related studies were undertaken in Australia, the US and China. Given that Australia has the most developed and adopted water markets in the world, and that it also has a significant portion of the academic literature, we turn to studying it in more detail to understand lessons from water markets in more detail.

4 Case study: Australia's Murry–Darling Basin

4.1 Water sharing in the southern Murray–Darling Basin

The MDB in south-eastern Australia has the majority of Australia's irrigation, food production and irrigation farm numbers. It is also a region of great importance for cultural purposes, environmental sites, tourism and recreation purposes. The Murray–Darling Basin Agreement describes how water is shared in the basin across three different states [South Australia (SA), Victoria (VIC) and New South Wales (NSW)] and one territory [Australian Capital Territory (ACT)]. There exists a long history of disputes between the three states regarding water management along the River Murray (e.g. Wheeler, 2014). Over time, the states realised that the success of their irrigation projects was dependent on the cooperation and compliance of the other relevant states. For example, there was a need for VIC and NSW to cooperate on building water storage and to settle water-sharing disputes along the shared border. The Federation Drought in 1902 compelled the states to come together and discuss those water-sharing disputes, which led to the first catchment-wide agreement between NSW, VIC and SA; the River Murray Waters Agreement in 1915 (Connell and Grafton, 2011; Quiggin, 2001). The agreement arranged for the building of a network of dams, weirs, locks and barrages, aiming to increase water supply security. The agreements' aim was to ensure water in the River Murray was managed fairly and to guarantee the downstream state SA with certain water security levels. A joint body, the River Murray Commission, was established alongside the agreement in order to assign responsibilities for managing the operation of infrastructure and distributing water to the states (Connell, 2007; Hallows and Thompson, 1995; NWC, 2011). The original water-sharing agreement has been updated a number of times since then to form today's Murray–Darling Basin

Agreement. In 2011, the new Basin Plan was implemented including Sustainable Diversion Limits (SDLs) which define water diversions for consumptive uses while maintaining environmental assets and ecosystem functions. The proposal and implementation phase of the Basin Plan was highly contentious, and irrigators showed a hostile reaction by publicly burning drafts of the Basin Plan (Bowmer, 2014; Quiggin, 2012). Several reasons have led to this reaction, such as a failure to balance social and environmental needs associated with water and a lack of consultation with stakeholder groups during the development phases (Crase, 2011). Figure 1 shows the allocation and shares of local water resources along the River Murray amongst the three different states and various river systems. In addition, the states developed individual water-sharing plans for their catchments.

In addition, another much discussed topic in Australia is the recognition of indigenous peoples' objectives and values for water and the integration in water sharing plans and environmental flow assessments. Only in recent years have studies increasingly highlighted and analysed the indigenous social, cultural and environmental objectives for water and their water needs (e.g. Jackson et al., 2019).

4.2 Water trading in Australia: lessons learnt

In the MDB, water markets were endorsed for water reallocation purposes since the 1960s and 1970s, in search of a solution for the emerging water scarcity problem. During this time, irrigators started to trade water informally. A wide

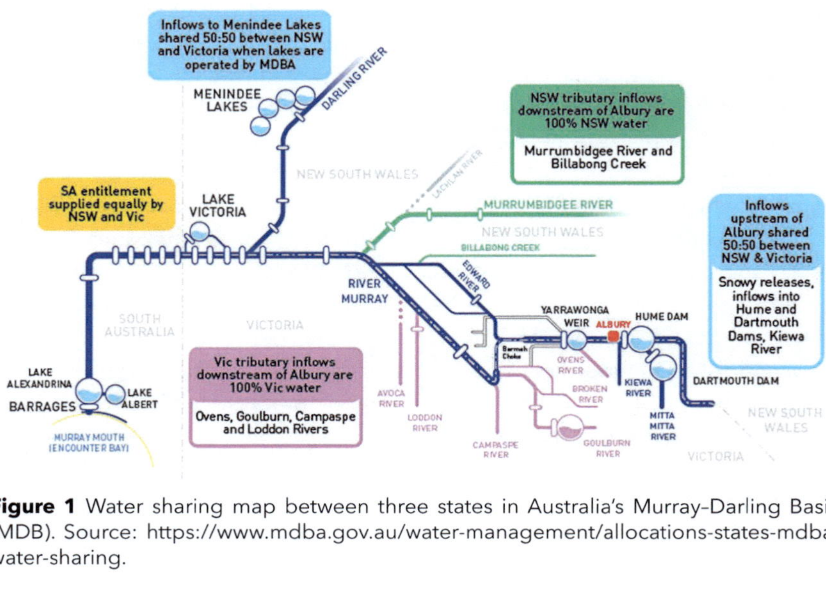

Figure 1 Water sharing map between three states in Australia's Murray–Darling Basin (MDB). Source: https://www.mdba.gov.au/water-management/allocations-states-mdba/water-sharing.

range of policy and institutional changes were initiated from the 1990s onward to facilitate formal water markets and the reallocation from consumptive water uses to environmental needs (e.g. Crase et al., 2004). Major reforms in 1994 and 2004 arranged for the separation of water rights from land rights and enabled the expansion of water markets across borders of the MDB (e.g. Turral et al., 2005; Haensch et al., 2021). Australian water policies have continuously evolved and adjusted to current issues (e.g. adjusting water recovery levels, limiting water extractions, registering foreign water ownerships with Australian Taxation Office) which are detailed in Wheeler et al. (2020b). Today, the MDB water market infrastructure is decentralised, with all states in the MDB having their own water registers. Water trading is arranged within and between various water trading zones, with varying rules applying to carry-over of unused water allocations, tagged trade and inter-valley trade restrictions across regions (Holley and Sinclair, 2016).

Within the MDB, the southern Basin is by far the largest water market, with the highest volumes/numbers of water traded (Table 1). Northern MDB water markets are much less developed, which Wheeler and Garrick (2020) attributed to relative illiquidity, lower storage, less hydrological connectivity and crop diversity, less regulated water rights, more homogenous agricultural production, far greater on-farm water storage and groundwater extraction. Table 1 shows the volume of water allocation (temporary water rights) trades in Australia during 2020–1, which has grown substantially over the last decade (BoM, 2022a). A record volume of water allocations were traded in 2020–1, increasing by 27% from the previous year, although the number of transactions was similar. Most research has concentrated on surface water trade issues, though there is an increasing focus on groundwater issues (e.g. de Bonviller et al., 2020; Wheeler et al., 2021).

Regarding volumes of water entitlements (permanent water rights) traded, 2547 gigalitres (GL) were traded nationally in 2020–1, a 30% increase compared to the previous year (Table 2). Once again, this record-high volume of water entitlement trades was primarily driven by increased trade in the southern MDB (BoM, 2022b).

In terms of monetary value of water traded, water markets in Australia had an estimated turnover of AUS$6 billion (see Tables 1 and 2), down from a record high of AUS$7 billion in 2019–20. With increased rainfall leading to improved water availability in 2020–1, there were record-high volumes of water allocations traded (7991 GL) and therefore prices paid decreased significantly from the previous year (BoM, 2022a,b).

Today Australia's water market is mature and ranks highest in terms of institutional foundations, economic efficiency, and environmental sustainability when compared with other water markets in the world (Grafton et al., 2011).

Table 1 Water allocation trade summary 2020–1

Region	Resource type	Transactions	Trades with market rate price reported[a] (%)	Volume (GL)	Estimated turnover[b] (AUS$ million)
Southern MDB	Surface water	29 890	57	7267	469
Northern MDB	Surface water	1035	34	350	30
Groundwater MDB	Groundwater	718	53	172	17
Rest of Australia	Surface water	2023	9	186	3
	Groundwater	286	19	16	1
Australia – total	Surface and groundwater	33 952	53	7991	520

[a]Allocation trade market rate price involved transactions with a reported price above $5/ML and below $10 000/ML. ML is megalitre.
[b]Price data have been cleansed to remove zero prices and outliers that are unlikely to be valid (see BoM, 2022a for details).
Source: BoM (2022a).

Table 2 Water entitlement trade summary 2020–1

Region	Resource type	Transactions	Trades with market rate price reported[a] (%)	Volume (GL)	Estimated turnover[b] (AUS$ million)
Southern MDB	Surface water	3836	51	1662	3930
Northern MDB	Surface water	448	41	174	450
Groundwater MDB	Groundwater	827	36	170	260
Rest of Australia	Surface water	2867	22	316	670
	Groundwater	1619	10	225	290
Australia – total	Surface and groundwater	9597	34	2.547	5600

[a]Entitlement trade market rate price involved transactions with a reported price above \$50/ML and below \$20 000/ML. ML is megalitre.
[b]Price data have been cleansed to remove zero prices and outliers that are unlikely to be valid (see BoM, 2022b for details).
Source: BoM (2022b).

Australia's water market guides many other nations in the world that experience similar water scarcity concerns (Wheeler et al., 2015).

Water trading has become an important tool for managing water scarcity and is widely adopted by irrigators as an adaptation strategy (Holley and Sinclair, 2016; Rafey, 2020; Wheeler et al., 2014a). The biggest driver of changes in water market prices has been found to be seasonal water scarcity factors (e.g. seasonal water allocations, drought, reduced rainfall and low water storages) (e.g. de Bonviller et al., 2019; Haensch et al., 2016; Wheeler et al., 2008; Zuo et al., 2019). Similarly, the water market provides irrigators an avenue for risk and uncertainty adjustments across seasons (e.g. Wheeler et al., 2008; Zuo et al., 2015, 2016). Sociological influences, for example, the influence of neighbours on water trade behaviour, has also been detected (Haensch et al., 2019).

Although adoption of water markets by irrigators was initially very slow, policy and institutional reform and water scarcity have significantly increased adoption over time. For example, less than 10% of irrigators in the southern MDB had conducted a water market trade in the year 2000, but by 2015–6, around half of all irrigators in the southern MDB had made at least one water right permanent trade, while 78% had conducted at least one water allocation trade (Grafton and Wheeler, 2018). In 2019–20, it was estimated that the value of permanent water rights on issue across Australia was AUS$26.3 billion, with an average annual water market turnover of around AUS$1.8 billion (ACCC, 2021). Trade has been much more accepted at the temporary seasonal level than at the permanent level (Bjornlund et al., 2011).

The Australian economics literature – both theoretical, applied and hydro-economic modelling focused – has outlined the overall net welfare gain to society from water markets. Wheeler (2014) provides an overview of the range of some of the studies that have been conducted, and Wheeler et al. (2016) outlined the financial private returns to irrigators from owning water. Rafey (2020) estimated that shutting down water markets in the southern MDB would be comparable to an 11% uniform decline in surface water allocations in terms of farm profits. ABARES (2021) estimated that water markets generated an average annual total benefit of AUS$117 million to irrigators in the southern MDB. Indeed, water markets will become increasingly important as climate change is predicted to worsen in the southern MDB with projected higher temperatures and more variable rainfall.

However, the impact of water trading and permanent water selling on the farm, families and the wider community is a much-discussed topic in Australia and has been the subject of a number of parliamentary enquires and a wide range of research. There has been a difference to date, in the findings by the academic literature, and the paid consultant/industry literature on these impacts, with the socio-economic negative impacts found to be much higher in the consultant literature (e.g. see commentary in Wheeler, 2014; 2022; Zuo et al., 2019). A recent broad-based assessment of social and economic conditions in the MDB (Sefton et al., 2020) attempted to assess the impacts of water recovery on communities and suggested that heterogeneous impacts existed across rural communities, with high-value irrigation areas benefiting more. Empirical studies using unit records from farmers have found little evidence for significant socio-economic impact by focusing on impacts on the farm level (Wheeler et al., 2014b; Wheeler et al., 2014c; Wheeler and Cheesman, 2013), while studies using broader area statistics also do not find significant associations (Haensch et al., 2021; Kirby et al., 2014; Wheeler et al., 2020a). However, within the literature, it is very rare to see any longitudinal analysis, because of the difficulty in obtaining data consistently measured over time, and the lags in available public data. Indeed, trying to control for all the local, national and international factors that can influence socio-economic changes in the MDB is incredibly difficult and remains an area that needs considerable future research attention.

The increasing unrest about the impacts of water markets and water recovery culminated in a review by the Australian Competition and Consumer Council (ACCC, 2021). Overall, the ACCC found that water markets in the MDB were working well and had provided many social benefits but that given the progress in their adoption, there was a need for institutional change. There were also market failures identified in terms of imperfect competition, externalities, and information asymmetry. However, most of the issues identified were not associated necessarily with water markets but rather other issues stemming from distributional issues, pecuniary externalities and institutional structures (Wheeler, 2022). Some of the institutional changes needed included hydrological issues, such as better understanding and measuring the impact of return flows and measuring water extraction (Grafton et al., 2018); improvements in terms of regulation, monitoring and compliance (especially in the northern parts of the basin); improvements in increased information and development of water registers, water accounting, water pricing and trade products.

Considering the operation of water markets in Australia, in particular, the ACCC (2021) review made the following key recommendations:

- Improved market integrity and conduct – especially by market intermediaries;
- A Water Markets Data Standards to improve market information and allow for more consistent reporting;
- A Water Market Information Platform;
- Strengthened water metering and monitoring of water extractions; and
- The establishment of an independent Water Markets Agency to undertake market surveillance, enforcement and reporting.

Others have suggested the need to consider substantial improvements to MDB water market data information (improving the quality, consistency and range of information – both pre-trade and post-trade) and to concentrate further research around computerised 'smart markets', ownership impacts, interactions between surface water and groundwater trade and blockchains and American-style centralised 'water banks' (Seidl et al., 2020a,b; Wheeler et al., 2021; Wheeler, 2022). What is clear is that well-designed marketplace rules and infrastructure will encourage participation, reduce strategic gaming, aggregate information and improve efficiency, liquidity and equity, which will facilitate more efficient and equitable allocation (Wheeler, 2022). Facilitating water sharing through water trading schemes is much more effective at conserving water than through more popular schemes such as subsidising irrigation infrastructure to 'save' water, which can often result in increased water consumption and reduced irrigation resilience (Grafton et al., 2018; Wheeler et al., 2020b).

5 Conclusion

This chapter reviewed the fundamentals and developments of water sharing and water trading schemes worldwide, with a particular focus on Australia's experience. Amongst the studies and experiences globally, three major themes can be identified as a basis for successful water allocation regimes: institutional/regulatory capacity, integration/stakeholder involvement and monitoring/control systems. Most studies highlight the relevance of regulations and restrictions to secure positive outcomes of water-sharing and water-trading systems and hinder adverse effects (e.g. increased water abstractions). It is also clear that there exists no 'one-size-fits-all' approach to water allocation regimes globally and that ongoing institutional developments and dialogues between all stakeholders are vital for the continued success of water allocation systems. It is also clear that there are many different types of trading systems that can be established, ranging from informal trade to formal trade between a couple of stakeholders and to formal trade between many different types of stakeholders. Trading has also been accepted more in temporary water allocation and sharing products, than in permanent trade in general.

In the light of future population growth, increasing socio-economic developments and growing water quantity and quality issues, functioning water sharing and water trading systems will be at the core of improving water management and delivering resilient irrigation communities. According to the UN Water, there has been substantial progress made in the different countries towards establishing transboundary water-sharing agreements since the monitoring started of the indicator 6.5.2 (SDG 6.5). It is important to continuously monitor this progress as well as the long-term management and compliance with those agreements and the increasing cost-effectiveness and useability of satellite data in measuring water consumption across regions, and farms will facilitate both tracking progress and monitoring and enforcement possibilities.

6 Where to look for further information

Areas of further information include:

- Wheeler's (2021) book of water markets, where 27 regions were assessed for their water governance and hydrological conditions, in readiness for implementation of economic incentives.
- IBNET database of water tariffs around the world (World Bank, 2022). The Global Water Tariff survey gathers water tariffs for 569 cities across 186 countries. Access to this database requires a subscription fee. IBNET covers 219 countries and territories and 2786 utilities, and access is free, and available here: https://tariffs.ib-net.org/sites.
- UN Water, and World Bank Databank provides information on water stress, renewable water resources, stakeholder share across all countries.
- History of water markets in Australia, see NWC (2011).
- Recent developments by the FAO in satellite measurement technology of water resources and productivity, see FAO & World Bank (2022).
- Outcomes from the World Water Week (https://www.worldwaterweek.org/), the UN 2023 Water Conference (https://sdgs.un.org/conferences/water2023) and the related Global Commission on the Economics of Water (https://watercommission.org/).

7 References

ABARES (2021). *A Model of Spatial and Inter-temporal Water Trade in the Southern Murray-Darling Basin*. ABARES technical report. Canberra.

ACCC (2021). *Murray-Darling Basin Water Markets Enquiry*. Final report. Canberra: Australian Competition and Consumer Commission (ACCC).

Ambec, S., Dinar, A. and McKinney, D. (2013). Water sharing agreements sustainable to reduced flows. *Journal of Environmental Economics and Management* 66(3): 639–655.

Avarideh, F., Attari, J. and Moridi, A. (2017). Modelling equitable and reasonable water sharing in transboundary rivers: the case of Sirwan-Diyala river. *Water Resources Management* 31(4): 1191–1207.

Barbier, E. (2019). *The Water Paradox: Overcoming the Global Crisis in Water Management*. Yale University Press.

Barlow, M. and Clarke, T. (2003). *Blue Gold: the Battle Against Corporate Theft of the World's Water*. London: Earthscan Publications Ltd.

Bjornlund, H. and McKay, J. (2002). Aspects of water markets for developing countries: experiences from Australia, Chile, and the US. *Environment and Development Economics* 7(4): 769–795.

Bjornlund, H., Wheeler, S. A. and Cheesman, J. (2011). Irrigators, water trading, the environment and debt: buying water entitlements for the environment. In: Connell, D. and Grafton, R. Q. (Eds). *Basin Futures - Water Reform in the Murray-Darling Basin*. Canberra: Australian National University Press, pp. 291–302.

BoM (2022a). *Water Allocation Markets in 2020–21*. Bureau of Meteorology (BoM). Canberra: Australian Government.

BoM (2022b). *Water Entitlement Markets in 2020–21*. Bureau of Meteorology (BoM). Canberra: Australian Government.

Boretti, A. and Rosa, L. (2019). Reassessing the projections of the World water Development Report. *NPJ Clean Water* 2(1): 15.

Bowmer, K. H. (2014). Water resources in Australia: deliberation on options for protection and management. *Australasian Journal of Environmental Management* 21(3): 228–240.

Browder, G. and Ortolano, L. (2000). The evolution of an international water resources management regime in the Mekong River basin. *Natural Resources Journal* 40(3): 499–531.

Brumbelow, K. and Georgakakos, A. (2007). Optimization and assessment of agricultural water-sharing scenarios under multiple socioeconomic objectives. *Journal of Water Resources Planning and Management* 133(3): 264–274.

Burness, H. S. and Quirk, J. P. (1979). Appropriative water rights and the efficient allocation of resources. *American Economic Review* 69(1): 25–37.

Chengot, R., Knox, J. W. and Holman, I. P. (2021). Evaluating the feasibility of water sharing as a drought risk management tool for irrigated agriculture. *Sustainability* 13(3).

Chong, H. and Sunding, D. (2006). Water markets and trading. *Annual Review of Environment and Resources* 31(1): 239–264.

Connell, D. (2007). *Water Politics in the Murray-Darling Basin*. Annandale, NSW: Federation Press.

Connell, D. and Grafton, R. Q. (2011). Water reform in the Murray-Darling Basin. *Water Resources Research* 47(12).

Connor, J. D., Franklin, B., Loch, A., Kirby, M. and Wheeler, S. A. (2013). Trading water to improve environmental flow outcomes. *Water Resources Research* 49(7): 4265–4276.

Crase, L. (2011). The fallout to the guide to the proposed basin plan. *Australian Journal of Public Administration* 70(1): 84–93.

Crase, L., Pagan, P. and Dollery, B. (2004). Water markets as a vehicle for reforming water resource allocation in the Murray-Darling Basin of Australia. *Water Resources Research* 40(8): 1–10.

Dai, X., Han, Y., Zhang, X., Chen, J. and Li, D. (2017). Development of a water transfer compensation classification: A case study between China, Japan, America and Australia. *Agricultural Water Management* 182: 151–157.

Davis, M. D. (2007). Integrated water resource management and water sharing. *Journal of Water Resources Planning and Management* 133(5): 427–445.

de Bonviller, S. and de Bonviller, A. (2021). *Water Markets in France: Appropriate Water Scarcity Management Mechanisms? Water Markets*. Cheltenham, UK: Edward Elgar Publishing.

de Bonviller, S., Wheeler, S. A. and Zuo, A. (2020). The dynamics of groundwater markets: price leadership and groundwater demand elasticity in the Murrumbidgee, Australia. *Agricultural Water Management* 239: 106204.

de Bonviller, S., Zuo, A. and Wheeler, S. A. (2019). Is there evidence of insider trading in Australian water markets? *Australian Journal of Agricultural and Resource Economics* 63(2): 307–327.

Dinar, A. (2004). Exploring transboundary water conflict and cooperation. *Water Resources Research* 40(5).

Dinar, A., Rosegrant, M. W. and Meinzen-Dick, R. (1997). *Water Allocation Mechanisms: Principles and Examples*.

Draper, S. E. (2007). Introduction to transboundary water sharing. *Journal of Water Resources Planning and Management* 133(5): 377–381.

Dridi, C. and Khanna, M. (2005). Irrigation technology adoption and gains from water trading under asymmetric information. *American Journal of Agricultural Economics* 87(2): 289–301.

Easter, K. W. and Huang, Q. (Eds.) (2014). *Water Markets for the 21st Century: What We Have Learned*. Boulder, CO: Westview Press.

Easter, K. W., Rosegrant, M. W. and Dinar, A. (Eds.) (1998). *Markets for Water: Potential and Performance*. Kluwer Academic Publishers.

Easter, K. W., Rosegrant, M. W. and Dinar, A. (1999). Formal and informal markets for water: institutions, performance, and constraints. *The World Bank Research Observer* 14(1): 99–116.

EBC, R. M. C. G., Marsden Jacob Associates, EconSearch, McLeod, G., Cummins, T., Roth, G. and Cornish, D. (2011). Community impacts of the guide to the proposed Murray-Darling Basin Plan. Volume 3. *Community Impacts*. Murray-Darling Basin Authority (MDBA Canberra), Australia.

FAO and World Bank (2022). Irrigating from space: Using remote sensing for agricultural water management. FAO, Rome.

Gardner, B. D. and Fullerton, H. H. (1968). Transfer restrictions and misallocations of irrigation water. *American Journal of Agricultural Economics* 50(3): 556–571.

Gómez, C. M., Pérez-Blanco, C. D., Adamson, D. and Loch, A. (2018). Managing water scarcity at a river basin scale with economic instruments. *Water Economics and Policy* 04(1): 1750004.

Grafton, R. Q. (2014). Economics. In: Grafton, R. Q., Wyrwoll, P., White, C. and Allendes, D. (Eds). *Global Water: Issues and Insights*. Canberra: Australian National University Press.

Grafton, R. Q., Adamowicz, W., Dupont, D., Adamowicz, W., Dupont, D., Nelson, H., Hill, R. J. and Renzetti, S. (2004). Water economics. In: Grafton, R. Q., Adamowicz, W., Dupont, D., Nelson, H., Hill, R. J. and Renzetti, S. (Eds). *The Economics of the Environment and Natural Resources*. Blackwell Publishing Ltd, pp. 161–192.

Grafton, R. Q., Horne, J. and Wheeler, S. A. (2016). On the marketisation of water: evidence from the Murray-Darling Basin, Australia. *Water Resources Management* 30(3): 913–926.

Grafton, R. Q., Libecap, G., McGlennon, S., Landry, C. and O'Brien, B. (2011). An integrated assessment of water markets: A cross-country comparison. *Review of Environmental Economics and Policy* 5(2): 219–239.

Grafton, R. Q. and Wheeler, S. (2015). Water economics. In: Halvorsen, R. and Layton, D. (Eds). *Handbook on the Economics of Natural Resources*. Cheltenham, UK: Edward Elgar Publishing.

Grafton, R. Q. and Wheeler, S. A. (2018). Economics of water recovery in the Murray-Darling Basin, Australia. *Annual Review of Resource Economics* 10(1): 487–510.

Grafton, R. Q., Williams, J., Perry, C. J., Molle, F., Ringler, C., Steduto, P., Udall, B., Wheeler, S. A., Wang, Y., Garrick, D. and Allen, R. G. (2018). The paradox of irrigation efficiency. *Science* 361(6404): 748–750.

Graham, N. T., Hejazi, M. I., Kim, S. H., Davies, E. G. R., Edmonds, J. A. and Miralles-Wilhelm, F. (2020). Future changes in the trading of virtual water. *Nature Communications* 11(1): 3632.

Griffin, R. C. (2006). *Water Resource Economics: the Analysis of Scarcity, Policies, and Projects*. Cambridge, MA, London: MIT Press.

Haensch, J., Wheeler, S. A. and Zuo, A. (2019). Do neighbors influence irrigators' permanent water selling decisions in Australia? *Journal of Hydrology* 572: 732–744.

Haensch, J., Wheeler, S. A. and Zuo, A. (2021). Explaining permanent and temporary water market trade patterns within local areas in the Murray-Darling Basin. *Australian Journal of Agricultural and Resource Economics* 65(2): 318–348.

Haensch, J., Wheeler, S. A., Zuo, A. and Bjornlund, H. (2016). The impact of water and soil salinity on water market trading in the southern Murray-Darling Basin. *Water Economics and Policy* 02(1): 1650004.

Hallows, P. J. and Thompson, D. G. (1995). *The History of Irrigation in Australia*. Mildura, Victoria: Australian National Committee on Irrigation Drainage (ANCID), First Mildura Irrigation Trust.

Hanemann, W. (2006). The economic conception of water. In: Rogers, P. P., Llamas, M. R. and Martinez-Cortina, L. (Eds). *Water Crisis: Myth or Reality?* Taylor Francis, pp. 61–91.

Hearne, R. R. and Easter, K. W. (1997). The economic and financial gains from water markets in Chile. *Agricultural Economics* 15(3): 187–199.

Holley, C. and Sinclair, D. (2016). Governing water markets: achievements, limitations and the need for regulatory reform. *Environmental and Planning Law Journal* 33(4): 301–324.

Howe, C. W. (2000). Protecting public values in a water market setting: improving water markets to increase economic efficiency and equity. *University of Denver Water Law Review* 3(2): 357–372.

Howe, C. W., Schurmeier, D. R. and Shaw, W. D. (1986). Innovative approaches to water allocation: the potential for water markets. *Water Resources Research* 22(4): 439–445.

Howitt, R. E. (1994). Empirical analysis of water market institutions: the 1991 California water market. *Resource and Energy Economics* 16(4): 357–371.

Jackson S., MacDonald D. and Bark R. (2019). Public Attitudes to Inequality in Water Distribution: Insights From Preferences for Water Reallocation From Irrigators to Aboriginal Australians. *Water Resources Research* 55: 6033–6048.

Jalilov, S.-M., Varis, O. and Keskinen, M. (2015). Sharing benefits in transboundary rivers: an experimental case study of Central Asian water-energy-agriculture nexus. *Water* 7(12): 4778–4805.

Kirby, M., Bark, R., Connor, J., Qureshi, M. E. and Keyworth, S. (2014). Sustainable irrigation: how did irrigated agriculture in Australia's Murray–Darling Basin adapt in the Millennium Drought? *Agricultural Water Management* 145: 154-162.

Knapp, K. C., Weinberg, M., Howitt, R. and Posnikoff, J. F. (2003). Water transfers, agriculture, and groundwater management: a dynamic economic analysis. *Journal of Environmental Management* 67(4): 291–301.

Levers, L. R., Skaggs, T. H. and Schwabe, K. A. (2019). Buying water for the environment: A hydro-economic analysis of Salton Sea inflows. *Agricultural Water Management* 213: 554–567.

Loch, A., Wheeler, S. A. and Settre, C. (2018). Private transaction costs of water trade in the Murray-Darling Basin. *Ecological Economics* 146: 560–573.

Lountain, S., Crase, L. and Cooper, B. (2021). When the genie is out of the bottle: the case of dynamic groundwater markets in West Bengal, India: water Markets. In: Cheltenham, UK: Edward Elgar Publishing.

Maestu, J. (Eds.) (2013). *Water Trading and Global Water Scarcity: International Perspectives*. Resources for the Future Press/Taylor and Francis/Routledge.

Miner, M., Patankar, G., Gamkhar, S. and Eaton, D. J. (2009). Water sharing between India and Pakistan: a critical evaluation of the Indus water Treaty. *Water International* 34(2): 204–216.

NWC (2011). *Water Markets in Australia: a Short History*. Canberra: National Water Commission (NWC).

NWC (2012). *Impacts of Water Trading in the Southern Murray-Darling Basin between 2006–07 and 2010–11*.

Ostrom, E. (2010). Beyond markets and states: polycentric governance of complex economic systems. *American Economic Review* 100(3): 641–672.

Ostrom, E., Stern, P. C. and Dietz, T. (2003). Water rights in the commons. *Water Resources Impact* 5(2): 9–12.

Palomo-Hierro, S., Gómez-Limón, J. and Riesgo, L. (2015). Water markets in Spain: performance and challenges. *Water* 7(12): 652–678.

Pérez-Blanco, C. D., Loch, A., Ward, F., Perry, C. and Adamson, D. (2021). Agricultural water saving through technologies: a zombie idea. *Environmental Research Letters* 16(11): 114032.

Quiggin, J. (2001). Environmental economics and the Murray–Darling River system. *Australian Journal of Agricultural and Resource Economics* 45(1): 67–94.

Quiggin, J. (2012). Why the Guide to the proposed basin plan failed, and what can be done to fix it. In: Quiggin, J., Mallawaarachchi, T. and Chambers, S. (Eds). *Water Policy Reform: Lessons in Sustainability from the Murray–Darling Basin*. Cheltenham: Edward Elgar Publishing.

Rafey, W. M. (2020). In: Droughts, deluges, and (river) diversions: valuing market-based water reallocation. *Essays in Environmental Market Design* Rafey, W. M. (Ed.): Ph.D Thesis. Massachusetts Institute of Technology, Department of Economics, pp. 7–110.

Randall, A. (1981). Property entitlements and pricing policies for a maturing water economy. *Australian Journal of Agricultural Economics* 25(3): 195–220.

Rolfe, J. (2008). Associated off-farm economic values of saving water and restoring pressure in the great artesian basin. Report for Dept of the Environment and Water Resources, Canberra.

Rouillard, J. and Rinaudo, J.-D. (2020). From State to user-based water allocations: an empirical analysis of institutions developed by agricultural user associations in France. *Agricultural Water Management* 239: 106269.

Sefton, R., Peterson, D., Woods, R., Kassebaum, A., McKenzie, D., Simpson, B. and Ramsay, M. (2020). *Final Report: Independent Assessment of Social and Economic Conditions in the Murray-Darling Basin*. Panel for Independent Assessment of Social and Economic Conditions in the Murray-Darling Basin, Melbourne.

Seidl, C., Wheeler, S. A. and Zuo, A. (2020a). High turbidity: water valuation and accounting in the Murray-Darling Basin. *Agricultural Water Management* 230.

Seidl, C., Wheeler, S. A. and Zuo, A. (2020b). Treating water markets like stock markets: key water market reform lessons in the Murray-Darling Basin. *Journal of Hydrology* 581: 124399.

Settre, C., Connor, J. and Wheeler, S. A. (2017). Reviewing the treatment of uncertainty in hydro-economic modelling of the Murray-Darling Basin, Australia. *Water Economics and Policy* 03(3): 1650042.

Settre, C. and Wheeler, S. A. (2016). Environmental water governance in the Murray-Darling Basin of Australia: the movement from regulation and engineering to economic-based instruments. In: Ramiah, V. and Gregoriou, G. N. (Eds). *Handbook of Environmental and Sustainable Finance*. UK: Academic Press, pp. 67–91.

Swain, A. (2011). Challenges for water sharing in the Nile Basin: changing geo-politics and changing climate. *Hydrological Sciences Journal* 56(4): 687–702.

Turral, H. N., Etchells, T., Malano, H. M. M., Wijedasa, H. A., Taylor, P., McMahon, T. A. M. and Austin, N. (2005). Water trading at the margin: the evolution of water markets in the Murray-Darling Basin. *Water Resources Research* 41(7).

UN and UNESCO (2021). *Progress on Transboundary Water Cooperation: Global Status of SDG Indicator 6.5.2 and Acceleration Needs*.

UNESCO and WWAP (2020). The United Nations World Water Development Report 2020: Water and Climate Change, Facts and Figures. United Nations Educational, Scientific and Cultural Organization World Water Assessment Programme (WWAP).

van Huynh, C., van Scheltinga, C. T., Pham, T. H., Duong, N. Q., Tran, P. T., Nguyen, L. H. K., Pham, T. G., Nguyen, N. B. and Timmerman, J. (2019). Drought and conflicts at the local level: establishing a water sharing mechanism for the summer-autumn rice production in Central Vietnam. *International Soil and Water Conservation Research* 7(4): 362–375.

Vaux, H. J. and Howitt, R. E. (1984). Managing water scarcity: an evaluation of interregional transfers. *Water Resources Research* 20(7): 785–792.

Ward, F. A. (2022). Enhancing climate resilience of irrigated agriculture: a review. *Journal of Environmental Management* 302(A): 114032.

Whaley, L. and Weatherhead, E. K. (2014). Competition, conflict, and compromise: three discourses used by irrigators in England and Competition, Conflict, and Compromise: three Discourses Used by Irrigators in England and Their Implications for the Co-Management of water Resources. *Water Alternatives* 8(1): 800–819.

Wheeler, S. A. (2014). Insights, lessons and benefits from improved regional water security and integration in Australia. *Water Resources and Economics* 8: 57–78.

Wheeler, S. A. (2021). *Water Markets A Global Assessment*. Cheltenham, UK: Edward Elgar Publishing Limited.

Wheeler, S. A. (2022). Debunking Murray-Darling Basin water trade myths. *Australian Journal of Agricultural and Resource Economics* 66(4): 797–821.

Wheeler, S. A. (2023). Informal and formal markets in meeting water needs. *Nature Sustainability*, 1–2. https://doi.org/10.1038/s41893-023-01178-6.

Wheeler, S. A., Bjornlund, H., Shanahan, M. and Zuo, A. (2008). Price elasticity of water allocations demand in the Goulburn-Murray Irrigation District. *Australian Journal of Agricultural and Resource Economics* 52(1): 37–55.

Wheeler, S. A., Carmody, E., Grafton, R. Q., Kingsford, R. T. and Zuo, A. (2020b). The rebound effect on water extraction from subsidising irrigation infrastructure in Australia. *Resources, Conservation and Recycling* 159: 104755.

Wheeler, S. A. and Cheesman, J. (2013). Key findings from a survey of sellers to the restoring the balance programme. *Economic Papers: a Journal of Applied Economics and Policy* 32(3): 340–352.

Wheeler, S. A. and Garrick, D. E. (2020). A tale of two water markets in Australia: lessons for understanding participation in formal water markets. *Oxford Review of Economic Policy* 36(1): 132–153.

Wheeler, S. A., Loch, A., Crase, L., Young, M. and Grafton, R. Q. (2017). Developing a water market readiness assessment framework. *Journal of Hydrology* 552 (Suppl. C): 807–820.

Wheeler, S. A., Loch, A., Zuo, A. and Bjornlund, H. (2014a). Reviewing the adoption and impact of water markets in the Murray–Darling Basin, Australia. *Journal of Hydrology* 518: 28–41.

Wheeler, S. A., Zuo, A. and Bjornlund, H. (2014b). Investigating the delayed on-farm consequences of selling water entitlements in the Murray-Darling Basin. *Agricultural Water Management* 145: 72–82.

Wheeler, S. A., Zuo, A. and Hughes, N. (2014c). The impact of water ownership and water market trade strategy on Australian irrigators' farm viability. *Agricultural Systems* 129: 81–92.

Wheeler, S. A., Rossini, P., Bjornlund, H. and Spagnoletti, B. (2016). The returns from investing in water markets in Australia. In: Ramiah, V. and Gregoriou, G. N. (Eds). *Handbook of Environmental and Sustainable Finance*. San Diego: Academic Press, pp. 371–384.

Wheeler, S. A., Xu, Y. and Zuo, A. (2020a). Modelling the climate, water and socio-economic drivers of farmer exit in the Murray-Darling Basin. *Climatic Change* 158(3-4): 551–574.

Wheeler, S. A., Zuo, A., Xu, Y., Haensch J. and Seidl C. (2020b). *Water Market Literature Review and Empirical Analysis*. Prepared for the Australian Competition and Consumer Commission (ACCC).

Wheeler, S. A., Zuo, A. and Kandulu, J. (2021). What water are we really pumping anyway? the nature and extent of surface and groundwater substitutability in Australia and implications for water management policies. *Applied Economic Perspectives and Policy* 43(4): 1550–1570.

Winpenny, J. (1994). *Managing Water as an Economic Resource*. London: Routledge.

Wolf, A. T. (2007). Shared waters: conflict and cooperation. *Annual Review of Environment and Resources* 32(1): 241–269.

World Bank (2022). *The International Benchmarking Network for Water and Sanitation Utilities (IBNET)*. Washington, DC: The World Bank Group. Accessed at: https://wbwaterdata.org/ibnet.

Young, R. A. (1986). Why are there so few transactions among water users? *American Journal of Agricultural Economics* 68(5): 1143–1151.

Young, R. A. and Haveman, R. H. (1985). Economics of water resources: a survey. In: Kneese, A. V. and Sweeney, J. L. (Eds). *Handbook of Natural Resources and Energy Economics*. Amsterdam: Elsevier, North-Holland, pp. 465–529.

Zuo, A., Nauges, C. and Wheeler, S. A. (2015). Farmers' exposure to risk and their temporary water trading. *European Review of Agricultural Economics* 42(1): 1–24.

Zuo, A., Qiu, F. and Wheeler, S. A. (2019). Examining volatility dynamics, spillovers and government water recovery in Murray-Darling Basin water markets. *Resource and Energy Economics* 58: 101113.

Zuo, A., Wheeler, S. A., Adamowicz, W. L., Boxall, P. C. and Hatton-MacDonald, D. (2016). Measuring price elasticities of demand and supply of water entitlements based on stated and revealed preference data. *American Journal of Agricultural Economics* 98(1): 314–332.

Chapter 12

Modernizing major and medium irrigation schemes in India: formulating a programme of change

Martin Burton, Water Resources and Irrigation Management, UK; M. G. Shivakumar, Asian Development Bank; India; Rahul Sen, Water Resources Policy and Institutions, India; Kuppannan Palanisami, International Water Management Institute, India; and Lance Gore, Asian Development Bank, The Philippines*

1 Introduction

India has more than 1700 major and medium irrigation (MMI) schemes. Measures to improve their performance to increase water use efficiency, production, farmer livelihoods, and water productivity will contribute to the saving of water that can be used to increase the irrigated area or transferred to other sectors. The Department of Water Resources, River Development and Ganga Rejuvenation (DoWR, RD&GR) has set a goal of improving water use efficiency by 2047. The Support for Irrigation Modernization Program (SIMP) has the potential to address these challenges using the framework and strategy developed under the programme. During Phase 1 of the programme, following submissions from the states, four initial irrigation schemes covering some 417 985 ha at a cost of US$562 million were identified for modernization over the next decade, with further schemes being brought in as the programme progresses.

The modernization programme comprises a broad range of inter-acting and complementary actions which contribute to the overall objectives of (i)

* Corresponding author: martinaburton@btinternet.com.

http://dx.doi.org/10.19103/AS.2023.0123.27

increasing farmer income, (ii) optimizing agricultural production from the scheme, (iii) raising water use efficiency to optimum levels, (iv) raising water productivity to optimum levels, and (v) sustaining the physical infrastructure over time. Phase 1 (Establishment) of SIMP has been completed and has now moved into phase 2 where Irrigation Modernization Plans (IMPs) are being formulated. This will be followed by phase 3 when Detailed Project Reports (DPRs) and tender documents will be prepared before letting contracts for implementation under phase 4. The DPRs are a government requirement for all irrigation development projects. They include detailed surveys, hydrological analysis, economic analysis, and environmental impact assessments to support the preparation of detailed designs, bills of quantities, cost estimates, and tender documents. Phases 2 and 3 will be implemented by state governments from their own resources or through loans. The SIMP has adopted a holistic approach to modernization with interventions classified into six action areas: (i) policy development and legislation, (ii) institutional reform, (iii) water resources identification and management, (iv) system modernization, (v) strengthening agricultural services and practices, and (vi) building knowledge and information systems. The sixth action area cuts across the five others.

This chapter outlines the establishment and progress of SIMP under phases 1 and 2. It outlines a typology developed to classify MMI systems as an aid to identifying systems for modernization. It also introduces the concept of 'tiered development' and the identification of over 85 possible modernization interventions at different levels in an irrigation scheme. From these, a 'cafeteria' approach can be used to select interlinked and complimentary components to include in the modernization of a particular scheme. Progress made during Phase 2 on preparation of the four IMPs is outlined and discussed. This step highlights an important change in the approach to rehabilitation and modernization projects in India by conceptualizing and identifying options for modernization prior to implementation of the full surveys carried out under the DPRs. Setting out the roadmap for modernization in the IMP then focusses the scope of the detailed surveys, investigations, and studies required to prepare the DPR. Finally, the chapter stresses the importance of understanding the linkages that exist between different components in an irrigation scheme, and how they interact together to produce the final product and outcomes.

2 Recognizing the need for irrigation project modernization in India

2.1 Drivers and enablers of change

By 2050, India's population is predicted to reach 1.64 billion (UN, 2019), an increase of 19% on the current population. The Food and Agricultural Organization (FAO) predicts that to meet food demand, agriculture will need to increase production by about one-third by 2050 (FAO, 2017). Irrigated agriculture is expected to play a major role in meeting this demand. Because of this rising population, the per capita water availability in India is decreasing, putting serious pressure on the irrigation sector to (i) produce more food per unit of water withdrawn and (ii) release water for other uses (principally domestic water supply and industry). With per capita water availability of 1545 m^3, India is technically 'water stressed' with studies showing that with rising population, the projected per capita water availability will decrease to 1401 m^3 and 1191 m^3 by 2025 and 2050, respectively (ADRI, 2023). In addition, climate change is anticipated to have multiple and unprecedented effects on precipitation and temperature regimes, impacting hydrological cycles, glacial meltwater, drought, and flood frequency (Mall et al., 2006; Gosain and Rao, 2003; Gosain et al., 2006; Guiteras, 2009; Lahiry, 2018; Palanisami et al., 2011; Palanisami et al., 2014; Pathak et al., 2014; Srinivasan, 2019).

Farmers will be at the centre of the processes required to adapt to these changing circumstances. They will likely be more knowledgeable, better informed, more orientated towards cash crops and profit, and be more productive and better-off than today. 'Protective' irrigation will give way to 'productive' irrigation as the focus changes to optimizing agricultural production from the resources used (land, water, labour, and finance). Irrigation schemes will need to be re-engineered to meet these challenges and move from the current supply-orientated to demand-orientated provision of irrigation water. The key drivers for change and modernization in India include the following:

1 Climate change impacts: Increasing global and regional temperatures are already being felt, and will adversely impact on water resource availability, crop growth, and production.
2 Growing population: The global population is predicted to rise to 9.8 billion by 2050 from the current 7.6 billion (United Nations, 2019). India will have the largest population with 1.64 billion with further growth anticipated beyond 2050 (ESA, 2023).
3 Reduction in available supplies: Global modelling studies by the International Institute for Applied Systems Analysis (IIASA) which factor in seasonal variations in availability and consumption of water indicate that *c*4.0 billion people (half of whom live in India and China) already live under conditions of severe water scarcity for at least 1 month per year. The prediction is that the number of people in Asia living in severely water scarce areas will increase from the current 1.2 billion to 2.1 billion, or about 40% of Asia's future population. The largest share of affected people will be in South Asia, followed by East Asia.
4 Changing demographics: India is experiencing growing urbanization as the rural youth drift to urban centres in search of employment

and different lifestyles. While this may be beneficial in rural areas by facilitating the creation of larger landholdings, it increases the demand for industrial and domestic water supplies.

5 Changing diets: Increasing urbanization and an expanding middle class is predicted to result in an increased and diversified food demand. This will drive diversified cropping and may provide opportunities for farmers with reliable irrigation to grow higher value crops.
6 Increasing food demand: To meet demand, agriculture worldwide will need to produce almost 50% more food, feed, and biofuel than in 2012 (FAO, 2017). In India, agriculture will need to produce almost 32% more food in 2050 than in 2015 (Economic Times, 2015). To meet this increase, yields will need to rise 0.9–1.2% annually (FAO, 2017).
7 Competing demands for finite water resources: Currently, irrigation accounts for around 80% of all water abstractions in India. With the projected population increase combined with demographic changes (urbanization), this balance is changing, with priority being given to satisfying domestic water supplies, industry, and power generation. Irrigated agriculture will have no alternative but to become more efficient and productive.
8 Falling groundwater levels: Groundwater in many parts of India is in crisis as farmers continue to over-abstract available groundwater supplies. Whilst groundwater can offer many opportunities for smallholder farmers, its continued abstraction above renewable levels is problematic. Groundwater recharge and community groundwater management are becoming increasingly important.
9 Attaining the Sustainable Development Goals (SDGs): Goal 6 of the 17 SDGs specifically relates to water. A further six SDGs are closely linked to water (Goal 2: Food Security and Agriculture; Goal 7: Energy; Goal 13: Climate; Goal 12: Sustainable Consumption and Production; Goal 15: Biodiversity and Ecosystems; and Goal 17: Implementation of SDGs through Technology, Capacity Development and Partnerships). These goals support the imperative for modernizing the irrigation sector. Target 6.4 (increasing water use efficiency) directly relates to the SIMP objectives.

On a more positive note, there are several enablers of change and modernization. These include improved communications, technology, and people. One of the biggest drivers and enablers of change are today's farmers. They are better educated, more informed, more knowledgeable, more aspirational, and more enterprising than ever before. They are better connected to the outside world than their forebears and more aware of the opportunities available to them.

The development of road networks and rural electrification has narrowed the divide between rural and urban communities, opening up previously inaccessible market opportunities. Markets are more accessible to areas of production and transportation costs and time to market have reduced markedly. The increase in the number and size of secondary towns is significant as they now provide more accessible inputs, off-farm employment, specialist services, and markets to the rural farming community.

The level, access to, and use of technology has also dramatically changed the picture since the days of the Green Revolution. Affordable mobile phones and associated services together with internet connectivity has been one of the most significant enablers of change for rural communities. Other beneficial developments include widespread use of smart phones, greater knowledge and use of computers and the Internet, remote sensing, management information systems, and improved plant technologies. A very recent innovation is the use of unmanned aerial vehicles (UAVs) or drones for a variety of activities ranging from canal inspections for Irrigation and Drainage (I&D) agency staff to helping farmers determine levels of fertilizer to apply to fields.

2.2 Irrigation development

Irrigation development has been a major sector of public investment in India since the first 5-year Plan (1951–1956). A summary of the expenditure on irrigation development and flood control from 1951 to 2012 is given in Table 1, which shows that since the first 5-Year Plan the major focus of investment in irrigation has been in MMI development. This public investment has been used to initiate construction of a total of 473 major and 1279 medium-sized irrigation projects. Of these, 295 major and 1018 medium irrigation projects have been completed up to the end of the 11th 5-Year Plan (March 2012).

The Ultimate Irrigation Potential (UIP) estimated for India by the Central Water Commission (CWC) is 139.89 million ha (Mha), of which 58.47 Mha (42%) is from MMI sources and 81.43 Mha (58%) is from minor irrigation sources (including ground water and surface water sources). Of this, up to end of the 11th 5-Year Plan, about 113.53 Mha (81%) of irrigation potential has been created, which includes 47.97 Mha (42%) from (MMI) sources and 65.56 Mha (58%) from minor irrigation sources (CWC, 2020).

Despite significant investment, irrigation development has not been spread equally amongst the states. Some of this is due to limitations on the areas that can be developed for irrigation, and some is due to the different pace of irrigation development in individual states. By the end of the 11th Five-Year Plan, more than 81% of the UIP from MMI schemes in the country had already been created. The states that have created more than 81% of their UIP from MMI schemes included Bihar, Chhattisgarh, Gujarat, Haryana, Maharashtra,

Table 1 Plan-wise expenditure on irrigation and flood control in India (Rs. billion)

Plan period	Major and medium irrigation	Minor irrigation and CAD	Flood control	Total (MMI + MI + CAD + FC)	% MMI/total
I plan (1951--1956)	3.76	0.66	0.13	4.55	82.64%
II plan (1956-1961)	3.80	1.62	0.48	5.90	64.41%
III plan (1961-1966)	5.76	4.43	0.82	11.01	52.32%
Annual plans (1966-1969)	4.30	5.61	0.42	10.33	41.63%
IV plan (1969-1974)	12.42	11.73	1.62	25.77	48.20%
V plan (1974-1978)	25.16	14.10	2.99	42.25	59.55%
Annual plans (1978-1980)	20.79	13.45	3.30	37.54	55.38%
VI plan (1980-1985)	73.69	41.60	7.87	123.16	59.83%
VII plan (1985-1990)	111.07	76.27	9.42	196.76	56.45%
Annual plans (1990-1992)	54.59	36.50	4.61	95.70	57.04%
VIII plan (1992-1997)	210.72	138.85	16.92	366.49	57.50%
IX plan (1997-2002)	492.89	137.60	30.38	660.87	74.58%
X plan (2002-2007)	836.47	164.59	43.44	1044.50	80.08%
XI plan (2007-2012)	1653.50	463.50	201.00	2318.00	71.33%
Total	**3508.92**	**1110.51**	**323.40**	**4942.83**	**70.99%**

Preparation of 5-Year Plan was halted by the Government of India following the preparation of the 12th 5-Year Plan.
Source: Presented in the 2019-2020 CWC Annual Report (CWC, 2020) using data from the Planning Commission.

Nagaland, Punjab, Rajasthan, Uttar Pradesh, Uttarakhand, and West Bengal. The major states where irrigation potential created was less than 81% of the UIP from MMI schemes included Jammu and Kashmir, Karnataka, Kerala, Madhya Pradesh, Orissa, and Tamil Nadu.

Currently, over 80% of the available water is used by the irrigation sector. In some states, such as Punjab, Rajasthan, and Uttar Pradesh, the development of irrigation has resulted in the full development of the available water resources, leading to critical water scarcity and shortages of water for other uses. As highlighted in the 12th 5-Year Plan, in these and other states where a large proportion of the UIP has been realized, the focus needs to change from construction of new schemes to more efficient and productive management of existing schemes (NWUEISP, 2014). It is also worth noting that in India, the term 'irrigation project' is used to describe irrigation schemes, whether they are under construction or have already been implemented. In this chapter a 'project' is taken to be a time-bounded activity, whilst a system describes the canal and drainage network and a scheme covers the entire irrigation activity (canals, drains, land, villages, roads, and groundwater) within the boundary of the scheme. Irrigation schemes are categorized as being small (Cultivable Command Area, CCA < 2000 ha), medium (CCA > 2000 ha and <10 000 ha), and major (CCA > 10 000 ha).

2.3 Changing dynamics of irrigated agriculture in India

The present trends in crop area and production reveal that agricultural growth can be attained through improving irrigation productivity rather than increasing the area of land under production. However, there are challenges that need to be addressed in the irrigation sector. The way forward could be through technological advancements and/or innovations with institutional support in the irrigation sector to bring systemic change in the water–food–energy nexus. This may include subsidized loans and incentives for modern irrigation equipment; the use of solar power for irrigation; buried pipe distribution systems; buried drainage on waterlogged and salinized land; modernized rental markets; increased use of mobile phone-based knowledge sharing applications; strengthened water users associations; and strengthened marketing. These changes at the farm level should be simultaneously accompanied by improvements in irrigation infrastructure, and its operation and maintenance at the main system level.

2.4 Climate change on agriculture and irrigation

India has been identified as one of the country's most vulnerable to climate change (Cruz et al., 2007). The projections of the Coupled Model Intercomparison Project (CMIP) for India show that the average climate will be warmer by +1.7 to +2.0°C by the 2030s and by +3.3 to +4.8°C by the 2080s, and that rainfall will increase by 5–6% and 6-14% by the 2030s and 2080s, respectively (Chaturvedi et al., 2012). This will affect the spatial and temporal distribution of rainfall runoff, soil moisture, groundwater reserves, and may increase the frequency and magnitude of droughts and floods. Though future changes in climate will have global impacts it is likely to be felt most severely in developing countries including India (Pathak et al., 2014). Irrigation modernization will therefore need to be designed to cope and adapt to these changing conditions.

2.5 Moves towards modernization

Most of the MMI schemes have been developed over the past century as protective systems to supplement rainfall allowing farmers to grow at least one crop annually. Prevailing climate, water, and food security concerns, along with an emphasis of increasing farmer's income, now require these schemes to be modernized to function as productive irrigation systems, providing year-round assured water supplies to farmers with flexible water deliveries allowing them to irrigate a wider range of crops (ADB, 2019). In addition, there is intense pressure on MMI schemes to not only meet food needs, but also, where feasible, to release water for other uses. To achieve this, MMI schemes need to address three key issues:

1 Low performance: Irrigation schemes are often not performing as well as they might with low efficiencies in terms of water use, and poor land and water productivity. In too many cases, there is over-abstraction, inequitable water delivery, and less than optimum levels of income to farming families. The loss of productivity due to build-neglect-rebuild cycle following inadequate funding for management, operation and maintenance is shown in Fig. 1.
2 Low water use efficiency (WUE): A study by the CWC in 2016 on 35 large-scale irrigation projects revealed that the overall project efficiency on these schemes ranged between 30% and 40% (CWC, 2016). The neglect of maintenance, failure of participatory irrigation management (PIM), and a lack of policy support had resulted in the deterioration of these irrigation systems needing immediate interventions to improve

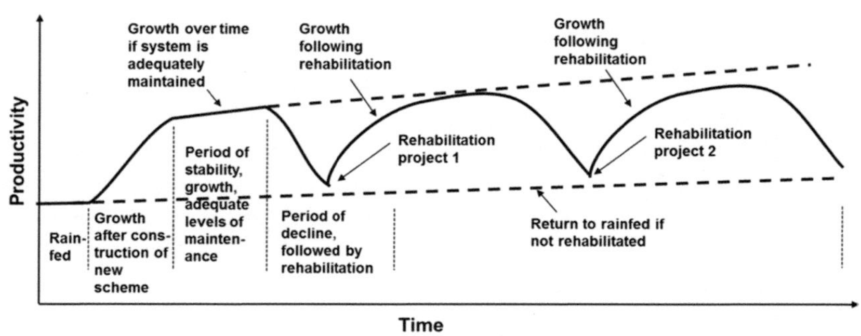

Figure 1 The depressing cycle of build-neglect-rehabilitate-neglect for irrigation schemes. Source: Adopted from: Burton (2010).

their performance. This low level of water use efficiency is a cause for serious concern. It is necessary to address this core issue and raise productivity levels, both for water allocation purposes and for those crops that are dependent on this water.

3 Low productivity of water (PoW): Allied to low water use efficiency is the low level of water productivity. Water use efficiency relates to the conveyance, distribution, and application of the water from the source to the crop root zone. The productivity of water relates more to the production obtained from the water supplied to the crop. Poor agronomic practices, poor quality seeds, a lack of fertilizer and pesticides, and poor irrigation scheduling practices are factors contributing to low PoW values.

To address these issues, the Government of India (GoI) has launched a variety of initiatives to support the states to increase the irrigation potential, increase farmers' income and provide funds for completion, rehabilitation, and modernization of irrigation projects. These include the Accelerated Irrigation Benefit Program (AIBP), the Command Area Development and Water Management Program (CAD&WM), Pradhan Mantri Krishi Sinchayee Yojana (PMKSY), the National Water Mission under the National Action Plan on Climate Change (NAPCC), the Repair, Renovation & Restoration (RRR) of Water Bodies, the Sahi Fasal (water use efficiency cropping), and the Har Khet Ko Paani (water to every farm) programmes. Further innovations to improve irrigation system performance in a sustainable manner are, however, still required.

3 Opportunities and options for modernization

3.1 Support for irrigation modernization programme

The Asian Development Bank (ADB) has been working with the GoI over the past decade to develop strategies to improve India's water use efficiency and crop water productivity. This support has included two major studies:

- The National Water Use Efficiency Improvement Support Program Scoping Study (NWUEISP, 2014); and
- Innovations for More Food with Less Water (MFLW) study (MFLW, 2015a,b).

Several other projects have also been supported, including:

- the Madhya Pradesh Irrigation Efficiency Improvement Irrigation Project (MPIEIP) – including the development of highly efficient pressurized 125 000 ha irrigation system;

- technical assistance for the modernization of the 200 000 ha Ara Canal Irrigation System in Bihar;
- Karnataka Integrated and Sustainable Water Resources Management Investment Program – modernization of 15 000 ha for higher water productivity and water savings; and
- proposed modernization of the 244 000 ha Tungabhadra Left Bank Canal Project in Karnataka with preparation of MIP for automated regulation and higher water use efficiency/water productivity.

To follow on from these studies and projects, the ADB and GoI formulated the SIMP to expand the modernization of MMI scheme in India in a more systematic and streamlined manner (ADB, 2019). The overall goal of the SIMP was to establish a robust framework to support states with preparing future irrigation modernization programmes for MMI schemes that sustain and enhance system performance. The core objectives are to produce more efficient, productive, financially beneficial, and sustainable MMI schemes. These objectives are founded on the process of modernization of irrigation (and drainage) systems as defined by ADB in the Concept Note (ADB, 2019): 'The process of upgrading infrastructure, operations and management of irrigation systems to sustain the water delivery service requirements of farmers and optimize production and water productivity'. Implementation of the SIMP will be a joint venture between central and state governments.

1 At the central level, a Central Irrigation Modernization Office (CIMO) has been established in the Central Water Commission (CWC) with the technical expertise to (a) build awareness and understanding at the national and state level of irrigation modernization processes, outputs, and outcomes; (b) strengthen design guidelines and streamline review and approval processes, and (c) support states in developing and implementing modernization programmes.
2 At the state level, State Irrigation Modernization Units (SIMU) are being established to provide technical and financial support to states to (a) prepare feasibility studies and irrigation modernization plans for selected schemes; (b) prepare detailed project reports (DPRs) to achieve identified modernization programme outputs and outcomes; (c) provide specialist advice during preparation of detailed designs and tender documents, and (d) assist in identifying financing for implementing modernization programmes.

The SIMP has four phases:

1 Phase 1 – CIMO Establishment (Completed; August 2021-December 2022). During phase 1, CIMO was established and 4 schemes totalling

Table 2 Summary details for each of the final shortlisted proposals

State	Scheme	CCA (ha)	Estimated cost	
			Rs. (crores)	US$ (millions)
Rajasthan	IGNP-II Lift, or	320 000	2810	380
Haryana	Loharu Canal Project	19 940	432	58
Maharashtra	Palked	65 045	182	25
Karnataka	Vani Vilasa Sagara Project	13 000	733	99
	Total	**417 985**	**4157**	**562**

1 Crore = 10 million.
Source: Adopted from: SIMP Final Report, Dec 2021.

some 418 000 ha in 4 states were identified and selected for modernization (batch 1). The total estimated cost for these schemes was Rs. 41 570 million (US$562 million). During this, phase selection procedures were developed together with procedures for preparing IMPs and options for modernization. An awareness raising and training programme on SIMP and modernization was also prepared. Table 2 summarizes the schemes selected for modernization.

2 Phase 2 – Operationalizing the CIMO (Ongoing; January 2023 - April 2024). During this phase consultants are working with CIMO, SIMUs and state Water Resources Departments to prepare IMPs (with costings) for each scheme. Awareness raising and training programmes will also be initiated.

Following discussions with the states and IMP surveys, draft IMPs have been prepared for four projects as shown in Table 3. The IGNP-II Lift project in Rajasthan was dropped as the state implemented this with its own funds. This project was replaced by another shortlisted project (Purna) in Maharashtra. The current planned modernization area is 284 519 ha at an estimated cost of Rs. 50.10 million (US$612 million). This represents an average cost of US$2152 per hectare.

3 Phase 3 – Supporting Project Readiness (Ongoing; September 2023–August 2024). During this phase, each state will prepare their DPRs, detailed designs and tender documents for the batch 1 schemes. Financing for these schemes will be identified by implementing states. Selection of batch 2 schemes will be initiated.

4 Phase 4 – Project Implementation (September 2024–August 2029). During this phase, work will commence on implementation of the modernization programmes identified for each scheme in the IMP and DPR. Financing for this phase will be identified by the implementing

Table 3 Summary details for each of the draft IMPs prepared (November 2023)

State	Scheme	CCA (ha)	Estimated number of beneficiaries	Estimated cost		Estimated economic rate of return (ERR)	Proposed interventions
			No.	Rs. (crores)	US$ (millions)		
Haryana	Loharu Canal Project	128 157	250 000	1413	172	14.0%	• Upgrade canal network (lining, SCADA, automation) • Groundwater banking • Reclaiming waterlogged areas • PIN in selective areas • Optimize energy spending • Agriculture • Institutions • Management
Maharashtra	Palked*	61 230*	60 000*	1270*	155*	n/a	• Upgrade canal network (lining, SCADA, automation) • Gravity PDN • Groundwater banking • Agriculture • Institutions • Management
	Purna	78 486	50 000	1700	207	17.8%	• Upgrade canal network (lining, SCADA, automation) • Gravity PDN and PIN • Groundwater banking • Agriculture • Institutions • Management

Karnataka	Vani Vilasa Sagara Project	16 646	11 000	627	78	14.8%	• Upgrade canal network (lining, SCADA, automation) • PIN for full CCA • Groundwater banking • Agriculture • Institutions • Management
	Total	**248 519**	**371 000**	**5010**	**612**	-	

SCADA, Supervisory Control and Data Acquisition; PDN, Piped Distribution Network; PIN, Piped Irrigation Network.
*Initial estimates, IMP not finalized (November 2023).
Source: Irrigation Modernization Plans, phase 2, SIMP.

states. Batch 2 schemes will be progressed, and a programme for inclusion of schemes under following batches identified and started.

3.2 Stages in irrigation development

Figure 2 outlines the possible stages in irrigation system development. This formulation was based on observation of irrigation development in several countries, notably the USA and Australia where irrigation systems have developed over time to meet changing demands, increased knowledge, and advances in technology. Moving from one level to another requires motivation, changes in attitude, practices and procedures, energy, and investment, with changes required not only in the 'hardware' (technology) but also in the 'software' (understanding, knowledge, skills, and attitudes).

Table 4 develops this framework and introduces the concept of 'tiers of irrigation development' for different components of irrigation management, operation and maintenance. Tier 1 is the 'basic' level where irrigation control is limited and efficiencies and production levels are low. Tier 4 is the 'advanced' level where irrigation control is extensive, and efficiencies and production levels are high. The aim of the SIMP modernization programme is to move

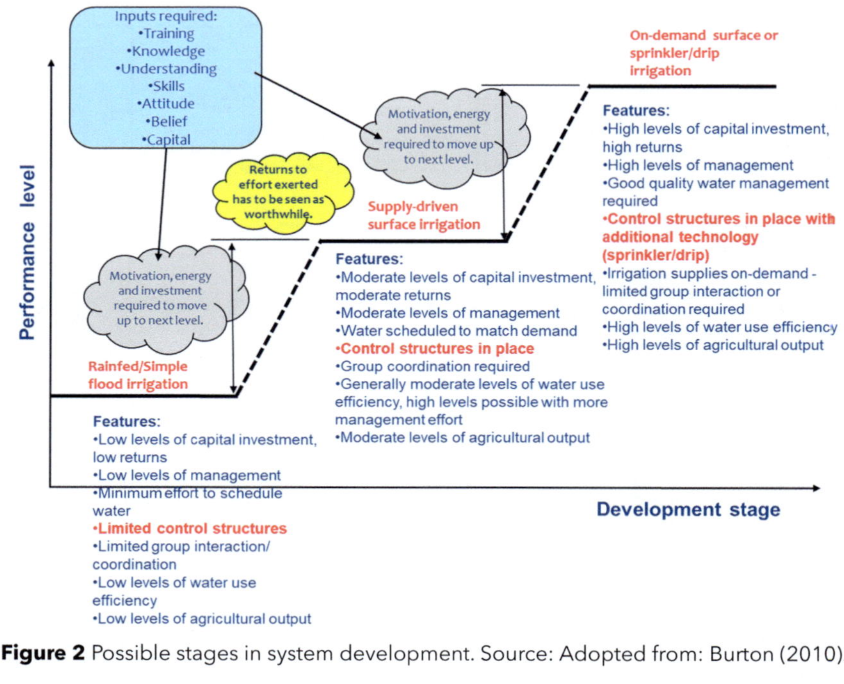

Figure 2 Possible stages in system development. Source: Adopted from: Burton (2010).

projects up from their existing level to a higher, more efficient, productive, and remunerative level. There are several considerations:

1 **Phased development:** It may not be sensible to move too rapidly from, say, tier 1 to tier 4 as all the necessary conditions for a functioning Tier 4 scheme may not be in place. It may be preferable to move in a phased manner from, say, tier 1 to tier 2 and then to tier 3 when conditions are suitable. One of the main constraints to rapid movement from tier 1 to tier 4 will be the knowledge, attitude, and skills of the farmers. Changing these 'soft' variables generally takes longer than changing the 'hard' variables such as physical structure.
2 **Different parts of a scheme may be in different tiers:** This is the case where farmers at the field level have adapted to unreliable, top-down, supply-orientated water delivery (tier 1) by developing groundwater resources. In this instance, the use of groundwater gives farmers on-demand irrigation (tier 4), topped up by periodic supplies from the main canal. In this case, automating the main canal to provide reliable on-demand irrigation would likely be successful.
3 **Cost implications:** There are cost implications in moving from one tier to another, higher, tier. In a large modernization programme, consideration should be given to whether it is preferable to move, say, 20 schemes from tier 2 to tier 3, rather than, say, 5 schemes from tier 2 to tier 4.

While moving from one tier to a higher tier may be beneficial on paper, it is essential to consider the inputs required to overcome constraints, particularly in terms of attitude, knowledge, behaviour, and skills of both I&D and agricultural agency staff and farmers.

3.3 Structuring the modernization programme

The team for SIMP phase 1 comprised two irrigation/water resources engineers, an institutions specialist, and an agricultural specialist. From the start, it was seen as imperative that any modernization programme comprised these three fundamental components - engineering, institutions, and agriculture. This led to the formulation of a holistic framework for modernization comprising six action areas (Table 5). The sixth action area cuts across the five others. Under this framework a series of modernization activities were then identified (Table 6).

These interventions were formulated to address the issues identified in the performance of MMI schemes as structured in the problem tree analysis (Fig. 3).

Table 4 Conceptual example of how different levels of physical setup and MOM impact levels of performance of irrigation schemes

Component	Tiers of development and performance							
	1		2		3		4	
	Main system	On-farm	Main system	On-farm	Main system	On-farm	Main system	On-farm
Physical	• Open channel (unlined) • Simple control structures • No measurement	• Open channel (unlined) • Ungated control structures • No formal turnouts	• Open channel (unlined) • Simple control structures • No measurement	• Open channel (unlined) • Some control structures	• Open channel (lined) • Good control structures • Measuring structures	• Open channel (lined) • Good control structures	• Piped system • Automated control • Flow meters	• Piped distribution systems • Full farmer-controlled turnouts with measurement
Management	• Low input • No seasonal plans • No service delivery contract • Low fee collection • Manual billing	• Low input • No seasonal plans • Farmers organize water distribution • Limited WUA involvement	• Moderate input • Seasonal plans • No service delivery contract • Mod. fee collection • Manual billing	• Moderate input • Seasonal plans • WUA organizes water distribution • WUA service fee charged	• High input • Seasonal plans • Service delivery contract • High fee collection • Manual billing	• WUA has O&M staff • WUA charges a fee	• Highly technical input • Seasonal plans submitted online • On-demand irrigation • Automated billing	• Low management input
Operation	• Manual • Simple rules • Limited scheduling	• Limited, farmers organize amongst themselves	• Manual • More complex rules • Scheduling based on norms	• WUA field oversee water distribution	• SCADA system • ET-based scheduling • Monitoring of supply versus demand	• WUA field staff oversee water distribution	• Automated system • Centralized control	• Limited due to pressurized on-demand system

Maintenance	• Limited work done • Manual vegetation and silt removal	• Limited work done • Manual vegetation and silt removal	• Work organized and done manually	• Work organized by WUA and done manually	• Maintenance scheduled and done • Machinery and equipment used • Contractors may be used	• Work organized by WUA and done with some machinery	• Limited but expensive and high tech when it's needed • Need reserve of spare parts. • Need good technical support	• Very little required until pipes need replacement or are damaged
Agriculture	• Supply-orientated • Fixed canal rotation pattern • No pre-season planning. Available water distributed in proportion to irrigable area	• Farmers plant crops to match historic water supply pattern • Farmers select low input, low initial cost crops • Low use of good quality seeds, fertilizer and pesticides	• Supply-orientated water delivery • Seasonal planning of estimated demand • Fixed rotation of main canals • Canal rotations published at start of the season	• Water supply more reliable, adequate and timely • Farmers plant higher risk and higher value crops	• Arranged-demand supply-orientated water delivery • Pre-season planning and in-season periodic adjustment to match requested demands	• Close liaison between scheme management and farmers • Water supply relatively closely matches irrigation demand • Farmers invest in seeds, fertilizer and pesticides	• Full demand-orientated supply to match actual irrigation water demand	• Farmers invest in high value seeds, fertilizers, and pesticides • Farmers plant crops to match market demand and price • Farmers able to pay for higher quality service provided
Level of service	**Low**		**Moderate**		**High**		**Optimum**	
Production	• Protective irrigation • Low yields • Low production • Low cropping intensity • Top-tail inequity		• Productive irrigation • Moderate production • Moderate cropping intensity • Some top-tail inequity		• Productive irrigation • High yields • High production • High cropping intensity • Limited top-tail inequity		• Productive irrigation • Optimum yields • Optimum production • High cropping intensity • No inequity	

Source: Authors.

Table 5 Action areas to achieve the SIMP core objective

SIMP core objective: More efficient, productive, financially beneficial, and sustainable MMI schemes					
Action area 1	Action area 2	Action area 3	Action area 4	Action area 5	Action area 6
Policy Development and Legislation to Frame the Modernization Process	Institutional Reform to Structure Management of the Process	Water Resources Identification, Planning and Management	System Modernization for Improved Performance	Strengthening Agricultural Services and Practices for Enhanced Production	Building Knowledge and Information Systems for Growth
National and State Level	National, State, and Scheme Level	Basin and Scheme Level	National, State, and Scheme Level	National, State, and Scheme Level	National, State, and Scheme Level

Source: Authors.

Table 6 Typical interventions required to support modernization of MMI schemes

#	Action area	Role	Typical interventions
1.	Policy and legislation	Required to frame, direct, and regulate the modernization program	• Legislation on water allocation and quality • Legislation on irrigation development and services • Legislation for Command Area Development (CAD) and Water Users Associations (WUAs) • Policy and legislation for Public-Private-Partnership (PPP) activities
2.	Institutional reform	Required to facilitate changes in attitudes and behaviours and, where needed, organizational restructuring	• Formation of WUAs • Restructuring of the WRD • Separation of water resources management from irrigation scheme management • Strengthening of the Agricultural Dept. • Strengthening of Water Resources Information Management and Decision Support Systems
3.	Water resources identification, planning, and management	Required to ensure that supplies to the scheme are adequate, reliable, and sustainable. Includes analysis of groundwater resources available to farmers	• Review of river basin plans (including groundwater) • Assessment of allocation and management of available water resources • Assessment of prioritization criteria for allocation of scarce water resources (e.g. during droughts or pressure from other uses) • Review of drought management plans
4.	System modernization	Modernization of the physical infrastructure together with the management, operation, and maintenance (MOM) procedures to improve control and measurement, reduce wastage of water and better match supply to demand	• Modernization of control and measurement structures • Modernization of the MOM procedures, including operating systems for conjunctive use of surface and groundwater resources • Introduction of on-demand irrigation water delivery • Development of asset management plans (AMPs)

(*Continued*)

Table 6 (*Continued*)

#	Action area	Role	Typical interventions
5.	Agricultural reform and modernization	Agricultural reform is required to optimize the benefits to be derived from reliable, adequate, and timely irrigation supplies	• Introduction of water-saving technology and techniques to farmers • Adoption of modern agronomic practices (seed dressing, tillage) • Supporting crop diversification • Improving storage • Strengthening of marketing
6.	Upgrading knowledge and information	A cross-cutting action area to strengthen stakeholders' understanding, knowledge, and skills to formulate, promote and support the modernization process	• Strengthening and modernizing teaching of irrigation in universities, colleges, and training institutes • Supporting study tours and exchange visits to experience modern irrigation practices • Strengthening adaptive research • Developing online education and training as part of a continuing education programme

Source: Authors.

3.4 Options for modernization

The interventions identified in Table 6 were further broken down into their component parts. For example, modernization of control and measurement structures was broken down into the type of control structure (manual/automated, localized control, remote control) and the measurement structure (weir, flume). In all some 85 possible modernization interventions were identified at scheme, main system, on-farm, field, and drainage levels. One quarter of the identified modernization interventions occur in action area 5 – Strengthening Agricultural Services and Practices. Action area 6 – Building Knowledge and Information Systems comprises specific activities but also spreads throughout the other five action areas. Examples of the analysis of the identified modernization options covered the benefits, constraints/limitations, point of application and likely costs, are shown in Table 7. Many options were identified, not all of which can be introduced under one modernization project. The issue of deciding which 'mix' of activities to choose is discussed in the following section.

Under the SIMP the crucial stage is when the 'mix' of interventions is decided during phase 2 when the Irrigation Modernization Plans (IMPs) have been prepared. For this exercise, a multidisciplinary team has been engaged to survey and assess the current status of the batch 1 schemes, following which

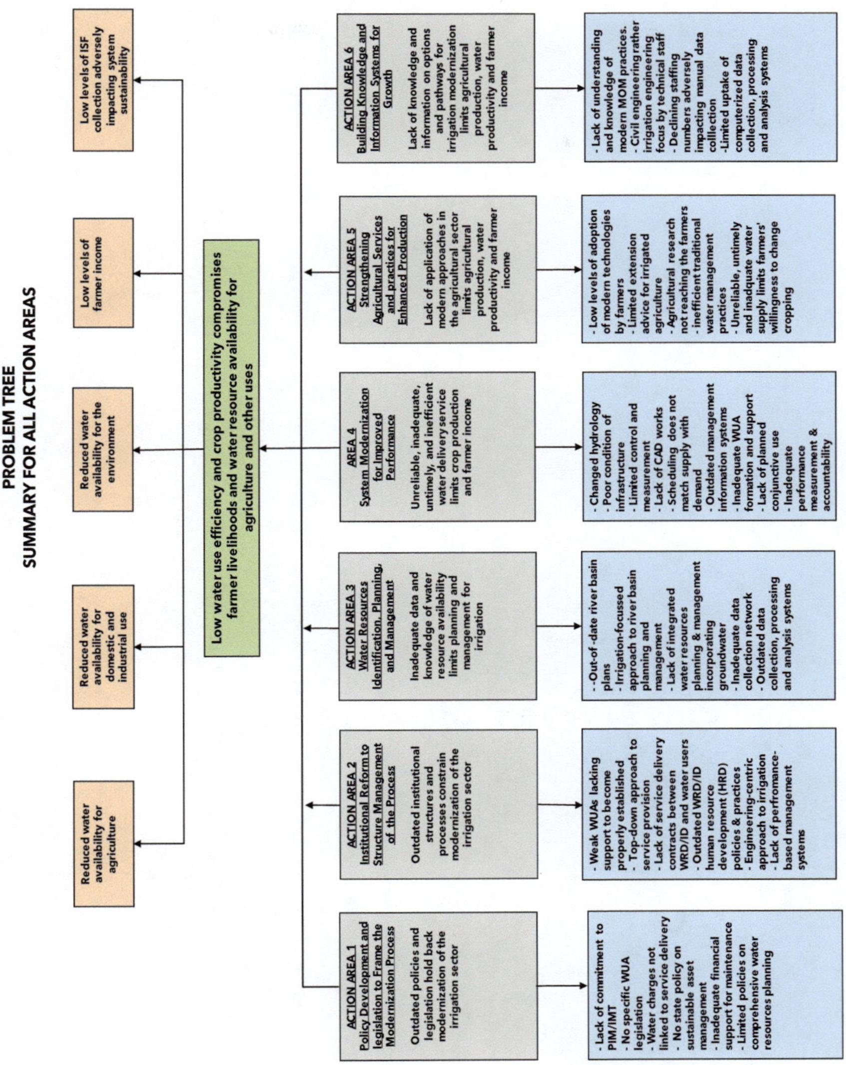

Figure 3 Example of a SIMP problem tree analysis for the six action areas. Source: Authors.

Table 7 Illustration of the range of identified modernization interventions for MMI schemes

Example	Modernization intervention	Benefits	Constraints/limitations	Where/when to apply/likely cost
Policy, legislation, and organizations				
 WUA Management Committee, Karnataka	Formation, support, and strengthening of Water Users Associations (WUAs)	Engages water users in the MOM of their irrigation systems. Improves transparency and accountability of water distribution Increases irrigation service fee (ISF) collection More reliable and equitable water distribution Improves maintenance of the system	Needs support and training at inception and in early stages to establish and strengthen WUA Need to avoid politicization	Beneficial in all systems Relatively low cost
River basin (water resources management)				
 Crop water productivity for rice, Karnataka	Use of remote sensing to determine key indicators (crop type, crop areas, actual evapotranspiration, biomass production, crop yields, crop water productivity)	Enables determination of key indicators over a large area Enables determination of crop type and area, ETa, PoW, biomass production Can look at current or historic data	Requires specialist skills and equipment	Where data of key parameters (crop area, crop type, crop ETa, CWP) are required over a large area Low cost compared to large scale field surveys

Main canal level				
 Canal automation, Australia	Canal automation	Increased crop production Water and labour savings Improved service to the farmers Reliable and equitable distribution of water Environmental protection (waterlogging and salinity minimized)	Initial cost is high Maintenance and servicing of the automation system Requires skilled and trained personnel to operate the system	Where improved control of main canal is required Water losses are high The efficiency of the project is low Where the structures can be secured from tampering or vandalism High capital cost recouped through efficiency gains
On-farm level				
 Smart irrigation-sensors to measures oil moisture, Morocco	Internet of Things (IoT)-Smart irrigation	Water supply can be matched to demand Facilitates monitoring by connecting sensors to smartphone or laptop Provides real-time updates and tracks environmental conditions (humidity, rainfall, temperature)	It requires good internet connectivity all the time Requires skilled professionals to monitor Capital cost and maintenance	Useful in irrigation schemes having good internet connectivity and availability of skilled professionals to manage Where farmers are ready to invest in the technology Relatively low cost

(Continued)

Table 7 *(Continued)*

Example	Modernization intervention	Benefits	Constraints/limitations	Where/when to apply/likely cost
Field level				
 Drip lines for vegetable crop, Karnataka	Drip irrigation for conventional planting, improved planting, with fertigation for system of sugarcane intensification (SSI)	Reduces wastage of water and fertilizer Better matches crop water requirements with supply Increases crop productivity	Cost of drip equipment Pumping costs to pressurize the drip system Not suitable where water has a high sediment load Emitters prone to clogging Limited to certain crop types	Where water is scarce Well suited to some high value orchard and row crops such as vegetables, tomatoes, sugar cane, and grapes Costly to purchase and operate. Pipes need periodic replacement
 Automated laser-guided grader, Punjab	Precision land levelling using laser technology	Conserves water and reduces time for irrigation Uniform distribution of water Less water consumption in land preparation Lesser weeds in the field Uniformity in crop growth Improved yield and water productivity	High cost of equipment Annual planing is needed to maintain the field grade	Where land or field plots are uneven Relatively expensive equipment cost but low costs per unit area when done by an agricultural contractor
Agricultural practices				
 West Bengal	System of Rice Intensification (SRI)	Yield increase and water saving	Needs training to develop planting and watering skills	Water scarcity and rice yield stagnation schemes Low cost - relatively simple change in agronomic practices

Table 8 Key studies and steps required in the preparation of an irrigation modernization plan

#	Study tool	Description
1	Benchmarking	Benchmarking is required to ascertain the current condition and performance of the irrigation scheme. Benchmarking will provide a systematic analytical approach to assess the current condition and performance of each scheme. The FAO developed RAP-MASSCOTE tools or similar approach with a team of specialists from different disciplines (engineering, agriculture, institution, remote sensing) which can be used to develop a preliminary vision for service-oriented management and a plan which can be followed up with detailed studies and investigations to confirm the findings.
2	Focus group discussions (FGDs)	FGDs can be used to develop a quick non-quantitative assessment of current farming systems, constraints and indicative responses to possible initiatives to modernization and efficiency initiatives. The FGDs would be used to develop a broad understanding of the key issues and opportunities.
3	Participatory rural appraisal (PRA)	PRAs will build on the FGDs and will consist of structured and detailed household and farmer surveys using questionnaires and enumerators to widen the sample base. The PRAs will allow some quantitative assessment of constraints, issues, and responses to possible initiatives to meet the modernization objectives, as well as soliciting opinions regarding approaches for cost recovery related to the management, operation, and maintenance (MOM) of the I&D scheme now and in the future (i.e. after modernization).
4	Medium-level remote sensing	A vegetation index analysis will be carried out, which will be based on a medium resolution 250m pixel analysis of the Normalized Difference Vegetation Index (NDVI). The gross command area will be surveyed to provide a first assessment of land use during the dry season and contribute to preliminary estimates of the productivity of water (PoW).
5	Detailed-level remote sensing	The high-detail analysis will provide an in-depth analysis of crop water productivity using satellite information with a 30-m resolution. The resulting time series and land cover map will be used to derive a crop biomass/water productivity map that can be upgraded to a crop yield/water productivity map using official crop statistics with some supporting fieldwork.
6	Basin/sub basin water balance	The sub-basin water balance will include an assessment of water availability and summary of withdrawals for irrigation and other water uses.
7	Institutional and technical analysis	The institutional and technical analysis will depend on the characteristics of each scheme as well as whether a RAP or PRA are programmed for each scheme. The institutional and technical analysis will build on the outputs of the RAP, FDGs, PRA, and key informant interviews and consultation (KII&C), together with scheme-specific data collection.

(*Continued*)

Table 8 (*Continued*)

#	Study tool	Description
8	Agricultural assessment	Using a farmer sample in each selected village cluster, the situation analysis showing the existing crop pattern, crop production and input use practices/technologies, and marketing will be carried out. The existing crop productivity, water use and income per ha of crop production will be assessed. A rapid key informant survey will also cover selected farmers in each cluster. Crop and technology improvement practices, marketing channels/methods suitable for each project will be examined and potential practices/technologies for inclusion in the modernization package will be shortlisted in consultation with the stakeholders. A constraint analysis will be done to assess farmers need for capacity building. Appropriate tools such as partial budget, cost curves, and production functions will be applied for the assessment.
9	Proposals for improved water management	Pre-feasibility plans will be prepared to assess options and present proposals to improve water management and agricultural production including proposals for conjunctive use and development of strategies to increase the productivity of water (POW). The plans incorporate assessment of current water availability, water allocations and consumptive use and identify and assess technical and institutional options for improving performance and meeting the modernization objectives, including innovative agricultural practices. These proposals will be discussed with stakeholders to gather their views and feedback.
10	Preliminary plans for modernization and efficiency improvements	An analysis of issues and presentation of preliminary plans to modernize and increase efficiencies will be prepared. The plans will build on the proposals to improve water management and agriculture production systems. The plans will examine the surface and groundwater systems together with outline assessments of investments to upgrade and modernize infrastructure, institutions and agricultural practices. The plans will examine necessary policy changes and requirements for institutional development and reforms. Preliminary economic assessments will be prepared. These plans will be discussed with the key stakeholders to gather their views and feedback.

FAO, Food and Agriculture Organization: RAP, Rapid Appraisal Process; MASSCOTE, Mapping System and Services for Canal Operation Techniques (FAO, 2007).
Source: SIMP Final Report, December 2021. Updated based on experience during phase 2 implementation.

proposals are developed for modernization. The key studies required for the preparation of an IMP are summarized in Table 8.

The guidelines for preparing the modernization plans are reviewed following implementation during Phase 2. The CWC is broadly in agreement with the approach followed in preparing the 4 IMPs in different states and settings and are also aware that it is not feasible to carry out detailed MASSCOTE studies in all MMI schemes in India due to a lack of trained human resources. However, a similar approach/study with specialists is being developed. The preparation of high-quality IMP calls for the involvement of a team of engineers with modernization experience and other specialists in agriculture, institutions, community development, and remote sensing. Experience during phase 2 has shown that it is possible to produce an IMP in 3 months, and that it is not as time-consuming and complex as initially envisaged. In fact, it saves time and guides the DPR preparation. Another key aspect is that it involves consultation with all key stakeholders. This ensures that broad agreement is reached on the core objectives and features of the modernization plan. The IMP preparation can also be seen as a part of DPR preparation and not a separate activity, i.e. it is akin to a pre-feasibility report.

4 Making the right choices

Typically, infrastructure (physical) costs dominate rehabilitation or modernization projects (Table 9). The agricultural, institutional, and management components, where included, form a fraction of the total costs (Table 10). While these three components might constitute the least in terms of the total cost, they often contribute significantly in terms of the beneficial outputs and outcomes from the project (Table 11).

The results of the Farmers Participatory Action Research Project (FPARP) conducted in 5000 locations across India has indicated that given the current rate of adoption and rate of return, the success rate of the water management technologies is only about 12% (Palanisami et al., 2015). The region-wise performance gap was calculated where, the percent gap in water saving was comparatively lower (32%) in the southern region whereas the percent gap in yield was higher in the eastern region (48%) (Table 12). Hence, it is important to focus on measures that could enhance the technology adoption at farm level. Promotion of digital agriculture and transformative investment options in climate smart agriculture could be explored and practiced to scale-up the adoption of the technologies. Agri-business models can be encouraged with private partners including seed providers, food companies, micro irrigation system manufacturers, and input suppliers who will be interested in pilot programmes to encourage and incentivize farmers for adopting specific

Table 9 Cost components of typical irrigation rehabilitation/modernization projects

Scheme/ country	Total project costs (US$ millions)	Target area (ha)	Target number beneficiaries	Component items and costs (US$ millions)			
				Physical works	Agriculture	Institutional	Water management
WMIP Kyrgyz Rep.	28	85 000	40 000	19.8	-	2.5	4.85
OIP-2 Kyrgyz Rep.	20	51 000	34 000	15.4	-	4.2	-
MPWSRP Madhya Pradesh	443	495 000	192 863	388.0	(included in inst. costs)	38.35	7.27
FIMPEgypt	180	80 000	140 000	153.5	14.1	-	-

WMIP, Water Management Improvement Project; OIP-2, Second On-farm Irrigation Project; MPWSRP, Madhya Pradesh Water Sector Restructuring Project; FIMP, Farm-level Irrigation Modernization Project.
Source: World Bank Project Appraisal Documents.

Table 10 Unit costs and share of total costs for components of typical irrigation rehabilitation and modernization projects

Scheme/ country	Total costs per unit area (US$/ha)	Total costs per beneficiary (US$/ha)	Component item costs per unit area (US$/ha, % of total)				
			Physical works	Agricultural	Institutional	Water management	Project management
WMIP Kyrgyzstan	331	703	233 (70%	–	29 (9%)	57 (17%)	12 (4%)
OIP-2 Kyrgyzstan	403	604	301 (75%)	–	83 (21%)	–	19 (5%)
MPWSRP Madhya Pradesh	895	2298	784 (88%)	–	77 (9%)	15 (2%)	11 (1%)
FIMP Egypt	2250	1286	1,919 (85%)	176 (8%)	–	–	–

WMIP, Water Management Improvement Project; OIP-2, Second On-farm Irrigation Project; MPWSRP, Madhya Pradesh Water Sector Restructuring Project; FIMP, Farm-level Irrigation Modernization Project.
Source: World Bank Project Appraisal Documents.

Table 11 Value addition from improved agricultural practices

Indicator	Soybean	Wheat	Paddy
Yield (q/ha)	5.7 increased to 8.1 (43%)	27.0 increased to 37.2 (38%)	27.8 increased to 38.6 (39%)
Productivity of water (kg/m^3)	0.9 increased to 1.6 (87%)	1.1 increased to 2.1 (82%)	4.8 increased to 7.7 (60%)
Productivity of water (Rs/m^3)	24.4 increased to 42.8 (76%)	17.5 increased to 31.6 (80%)	67.3 increased to 107.8 (60%)
Net margin (Rs/ha)	8193 increased to 13 986 (71%)	36 159 increased to 51 112 (41%)	32 444 increased tc 47 819 (47%)

Source: College of Agricultural Engineering, Jabalpur adaptive trials reported in MPWSRP, 2015.

climate smart practices. Transformative policies both at state and national level that will promote technology spread and adoption are also needed.

Institutional dimensions influence not only the policy and regulatory environment and framework within which a modernization programme is planned and implemented, but they also determine how the scheme is managed, operated, and maintained and influence the understanding and knowledge of both agency staff and farmers. As with the agricultural dimensions, they are an integral part of an irrigation scheme, and its modernization. The different levels of institutional dimensions that come into play in irrigation management range from irrigation governance (entitlement, tariff, organizational setup/mandates) to irrigation management (demand–supply estimation, allocation, scheduling, data generation and analytics systems) to irrigation service (service delivery planning, cost recovery mechanism, performance measurement, engagement/communication process with water users, staff capacity development).

The various instruments, structures and tools that facilitate these institutional dimensions of irrigation management include

- policy-legislation-regulatory provisions at the governance level;
- statutory processes for capacitated organizational setup for decision-making;
- service delivery and water users' engagement;
- creating a cadre of knowledgeable and trained staff;
- establishing hydro-meteorological observation systems;
- data analysis and decision support systems at the management level;
- statutory systems and tools for participatory planning of irrigation service delivery;
- procedures for estimating, invoicing and collecting the IFS;
- empowering and strengthening WUAs; and
- benchmarking and water auditing scheme performance.

Table 12 Comparison of technologies at research station and farmers' fields

Region	Increase in water saving (%)			Increase in yield (%)			Internal rate of return (%)		
	Research station	FPARP	Gap (%)	Research station	FPARP	Gap (%)	Research station	FPARP	Gap (%)
Southern	26.58	18.12	31.84	24.75	8.03	67.54	22.25	10.44	53.06
Western	38.34	11.98	68.75	24.63	10.79	56.21	25.00	10.86	56.57
Northern	30.95	12.23	60.49	20.86	10.72	48.60	19.08	11.44	40.01
Eastern	29.40	15.29	48.00	19.93	8.60	56.85	15.50	9.25	40.32
North-eastern	42.50	27.53	35.22	80.33	12.80	84.07	17.00	14.00	17.65
Average	**32.73**	**14.85**	**54.61**	**26.21**	**9.72**	**62.92**	**20.96**	**10.80**	**48.48**

Source: Palanisami et al. (2015).

Table 13 Institutional change initiatives under current irrigation modernization programmes in India

Institutional dimension	Type of institutional change initiatives
Policy-legislation-regulation	Karnataka has issued its new State Water Policy 2022 with specific irrigation modernization related strategies amongst others such as: • Managing the State's water infrastructure for irrigation • Modernizing irrigation • Participation of the community and capacity building
	Maharashtra has enacted the Maharashtra Water Resources Regulatory Act establishing an independent regulatory authority for regulation of water resources in the state
	A number of states have WUA Laws for the formation and empowerment of WUAs for participatory irrigation management
Organizational setup	Karnataka has reorganized its Water Resources Department into the Water Resources Development Organization and the Irrigation Corporations (Nigams) separating the water management functions from the irrigation development and service delivery functions
	Andhra Pradesh had reorganized the I&CAD Department into the Irrigation construction wing and the Command Area Development wing for irrigation scheme MOM
	Maharashtra has established a separate wing in the WRD as the Chief Auditor, Water and Irrigation Maharashtra State in Aurangabad
Data management systems	Karnataka has established the ACIWRM which is designing and developing the KWRIS as the single window repository of all water resources and irrigation service data for the state
	Andhra Pradesh has established the APWRIMS which include the state WRIS and WRMS providing both observational data and analyzed reports on various water resources and irrigation parameters
Performance management and DSS	Maharashtra WRD has been carrying out water audit and benchmarking studies of its irrigation schemes and publishing annual reports for the last 2 decades now covering a total of 2621 irrigation schemes (65 major, 239 medium and 2317 minor tanks)
	I&CAD Department, Andhra Pradesh had designed and developed a number of digital apps, tools and dashboards to facilitate planning and performance management of irrigation schemes by transferring a lot of the back-end data analysis processes to IT-enabled systems and enable better planning, implementation, and management decision-making. Some of these digital apps and tools included: • MIS for irrigation water demand and supply; • reservoir storage and canal network flow monitoring system; • work tracking system for irrigation O&M works; • satellite remote sensing-based estimation of irrigated area; • satellite remote sensing-based tank hydrology re-assessment; • satellite remote sensing-based drainage congestion identification; • comprehensive irrigation scheme performance assessment; • participatory self-assessment by water user associations; and • participatory ground water management planning and monitoring.

(*Continued*)

Table 13 (*Continued*)

Institutional dimension	Type of institutional change initiatives
Capacity building and training	ACIWRM, Karnataka is carrying out extensive training for WRD staff on topics such as IWRM, MASSCOTE, water accounting, water productivity, remote sensing, and GIS for water resources management
	WALAMTARI had setup field training centres with requisite training staff in all 18 irrigation circles covering all MMI schemes in Andhra Pradesh for both WUA and irrigation engineers training and support

ACIWRM, Advanced Centre for Integrated Water Resources Management; KWRIS, Karnataka Water Resources Information System; I&CAD, irrigation and Command Area Development Department; MOM, Management, operation and maintenance; WRD, Water Resources Department; APWRIMS, Andhra Pradesh Water Resources Information Management System; WRIS, Water Resources Information System; WRMS, Water Resources Management System; GIS, Geographical Information System; IWRM, Integrated Water Resources Management; WALAMTRI, Water and Land Management Training and Research Institute; WUA, Water Users Association.
Source: Authors.

Over the last couple of decades, various states in India have implemented different institutional change initiatives under their respective irrigation modernization programmes. Details of some of these initiatives are briefly summarized in Table 13.

Whilst most of these institutional change initiatives listed in Table 13 (and a number of others not listed) have delivered good results in supporting their respective irrigation modernization programmes during implementation, a common concerning feature shared amongst them is that a lot of them became dysfunctional soon after the closure of the multilateral development bank funded project under which they were initiated or with the change in the administrative leadership in the WRD or political leadership in the government. Institutionalization of the changing systems in the institutional dimensions of the irrigation sector is perhaps the most challenging aspect of irrigation modernization. Unless a lasting solution is found to this issue, any irrigation modernization programme, including SIMP, will remain confronted with achieving the desired management outcomes of effectiveness and sustainability.

5 Linking project activities to achieve the desired objectives

Using the logical framework (European Commission, 2004) approach there is a logical relationship between the various components of a modernization project. This is set out in the IF-THEN hierarchy of the logical framework. As shown in the problem tree (Fig. 3) a complex project may have several strands (engineering, agriculture, and institutional) each leading to distinct outputs which in turn combine to achieve the overall objective. The logical framework's IF–THEN logical hierarchy is based upon the premise that the conditions and achievements specified for each level are both necessary and sufficient to attain the next highest level. Thus, each causal linkage must be planned and executed to ensure that the required conditions not only exist but are sufficient to achieve the next level. Thus, the IF–THEN relationship not only relates to progression up the hierarchy from inputs to the development objective, but also applies between activities where one activity might be a necessary input to another part of the chain of activities. This thinking then structures how the modernization project is designed and which components are included into the project. By thinking about the inter-linkage of activities those planning modernization projects are required to consider the following:

1 What are the linkages between project activities?
2 Are linkages sequential or can they be carried out in parallel?
3 How does a given linkage work? Is Activity A a pre-requisite for Activity B? Can Activity A achieve its required output without Activity B? Is the completion of
4 Activity A an input to Activity B?
5 How do the outputs of the different chains of activities combine to provide the desired outcome(s)?
6 What assumptions are made in framing the activities and their outputs?
7 What is the importance of each activity?

If the objective is to increase agricultural production and income to farmers, then a typical logical framework might be as shown here:

IF the water supply from the main canal is transparent, reliable, timely, adequate, and equitable, and

↓

IF the water supply at the on-farm (chak) level is also transparent, reliable, timely, adequate, and equitable, and

↓

IF there is a market for the agricultural products

↓

THEN farmers may consider planting crops for productive rather than protective agriculture

Behind this logical chain, there are several inherent assumptions. For the main canal, the assumption is that there is control and measurement of the flow (gates, measuring structures) and the canal section is in good condition. There is also the assumption that the Irrigation Agency has the staff, knowledge, skills, and resources to schedule and regulate the irrigation water to match farmers' needs. For the on-farm system, the assumption is that there are adequate control and distribution structures, and that channels are in a reasonable condition. In addition, there is the assumption that there is an organization (e.g. Water User Association) to schedule and regulate the irrigation to match farmers' daily needs. There is a further assumption that there is 'social control' – that is the community trusts the on-farm organization to be honest and fair in organizing the water distribution. For the market, the assumption is that traders can be trusted and offer a fair price. Finally, the location of the market and means of transport of agricultural production to the market needs to be considered.

It is important to recognize that irrigation is a complex activity, with many different components interacting in many different ways, and not always in the same way in different schemes. Table 14 sets out the key activities and

Table 14 Basic activities in a modernization project

Project activity	Process	Output	Assumptions/ necessary conditions
Modernize institutional arrangements	Frame appropriate policy, laws and regulatory statutes and capacity building of decision makers, service providers, and water user organizations	Appropriate policy-legal-regulatory ecosystem and capacitated decision makers, service providers, and water user organizations established	Across the board consensus on the modernization programme amongst all stakeholders and availability of necessary data/information and analysis to support decision-making, providing services and water use
Modernize main canal with automated gates	Full control of the available water supplies	Timely, reliable, and adequate supplies at all locations	Knowledgeable, capable, and motivated O&M staff
Modernize on-farm system with control structures and lining	Full control of available water supplies	Timely, reliable, and adequate supplies to all landholdings	Knowledgeable, capable, and motivated WUA with field staff
Change farming practices from protective to productive irrigation	High-yielding crops grown using dressed seed, fertilizer, and pesticides	High yields of marketable crops	Inputs of seeds, fertilizer, and pesticide available. Farmers' knowledgeable and skilled in growing high-yielding varieties dependent on reliable, adequate, and timely irrigation. Market available for agricultural produce

processes with their desired outputs and assumptions/necessary conditions. In some cases, the necessary conditions may be outside the control of the project (e.g. input and market prices, transport to market) but in others they may be within the control of the project. This might be the case with 'capable and motivated O&M staff' where a new project activity can be created to train and build capacity of the staff.

6 Conclusion

Modernization of MMI schemes in India is challenging due to the sheer number of schemes needing modernization, the context of irrigated agriculture which has changed significantly from protective to productive irrigation, multiple and competing uses of water, conjunctive use of surface and groundwater, and climate change. The framework developed under the SIMP for modernizing MMI schemes with a stepwise approach in preparing the IMPs has been tested successfully in four schemes in India. This framework is expected to pave the way for taking up more schemes in the years to come in many states of India, ultimately leading to the modernization of all the MMI schemes.

The main conclusion emerging from our work in India is that modernization of I&D schemes is a multifaceted activity with engineering, agriculture, and institutional components at its core. Addressing only the engineering aspects, as has been the case in many rehabilitation programmes in the past, will not produce the required efficiency, productivity and remunerative improvements required. Modernization needs to be a holistic process and must include modernization of scheme management along with empowering the farmers to lead the process and moving the focus of the irrigation authorities towards a service delivery orientation as key driving forces in spearheading sustainable modernization efforts.

I&D schemes are not all identical, each scheme differs from others in some way or another. Some aspects are generic, such as the standard design of canals and structures, but the way these assets are managed varies widely, as does the way farmers use the water delivered. Rather than adopting a blueprint approach to modernization, it is recommended that a 'cafeteria' approach is adopted, applying the technology and support most appropriate to each scheme, taking account of the key variables. Particularly important will be the need to adopt a phased approach in moving from one tier to another, higher, tier, consideration of farmers' ability and willingness to change and achieving progressive capacity development of key personnel involved in modernization.

7 References

ADB. 2019. Concept Note, Support for Irrigation Modernization Program. Asian Development Bank, Manila, Updated 11 November.

ADRI. 2023. India Water Facts. Water Resources of India, Asian Development Research Institute. Available at: https://www.adriindia.org/adri/india_water_facts, accessed October 2023.

Burton, M. 2010. *Irrigation Management: Principles and Practices*. CAB International, Wallingford.

Chaturvedi, R. K., Joshi, J., Mathangi Jayaraman, G. and Bala, R. N. H. 2012. Multi-model climate change projections for India under representative concentration Pathways. *Curr Sci* 103(7), 1–12.

Cruz, R. V., Harasawa, H., Lal, M., Wu, S., Anokhin, Y., Punsalmaa, B., Honda, Y. and Jafari, M. 2007. Asia, climate change 2007: impacts, adaptation and vulnerability. In: Parry, M. L., Canziani, O. F., Palutikof, J. P., Van der Linden, P. J. and Hanson, C. E. (Eds) *Contribution of Working Group II to the Fourth Assessment Report of the Intergovernmental Panel on Climate Change*. Cambridge University Press, Cambridge, pp. 469–506.

CWC. 2016. *Summary Report on Water Use Efficiency Studies for 35 Irrigation Projects*. Central Water Commission, New Delhi.

CWC. 2020. *Annual Report (2019–20)*. New Delhi. Available at: http://cwc.gov.in/sites/default/files/arcwc2019-20.pdf.

Economic Times. 2015. FICCI-Grant Thornton Report, New Delhi. *The Economic Times*, December 3.

European Commission. 2004. *Project Cycle Management Guidelines, Vol. 1*. European Commission, Brussels, March.

European Space Agency. 2023. *Department of Economic and Social Affairs, Population Division, World Population Prospects 2022*. United Nations, New York. Available at: https://esa.un.org/unpd/wpp/, accessed October 2023.

FAO. 2007. Modernizing irrigation management – the MASSCOTE approach. In: Renault, D., Facon, T. and Wahaj, R. (Eds) *IDP 63*. Food and Agriculture Organisation of the United Nations, Rome.

FAO. 2017. *The Future of Food and Agriculture – Trends and Challenges*. Food and Agricultural Organization, Rome.

Gosain, A. K. and Rao, S. 2003. In: Shukla, P. R., Sharma, S. K., Ravindranath, Garg, A. and Bhattacharya, S. (Eds) *Climate Change and India: Vulnerability Assessment and Adaptation*. Universities Press (India) Pvt Ltd, Hyderabad, p. 462.

Gosain, A. K., Rao, S. and Basuray, D. 2006. Climate change impact assessment on hydrology of Indian River basins. *Curr Sci* 90, 346–353.

Guiteras, R. 2009. *The Impact of Climate Change on Indian Agriculture*. Report, University of Maryland, USA.

Lahiry, S. C. 2018. *Climate Change and Its Impact on Water Resources*. India Water PortalArghyam Foundation, Bengaluru, India.

Mall, R. K., Gupta, A., Singh, R., Singh, R. S. and Rathor, L. S. 2006. Water resources and climate change: an Indian perspective. *Curr Sci* 90(12, 1610–1626).

MFLW. 2015a. India: innovations for more food with less water. Task 2. Report prepared by Lahmeyer International in association with Lahmeyer International India

BETS Consulting Services, Centre for Environment and Development, and Total Management Services. Project No. 45072-001, Regional Technical Assistance TA7967-REG. Asian Development Bank, Manila, April.

MFLW. 2015b. Innovations for more food with less Water. Task 1: Final Report. Report prepared by the International Water Management Institute. Project No. 45072, Regional Technical Assistance TA-7967-REG. Asian Development Bank, Manila, May.

MPWSRP. 2015. Implementation completion and results report. Madhya Pradesh Water Sector Restructuring Project. World Bank, Washington, DC.

NWUEISP. 2014. Scoping study for a national water use efficiency improvement support program. Report prepared by Martin Burton and Amarjit Dhingra. Project No. SC 100903 (IND) and SC 100905 (IND). Asian Development Bank, Manila.

Palanisami, K., Ranganathan, C. R., Kakumanu, K. R. and Sekhar, U. N. 2011. A hybrid model to quantify the impact of climate change on agriculture in Godavari Basin, India. *Energy Environ Res* 1(1, 32-52).

Palanisami, K., Ranganathan, C. R., Nagothu, U. S. and Kakumanu, K. R. 2014. *Climate Change and Agriculture in India: Studies from Selected River Basins*. Routledge, New Delhi, p. 101.

Palanisami, K., Suresh Kumar, D., Malik, R. P. S., Raman, S., Kar, G. and Mohan, K. 2015. Managing water management research: analysis of four decades of research and outreach programmes in India. *Econ Polit Wkly* 1(26 & 27), June, 33-44.

Pathak, H., Pramanik, P., Khanna, M. and Kumar, A. 2014. Climate change and water availability in Indian agriculture: impacts and adaptation. *Indian J Agri Sci* 84(6), 671–679.

SIMP. 2021. Phase 1 Final Report prepared by Martin A. Burton, M. G. Shivakumar, Rahul Sen and K. Palanisami. Support for Irrigation Modernization Program, Asian Development Bank, Manila.

Srinivasan, V. 2019. Climate adaptation in the water sector in India. In: Dubash, N. K. (Ed.) India in a Warming World: Integrating Climate Change and Development. Oxford Scholarship Online: December 2019. DOI: 10.1093/oso/9780199498734.001.0001.

United Nations. 2019. Available at: https://population.un.org/wpp/Publications/Files/WPP2019_Volume-II-Demographic-Profiles.pdf.

Part 4

Addressing future water and climate risks

Chapter 13

PRECIMED: development of a decision support system (DSS) for precision irrigation in Mediterranean agriculture

María F. Ortuño[1], Andrés Parra, Juan J. Alarcón and María J. Gómez-Bellot, Centro de Edafología y Biología Aplicada del Segura (CEBAS-CSIC), Spain; Antonio Skarmeta, Manuel Mora and Rafael Marín, Odin Solutions S.L. (ODINs), Spain; Nikolaos Katsoulas, Sofia Faliagka and Dimitris K. Papanastasiou, University of Thessaly (UTH), Greece; Mohammed Semiani, Farouk Eddine Belkhiri and Smadhi Dalila, National Institute of Agronomic Research of Algeria (INRAA), Algeria; and Khouloud Boukadi, Rima Grati and Bassem Bouaziz, University of Sfax (OPTIM), Tunisia

1 Introduction

In the twenty-first century, some problems such as water quality degradation, groundwater depletion, demographic unbalances between rural and urban areas and soil salinization processes have become evident in both arid and semi-arid areas (WWAP, 2015). Some of these problems are especially relevant in the Mediterranean region; in fact, currently large areas in the Mediterranean region are facing recurrent water scarcity and drought events (Ali et al., 2022). In the Mediterranean region, irrigated agriculture contributes 75% to total crop production (Wani et al., 2009; Daccache et al., 2014). There is therefore a need for innovative technologies that increase water and nutrient use efficiency (WUE and NUE) and make non-conventional water sources available for fertigation,

[1] Corresponding author: mfortuno@cebas.csic.es.

http://dx.doi.org/10.19103/AS.2023.0123.24

thereby decreasing water scarcity and the discharge of water and nutrients to the environment (Nikolaou et al., 2020). In this sense, the optimization of the sustainability of agriculture by using precision irrigation techniques, adopting and implementing new water and nutrient management practices, takes on special relevance.

Although more complex definitions exist, the simple description of precision agriculture is a way to 'apply the right treatment in the right place at the right time' (Gebbers and Adamchuk, 2010). Precise irrigation methods are developing rapidly in order to save water, while improving yields and fruit quality (Conesa et al., 2021; Zhang et al., 2022; El Chami et al., 2023). Although irrigation has been practiced for centuries, precision irrigation is a relatively new concept, as the sector had to respond to societal demands for reductions in water allocation and improvements in efficiency. Certain irrigation strategies have been proven to successfully increase WUE by reducing water use while increasing yield (Costa et al., 2007; Tejero et al., 2011; Du et al., 2015). Thus, the current trend of irrigation and fertigation management implies (i) precision crop irrigation and fertigation, (ii) the use of crop-based information (crop indicators/descriptors), (iii) enhanced analysis, interpretation and valorization of field data and (iv) development of systems to empower growers to control fertigation more effectively. Therefore, the contribution of knowledge, know-how and tools related to efficient and smart applications to support irrigation and fertigation control is essential.

Precision irrigation is technologically feasible through the use of environmental sensors, probes and decision support systems (DSSs) (NEEA, 2015). Nowadays, there is a wide variety of possible solutions available on the market ranging from the simplest based on hand sensing (Conejero et al., 2011; Gómez-Bellot et al., 2015; Bañón et al., 2020) to the more complex that uses satellite imagery (Yang, 2018; Ihuoma et al., 2021; Ramírez-Cuesta et al., 2022). In addition, the 'Internet of Things' (IoT) is a highly promising family of technologies which is capable of offering many solutions towards the modernization of agriculture (Tzounis et al., 2017; Vladisavljević et al. 2019; Zamora-Izquierdo, et al., 2019; Xu, et al., 2022). Recently, the IoT world has offered new LPWAN (low power wide-area network) such as Sigfox, LoRa and 5G (LTE-M and NB-IoT), to allow wireless and autonomous devices to be monitored and controlled without expensive routers or repeaters. LPWAN technologies enable internet connection in remote and dispersed places without traditional connection based on 2G/3G/GPRS/GSM technologies. To integrate legacy sensors/actuators, a new IoT paradigm called the cyber-physical system (CPS) is being developed to enable interaction with crop facilities to perform precise data acquisition (for plant, soil and environment) and also to enhance remote actuations for fertilizer/nutrition pumps, irrigation valves and greenhouse actuators (e.g. ventilation or lighting). Finally, the analysis of IoT data ('Big Data') allows interpreting and valuing the collected data using complex processes, including scenario prediction and via DSSs (Cravero and Sepúlveda, 2021; Saggi and Jain, 2022; Ben Abdallah et al., 2023).

Both intensive soil and soilless greenhouse production and open-field cropping systems experience multiple issues caused by excessive irrigation and nutrient supply. Growers tend to over-irrigate without taking into consideration actual crop water needs (Valenzano et al., 2008; Rallo et al., 2011; Katsoulas et al., 2017) by applying standard practices based on their own perspective rather than more objective science data, resulting in water and nutrient losses (Fernández et al., 2007; Thompson et al., 2007; Navarro-Hellín et al., 2016). This makes the development and further adoption of user-friendly tools for fertigation management a necessity for Mediterranean farming systems. Model-based DSSs can contribute to the preparation of plans that are more suitable for fertigated greenhouse crops as demonstrated by multiple studies (Giménez et al., 2013; Gallardo et al., 2016, 2020; Tei et al., 2020), however, at a lower frequency for open-field crops.

There are solutions available on the market covering a range of individual operations in the agricultural sector to optimize irrigation, among others, tree pruning, and pest management (Sciarretta et al., 2019). On their own, these solutions are providing value to farmers, either by enabling reduction of production costs and/or more efficient execution of targeted operations. However, most of these solutions are built around proprietary protocols and are the so-called point solution, i.e. addressing a single farm operation and implemented as an isolated solution, despite using modern cloud-based technologies as the IoT concept. The main consequence of this situation on the market is that a single farm must use multiple solutions. For this reason, an innovative DSS termed PRECIMED DSS was proposed, which will alleviate these problems by validating the real-life potential and market readiness of advanced interoperability between agricultural devices, for different production sectors such as greenhouse and outdoor crops.

2 PRECIMED approach

2.1 Overview of PRECIMED project

The PRECIMED DSS has been developed within the framework of the project 'Precision irrigation management to improve WUEand NUE in the Mediterranean region' (PRECIMED), a research and innovation (R&I) project co-funded under the PRIMA 2018 programme (Partnership for Research and Innovation in the Mediterranean Area), which seeks to support initiatives in the Mediterranean area to improve agri-food and water management systems (https://prima-med.org/). The PRECIMED consortium was comprised of three entities from two European countries (Spain and Greece) and two entities from two African countries (Algeria and Tunisia), from which one was a SME (ODIN Solutions S.L., ODIN), and four were RTDs (Spanish National Research Council, CEBAS-CSIC; University of Thessaly in Greece, UTH; Algerian National Institute of Agricultural Research, INRAA; and University of Sfax in Tunisia, OPTIM) (Fig. 1). In order to bring excellence and a broad reach, the consortium brings together four different Mediterranean regions with different typologies with respect to potential end-users of the PRECIMED DSS. The system has been tested in Spain, Greece, Tunisia and Algeria. This allowed the PRECIMED DSS to evolve by receiving inputs from different types of users, located in different geographical environments. In this way, on the one hand, the project delivers a technological solution with the desired functionalities and quality, and on the other hand, it is possible to set the basis for the exploitation of the PRECIMED DSS at European and African levels, specifically in the Mediterranean basin, ensuring its scalability in terms of reaching the largest number of possible end-users.

2.2 PRECIMED objectives

The PRECIMED project has researched, developed and validated a standards-based DSS including irrigation and fertigation models for analysis of real-time crop and meteorological status data to improve the efficient use of water, nutrients and energy. For this reason, the consortium has integrated knowledge of fertilizers and irrigation of Mediterranean crops with innovative information and communication technologies (ICT) to develop a solution that will be

Figure 1 Geographical distribution of the PRECIMED partners and pilot sites.

respectful with the environment and economically profitable. In the PRECIMED project, the DSS has been developed for an end-user that can easily access and manage through web interfaces from any location with Internet access either through a mobile phone device, tablet or personal computer. The DSS platform can collect and ingest a large amount of crop data, which is processed and analyzed to provide information to the end-user regarding crop water needs and real-time recommendations regarding the best irrigation and fertilization practices. The DSS offers management services and remote actuation to improve sustainability while saving water and fertilizers in a region with significant problems relating to both water resource stress and groundwater quality.

To achieve the overall aim of the project, four key objectives were identified:

1. To improve WUE and NUE in the Mediterranean region by using intensive ICT solutions;
2. To facilitate the interchange of technology and good agricultural practices between EU and non-EU Mediterranean countries;
3. To develop and validate (on different demonstration farms) a standards-based DSS for data-driven irrigation and fertilization management that evaluates the medium-term evolution of crop nutritional status, soil salinity, yield and fruit quality, optimizing the water and fertilizers needs and the energy costs at farm level; and
4. To ensure that the project activities and outcomes would reach and impact relevant target groups, especially end-users (farmers), thus enhancing the market uptake and potential of PRECIMED's solutions.

2.3 PRECIMED concept

2.3.1 Methodology

The implementation of the PRECIMED DSS was carried out in different regions in Algeria, Greece and Spain, under both open-field and greenhouse crops, by adopting these steps:

1. Determination of the requirements that farmers have recognized. This has been done through a continuous process to engage farmers during the design, development, testing and validation of the PRECIMED tool, and by the agronomic assessment of their crops and soils where PRECIMED was to be deployed, and based on scientific results from previous projects and research studies.

2 Design and development of a cost-effective DSS system for precise irrigation and fertilization. For this reason, the cost-effective cyber-physical system (CPS) gateways, standardized IoT platform, irrigation models and Big Data techniques were defined. The proposed gateways enable cost-effective data acquisition, using novel IoT technologies such as CPS for sensors/actuators integration and LPWAN networks for long-range wireless communication. In addition, the proposed DSS platform was based on FIWARE, a European IoT platform for data acquisition, information processing and end-users' services. This platform integrates fertilization and irrigation models obtained in several research EU projects in order to improve crop productivity and water sustainability. The integrated models are developed using Big Data information processing techniques to offer irrigation services with end-users' interfaces based on the Software as a Service (SaaS) approach.
3 Deployment and validation of the PRECIMED DSS prototype in the different scenarios, working together with end-users.

This methodology allowed end-users (farmers, technicians and other stakeholders) to participate directly and actively not only in the preparation and development of the PRECIMED DSS but also during the test and validation phases of the tool's implementation.

2.3.2 Pilot farm assessment

The assessment of the pilot farms for the design of the PRECIMED DSS focused on identifying the requirements of both the farms and the farmers, analyzing the management of nutrient and water availability in the context of climate change taking into account the impact of irrigation technologies on the productivity of water and fertilizers use efficiency, and defining the most interesting system suitable for each pilot farm included in the assessment, concerning the type and the number of sensors required to manage water and fertilizers in each case study. At the beginning of the project, several meetings were held with farmers, cooperatives and agricultural partners from the area where the farms were located. The findings of these meetings are summarized as follows:

For greenhouse production, growers currently use indices to help them monitor the microclimate conditions and fertigation management practices. These indices include the following:

- Solar radiation inside or outside greenhouse;
- Air vapour pressure deficit and if not available, the greenhouse air temperature and relative humidity;

- Volumetric water content (VWC) of the crop substrate;
- pH and electrical conductivity (EC) values of the nutrient solution in the substrate and in the drainage (leachate) solution from the substrate; and
- Ratio of the volume of drained leachate to the volume of nutrient solution applied.

Usually, based on these indices, growers modify the dose and frequency of their nutrient application solution. However, they currently have no tools or indices to modify the composition of the applied nutrient solution and thus, they usually apply a standard nutrient solution depending on the growth stage of the crop (e.g. four stages for tomato, two stages for cucumber crops). Nevertheless, they pointed out that it would be ideal if they could modify the composition of the nutrient solution applied on a more frequent weekly or bi-monthly basis, so that the applied nutrient concentrations could be more closely related to the fertilization needs of the crop.

Regarding the outdoor field crops, it was concluded that to develop a DSS to improve the WUE through precise irrigation scheduling, each pilot needed the following:

- An agro-meteorological station to monitor, at least, rainfall events and crop evapotranspiration (ET_c) rates;
- Some devices to monitor changes in soil water content and temperature;
- An initial physical-chemical analysis to characterize the local soils, i.e. soil texture, organic matter content and mineral composition;
- To analyze the irrigation water quality through electrical conductivity and pH; and
- To describe the crop coefficient (K_c) for each phenological state of the plant.

According to the main requirements identified in the different countries, the pilot leaders configured the most appropriate approach for the deployment of the PRECIMED system in each pilot farm; a pomegranate field in Spain, citrus and potato farms in Algeria, and a soilless-greenhouse cucumber and tomato production in Greece were selected.

2.3.3 PRECIMED platform

This section addresses the cost-effective data acquisition and remote actuation with wireless gateways connected to an IoT-standards-based platform through the internet. Within the framework of the PRECIMED project, a FIWARE (EU FIWARE)-based platform (https://www.fiware.org/) supporting edge/cloud computing and the integration of three subsystems (i.e. data acquisition, data processing and end-users services) was developed in the context of precise irrigation and fertigation management. Moreover, the FIWARE-based platform employs and integrates outcomes of the EU FIWARE project, including the ORION Context Broker (Cristescu, 2015), open and standardized lightweight IoT-data protocols (Haller, 2010; Al-Sakran, 2015) such as MQTT (Hunkeler et al., 2008), CoAP or even REST (Shelby et al., 2014) services based on HTTP, as well as NGSI & NGSI-LD (NGSI, 2019) to facilitate the acquisition, integration, and exchange of large datasets with CPS (Wolf, 2009) gateways.

The analysis of the IPv6 and 6Lowpan protocols and different LPWAN (Song et al., 2017) networks (Sigfox, LoRa, and NB-IoT/LTE-M 5G) was carried out to consider the best wireless communication option in terms of long-range, low consumption, and low cost for the CPS systems (Wolf, 2009). Considering the analysis of results, the separation of the communications module in the design of the printed circuit boards for the CPS was suggested as a more scalable and flexible solution, to integrate different communication technologies through commercial LPWAN modules.

Regarding the PRECIMED IoT platform, its design was carried out based on open standards (The Internet Engineering Task Force) of IoT, MQTT, and EU FIWARE) to ensure interoperability, with a multilayer architecture to ensure scalability and support for edge and cloud computing and following the distribution model of SaaS for its development. During its development, the backend and frontend were developed and updated. Different connectors (IoT agents) were created to integrate external data sources, such as the deployed IoT devices, weather services (Weatherbit), and multispectral satellite imagery (Sentinel-2). ODIN worked on different solutions (advanced graphics, data files, REST API, and the Context Broker) to cover the end-user needs identified by the partners and offer the data registered on the platform for its exploitation by users or by other services (i.e. DSS services). In addition, different agronomic services (irrigation planning, rules management, plot, and agronomic data management) were included together with needed HMI (human machine interfaces) (Fig. 2).

Different open-source web technologies have been used for development (Fig. 3). The frontend was developed using Vue.js (Vue.js), a framework based on HTML, CSS, and JavaScript, and using OpenLayers, a JavaScript library for displaying interactive maps. The backend was developed on top of JavaScript and integrates several components:

- A MQTT Broker (Mosquitto, https://mosquitto.org/) that receives data from MQTT-compliant sources to be processed by the IoT agents.
- The IoT FIWARE JSON Agents, a software that translates information received from the MQTT Broker and other data sources into a superior

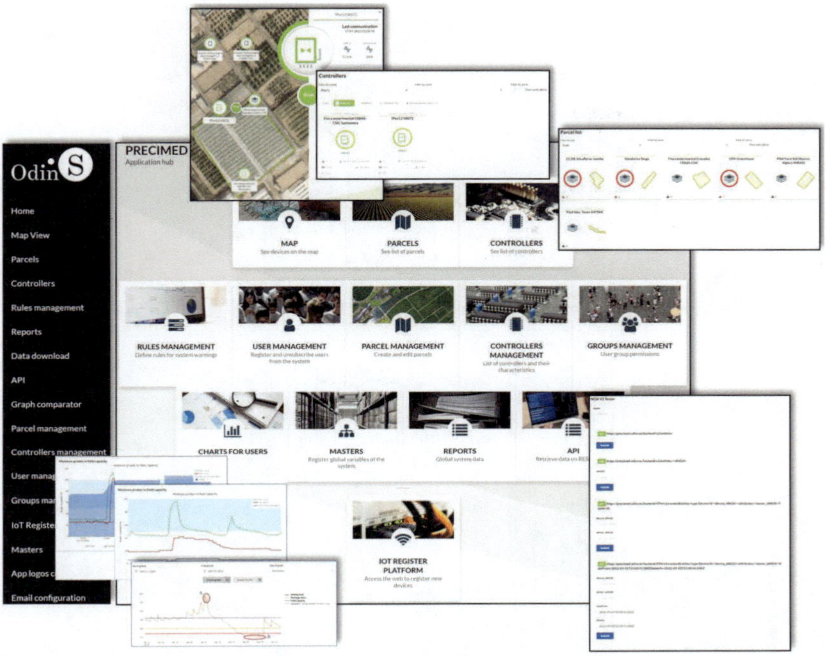

Figure 2 PRECIMED DSS dashboard showing the end user HMI.

JSON (Org, 2009)-based data model to record them in the internal DB as historical data.

- An internal document-oriented NoSQL JSON database (MongoDB) where the IoT agents record the data.
- A REST API based on the FIWARE NGSI data model to give access to the registered information for exploitation.
- In addition, the possibility of deploying a FIWARE Orion-LD Context Broker that supports the FIWARE NGSI-LD data model and offers an API REST has been tested, to access the data registered by publishing/subscribing.

2.3.4 Irrigation/fertilization approaches

DSS models and their application in precision agriculture need to be tailormade to specific locations or management objectives. For this reason, different approaches, which are described below, were developed during the project according to the first assessment. In addition, two models based on the most recent advances in machine learning and neural networks were developed and tested at the laboratory of OPTIM (Tunisia), using data from the pilot farms established during the project (described in Section 2.3.5).

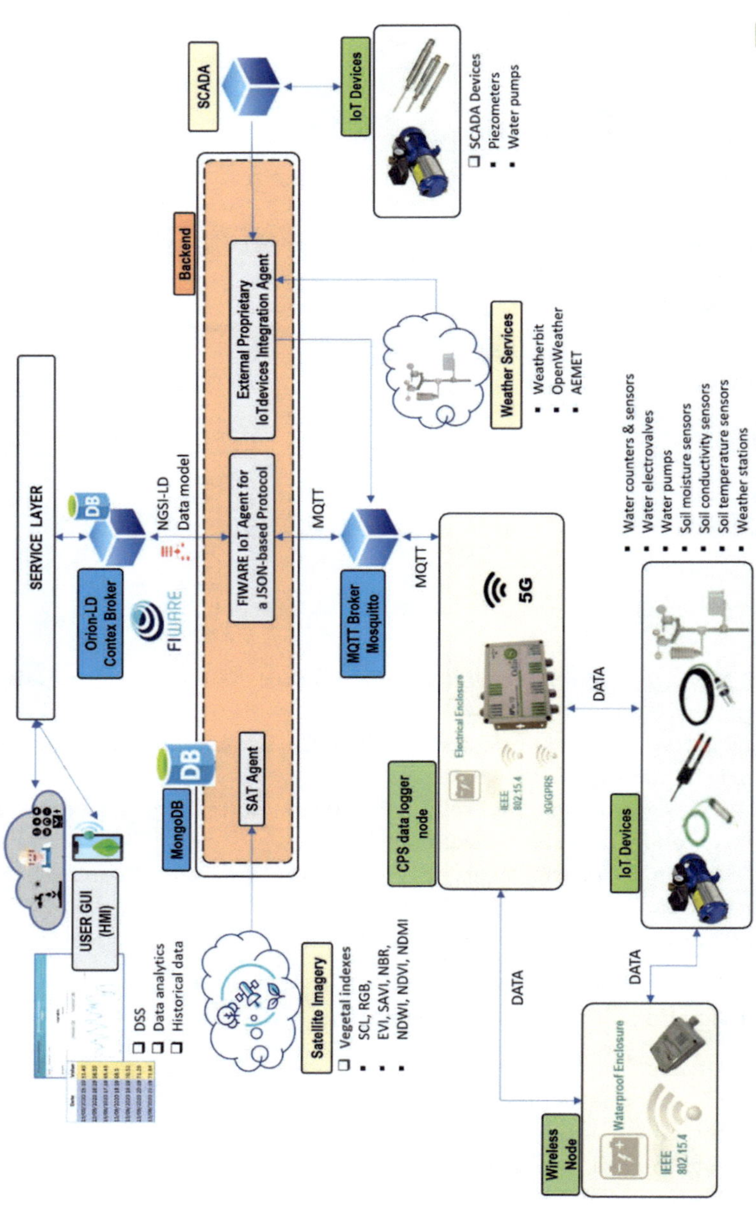

Figure 3 Conceptual deployment of the different elements within the PRECIMED DSS platform.

2.3.4.1 Greenhouse conditions

The irrigation scheduling in soil and soilless greenhouses should consider three significant parameters: (i) the dose, which equals the water irrigation volume that is transpired by the crop, (ii) the leachate fraction, and (iii) the salt accumulation in the root zone (Pardossi et al., 2011). Extensive research has been conducted over the years to estimate the crop transpiration and investigate the effects of climatic parameters on crop water uptake, in an attempt to match the crops' irrigation needs with daily transpiration fluctuations. This has led to the development of several transpiration models (Katsoulas and Stanghellini, 2019) based on the Penman–Monteith approach. In the context of the 'fourth agricultural revolution,' sensing, smart and sustainable technologies have been developed to improve soilless culture systems using emerging ICT (Rose and Chilvers, 2018; Miranda et al., 2019). As part of the PRECIMED project, fertilization simulation models to minimize water and nutrient applications in the greenhouse emphasized the 'feed-forward irrigation control systems' for more accurate results. Running simulation models for on-time fertigation management could support growers to make more informed water and nutrient plans over a shorter period, whereas off-line analysis that uses historical data and weather forecasts could provide information for longer periods. The concept of fertigation management for greenhouse crops can be summarized into the following four steps:

1. In the first stage, the crop nutrient needs in relation to dry matter production (DMP) are estimated. Any changes in DMP as the plant grows or when there is biomass removal due to harvesting or leaves and stems pruning, are taken into consideration daily.
2. In the second stage, crop water needs, as a function of crop transpiration, are estimated as the difference between the nutrient solution supplied to the crop (irrigation) and the nutrient solution drained from the crop (drainage) that are correlated with the climate conditions (solar radiation and vapor pressure deficit) observed in the greenhouse, in a regular basis. Thus, using the recent historical data (e.g. last 10 days) the system estimates the coefficients used in the model in the greenhouse under consideration. Any change in the crop management strategy or crop development is automatically taken into account in the coefficients.
3. The amount of fertilizer and water that needs to be applied to the crop is then estimated and used to adjust the nutrient solution recipe based on the crop's daily needs.
4. Lastly, the fertigated solution recipe is adapted based on any changes in the crop biomass and/or climatic conditions and weather forecast.

The PRECIMED simulation model was developed to improve both WUE and NUE in a hydroponic greenhouse in the Mediterranean region, promoting precision agriculture for better resource management. The model considers several climatic inputs such as the daily maximum and minimum temperature and solar radiation and simulates the crop DMP and crop transpiration. The optimal ion (NO_3^-, K^+, PO_4^{3-}, and Ca^{2+}) concentration in the fertigation nutrient solution that has to be supplied to the plants is then estimated considering the nutrients' content in the dry matter and the nutrients' absorption concentration. The model was evaluated in soilless tomato and cucumber greenhouse crops cultivated either during an autumn–winter or spring–summer period to assess the efficiency of the model through the year.

To assess the effectiveness of the model, measurements of crop DMP, macronutrients concentration in the dry matter, water, and nutrients consumption and crop yield were collected and analyzed. The purpose of this work was to provide the end-users with a tool that would estimate on a weekly time-step the optimal composition of the irrigation nutrient solution based on the growth stage of the crop and local weather conditions, as an alternative to the current practice where the composition of the nutrient solution only changes about two or three times during the entire growth period.

2.3.4.2 Open-field conditions

Regarding the outdoor field crops cultivated in the Algerian pilot farms, a water balance model was proposed to generate an irrigation scheduling programme for citrus and potato crops. Due to the high cost of sensors, including Drill and Drop (https://sentektechnologies.com/products/soil-data-probes/drill-drop/), that can be used for *in situ* soil moisture monitoring and irrigation scheduling, this approach was proposed to allow a larger group of farmers that can also access a limited number of weather stations. The main steps for calculating the daily soil moisture balance using evapotranspiration data can be summarized as follows:

- Determination of the daily reference evapotranspiration (ET_0) according to the Hargreaves method (Hargreaves and Samani, 1985);
- Determination the appropriate daily K_c according to the FAO Penman-Monteith equation (Allen et al., 1998) related to the numerical calculation of K_c:

$$kci = kc\,prev + \left[i - \Sigma\left(L\,prev\right) / L\,stage\right]\left(kc\,next - kc\,prev\right)$$

where:

i is the day number within the growing season (length of the growing season),

K_c i is the crop coefficient on day *i*,

L_{stage} is the length of the stage under consideration (days)

- $\sum(L_{prev})$ is the sum of the lengths of all previous stages (days);
- Calculation of the daily crop water requirements;
- Calculation of the net irrigation taking into account the irrigation system efficiency;
- Estimation of the effective rainfall;
- Calculation of the daily change in water balance;
- Estimation of the remaining availability water in the soil, and;
- When the remaining available water of the soil reaches a third of the total available water (TAW), an irrigation event is triggered to bring the soil profile back to field capacity (FC).

The Diviner 2000 (https://sentektechnologies.com/products/soil-data-probes/diviner-2000/) and the Drill and Drop probe were then used to verify the reliability of the soil water content calculated by the use of water balance method. In addition to climate data, this method also requires defining some soil characteristics, in particular the determination of the soil water content at permanent wilting point (PWP) and FC. The water balance method provides the option to choose a desired level of soil water depletion, taking into account the sensitivity of the crop to water stress, the acceptable yield losses, and the quality of production. In the framework of the PRECIMED project, an evaluation of farmer irrigation practices was made by comparing their irrigation scheduling based on the water balance method during the growing seasons of 2021 and 2022.

At the CEBAS-CSIC experimental farm in Spain, a classical approach according to the Penman-Monteith model was used to irrigate the crop during the first growing season (2021). Different deficit irrigation strategies were also established, aiming to obtain a larger dataset that could then be used in the deep learning modelling of soil moisture content (LSTM) to predict the irrigation scheduling components (Section 2.3.5). In addition, another irrigation protocol (termed CEBAS-PRECIMED DSS) was adopted leveraging the functionality of the PRECIMED platform and the multilayer soil sensors (Drill and Drop) deployed on the farm. This protocol was based on soil moisture measurements obtained from these sensors at different depths (Fig. 4).

However, to schedule irrigation, attention was focused on two specific depths (40 cm and 60 cm) based on the evidence that for drip-irrigated woody

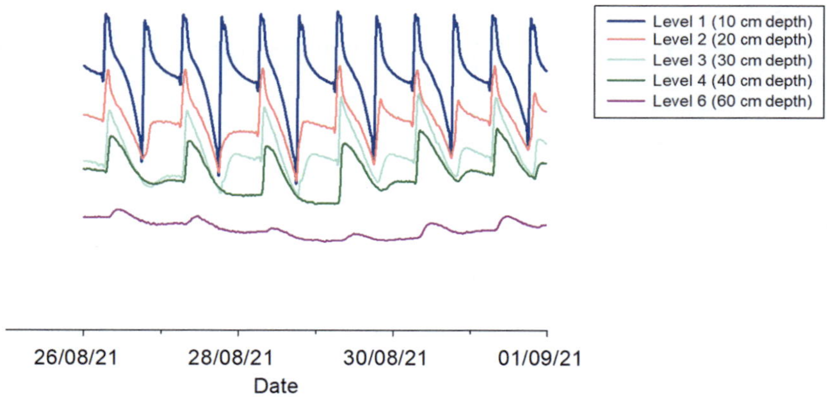

Figure 4 Example outputs from soil moisture sensors operating at five different depths in the CEBAS-PRECIMED DSS.

crops, the more active roots are located in the top 50 cm of the soil profile (Intrigliolo et al., 2013). The information obtained from the soil probes was used to ensure that the soil water content was maintained within an optimal range between these two depths, preventing drainage beyond 60 cm and ensuring adequate and constant soil moisture at 40 cm depth with each daily watering. To do this, the dynamics of soil moisture at these depths were followed after each irrigation event. In summary, the DSS monitors whether the soil moisture is higher one hour after the end of an irrigation event (F) compared to that before starting an irrigation application (I) 40 cm and 60 cm depths, and based on the results, the system executes the following commands:

At 40 cm depth:

- F40 - I40 > 1 →Irrigation is maintained
- F40 - I40 < 1 →Irrigation run time is increased by 15% for the next day[2]

At 60 cm depth:

- F60 - I60 > 0.3 →Irrigation is reduced the next day by 15%
- F60 - I60 < 0 →There is no water drainage and irrigation is maintained

2.3.5 Big data techniques and services

The algorithms corresponding to the previous models were developed and integrated into the PRECIMED DSS platform. End-user HMI interfaces were designed and implemented to provide the necessary dashboards to manage DSS services once integrated with the platform. The interfaces were designed

[2] The values of irrigation differences and percentages of increase and decrease in irrigation (15%) were established based to previous studies.

to be simple, easy to use and responsive. In parallel, two models based on the most recent advances in machine learning (Ben Abdallah et al., 2022a,b) and neural networks were developed and validated in the laboratory. The first model was intended for outdoor crops (Fig. 5), while the second was intended for greenhouses (Fig. 6).

Several studies have noted the relevance of soil moisture prediction in irrigation environments for outdoor crops. For this purpose, the approach adopted was to predict the soil moisture value followed by the actual estimation of net irrigation from this predicted value (Bamory et al., 2023). The prediction of soil moisture could capture potential ground-water variations that would be otherwise difficult to observe, thus avoiding inefficient irrigation decisions. Several stand-alone data-driven models for predicting future soil-moisture values (Dubois et al., 2021), have compared different approaches including feed-forward neural networks, random forest (RF), and support vector machine learning for predicting weekly soil water potential. However, a novel strategy was used in this study. At first, the well-known time series prediction model, LSTM, was employed in different experiments to find the optimal length and number of time-steps required to predict the next day's soil moisture level given the data at hand. Furthermore, because soil moisture exhibits spatio-temporal variability (Song et al., 2016), the appropriateness of a hybrid CNN-LSTM model was also evaluated.

The estimation of the net irrigation amount (depth applied, mm) from the predicted soil moisture was made using the formula proposed by INRAA. Hence, the outdoor cropping approach could be summarized as follows:

1 Integration of data collected from the CEBAS Controller (climate and soil variables): The data was cleaned and preprocessed by OPTIM for correct use by the model. Some of these preprocessing techniques included interpolating missing values, by estimating missing values based on adjacent values.
2 Development of a deep learning model for soil moisture prediction: Soil moisture prediction is a time-series forecasting problem, and therefore there is time-dependency between soil moisture observations. However, past studies dealing with soil moisture prediction have largely used the feed forward type of Artificial Neural Networks (ANN) (Sanuade et al., 2020), which ignores the temporal-dependency nature of time series data. In contrast, recurrent neural networks are a special type of ANN that has the ability to store neurons' outputs between different time steps, making them more suitable for time series processing (Adeyemi et al., 2018; Marini et al., 2020). This specific characteristic of RNNs leads to their adoption in some soil moisture prediction studies. For this

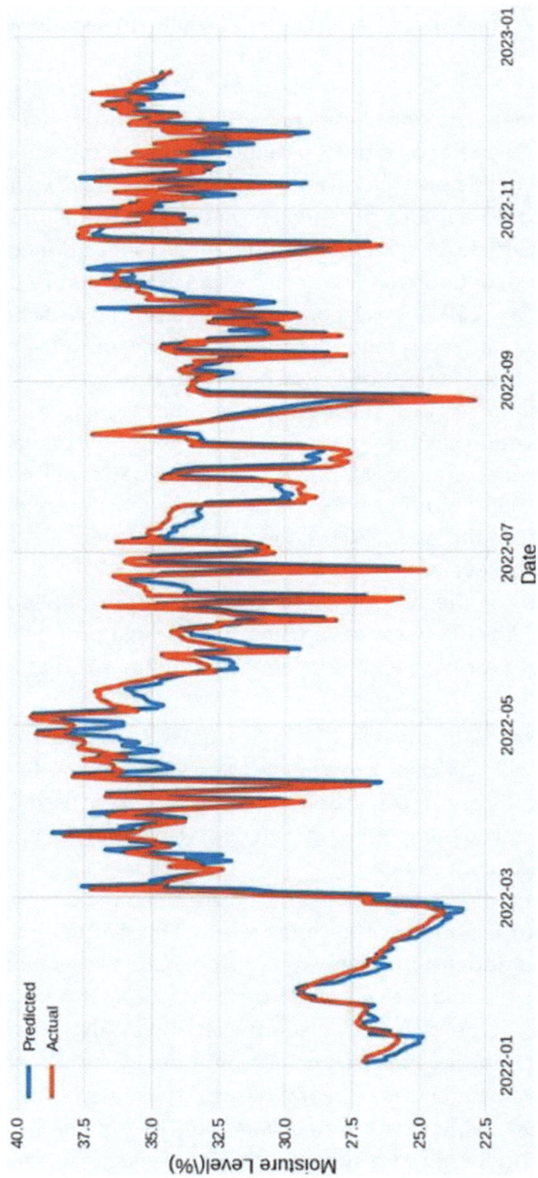

Figure 5 Comparison of actual versus predicted (modelled) soil moisture as part of the validation of the PRECIMED model for an outdoor field crop.

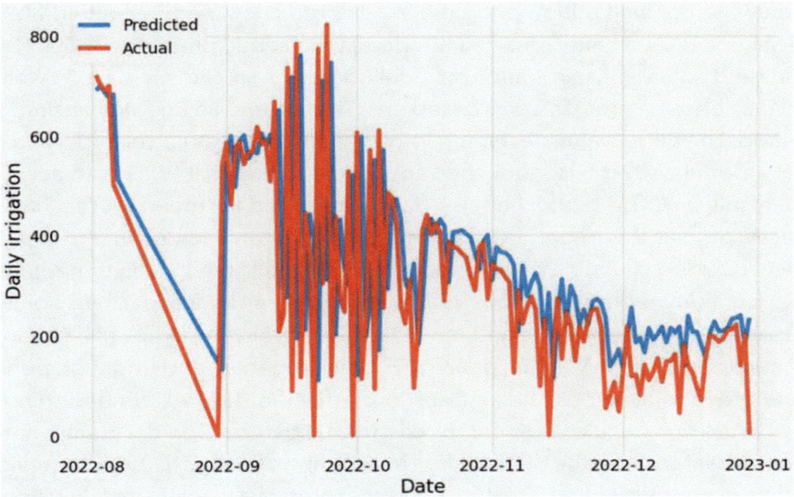

Figure 6 Comparison of actual versus predicted (modelled) soil moisture as part of the validation of the PRECIMED model in greenhouse environment.

reason, OPTIM investigated the suitability of RNN based models namely LSTM, Bi-LSTM, and CNN-LSTM.

3 Model validation: A sample of the original CEBAS dataset was used by OPTIM to evaluate the modelled results. This evaluation process consisted of comparing recorded soil moisture values from the CEBAS dataset to the predictions made by the model.
4 Prediction of soil moisture using the developed model: The model was deployed and made available over the internet so that it can be used to make soil moisture predictions.
5 Computation of the irrigation requirements based on the predicted soil moisture value: Some additional parameters were also taken into account to compute the net irrigation, which was the end goal of the process.

Regarding greenhouses, despite the various solutions proposed in the literature to provide an accurate model for estimating future crop transpiration and daily water need values, few studies have been performed based on modern mathematical algorithmic techniques, notably machine learning (Ge et al., 2022). Recently, Ge et al. (2022) compared eight regression algorithms, such as XGBR, SVR, and KNR, to model drip-irrigated greenhouse tomato evapotranspiration (ET) and analyzed the model prediction results. Li et al. (2020) developed a model to predict plant transpiration that used RF regression with plant and environmental parameters. Even though these models can produce reliable and accurate predictions, they still require substantial data pre-processing as they are not well adapted to time series prediction. As highlighted above, LSTM, as a RNN architecture that supports time-series prediction, could be a promising solution for irrigation scheduling based on climatic data as well as previous observations. This was based on UTH data and an irrigation estimation method. The target value, that is the irrigation amount, was calculated using the UTH's formula, which also took the vapour pressure deficit (VPD) into account. As a result, an LSTM model has also been developed for greenhouses that can directly predict the irrigation amount. This study considered data from UTH, which included climatic and plant observation data (greenhouse temperature, AVG Humidity, Solar radiation, VPD, VWC, water nutrient leachate solution supplied to the crop by irrigation and drained from the crop, pH, electrical conductivity (EC), daily actual drain ratio, daily irrigation, and irrigation period. In the greenhouse crops, the aim was to predict the daily water needs (DWN), which were first calculated to complete the UTH dataset. The daily water needs were calculated using the transpiration rate (T_r) and drainage (D_r). While previous studies used a simplified T_r estimation by expressing it solely as a function of solar radiation, the equations used in the PRECIMED project included VPD to improve the model performance.

3 PRECIMED decision-support system case studies

The PRECIMED DSS prototypes were deployed and validated in different countries and cropping scenarios to showcase fully functional DSS applications that would encourage business opportunities in the exploitation of the project's results, with an emphasis on crop productivity, WUE, and end-user experience. Nevertheless, the deep learning models were integrated in PRECIMED platform although only validated under laboratory conditions, as by the time they were fully developed the field trials were concluded.

3.1 Pomegranate case study

3.1.1 Experimental design

The experiment was carried out on the experimental farm at CEBAS-CSIC in Santomera (Murcia, Spain) (Lat 38°06′ N, Long 1°02′ W, Alt 112 m). The plant material was own rooted 15-year-old pomegranate trees (*Punica granatum* (L.) cv. Mollar de Elche) in a 3 m × 5 m spacing pattern with a total cultivated area of 0.8 ha. The sandy clay loam soil of the experimental site was characterized by a high stone content (39% by weight) and a bulk density of 1.37 g/cc. The volumetric soil water content at saturation, FC, and PWP was 49%, 29%, and 18%, respectively. Irrigation water had an EC of 0.8–1.0 dS m^{-1}. Pest control and fertilization practices were those usually applied by local growers, while no weeds were allowed to establish within the orchard. The trees were lightly pruned every year, while sprouts and suckers were removed as they appeared, in order to encourage fruit production. Each year, four different irrigation treatments were performed to explore crop responses to both the timing and irrigation regime. During the first season (April–October 2021), four treatments provided data for the model described in Section 2.3.5:

- Control(C), irrigation was set at 120% of the estimated crop evapotranspiration (ET_c) (Tabari, 2010) avoiding limiting soil water conditions. ET_c was calculated as a product of the ET_0 (Allen et al., 1998) and crop coefficient (K_c).
- Sustained deficit irrigation (SDI), where water was applied at 50% of the water requirements of the crop through the whole season.
- Regulated deficit irrigation (RDI_{fg}), where irrigation was applied at 50% of the water requirements of the crop during July and August (linear fruit growth phase) while the rest of the season 120% ET_c was applied.
- Regulated deficit irrigation (RDI_r), where irrigation was reduced a 75% of the water requirements of the crop during the ripening phase (3 weeks before harvest) while the rest of the season 120% ET_c was applied.

During the second season (April–October 2022), the irrigation treatments were modified, and the four treatments were as follows:

- Control (C), where irrigation is scheduled to replace 100% of the ET_c.
- SDI, where irrigation is applied at 50% of the control treatment through the whole season.
- $C_{PRECIMED}$ treatment[3], where the plants were irrigated according to the CEBAS DSS protocol.
- $SDI_{PRECIMED}$[4], where irrigation is applied at 50% of the $C_{PRECIMED}$ treatment through the whole season.

To validate the $C_{PRECIMED}$ and $SDI_{PRECIMED}$ treatments that could be used as a reference for further DSS development, a series of measurements were performed during the season followed by analysis of the harvested fruits. Regarding plant water status, different indicators were used, such as the midday stem water potential, leaf gas exchange and crop water stress index.

[3] The $C_{PRECIMED}$ treatment was an irrigation protocol developed on the frame of the project (CEBAS-PRECIMED DSS protocol as described in Section 2.3.4) and was based on the information recorded from the multi-layer Drill and Drop soil probes installed on-farm.

[4] One of the reasons to include deficit irrigation protocols in the trials was to obtain a broader set of results when applying different irrigation strategies taking advantage of the soil probes installed on-farm. In this way, not only data about soil moisture and temperature were obtained for model input, but also to support the user in real-time, as during the 2 years of the pilot, a close monitoring of the different irrigation treatments was done through the platform, which allowed not only to interpret each irrigation event to the end-users, but also to discover irrigation anomalies remotely.

3.1.2 Deployment and integration of the DSS with the PRECIMED IoT platform

Agro-meteorological data were recorded by an automatic weather station located in the CEBAS-CSIC experimental field station, which recorded actual values every 5 min and average values every 15 min. From these data, ET_0 was calculated hourly using the Penman–Monteith equation (Allen et al., 1998). Soil humidity was recorded by the PRECIMED platform at four points on the farm and at six different depths (10 cm, 20 cm, 30 cm, 40 cm, 50 cm, 60 cm) using the Drill and Drop soil probes, to allow soil monitoring in real time. These devices, as well as the electro-valves to control the irrigation, were connected through the use of CPS gateways with a wireless LPWAN communication to the IoT platform, where data was then transmitted by the datalogger and stored for further exploitation, similarly to that done by the CEBAS-PRECIMED DSS protocol.

The DSS module retrieves historical data from the IoT platform for the electro valves that are linked to the irrigation system to find the last irrigation period (between the last on/off interval). With this information, the DSS module retrieves the historical data from the two soil moisture sensors to estimate the irrigation needed for the next irrigation event. The model also needs to be parametrized before doing its calculations. This is done by setting certain agronomic information relating to the crop. Finally, the DSS model returns the recommendations for the next irrigation using the following parameters (Fig. 7):

- Irrigation increment defined as the irrigation water variation, as a percentage (%), comparing the new recommendation with the last irrigation event.
- Irrigation duration (minutes).
- Number of irrigations to be applied.

3.1.3 Validation of the CEBAS-PRECIMED DSS prototype

As shown in Table 1, the use of the CEBAS-PRECIMED DSS estimated water savings of around 30% with a slight increase in crop production (14%) compared with the conventional irrigation protocol followed by the farmers in the area. Considering that energy consumption in drip irrigation system is typically 1.0 kW h/m^3, an energy saving of around 30% has been achieved. In relation to the physiological key performance indicators (KPIs), the increase in the gas exchange parameters and reduction in the canopy temperature and the crop water stress index (CWSI) for the plants that were irrigated based on the CEBAS-PRECIMEDS DSS showed significant benefits in terms of crop water status than those irrigated at 100% ET_c. In addition, regarding fruit quality, the total soluble solids increased, without observing an increase in fruit acidity, which can be

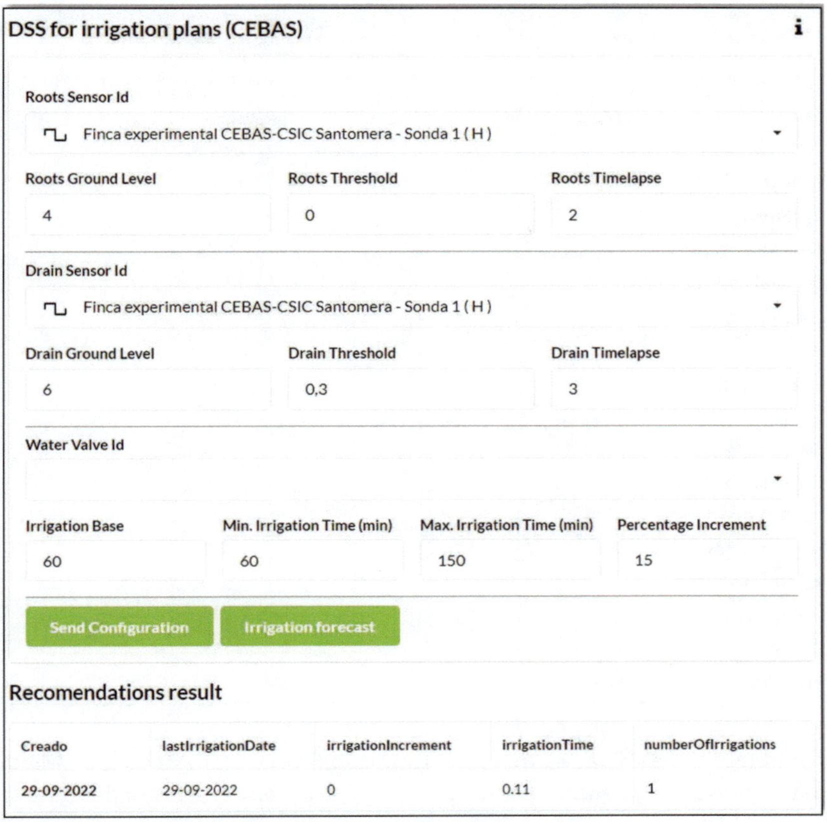

Figure 7 User interface for the CEBAS-PRECIMED DSS protocol.

perceived positively by consumers in terms of product quality. Also, the CEBAS-PRECIMED DSS irrigation protocol resulted in higher content of amino acids and sugars in the fruits, enhancing fruit flavour. An increase of 9% in the total phenolic compounds was observed in fruits from trees irrigated with CEBAS-PRECIMED DSS protocol, which is considered the most important in terms of health-promoting potential.

3.2 Citrus and potato case study

3.2.1 Experimental design

This trial was carried out between 2021 and 2022 on a farm located in the Mitidja region, central Algeria (Lat. 36.59 N, Lon. 3.094 E and Alt. 60 m) and in Sebain, in the Tiaret region of West Algeria (Lat. 35.44 N, Lon. 1.62 E and Alt. 937 m), respectively. The Mitidja region farm was a specialized orchard farm

Table 1 Key performance indicators achieved during the validation of the CEBAS-PRECIMED DSS protocol in the outdoor field farming scenarios using pomegranate crops

Key performance indicator	Estimated change (decrease (−) or increase (+))
Crop	
Water savings	+30%
Production improvement	+14%
Physiology	
Stem water potential	-3.5%
Stomatal conductance	+10%
Net photosynthesis	+2%
Intrinsic WUE	+20%
Canopy temperature	−12%
CWSI	−8%
Fruit quality	
Fruit weight	+15%
Total soluble solids	+8%
Maturity index	+2%
Amino acids	+38%
Organic acids	−21%
Sugars	+15%
Total phenolic compounds	+9%

growing fruit trees. Orange trees (*cv.* Thompson) were cultivated in the Mitidja area using drip irrigation as the main irrigation method. The orchard had a density of 5 m x 5 m, corresponding to 400 orange trees per hectare. Each row of trees included two lateral dripper lines at a distance of 2.5 m. Four drippers were used per tree (2 drippers per drip line) with each having a flow rate of 8 l h^{-1}. The distance between the drippers was 1 m.

On the other hand, the Sebaine Tiaret farm was specialized in vegetable crops, and in particular potato cultivation. This trial was conducted using potato (*Solanum tuberosum* cv. Spunta). The plant density was 47 619 plants per ha. This density corresponds to distances of 70 cm and 30 cm between rows and plants of potato, respectively. In contrast to the orange experiment, the potato trial was conducted using a sprinkler irrigation system. The trial aimed to compare the effects of two levels of nitrogen (120 kg N ha^{-1} and 250 kg N ha^{-1} combined with three irrigation regimes (Treatment 1, irrigation started at soil water content at about 100 mm by 0.40 m depth; Treatment 2, trigger irrigation when soil water content reaches about 80 mm by 0.40 m depth; and Treatment 3, trigger irrigation when soil water content reaches about 60 mm by 0.40 m depth).

3.2.2 Deployment and integration of the DSS system with the PRECIMED IoT platform

Soil moisture sensors and agrometeorological stations were installed on the two pilot farms. An automatic weather station type 'Imetos' and a Diviner 2000 probe were used in the two pilot sites during 2021 and 2022. The device installed in the Mitidja region was completed by the installation of a Drill and Drop Triscan 120 cm probe in July 2022. The weather station installed in the Mitidja farm recorded data every hour while the second weather station installed in Sebaine, Tiaret farm recorded data every 2 h. The two automatic weather stations were linked to the PRECIMED platform to transmit data. The Drill and Drop probes, for the control of soil water status to validate the reliability of the model were not yet integrated into the platform.

The DSS model for irrigation scheduling was based on crop evapotranspiration data and the status of soil water storage to generate/plan each irrigation event (when to irrigate and the depth of irrigation). The model used data from Imetos weathers stations and other IoT sensors like Drill and Drop (Fig. 8), and other data sources from the end-users, such as information on soil proprieties in relation to water retention (soil water at FC, PWP, and the total available water) and some references related to crop coefficients and the length of crop stages. The Drill and Drop probe and Diviner 2000 were used to validate the model. The main parameters used to perform the calculations associated with the model were summarized in section 2.3.4. The DSS model was not integrated with the PRECIMED IoT platform, but it is available on the INRAA computing infrastructure.

Figure 8 The in-field monitoring devices installed on the Algerian case study farm.

3.2.3 Validation of the INRAA-PRECIMED DSS prototype

The simulation of the soil moisture content and defining of the dates of irrigation were calculated during two growing seasons, 2021 and 2022, using climatic data recorded from the Imetos weather station installed in the citrus crop (*cv.* Thompson) field of the at the Mitidja farm. The comparison between the irrigation practices followed by farmers and the results of the predicted soil water storage and the irrigation dates by the model showed the following:

1 Based on irrigation scheduling using crop evaporation data, the modelled water requirements showed an estimated waste of water at farm level of 15.2% and 13.2%, in 2021 and 2022, respectively.
2 The soil water depletion allowed in the case of the model was one-third of the total available water (TAW). This corresponds to 47 mm. Each time, when the root depletion level is reached, irrigation is needed to refill the soil water storage back to FC. In the case of farmers, two refill points had been adopted: two-thirds of TAW during cell expansion stage and one-third of TAW during the full development of fruits.
3 The interpretation of the Diviner 2000 measurements showed that the active root zone, under the farmer irrigation technique was limited at 0.90 m depth.
4 During 2021, 7 irrigation events were estimated using the modelling approach and 7 based on farmer practices. In 2022, 7 irrigation events were predicted by the model, and 17 irrigation events based on the farmer irrigation practices.
5 The soil water content between the tree rows was better correlated with the predicted results by the irrigation scheduling using crop evapotranspiration data as opposed to the irrigation practices followed by the farmers.

3.2.4 Fertilization and irrigation of potato

Potato yields are strongly affected by water stressed conditions. On average, the yield losses compared to the optimum conditions of water supply vary from 24.3% to 35.4% for moderate and severe water stress regimes, respectively. On average, no differences between the two nitrogen rates tested were observed. Statistics indicate a yield of 36.2 tons ha^{-1} and 35.5 tons ha^{-1} for the nitrogen rates of 120 kg and 250 kg N ha^{-1}, respectively. The water stress affected strongly the NUE. The importance of this negative effect increases with the severity of the water stress. On average, the results of NUE were 264.09 kg ha^{-1}, 217.42 kg ha^{-1}, and 178.68 kg ha^{-1} for the best, the moderate and the severe stress conditions, respectively. However, the NUE of the nitrogen rate of the 250 kg N ha^{-1} treatment was lower than the NUE obtained when applying the

120 kg N ha^{-1} treatment. On average, the results of NUE for 250 kg N ha^{-1} and 120 kg N ha^{-1} treatments were 144.68 kg ha^{-1} and 295.44 kg ha^{-1}, respectively. Hence, for the same yield of potato tubers, the NUE under the nitrogen rate of 120 kg N ha^{-1} was higher than the NUE under the 250 kg N ha^{-1}.

Similar results were obtained in the framework of an experiment implemented in the same site using the isotopic nitrogen technique. The results showed that (although with a statistically weak significance ($p < 0.05$), the % NUE decreases with an increase in nitrogen fertilizer application rates. The NUE was 33.7%, 17.4%, 25.22%, and 20.4%, for the fertilizer rates of 120 kg N ha^{-1}, 170 kg N ha^{-1}, 250 kg N ha^{-1}, and 292 kg N ha^{-1}, respectively. The results showed that the most efficient nitrogen rate with the best fresh yields was that of 120 kg N ha^{-1}. These results showed that beyond the rate of 120 kg N ha^{-1} in the soil conditions of the pilot farm, the yield was not affected by the additional nitrogen (Semiani, 2022).

3.3 Greenhouse case study

3.3.1 Experimental design

Two different farming scenarios were tested using a commercial greenhouse facility called 'Kalliantzis Farm' located in Pirgetos, Larissa, in northern-central Greece (Lat. 39°55′12″N, Lon. 22°35′27″E, Alt. 117 m) where cucumber plants were grown hydroponically, and in the Greenhouse Park at the University of Thessaly facilities in the area of Velestino, Greece (Lat 39°44′, Long. 22°79′, Alt. 85 m), where tomato plants were grown in soilless substrate to assess the effectiveness of the PRECIMED DSS regarding plant growth and productivity. Plants were irrigated either through a standard nutrient solution (NS) comprising the conventional treatment (CT) or through the NS provided by the PRECIMED DSS comprising the PRECIMED treatment (PT). In contrast to the CT, the NS corresponding to the PT was produced once per week based on the nutrient uptake forecast provided by the model. Both experiments were performed in separate greenhouse compartments with a ground area of 240 m^2. Each compartment consisted of six hydroponic channels (three channels/treatment) above which perlite slabs were placed and used as the growing medium of the plants. Three of these non-adjacent hydroponic lines were used as the control treatment, comprising three replicates per treatment, where the CT were applied, whilst the remaining lines of the greenhouse compartments were used to assess the efficiency of the PRECIMED treatment. During both periods, 10 destructive samplings were performed where the plants were then cut, separated into leaves, stems, and fruits, and used for DMP determination and nutrient uptake analysis.

3.3.2 Deployment and integration of the DSS system with the PRECIMED IoT platform

Many different IoT devices were deployed in the Greek greenhouse pilot such as air temperature, relative humidity, and solar radiation as well as irrigation and drainage volume sensors. In addition, using data recorded during the experimental season, variables such as VPD, crop transpiration, water/nutrient solution supplied, water/nutrient solution drained, pH, and EC were estimated daily. These devices are integrated in the PRECIMED IoT platform, where data is retrieved from a data base deployed in the Greek pilot cloud, where data from devices is registered together with other processed relevant information including calculations using the sensors raw data (i.e. aggregations, average values, etc.). This data is then registered in the IoT platform for further exploitation, as done by the UTH DSS model. A user interface in the PRECIMED platform is also available to configure and run the UTH-PRECIMED DSS model. This model uses historical input data from four different IoT devices deployed in the greenhouse, including (i) air temperature, (ii) solar radiation, (iii) irrigation water volume, and (iv) drainage water volume. The data from these sensors was processed and registered for exploitation in a database at the pilot infrastructure side. This data was then retrieved from the pilot side by the IoT PRECIMED platform to be registered in its database. Finally, the DSS module retrieves the historical processed data from the four sensors via the IoT platform to estimate the nutrient recommendations needed for a given time period. The model also needs to be parametrized before performing the calculations. This is done by setting information related to the case study site. The DSS model then produces a weekly recommendation to the end-user comprising a nutrient recipe with the following nutrient concentrations (mmol L^{-1}): NO_3^-, $H_2PO_4^{3-}$, K^+, Ca^{2+}, Mg^{2+}, SO_4^{2-}.

The DSS model is integrated with the IoT platform via REST APIs. The DSS REST API allows the platform (working as a client) to interact with it, at the same time that the IoT platform REST API for data exploitation allows the DSS model (when works as a client) to retrieve the needed sensors' data to feed the model. The data exchanged is modelled using standard data models (i.e. EU FIWARE). Moreover, the DSS results are shown in the user frontend in the PRECIMED platform. The data exchanged by the REST APIs to integrate the DSS model is represented using the EU FIWARE NGSI/NGSI-LD data-models format.

3.3.3 Validation of the UTH-PRECIMED DSS prototype

The DSS prototype was initially operated using MS Excel based on the simulation model developed from a previous project deliverable (*D3.2: Integrated models of irrigation and fertilization for open-field crops and greenhouses*). Afterwards, it was provided to ODIN for the development of the model based on DSS aiming to predict crop nutrient and water needs and provide the end-user with a useful nutrient recipe to fertigate a soilless cucumber and tomato cultivation in the Mediterranean region. The software calculates the daily crop needs in K, Ca, P, and N and the daily increment of the DMP considering the water needs of the crop. The web-based platform where the PRECIMED DSS is hosted was tested and validated using all algorithms provided from UTH to ODIN in a two-period experiment with cucumber plants. To date, the DSS has been also tested and validated in a two-period experiment conducted with tomato plants. In general, the validation of the DSS prototype in real-case scenarios in the Mediterranean region, and in particular in Greece, under soilless greenhouse crops performed well.

As shown in Table 2, application of the PRECIMED DSS in greenhouse crops showed significant benefits in terms of productivity, water and NUE, and for the environment. In particular, the use of such DSS in greenhouse soilless crops resulted in water reduction by 10% with only a 9% reduction in cucumber productivity. Moreover, the DSS resulted in a significant reduction

Table 2 Key performance indicators achieved during the validation of the PRECIMED DSS in the greenhouse real-case farming scenarios using tomato and cucumber soilless crops

Key performance indicator	Decrease
Water	9%
Production improvement	9%
Fertilizer	24–36%
Reduction of nitrogen contamination (underground water)	33%
Global warming	25%
Stratospheric ozone depletion	35%
Ionizing radiation	25%
Ozone formation, impacts on human health	28%
Fine particulate matter formation	19%
Ozone formation, impacts on terrestrial ecosystems	28%
Terrestrial acidification	24%
Freshwater eutrophication	11%
Marine eutrophication	35%
Terrestrial ecotoxicity	31%
Freshwater ecotoxicity	33%
Marine ecotoxicity	31%
Human carcinogenic toxicity	16%
Human non-carcinogenic toxicity	34%
Land use	36%
Mineral resource scarcity	35%
Fossil resource scarcity	18%
Water consumption	30%

in the ozone formation, impacts on human health or terrestrial ecosystems, global warming, ionizing radiation, terrestrial acidification, fine particulate matter, fossil resource scarcity formation, human carcinogenic toxicity, and freshwater eutrophication by 10–30%. In addition, a significant reduction by 30–36% was observed in terms of fertilizer use in greenhouse crops, land use, stratospheric ozone depletion, marine eutrophication, mineral resource scarcity, human non-carcinogenic toxicity, reduction of nitrogen contamination (underground water), freshwater, terrestrial, and marine ecotoxicity, as well as water consumption.

4 Conclusion and future trends

The PRECIMED IoT DSS platform, based on open standards (like EU FIWARE) to promote its interoperability and integrate different agronomic services as well as the developed agronomic DSS with their HMI interfaces, has been developed and deployed for the four agricultural scenarios in the frame of the PRECIMED project.

The PRECIMED DSS supports better resource planning for suppliers of fertilizers and irrigation communities and offers a timely and guaranteed delivery of agricultural products, due to a greater predictability of yields (both in quantity and quality) that benefit food producers and the productivity of their farm. In addition, the PRECIMED DSS has an important impact in mitigating climate change due to its effects expected in reducing the use of water and fertilizers. Therefore, overall, the PRECIMED DSS provides a valuable contribution to society, providing tools to improve the quality and productivity of agriculture in areas affected by climate change (arid and semi-arid areas).

Regarding the development of the PRECIMED DSS, the technology used in the IoT hardware solutions should evolve in different ways, for example by integrating new communication technologies for scenarios where others are not available (i.e. like satellite communication), updating hardware designs to continuously optimizing energy consumption; configuring IoT devices with enough computing power to support embedded AI algorithms for Edge solutions (i.e. supporting critical controls in greenhouses), and by incorporating security technologies (i.e. MQTT/TSL and Blockchain) to secure data. The IoT DSS platform will have to evolve too, not only by integrating the new services that the clients may need and demand but also others that could make it more valuable. However, the platform is ready to be tested by the first customers, having now a technology readiness level of TRL7 with a status of PMV so that after the end of the project it could start its commercialization using a business model based on the cloud SaaS software distribution model (i.e. Freemium, pay-as-you-go, subscription, or tiered fixed fee).

To date, data-driven predictive models for irrigation recommendations, both in open-field and greenhouse, developed by OPTIM have been distributed via REST API to integrate into the PRECIMED platform. However, they have not been validated in the field, so there is a commitment to do so, both in open-field and greenhouse crops in the coming years. This fact and the partners' interest in improving the PRECIMED platform to validate the PRECIMED DSS in more crops and in other regions will keep the consortium collaborating once the project is completed.

5 Where to look for further information

The PRECIMED website provides a good overview of the project:

- https://precimed-prima.org/.
- https://www.facebook.com/precimed.prima/.
- https://www.linkedin.com/company/precimed-prima/.

For commercialized soil and climate sensors, further information is available at:

- Drill and Drop probe (https://sentektechnologies.com/).
- Imetos weather station (https://metos.at/en/contact/).

Key research in this area can be found in the following organizations:

- ODIN Solutions S.L. (https://www.odins.es).
- University of Thessaly (https://www.facebook.com/uth.lacec, https://www.uth.gr/).
- Centro de Edafología y Biología Aplicada del Segura (CEBAS-CSIC) (https://www.cebas.csic.es).
- University of Sfax. Miracl. (http://www.miracl.rnu.tn/).
- National Institute of Agronomic Research of Algeria (https://www.inraa.dz/).

6 Acknowledgements

The work has been carried out in the frame of the PRECIMED project that is funded by the AEI of Spain (PCI 2019-103608), the GSRT of Greece (155331/I4/19.09.18), CDTI of Spain (IDI-20190230), DGRSDT of Algeria (332-24/10/2019), and MESRS of Tunisia under the PRIMA Programme. PRIMA is an Art.185 initiative supported and co-funded under Horizon 2020, the European Union's Programme for Research and Innovation.

7 References

Adeyemi, O., Grove, I., Peets, S., Domun, Y. and Norton, T. (2018). Dynamic neural network modelling of soil moisture content for predictive irrigation scheduling. *Sensors* 18(10), 3408.

Ali, E., Cramer, W., Carnicer, J., Georgopoulou, E., Hilmi, N. J. M., Le Cozannet, G. and Lionello, P. (2022). Mediterranean region. In: Pörtner, H.-O., Roberts, D. C., Tignor, M., Poloczanska, E. S., Mintenbeck, K., Alegría, A., Craig, M., Langsdorf, S., Löschke, S., Möller, V., Okem, A. and Rama, B. (Eds). *Climate Change*: Impacts, Adaptation and Vulnerability. Contribution of Working Group II to the Sixth Assessment Report of the Intergovernmental Panel on Climate Change. Cambridge University Press, Cambridge, UK and New York, pp. 2233–2272.

Allen, R. G., Pereira, L. S., Raes, D. and Smith, M. (1998). Crop evapotranspiration: guidelines for computing crop water requirements. FAO Irrigation and Drainage Paper 56, Food and Agriculture Organisation, Rome, Italy.

Al-Sakran, H. O. (2015). Intelligent traffic information system based on integration of Internet of Things and Agent technology. *Int. J. Adv. Comput. Sci. Appl.* 6(2), 37–43.

Bamory, A. T. K., Grati, R., Bouaziz, B. and Boukadi, K. (2023). A new long short-term memory-based approach for soil moisture prediction. *J. Ambient Intell. Smart Environ.*, Pre-press, 15(3), 255–268.

Bañón, S., Ochoa, J., Bañon, D., Ortuño, M. F. and Sánchez-Blanco, M. J. (2020). Assessment of the combined effect of temperature and salinity on the outputs of soil dielectric sensors in coconut fiber. *Sustainability* 12(16), 6577. doi:10.3390/su12166577.

Ben Abdallah, E., Grati, R. and Boukadi, K. (2023). Towards an explainable irrigation scheduling approach by predicting soil moisture and evapotranspiration via multi-target regression. *J. Ambient Intell. Smart Environ.* 15(1), 89–110.

Ben Abdallah, E., Grati, R., Fredj, M. and Boukadi, K. (2022a). A machine learning approach for a robust irrigation prediction via regression and feature selection. In: Barolli, L., Hussain, F. and Enokido, T. (Eds). *Advanced Information Networking and Applications. Lecture Notes in Networks and Systems*, Vol. 449. Springer, Cham, pp. 491–502.

Ben Abdallah, E., Grati, R. and Boukadi, K. (2022b). A machine learning-based approach for smart agriculture via stacking-based ensemble learning and feature selection methods. In: International Conference on Intelligent Environments (IE), Biarritz, France, pp. 1–8.

Conejero, W., Mellisho, C. D., Ortuño, M. F., Moriana, A., Moreno, F. and Torrecillas, A. (2011). Using trunk diameter sensors for regulated deficit irrigation scheduling in early maturing peach trees. *Environ. Exp. Bot.* 71, 409–415.

Conesa, M. R., Conejero, W., Vera, J. and Ruiz-Sánchez, M. C. (2021). Soil-based automated irrigation for a nectarine orchard in two water availability scenarios. *Irrig. Sci.* 39(4), 421–439.

Costa, J. M., Ortuño, M. F. and Chaves, M. M. (2007). Deficit irrigation as a strategy to save water: physiology and potential application to horticulture. *J. Integr. Plant Biol.* 49(10), 1421–1434.

Cravero, A. and Sepúlveda, S. (2021). Use and adaptations of machine learning in big data—applications in real cases in agriculture. *Electronics* 10(5), 552.

Cristescu, G. A. (2015). Development of a context management platform for the Internet of Things based on OMA NGSI-9 and NGSI-10 standards. Bachelor's Degree Final Project, University of Cantabria, Spain.

Daccache, A., Ciurana, J. S., Rodríguez Díaz, J. A. and Knox, J. W. (2014). Water and energy footprint of irrigated agriculture in the Mediterranean region. *Environ. Res. Lett.* 9(124014).

Du, T., Kang, S., Zhang, J. and Davies, W. J. (2015). Deficit irrigation and sustainable water-resource strategies in agriculture for China's food security. *J. Exp. Bot.* 66(8), 2253–2269.

El Chami, A., Cortignani, R., Dell'Unto, D., Mariotti, R., Santelli, P., Ruggeri, R., Colla, G. and Cardarelli, M. (2023). Optimization of applied irrigation water for high marketable yield, fruit quality and economic benefits of processing tomato using a low-cost wireless sensor. *Horticulturae* 9(3), 390.

Fernández, M. D., González, A. M., Carreño, J., Pérez, C. and Bonachela, S. (2007). Analysis of on-farm irrigation performance in Mediterranean greenhouses. *Agric. Water Manag.* 89(3), 251–260.

Gallardo, M., Elia, A. and Thompson, R. B. (2020). Decision support systems and models for aiding irrigation and nutrient management of vegetable crops. *Agric. Water Manag.* 240(3), 106209.

Gallardo, M., Fernández, M. D., Giménez, C., Padilla, F. M. and Thompson, R. B. (2016). Revised VegSyst model to calculate dry matter production, critical N uptake and ET_c of several vegetable species grown in Mediterranean greenhouses. *Agric. Syst.* 146, 30–43.

Ge, J., Zhao, L., Yu, Z., Liu, H., Zhang, L., Gong, X. and Sun, H. (2022). Prediction of greenhouse tomato crop evapotranspiration using XGBoost machine learning model. *Plants* 11(15), 1923.

Gebbers, R. and Adamchuk, V. I. (2010). Precision agriculture and food security. *Science* 327(5967), 828–831.

Giménez, C., Gallardo, M., Martínez-Gaitán, C., Stöckle, C. O., Thompson, R. B. and Granados, M. R. (2013). VegSyst, a simulation model of daily crop growth, nitrogen uptake, and evapotranspiration for pepper crops for use in an on-farm decision support system. *Irrig. Sci.* 31(3), 465–477.

Gómez-Bellot, M. J., Nortes, P. A., Sánchez-Blanco, M. J. and Ortuño, M. F. (2015). Sensitivity of thermal imaging and infrared thermometry to detect water status changes in *Euonymus japonica* plants irrigated with saline reclaimed water. *Biosyst. Eng.* 133, 21–32.

Haller, S. (2010). The things in the internet of things. Internet of Things Conference, Tokyo, Japan, pp. 26–30.

Hargreaves, G. H. and Samani, Z. A. (1985). Reference crop evapotranspiration from temperature. *Appl. Eng. Agric.* 1(2), 96–99.

Hunkeler, U., Truong, H. L. and Stanford-Clark, A. (2008). MQTT-S – a publish/subscribe protocol for Wireless Sensor Networks. In: Proceedings of the 3rd International Conference on Communication Systems Software and Middleware and Workshops (COMSWARE'08), Bangalore, India. IEEE, New York, pp. 791–798.

Ihuoma, S. O., Madramootoo, C. A. and Kalacska, M. (2021). Integration of satellite imagery and *in situ* soil moisture data for estimating irrigation water requirements. *Int. J. Appl. Earth Obs. Geoinf.* 102, 102396.

Intrigliolo, D. S., Bonet, L., Nortes, P. A., Puerto, H., Nicolás, E. and Bartual, J. (2013). Pomegranate trees performance under sustained and regulated deficit irrigation. *Irrig. Sci.* 31(5), 959–970.

Katsoulas, N., Bartzanas, T. and Kittas, C. (2017). Online professional irrigation scheduling system for greenhouse crops. *Acta Hortic.* 1154, 221–228.

Katsoulas, N. and Stanghellini, C. (2019). Modelling crop transpiration in greenhouses: different models for different applications. *Agronomy* 9(7), 392.

Li, L., Chen, S., Yang, C., Meng, F. and Sigrimis, N. (2020). Prediction of plant transpiration from environmental parameters and relative leaf area index using the random forest regression algorithm. *J. Clean. Prod.* 261, 121136.

Marini, A., Termite, L. F., Garinei, A., Marconi, M. and Biondi, L. (2020). Neural network models for soil moisture forecasting from remote sensed measurements. *Acta Imeko* 9(2), 59–65.

Miranda, J., Ponce, P., Molina, A. and Wright, P. (2019). Sensing, smart and sustainable technologies for agri-food 4.0. *Comput. Ind.* 108, 21–36.

Navarro-Hellín, H., Martínez-del-Rincon, J., Domingo-Miguel, R., Soto-Valles, F. and Torres-Sánchez, R. (2016). A decision support system for managing irrigation in agriculture. *Comput. Electron. Agric.* 124, 121–131.

NEEA (2015). *Agricultural Irrigation Initiative: The Future of Agricultural Irrigation*. Document prepared by Marshall English. Northwest Energy Efficiency Alliance (NEEA), Portland.

Nikolaou, G., Neocleous, D., Christou, A., Kitta, E. and Katsoulas, N. (2020). Implementing sustainable irrigation in water-scarce regions under the impact of climate change. *Agronomy* 10(8), 1120.

Org, J. S. O. N. (2009). Introducing JSON. Available at: https://www.json.org/json-en.html.

Pardossi, A., Carmassi, G., Diara, C., Incrocci, L., Maggini, R. and Massa, D. (2011). *Fertigation and Substrate Management in Closed Soilless Culture, Project Efficient Use of Imputs in Protected Horticulture, KBBE-2007-1-2-04*. Dipartimento di Biologia delle Piante Agrarie, Università di Pisa, Italy, pp. 1–63.

Rallo, G., Mario, M. and Provenzano, G. (2011). Comparison of SWAP and FAO agro-hydrological models to schedule irrigation of wine grapes. *J. Irrig. Drain. Eng. ASCE* 138(7), 581–591.

Ramírez-Cuesta, J. M., Ortuño, M. F., Gonzalez-Dugo, V., Zarco-Tejada, P. J., Parra, M., Rubio-Asensio, J. S. and Intrigliolo, D. S. (2022). Assessment of peach trees water status and leaf gas exchange using on-the-ground versus airborne-based thermal imagery. *Agric. Water Manag.* 267, 107628.

Rose, D. C. and Chilvers, J. (2018). Agriculture 4.0: broadening responsible innovation in an era of smart farming. *Front. Sustain. Food Syst.* 2, 87.

Saggi, M. K. and Jain, S. (2022). A survey towards decision support system on smart irrigation scheduling using machine learning approaches. *Arch. Comput. Methods Eng.* 29(6), 4455–4478.

Sanuade, O. A., Hassan, A. M., Akanji, A. O., Olaojo, A. A., Oladunjoye, M. A. and Abdulraheem, A. (2020). New empirical equation to estimate the soil moisture content based on thermal properties using machine learning techniques. *Arab. J. Geosci.* 13(10), 377.

Sciarretta, A., Tabilio, M. R., Amore, A., Colacci, M., Miranda, M. Á., Nestel, D., Papadopoulos, N. T. and Trematerra, P. (2019). Defining and evaluating a decision support system (dss) for the precise pest management of the Mediterranean fruit fly, *Ceratitis capitata*, at the farm level. *Agronomy* 9(10), 608.

Semiani, M. (2022). Evaluation of nitrogen fertilizers use efficiency for potato (*Solanum tuberosum*), using the 15N technique. International Symposium on Managing Land and Water for Climate Smart Agriculture, Vienna, Austria, p. 91.

Shelby, Z., Hartke, K. and Bormann, C. (2014). The constrained application protocol (CoAP). *IETF RFC* 7252, 1–112.

Song, X., Zhang, G., Liu, F., Li, D., Zhao, Y. and Yang, J. (2016). Modeling spatio-temporal distribution of soil moisture by deep learning-based cellular automata model. *J. Arid Land* 8(5), 734–748.

Song, Y., Lin, J., Tang, M. and Dong, S. (2017). An Internet of energy things based on wireless LPWAN. *Engineering* 3(4), 460–466.

Tabari, H. (2010). Evaluation of reference crop evapotranspiration equations in various climates. *Water Resour. Manag.* 24(10), 2311–2337.

Tei, F., De Neve, S., de Haan, J. and Kristensen, H. L. (2020). Nitrogen management of vegetable crops. *Agric. Water Manag.* 240, 106316.

Tejero, I. G., Zuazo, V. H. D., Bocanegra, J. A. J. and Fernández, J. L. M. (2011). Improved water-use efficiency by deficit-irrigation programmes: implications for saving water in citrus orchards. *Sci. Hortic.* 128(3), 274–282.

Thompson, R. B., Martínez-Gaitan, C., Gallardo, M., Giménez, C. and Fernández, M. D. (2007). Identification of irrigation and N management practices that contribute to nitrate leaching loss from an intensive vegetable production system by use of a comprehensive survey. *Agric. Water Manag.* 89(3), 261–274.

Tzounis, A., Katsoulas, N., Bartzanas, T. and Kittas, C. (2017). Internet of Things in agriculture, recent advances and future challenges. *Biosyst. Eng.* 164, 31–48.

Valenzano, V., Parente, A., Serio, F. and Santamaria, P. (2008). Effect of growing system and cultivar on yield and water-use efficiency of greenhouse-grown tomato. *J. Hortic. Sci. Biotechnol.* 83(1), 71–75.

Vladisavljević, R., Soleša, D. and Puvača, N. (2019). Internet of Things (IoT) in a function of smart agriculture development. *J. Agron. Technol. Eng. Manag.* 2(4), 287–294.

Wani, S. P., Sreedevi, T. K., Rockstrom, J. and Ramakrishna, Y. S. (2009). Rainfed agriculture: past trends and future prospectives. In: Wani, S. P., Rockstrom, J. and Oweis, T. (Eds). *Rainfed Agriculture: Unlocking the Potential*. CAB International, Wallingford, pp. 1–35.

Wolf, W. (2009), Cyber-physical systems. *Computer* 42(3), 88–89.

WWAP (2015). United Nations World Water Assessment Programme. The United Nations World Water Development Report 2015: Water for a Sustainable World. UNESCO, Paris.

Xu, J., Gu, B. and Tian, G. (2022). Review of agricultural IoT technology. *Artif. Intell. Agric.* 6, 10–22.

Yang, C. (2018). High resolution satellite imaging sensors for precision agriculture. *Front. Agr. Sci. Eng.* 5(4), 393–405.

Zamora-Izquierdo, M. A., Santa, J., Martínez, J. A., Martínez, V. and Skarmeta, A. F. (2019). Smart farming IoT platform based on edge and cloud computing. *Biosyst. Eng.* 177, 4–17.

Zhang, F., Zhang, Y., Lu, W., Gao, Y., Gong, Y. and Cao, J. (2022). 6g-enabled smart agriculture: a review and prospect. *Electronics* 11(18), 2845.

Chapter 14

Integrating biophysical and hydraulic models to assess the agronomic and environmental impacts of precision irrigation

A. Daccache, University of California, USA; and J. W. Knox and T. M. Hess, Cranfield University, UK*

1 Introduction

1.1 Role of irrigation for crop production and food security

Over the last five decades, major changes in agricultural land management have underpinned increases in crop productivity. Yield gains have arisen mainly from so-called Green Revolution technologies, including the introduction of high-yielding cultivars, and dependence on chemical fertilisers, pesticides, farm mechanisation and, importantly, irrigation (Foley et al., 2005). As a consequence, there have been significant increases in global fertiliser use and the total areas irrigated. Although development has slowed recently, irrigated croplands have increased from 100 M ha in 1950 to 367 M ha in 2018 (Lal, 2009; Siebert et al., 2013; Meier et al., 2018) and are reported to be twice as productive as non-irrigated or rainfed cropland. With more than a third of the world's food is produced on irrigated lands (Goodwin and O'Connell, 2008),

* Corresponding author: adaccache@ucdavis.edu.

http://dx.doi.org/10.19103/AS.2023.0123.23

global food supply, in terms of quantity, quality and diversity, is sustained by irrigation water. However, meeting future food demands will be increasingly challenging, given the rising population with changing dietary needs and competing demands for freshwater resources, especially in arid and semi-arid regions, where irrigated production is concentrated.

Globally, irrigation accounts for >70% total freshwater withdrawals and >90% of total consumptive water use (Siebert et al., 2010) but allocations of water for agriculture and the need to increase irrigation efficiency are under increased scrutiny, driven by public and regulatory concerns regarding the consequences of over-abstraction on the environment and water resources sustainability (Knox et al., 2012; Lankford et al., 2020). Rising competition for water resources and demands for greater environmental protection means that irrigated production is under severe pressure to improve water use efficiency and reduce its environmental footprint. The major yield benefits of irrigation over the last few decades have often been overshadowed by the adverse effects on natural resources, threatening its long-term productive potential. For example, land degradation and large-scale salinisation of irrigated areas are often linked to irrigation development and numerous examples in the scientific literature describe the negative impacts of unsustainable abstraction for irrigation on the aquatic environment (e.g. Tilman, 1999).

Many measures to increase crop productivity and water use efficiency in irrigated agriculture have therefore been proposed (e.g. Fereres et al., 2011) mainly in response to food security concerns. Management improvements and implementation of new technologies are reported to offer the greatest potential (e.g. Sadler et al., 2005). However, achieving the sustainable intensification of agriculture represents a major industrial and scientific challenge (Beddington, 2010). Agricultural intensification will require farmers to become more specialised and for many, investment in irrigation and innovative technologies will inevitably provide the basis for increasing profitability. Rising production costs, coupled with consumer demands for quality produce, mean that farmers and growers will also need to consider the impacts of irrigation practices on resource use (including land, energy, water resources and fertilisers) and crop production. Crop assurance schemes are also promoting advanced irrigation technologies, requiring growers to manage and audit their irrigation practices and to link them to traceability and accountability in support of public health and environmental regulation (Monaghan et al., 2013).

1.2 Defining 'precision irrigation' in agricultural modelling

Despite widespread international use of the term in the agri-food industry, precision irrigation (PI), as a scientific concept, is still relatively ambiguous in the research community (Smith and Baillie, 2009). This is evident through the remarkably limited number of articles in the scientific literature explicitly focusing on 'precision irrigation'. In the agri-food industry, PI is widely and loosely used to refer to micro (drip or trickle) irrigation. Other definitions refer to accurate irrigation management where the amount of water applied matches crop water demand without taking into consideration the spatial in-field variability. In this chapter, PI refers to applying precise amounts of water to the crop at precise times (to match crop water demand) and at precise locations (taking into consideration field spatial variability). The term is also used by large irrigation systems manufacturers (e.g. centre pivot and linear moves) and those providing technologies for monitoring *in situ* soil moisture or plant water status using UAV's, airborne and satellite remote sensing. There is, however, no commonly accepted scientific definition or conceptualisation of the term, and no framework to guide research or inform integrated model development. In order to embed PI into agricultural systems modelling, it is necessary to have a clear definition of its unique attributes.

From an agricultural systems perspective, the technological aspects related to (i) mapping field spatial variability, (ii) monitor soil and crop water status and (iii) irrigation control with variable rate application have seen rapid development over the last decade. Regardless of this technological advancement, a number of questions related to management complexity and economic justification still represent a hurdle for the wider uptake and adoption of PI technologies. The challenge is that PI benefits are very site and crop specific (Daccache et al., 2015) and therefore a rigorous modelling framework coupling irrigation system performance under variable rate irrigation (VRI) with biophysical yield prediction models is needed for precise irrigation management and importantly for economic and environmental quantification of the benefits. Whilst some of these modelling aspects have been evaluated (e.g. González Perea et al., 2017), the transferability of integrated PI modelling approaches between different crop sectors, production methods and irrigation application systems is still a major challenge.

The conventional management of an irrigation system generally aims to apply water in a 'uniform' manner, recognising agroclimatic, soil and engineering constraints and to minimise the negative impacts that any irrigation non-uniformity might have on crop growth. In contrast, PI involves the 'differential' application of water. PI has been defined in terms of applying water in the right place with the right amount (Al-Karadsheh et al., 2002), the accurate and precise application of water to meet specific requirements of individual plants or management units to minimize adverse environmental impact (Raine et al., 2007), or the application of water to a given site and timing to support optimum crop production, profitability or some other management objective (Camp et al., 1998). In this chapter, we refer to a review by Smith and Baillie (2009) who defined PI as a 'concept' rather than a specific technique and one that differs substantially from current practice. They described the PI concept as being holistic, adaptive and applicable to any method of irrigation, including surface, overhead or micro, rather than being linked to a specific application method.

1.3 Conceptualising precision irrigation for agricultural modelling

Despite these differences in interpretation, there is one common aspect to all PI definitions, which is the 'differential' component – both in terms of recognising the need for variable irrigation need across a field and then delivering the variable application of irrigation (Smith et al., 2010). PI thus focuses on small quasi-homogenous areas within a field, while conventional irrigation management assumes a whole-field approach aiming to apply water as uniformly as possible across the field. Not surprisingly, most PI definitions and studies in the scientific literature have focussed on VRI, now commonly used on pressurised overhead systems (including centre pivot or linear move systems) which apply water variably in response to feedback from spatial information delineating areas reflecting different stages of crop development and/or drought stress, soil type, topography or land use (e.g. El Nahry et al., 2004). However, it is important to recognise that VRI technology represents only one aspect of PI and there are other ways to manage and implement precision irrigation.

The concept of PI proposed by Smith and Baillie (2009) also includes the concept of adaptive control, implying that an irrigation system can be managed to achieve a multiple set of objectives, whether it be maximising water use efficiency, maximising crop yield and quality or farm profitability. This requires a modelling approach that can cope with large multiple datasets on water application, possibly at high resolution, simulate crop responses at a scale that reflects the approaches for field management, and then be used to evaluate decision-making strategies to improve farm and field-scale irrigation management. For example, through model simulation, spatial changes in soil moisture and crop response could be modelled and validated against *in situ* or direct field measurement, with the data then used as feedback to determine the timing and spatial amounts needed for irrigation (Smith et al., 2010). From this, it is evident there are a number of issues that are fundamental to the need to spatially model precision irrigation:

1. The requirement for water will vary spatially due to spatial differences in soil, crop development and micro-climate;
2. The scale of spatial variability may be very small in heterogeneous fields;

3 *In situ* soil moisture monitoring is necessarily limited by cost and practicality to a very few points within a field, so, there is a need to model water distribution to be able to estimate what is happening at a sufficiently fine spatial resolution;
4 Financially optimum water application depends on final crop yield and quality which has to be extrapolated from the current condition; and
5 Water distribution cannot be controlled precisely. Current overhead irrigation systems cannot apply water differentially to a high degree of accuracy because they rely on overlapped wetting patterns to deliver high uniformity. Distribution from overhead systems is also affected by variable wind and operating pressure, and there is a need to optimise equipment performance and energy use in conjunction with water application.

Finally, it is worth briefly highlighting why PI has been largely ignored within the modelling domain of precision agriculture. In a comprehensive review by Zhang et al. (2002), the spatial and temporal effects of 'irrigation pattern' on crop production were lumped within 'management variables' along with tillage, fertiliser management, seed rates and crop rotation. This is likely due to the spatial and temporal complexity that exists with irrigation and the confounding effects that spatial and temporal variabilities in soils, agroclimate and nutrient management can have on irrigation management and hence crop yield. The lack of mathematical models capable of simulating the spatial distribution of irrigation water (without extensive model calibration) and the costs of integrated model development are inevitably contributory factors. Indeed, the lack of such a Decision Support System (DSS) was identified by Smith et al. (2010) as being the most likely factor that has delayed significant commercial application of precision irrigation. More recently, there have been attempts to develop DSS for precision irrigation but their uptake by farmers has generally been limited.

2 Biophysical and ballistics modelling

2.1 Crop growth models

With careful calibration, crop simulation models provide reasonable estimates of potential yield for a given situation based on the wide range of physiological relationships that govern plant growth and development (Cassman, 2008). The development and application of crop growth models for agricultural decision support is a mature and well-established research domain. A relatively limited number of such models now dominate agricultural applications research, with the most widely used being the Decision Support System for Agrotechnology

Transfer (DSSAT) program (Jones et al., 2003). The diversity in application for DSSAT and other such crop models is undoubtedly due to the significant international interest in climate change impacts on crop production (White et al., 2011). Such models can help simulate the integrating effects of soils, weather, crop management, genetics and pests on plant growth and can be used to gain some (albeit still limited) insight into spatial land and crop management variability (Thorp et al., 2008). The majority of crop models were, however, developed for use in simulating crop growth and development for a single, homogenous field unit, and lack operational flexibility for use in evaluating precision agriculture, where multiple units or management zones with a single field need to be simulated. Of course, this can be overcome by running such models in batch mode (Barros et al., 2022) defining each different management unit as an individual point, but this is cumbersome and inelegant from a modelling perspective, and later constrains any attempt to import data into a GIS for spatial analysis.

Thorp et al. (2008) highlighted the limitations of such models and the need for additional procedures to simulate crop growth spatially across management zones at the field-scale. This is especially pertinent given the increasing availability of high resolution and often frequent time-step data emerging from the widespread uptake of information technologies and remote sensing in agriculture (Thenkabail et al., 2012). Their solution - a prototype precision farming DSS termed Apollo - represents a major development in functionality for crop simulation in precision agriculture. Although Apollo has been used for evaluating precision nitrogen management strategies (Thorp et al., 2006) and to some extent spatially variable irrigation (DeJonge et al., 2007), the approaches for dealing with spatial heterogeneity in water application are still limited. This may not necessarily be a problem when modelling regional level impacts but does represent a major limitation at the field scale. Other integrated models suffer similar constraints, particularly in evaluating sub field-level irrigation wetted patterns and their spatial impacts on crop development.

2.2 Irrigation water distribution models

Mapping water application using sprinkler irrigation systems is performed using two main modelling methods; firstly, empirical or semi-empirical approaches, involving extrapolation from measured single water distribution pattern data for various wind conditions (speed and direction) for a defined irrigation sprinkler nozzle, pressure and trajectory angle, and, secondly, physically described models, which apply ballistic theory to calculate the flight trajectories of individual water droplets, taking into account their size, velocity and atmospheric effects on the break up of the jet (Smith et al., 2008). Richards and Weatherhead (1993) developed a semi-empirical model to simulate water distribution pattern of large sprinklers (rainguns). The model was later refined by Al-Naeem (1993) to predict the effect of wind speed and direction on the water application pattern of mobile overhead irrigation system (hosereel with a travelling raingun, as used widely in northern Europe and Australia). For these large sprinklers/rainguns, the jet of water entrains a surrounding airflow that carries the water further than a simple ballistic approach (described later) would predict. The water travels in 'packets', with the leading droplets slowing due to air resistance and colliding with the following droplets which are carried partly by the entrained wind. Cross winds disrupt the entrained wind, shortening the throw. These interactions are not yet susceptible to deterministic modelling and therefore the model uses a complex series of algorithms and empirical parameters to convert a measured 'no-wind' pattern into a 'wind-distorted' pattern. Field data for model calibration include a full-circle pattern or radial leg under still (zero wind) conditions, and two full-circle wind distorted patterns for different wind conditions (Smith et al., 2008).

Other researchers started to focus on developing ballistics models for overhead application systems that could be modified and used for precision irrigation. These include fixed (set) sprinkler systems, mobile hosereels fitted with booms and centre pivots (Molle and Gat, 2000), systems which are all widely used in field-scale agriculture across the world.

There is a quite separate set of research where physics laws are applied at the droplet level and then extrapolated to sprinkler and to system level. The ballistic model used to simulate the effects of wind (speed and direction) on sprinkler irrigation performance was first developed by Fukui et al. (1980). This was then developed and improved by researchers including Von Bernuth and Gilley (1984), Vories et al. (1987), Von Bernuth (1988), Seginer et al. (1991), Han et al. (1994) and Tarjuelo et al. (1994). SIRIAS model developed by Carriòn et al. (2001) probably represents the most complete computational software for simulating the trajectory of water droplets discharged by a single sprinkler. The SIRIAS model was calibrated for several manufactured sprinkler types and is designed to simulate the wetted patterns under different wind conditions and nozzle configurations (Montero et al., 2001; Playan et al., 2006). The ballistic theory assumes that droplets are formed at the sprinkler nozzle and travel independently (droplets do not fuse or disintegrate) until they reach the soil or crop canopy. They also assume that each droplet can be modelled independently, ignoring any entrained wind effects and droplet collisions and break-up during flight. In the absence of wind, the trajectory of each droplet is subjected to the initial velocity of the droplet at the nozzle, a gravity force in the vertical direction and a resistance force, which opposes the relative movement of the droplet in the air (Vories et al., 1987; Seginer et al., 1991). Under windy conditions, the droplet velocity with respect to the ground is equal to the droplet velocity with respect to the air plus the wind vector.

The water distribution pattern for a single sprinkler operating under 'no wind' conditions and obtained using this procedure is circular. However, in reality, wind distortion consists of a narrowing of the water profile in the perpendicular direction to the wind and shortening windward and lengthening of the profile in the leeward direction (von Bernuth and Seginer, 1990). To achieve this deformation, Tarjuelo et al. (1994) introduced two correction coefficients to the air drag coefficient. These coefficients are fundamental to achieving a good fit between simulated sprinkler profiles and field (observed) measurements. However, the main limitation of the SIRIAS model is that it simulates the performance of only a single sprinkler under a well-defined operating (pressure and wind) conditions. It does not provide scope for analysing the performance of an entire sprinkler irrigation system operating under dynamic conditions where both the operating pressure and ambient wind conditions (wind speed and direction) are continuously changing, which reflects actual field practice. The SIRIAS model also works independently as a standalone ballistic model and cannot be used in conjunction with a crop growth model to assess the impacts of irrigation heterogeneity on crop yield.

Despite the significant research effort made in developing such models over the last two decades, there is a surprising lack of published studies describing their application for assessing irrigation heterogeneity to impacts on crop production. Ruelle et al. (2003) integrated a semi-empirical modelling approach with the STICS crop model to assess the impacts of irrigation heterogeneity on maize yield and nitrate leaching risks in shallow soils in France. In Australia, Smith et al. (2008) developed a DSS for evaluating the performance of overhead irrigation in the sugarcane and horticultural sectors using a crop model and the Richards and Weatherhead (1993) approach for simulating water application. Whilst both considered the effects of irrigation heterogeneity on crop yield at the field-scale they focused on simulating irrigation performance for equipment (a hosereel fitted with a raingun) that could not easily be managed according to principles of precision irrigation.

3 Conceptual framework for modelling overhead precision irrigation impacts on crop yield

As part of a broader research programme investigating the potential for implementing precision irrigation techniques into UK outdoor field-scale horticulture, the authors have developed a framework for integrating hydraulic and physical based ballistic irrigation simulation with a biophysical yield prediction model to evaluate the impacts of irrigation heterogeneity on crop growth and yield. . The approach allows comparison of conventional irrigation management practices versus alternative precision irrigation strategies and provides a framework for evaluating the agronomic (yield), water resource (irrigation use and water efficiency), energy (use, cost, footprint) and environmental (nitrate leaching, drainage) impacts for a range of different equipment and management scenarios. It also provides the potential for including feedback loops into defining the management zones and required application depths for the next scheduled irrigation. The approach has been developed for supplemental irrigation using a mobile hosereel fitted with a boom on high-value outdoor field-scale crops in a temperate climate, but it could readily be adopted for use on other crop production systems and agroclimatic conditions.

The framework consists of two components, or modules, including (i) an irrigation system module and (ii) a crop yield module. By 'irrigation system', we refer to the combined engineering, hydraulic and management components that are necessary to apply water via an overhead irrigation system. A brief description of each, including the input datasets and modelling architecture for data integration, together with sample output, is given below.

3.1 Irrigation system module

The irrigation system module combines data and information on field management units, the engineering hydraulics and characteristics of the irrigation system and ambient short-term changes in local weather conditions to dynamically simulate the performance of the irrigation system as it moves along the field, generating an overlapped wetted pattern. The model was developed using Microsoft VB.Net and a MapWindow plug-in (Ames, 2007) in order to incorporate a number of spatial datasets and to harness the GIS mapping functionality. The module contains eight discrete components which are combined to model and map the spatial distribution of water applied onto a crop but affected dynamically by spatial and temporal changes in wind (speed and direction), the irrigation system's operating pressure (as this influences flow rate and hence discharge) and field topography (as this influences operating pressure and wetted patterns).

The individual components and their inter-dependencies are shown in Fig. 1 and described below.

1. *Sprinkler data*: this includes experimental data relating to the measured water distribution patterns for an individual sprinkler. Data is also required on the height of the sprinkler from the ground (to estimate wetted area), the hydraulic operating characteristics of the sprinkler (design pressure to estimate droplet ballistics) and adjustment coefficients (for ballistic model calibration).
2. *Irrigation system*: this includes information and data relating to the hydraulic and engineering characteristics of the equipment (hosereel fitted with boom) and how it operates under normal conditions,

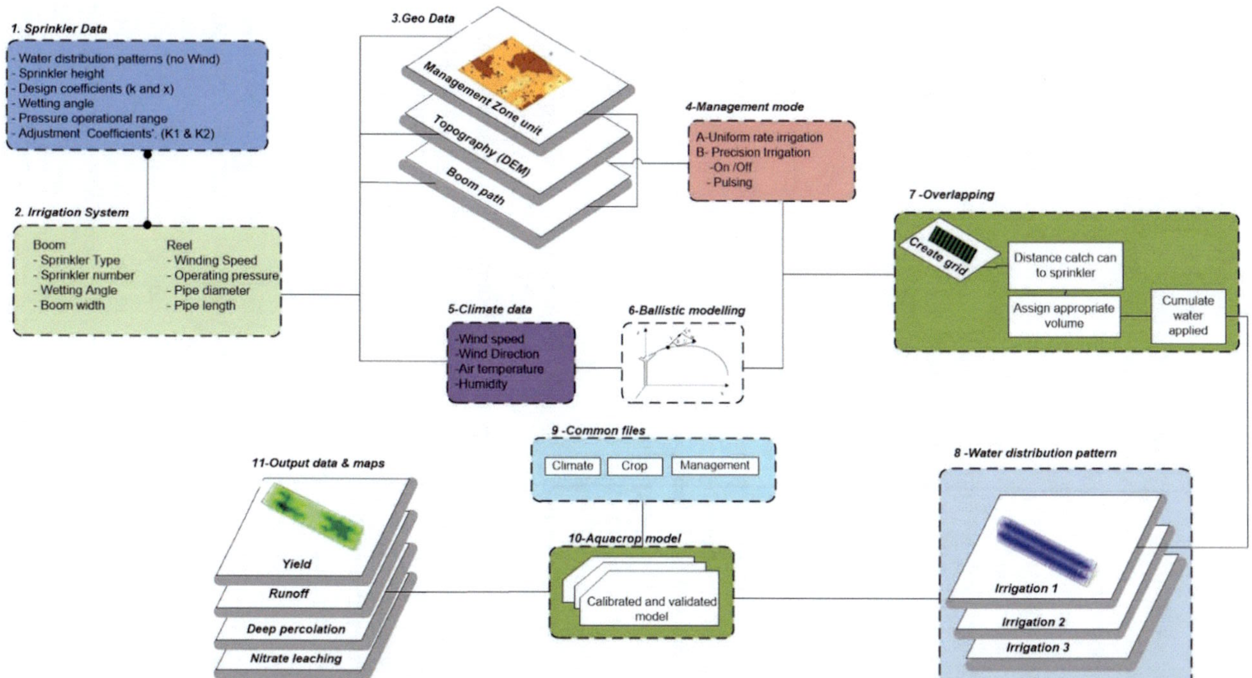

Figure 1 Schematic showing an integrated approach for combining a ballistic irrigation system model with a crop model (Aquacrop) to assess the agronomic and environmental impacts of precision irrigation.

including the number of installed sprinklers and their type, the wetted width for boom (instantaneous wetted area), the operating pressure and travelling speed details on the reel system hydraulics (pipe diameter and length to calculate pressure head and energy losses).

3. *Geodata*: information describing the configuration of the field in terms of topography (usually from a digital elevation model, DEM), as well as spatial datasets relating to in-field soil variability) presented as a map (shapefile) of homogeneous management zones (HMZs). The HMZs need to be large enough to be practically and economically manageable and independent from other field attributes (Dennis and Nell, 2002). The most common techniques used to delineate management zones are based on either (i) historical yield monitoring, (ii) real-time crop water stress sensing (i.e. satellite, aerial or UAV remote sensing) or (iii) soil mapping using different techniques (i.e. systematic soil sampling, EMI, electrical resistivity, gamma ray or microwave radiometry) (Fig. 2) In some agricultural systems, the importance of sequential planting dates and varietal boundaries for commercial cropping (e.g. salads)

Figure 2 An example of different methods used to map in-field soil spatial variability at high spatial resolution. From top left and clockwise: Electro-magnetic induction (AEMP 14), electrical resistivity (Veris U series), microwave (L-band) dual polarisation (TerraRad Tech) and on-the go soil spectroscopy.

is also an important issue for defining management zones (e.g. Hedley and Yule, 2009). The importance of field topography is that changes in elevation influence the operating pressure and hence the system capacity to provide the correct pressure at the sprinklers, with consequent impacts on the application rate, uniformity, soil erosion and runoff.

4. *Irrigation management*: Criteria for defining conventional management (the schedules for defining irrigation amounts, timing and frequency) or for precision irrigation (including rules for system management, sprinkler pulsing, switching on/off sprinklers in different management zones or changes in the travelling speed for continuously moving overhead systems).
5. *Climate data*: Hgh frequency local climate data for wind speed and direction (usually on a 15 min interval), temperature and humidity affecting wind drift and evaporation.
6. *Ballistics modelling*: Application of ballistic theory to simulate irrigation water distribution under initially 'no wind' and then with changing wind conditions (speed and direction) to generate gridded datasets showing wetted patterns.
7. *Overlapping* involves algorithms to generate patterns of wetted sprinkler overlap for each irrigation event, taking into account the direction and speed on the irrigation machine moving along the field, the effects of changing wind on water application rates, topography and operating pressure conditions on water distribution.
8. *Water distribution pattern*: Derived gridded datasets for the field showing the resulting spatial heterogeneity in irrigation water application and reflecting spatial and temporal changes in sprinkler operation (uniform or variable rate application), weather (mainly wind speed and direction) and hydraulic (Pressure changes) conditions.

An example of the typical output generated by the irrigation performance model is shown in Fig. 3, where the simulated water distribution along a field is shown, under both 'no wind' and typical wind conditions. The ballistic model was then used to simulate the effects of changes in wind speed and direction on the simulated 'no-wind' wetted distribution pattern. The operational mode of each sprinkler is defined to match the irrigation needs in each management zone to reflect differences in soil type, available moisture and crop development. Using an overlapping algorithm, data and map outputs are generated to show the spatial distribution of water applied during a precision irrigation event as affected by wind, pressure and topography and in response to decision rules regarding specific management zones. An example is shown in Fig. 4 where the irrigation uniformity under a system of conventional irrigation management is

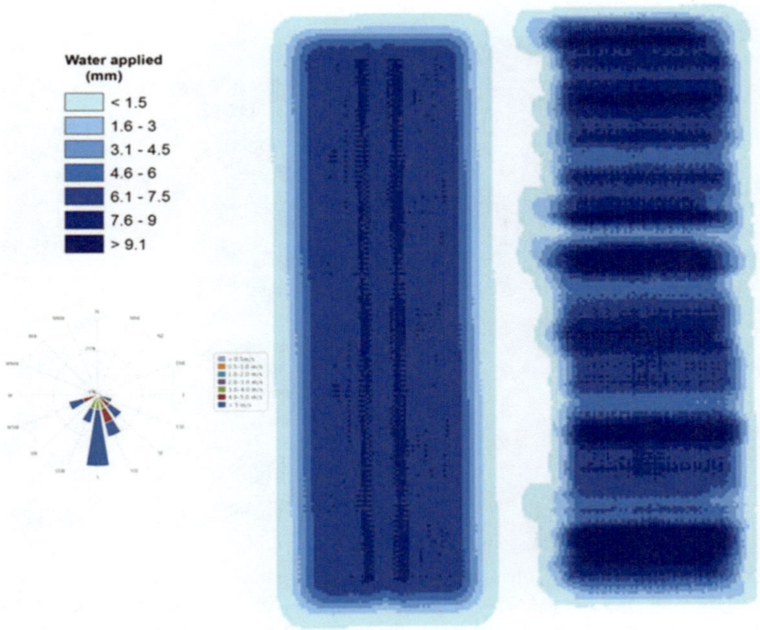

Figure 3 Simulated water distribution pattern from an overhead irrigation system (hosereel fitted with a boom) operating under (a) no wind conditions and (b) real-time wind conditions (wind rose highlights the wind speed frequency and direction).

compared against a schedule for precision irrigation, where application depths are modified according to management zones. The output data from these model runs were then exported to the crop module.

3.2 Crop yield module

Previous integrated models to assess spatial crop yield impacts have used the DSSAT crop modelling program (e.g. Thorp et al., 2010) and STICS model (Ruelle et al., 2003). In this study, the FAO Aquacrop model (Raes et al., 2009; Steduto et al., 2009) was used as it is particularly suited to address production conditions where water is a key limiting factor. The daily soil water balance model is combined with growing degree days (GDD) to predict plant growth stages. Plant development depends on temperature, with biomass accumulation depending on plant transpiration. The FAO Aquacrop model has been used extensively for research purposes, for example, it has been used to predict wheat yield response to different water regimes (Andarzian et al., 2011; Salemi et al., 2011), rapeseed yield (Zeleke et al., 2011) as well as assessing maize, sugar beet and sunflower production under contrasting agroclimatic conditions

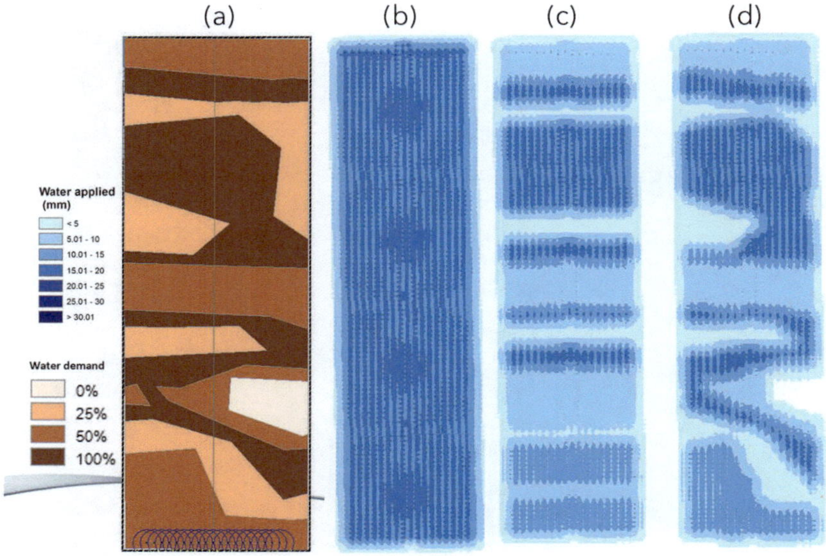

Figure 4 Simulated model output for an overhead irrigation system (hosereel fitted with a boom) showing the soil management zones (a), the water applied under conventional irrigation management (b), variable pulling in speed (c) and precision irrigation with individually controlled sprinklers in response to defined management zones (d).

(Stricevic et al., 2011). In this study, the FAO Aquacrop model was calibrated for an onion crop and then validated against experimental and field data.

Rather than modelling a single point for yield assessment, in this work the crop model has been linked to the irrigation system module to spatially simulate crop growth at a sub-field scale. To do this, the field area was first divided into a gridded mesh based on geodata (soil, topography management zones). It is important to set the grid mesh resolution to a scale that is appropriate for the modelling assessment. For example, modelling of small sprinklers is typically at a 1-m grid scale, whereas Richards and Weatherhead (1993) initially used an 8 m grid for collecting raingun pattern data. Spatially gridded data on soil characteristics and the volume of water applied during each irrigation event were then used to simulate crop growth and development. The Aquacrop model can be run in batch mode to simulate crop growth and yield, runoff, deep percolation and crop evapotranspiration. The outputs are generated as a grid file that can then be analysed statistically or imported into a GIS for spatial analysis. This approach combining the recent mathematical advances in ballistics modelling coupled with biophysical simulation within a framework for precision irrigation assessment on crop yield represents important progress in integrating irrigation ballistics with crop modelling.

4 Case study – assessing impacts of precision irrigation on onions in a humid climate

To demonstrate the potential of integrating hydraulic and biophysical crop growth models as a tool to assess the economic and ecological benefits of adopting precision irrigation and VRI techniques, we selected a site located in 'the Fens' in Eastern England. Originally known as 'Marchlands', the Fens were drained centuries ago to create highly fertile agricultural lands that are still supported by a system of drainage channels (dykes and drains) with automated pumping stations to control in-field water levels. As such, the soil is highly rich in organic content and characterised by black silts and clay deposition making the site one of the most agriculturally fertile regions in the UK but also one of the most variable. The field was cropped with onion (*cv* Arthur) and irrigated with an overhead boom irrigation system, as described below.

4.1 Soil spatial variability

To obtain an insight of the spatial distribution of soil properties such as texture, soil moisture content and salinity, we used a non-invasive EMI (electro-magnetic induction) scanning technology. The Geonics EM38 is characterised by a fixed distance (1 m) between the receiver coil and the transmitter and by a single operational frequency (14.5 KHz). Thus, providing an apparent electrical conductivity (ECa) measurement from an effective depth around 1.5 m when positioned in the vertical dipole orientation and around 0.75 m when in horizontal dipole orientation (Fig. 5a). To convert the ECa map into its constituent hydraulic properties, we collected and analysed 50 soil samples for their texture, organic and moisture content. Accordingly, three homogeneous zones were delineated based on texture (Fig. 5b) with water holding capacities of 80, 120 and 150 mm/m for a loamy sand, sandy loam and clay soil, respectively.

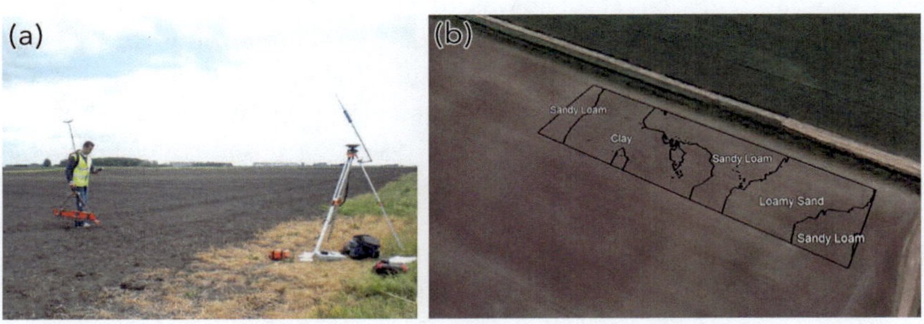

Figure 5 Soil scanning using the dual EM38 system in horizontal dipole orientation connected to an RTK GPS recorder (a), and the field study site showing the delineated irrigation management zones based on soil texture (b).

4.2 Biophysical crop modelling

In this study, the AquaCrop model was used to simulate onion yield response to uniform and VRI taking the hydraulic characteristics of the irrigation systems uses. The model was designed to help farmers, researchers and policymakers make informed decisions about crop production, irrigation management and water resource planning. A detailed description of AquaCrop model can be found in Raes et al. (2009) and Steduto et al. (2009). The FAO Aquacrop model was calibrated and validated using experimental data obtained from a set of irrigation trials on onion (*cv* Arthur) conducted in a polytunnel environment at Broom's Barn Research Centre (latitude 52.61°N; longitude 0.56°E; altitude 75 m asl), Suffolk. A detailed description of the trials is given in Lacey and Ober (2011, 2012), and for the model calibration and validation, readers are referred to the work reported by Perez-Ortola et al. (2015).

4.3 Hydraulic modelling

Mobile overhead boom irrigation systems are popular for use on field-scale vegetables and high-value crops cultivated under both humid and semi-arid climates on flat or undulating terrain. This efficient way of irrigating large fields offers multiple benefits in terms of water conservation (ability to apply low application rates), consistency in crop yields, equipment cost (lower energy demands than higher pressurised raingun systems), practical flexibility and labour savings. The main component of the boom system is a large mechanical arm (boom) fitted with sprinklers which provide even discharge and coverage across the entire wetted span to provide high application uniformity under low wind conditions (Fig. 6). The boom is typically mounted on a four-wheeled structure and pulled by a winding hosereel powered by mainly a hydraulic pump. For this study, the boom system was assumed to have seven sprinklers (Nelson S3000) on a 3.75 m spacing with the boom pulled by a hosereel at a fixed wind in speed of 30 m/h.

A GIS-based mapping routine was then developed in MapWindow and VB .net to simulate the operation of the boom system under both variable wind conditions and changes in operating pressure. The model used a combination of hydraulic (to simulate pressure changes along the system and between sprinklers) and ballistic (to simulate wind distortion of the sprinklers wetted profiles) modules to generate water distribution patterns for each irrigation event. A detailed description of the model is presented in González Perea et al. (2017). The hydraulic characteristics of the sprinklers (Nelson S3000) including the relationship between pressure (H) and discharge (Q), the individual sprinkler water profiles at different operating pressures (tested indoors under 'no wind' conditions), the engineering design of the boom system (pipe size, length of the boom and distance between sprinklers), the wind in speed and

Figure 6 Boom irrigation being used on salad crops in Eastern England.

the environmental conditions (i.e. wind speed and direction, temperature and relative humidity) were all used as input variables into the model to create an irrigation application profile along the entire field. The model was also designed to adjust the amount of water applied based on the delineated management zones (Fig. 5b) imported as an input file in GIS shapefile format. Water application adjustment can be achieved by adjusting the wind in speed of the hosereel or by controlling the operation of each individual sprinkler (on/off mode or pulsing at different rates) (Fig. 4c,d).

The boom system in this study was evaluated under both 'no wind' condition and assuming uniform rate irrigation (URI). Accordingly, the average volume of water applied (assuming a 50% adequately irrigated area) under the uniform irrigation was 19 mm per irrigation event (at 30 m/h) with a coefficient of uniformity and distribution uniformity equivalent to 89% and 81%, respectively (Fig. 7).

4.4 Irrigation management

Onion irrigation in Eastern England typically starts 50 days after planting with irrigation applied using the boom system (as described above) with a schedule that assumed a fixed application of 18 mm at a fixed soil moisture deficit of 20 mm. The 2 mm deficit left after each irrigation event is recommended practice in humid climate to accommodate for any summer rainfall storage and to reduce runoff or deep percolation losses. Two types of irrigation mode were simulated and compared, including (1) a URI where all the sprinklers operate at full capacity simultaneously (Fig. 4b) and (2) a VRI where each sprinkler has

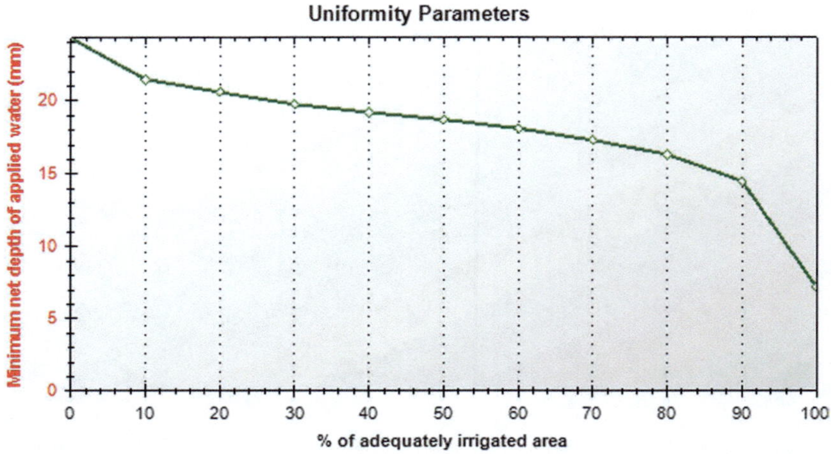

Figure 7 Relationship between the minimum depth of water applied (mm) and the proportion (%) of the area adequately irrigated.

the ability to be operated independently and at different pulsing rates (Fig. 4d) to respond to the demand of each management zone (Fig. 4a). In both cases, irrigation is triggered whenever the soil water deficit in the driest part of the field (corresponding to the loamy sand zone (Fig.5b)) reaches a deficit of 20 mm.

4.5 Integration of hydraulic and crop models

The hydraulic/ballistic modelling described previously aims to represent the spatial distribution of water application across the field under either a URI or VRI scenario. The Aquacrop model can be used to simulate plant growth and yield response to water. Therefore, there is a need to integrate both these models to estimate crop productivity, irrigation need and any losses taking into account the spatial variability of field soil characteristics and irrigation water non-uniformity. This was achieved using the batch version of the Aquacrop model operating at the pixel level whereby each pixel included the relevant soil characteristics, the amounts and timing of irrigation applied and the changing plant stress level. A schematic representation of the integrated hydraulic/ballistic irrigation and biophysical crop (Aquacrop) modelling framework used in this case study to assess the agronomic and environmental impacts of precision irrigation is shown in Fig. 1.

4.6 Results

Based on the biophysical modelling and local agroclimate conditions, eight irrigation events were required to satisfy the onion crop water demand in the driest management zone in the field (loamy sand zone Fig. 5b). For each irrigation event, the ballistic/hydraulic model generates an irrigation application map that was then used as a spatially distributed input dataset for the biophysical crop model. From this latter a set of spatially distributed variables (maps) for the each of the main Aquacrop model outputs (i.e. yield, run off, drainage, transpiration and drainage) were generated (Fig. 8). The effects of wind on the sprinkler water distribution patterns were eliminated by assuming zero wind speed conditions, in order to avoid any bias being introduced into the comparative analysis between the URI and VRI scenarios.

In this case study, the adoption of a precision irrigation approach based on VRI in response to spatial soil variability would result in 12.3% water savings and a 1.5% reduction in yield compared to conventional uniform rate application (URI). In terms of water use efficiency, the 6.5 t of dry matter (DM)/m^3 for the URI scenario increases to 7.1 t DM/m^3 under the VRI scenario which corresponds

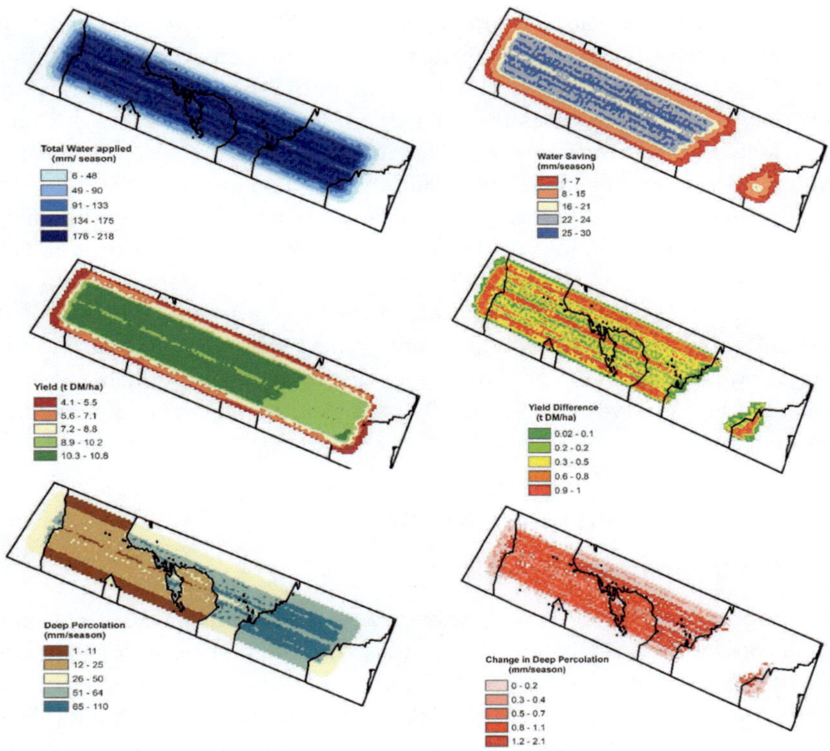

Figure 8 Summary mapping outputs from the integrated ballistic-biophysical modelling. Left panel shows the spatial distribution in total amount of water applied (mm/season), yield (t DM/ha), deep percolation losses (mm/season) for URI. Right panel shows the potential water savings (mm/season), yield differences (t DM/ha) and changes in deep percolation when the VRI precision irrigation approach is compared to the URI.

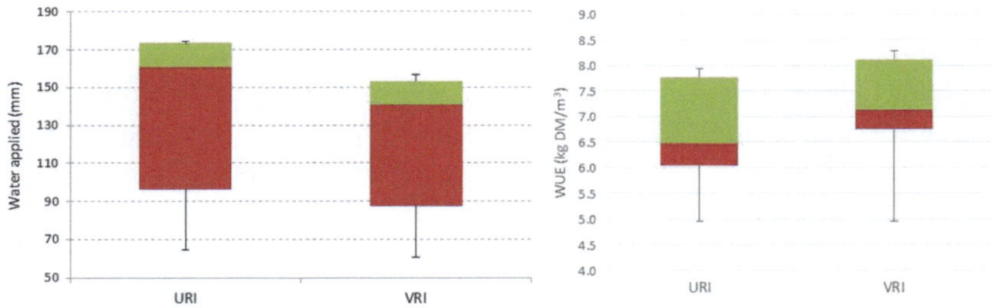

Figure 9 Comparison of the total water applied (mm/season), and water use efficiency (WUE) (kg DM/m^3) between the conventional uniform rate irrigation (URI) and variable rate irrigation (VRI) approach.

to a approximately 10% improvement in WUE (Fig. 9). There is clear evidence that VRI can deliver noticeable water saving with limited losses in yield. From the maps presented in Fig. 8, we can notice that under URI, both crop yield and deep percolation are far from uniform with relatively low yield and high infiltration losses on the coarser textured soils compared to the fine textured ones. On those relatively heavy soils, the benefits of PI in terms of water savings and deep percolation losses (hence nutrient leaching potential) are clearly more relevant. However, recommendations for VRI adoption must be evaluated economically to ensure it delivers a sufficient return on investment.

By extracting all the values for each pixel from the field yield and irrigation maps (Fig. 8), it is possible to develop a yield response curve to irrigation for all the different soil types (Fig. 10). Whilst this yield response curve is site specific to the local pedo-climatic conditions present in the case study site, such an approach would allow irrigation managers to optimize their water use taking into account in-field heterogeneity, rather than simply maximising production which does not necessarily represent the most economic and/or environmentally sustainable crop production system.

5 Conclusion and future trends

5.1 Model application

The approach described here provides a modelling environment and framework for application across a variety of agricultural science disciplines. In addition to assessing the impacts of irrigation heterogeneity on crop yield, the integrated modelling approach could be used for (i) conducting a Monte Carlo type simulation of the impacts of in-field soil variability on yield, (ii) simulating yield impacts of conventional irrigation management versus precision irrigation under a range of contrasting equipment and management scenarios, (iii) conducting investigations to assess different field management strategies to minimise nitrate leaching risks and deep drainage under supplemental

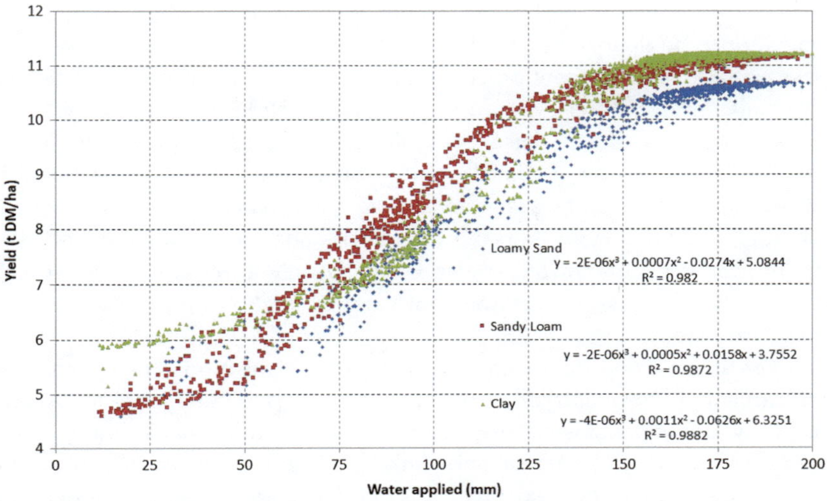

Figure 10 Yield (t DM/ha) response to water applied (mm) curve for the three contrasting soil types in the case study.

irrigation, (iv) evaluating management strategies to make more effective use of rainfall (free irrigation), and (v) assessing the impacts of conventional and precision irrigation management strategies on water and energy footprints, with consequences for reducing greenhouse gas emissions. By incorporating projections of future climate change, spatial differences in yield between rain-fed and irrigated production could be evaluated, as well assessing impacts on field-scale irrigation demand, including changes in the timing of peak (seasonal) demand and consequences for water and energy management.

5.2 Methodological limitations

The approach does, however, have several methodological limitations which need to be recognised, and there remains scope for further refinement in the methodological approach. These challenges include issues such as geodata availability, defining the appropriate scale of model application, ensuring transferability for use with other crop models, and integrating these approaches with current modelling developments in precision agriculture.

Adam et al. (2010) identified the challenges in coupling different crop models within an integrated modelling framework, both for modellers and end users; many of those issues are also pertinent here. Identifying and selecting the most appropriate scale for conducting the irrigation and crop modelling at the sub-field level is a major issue and needs to be strongly guided by the intended purpose and application of the output data. The scale of enquiry will also be influenced by the availability, format and resolution of the geodata. The rationale and delineation of management zones or units within a single field also needs to be defined carefully so that the minimum size suits and does not constrain farm management practices. For example, in the ballistic modelling, the simulation of different wetted patterns for individual management zones needs to consider the practical feasibility of physically applying the water, otherwise there is a risk of spatially modelling at a finer scale than is possible using equipment under the assumed farming practices. However, it is important to model at a finer scale than the intended management intervention. The current modelling framework is also currently only applicable for overhead (spray) irrigation systems and does not cater for micro (drip) or surface (e.g. furrow) irrigation.

There is also a need for careful documentation of the modelling approaches and particularly how datasets are pre-processed prior to model input, and then how derived datasets are passed between individual models. Great care therefore has to be taken when linking models, as errors in one can be easily propagated to the next. Errors may become exacerbated or attenuated through model integration. There is hence a risk of introducing additional modelling uncertainty, particularly where datasets of different provenance, scale and integrity are integrated. The use of an uncertainty matrix should be used to identify sources of uncertainty both within the irrigation ballistics and crop modelling components, and then used to inform the interpretation of the crop modelling outputs.

Though several techniques are now available for mapping soil variability, converting measured parameters (i.e. electrical conductivity, soil resistivity, surface brightness, reflectance) to meaningful soil physical values (i.e. water holding capacity, moisture content) requires detailed soil sampling for validation and careful calibration and validation. In addition, advanced geospatial techniques will be needed for accurate point measurement interpolation and homogeneous soil delineation. As such, a good level of soil mapping and geospatial analysis expertise is needed and may not always be available.

Finally, the effects of wind (although explicitly ignored in this study) on the distribution pattern of sprinkler system still represents a major challenge when irrigating based on well-defined homogeneous zones. Water droplets drift and evaporation may compromise the performance of the VRI system and in severe conditions may fail to apply the right depth of water in the right place. In this study, we assumed water was the only limiting factor affecting crop yield. In reality, crop production will also be affected by pests and disease risks, plant genetics, seeding/planting density, soil topography and germination rates, thereby creating additional sources of spatial variability. In some cases, homogeneous management zones may need to be dynamic (revised to account for spatial crop variability) rather than fixed and limited to only the local soil properties.

5.3 Future challenges and opportunities for precision irrigation uptake

The modelling framework described here is being used to assess the agronomic, economic and environmental impacts of different irrigation management strategies (conventional versus precision irrigation) for commercial field-scale vegetable production in the UK, and elsewhere internationally. The approach will provide a basis for transfer into other high-value horticultural crop sectors where there is an industry and scientific need to understand the impacts of irrigation heterogeneity on crop yield, but more importantly, crop quality, and to identify strategies that can be used to reduce the 'non-beneficial' water losses, to improve water (and energy) use efficiency and to reduce the environmental impacts associated with supplemental irrigation. Integrating biophysical and engineering models to advance our knowledge of these interactions will go some way to addressing these knowledge gaps.

Further work is needed to incorporate suitable farm economic models to assess the financial impacts of different irrigation management strategies on crop production (to inform discussions on the merits of precision irrigation), to evaluate the use of remote-sensed crop cover data for improving crop model calibration, to assess how rapid soil monitoring techniques such as EMI scanning could be used effectively to define soil and irrigation management zones and finally how integrated modelling frameworks such as the approach described in this chapter can be used to support training, extension and knowledge transfer to highlight the agronomic, water resource and environmental benefits of improved water management in food crop production.

6 Acknowledgement

The authors acknowledge UK Government (Defra) for funding this research (HL0196) through their Hortlink programme, the farm businesses (Elveden Estate, PDM Produce, Gs) and industry partners (Briggs Irrigation, Wroot Water, Adcon Telemetry, Plantsystems) who actively supported this research.

7 References

Adam, M., Corbeels, M., Leffelaar, P. A., Van Keulen, H., Wery, J. and Ewert, F. (2010). Building crop models within different crop modelling frameworks. *Agricultural Systems* 113: 57–63.

Al-Karadsheh, E., Sourell, H. and Krause, R. (2002). Precision irrigation: new strategy irrigation water management. Conference on International Agricultural Research for Development. In: Deutscher Tropentag 2002 Witzenhausen, 9–11 October 2002. Edited by Andreas Deininger - Kassl Univ. Press (2002).

Al-Naeem, M. (1993). A hosereel raingun irrigation system computer simulation to predict water distribution and crop yield: optimising trajectory angle, sector angle, sector position and lane spacing in different wind conditions. PhD thesis, Silsoe College, Cranfield University.

Ames, D. P. (2007). *Mapwin GIS Reference Manual*. Geospatial Software Lab, Idaho State University, Idaho Falls.

Andarzian, B., Bannayan, M., Steduto, P., Mazraeh, H., Barati, M. E., Barati, M. A. and Rahnama, A. (2011). Validation and testing of the AquaCrop model under full and deficit irrigated wheat production in Iran. *Agricultural Water Management* 100(1): 1–8.

Barros, V. D., Waltner, I., Minoarimanana, R. A., Halupka, G., Sándor, R., Kaldybayeva, D. and Gelybó, G. (2022). SpatialAquaCrop: an R package for raster-based implementation of the AquaCrop model. *Plants* 11(21): 2907.

Beddington, J. (2010). Food security: contributions from science to a new and greener revolution. *Philosophical Transactions of the Royal Society of London. Series B, Biological Sciences* 365(1537): 61–71.

Camp, C. R., Sadler, E. J., Evans, D. E., Usrey, L. J. and Omary, M. (1998). Modified centre pivot system for precision management of water and nutrients. *Applied Engineering in Agriculture* 14(1): 23–31.

Carrión, P., Tarjuelo, J. and Montero, J. (2001). SIRIAS: a simulation model for sprinkler irrigation. *Irrigation Science* 20: 73–84. https://doi.org/10.1007/s002710000031.

Cassman, K. G. (2008). Ecological intensification of cereal production systems: yield potential, soil quality, and precision agriculture. *Proceedings of the National Academy of Sciences of the United States of America* 96(11): 5952–5959.

Daccache, A., Knox, J. W., Weatherhead, E. K., Daneshkhah, A. and Hess, T. M. (2015). Implementing precision irrigation in a humid climate – recent experiences and on-going challenges. *Agricultural Water Management* 147: 135–143.

DeJonge, K. C., Kaleita, A. L. and Thorp, K. R. (2007). Simulating the effects of spatially variable irrigation on corn yields, costs, and revenue in Iowa. *Agricultural Water Management* 92(1–2): 99–109.

Dennis, H. J. and Nell, W. T. (2002). Precision irrigation in South Africa. In: The 13th International Farm Management Congress, Wageningen, The Netherlands, 7–12 July 2002. DOI: 10.22004/ag.econ.7023.

El Nahry, A. H., Ali, R. R. and El Baroudy, A. A. (2004). An approach for precision farming under pivot irrigation system using remote sensing and GIS techniques. *Agricultural Water Management* 98: 517–531.

Fereres, E., Orgaz, F. and Gonzalez-Dugo, V. (2011). Reflections on food security under water scarcity. *Journal of Experimental Botany* 62(12): 4079–4086.

Foley, J. A., Defries, R., Asner, G. P., Barford, C., Bonan, G., Carpenter, S. R., Chapin, F. S., Coe, M. T., Daily, G. C., Gibbs, H. K., Helkowski, J. H., Holloway, T., Howard, E. A., Kucharik, C. J., Monfreda, C., Patz, J. A., Prentice, I. C., Ramankutty, N. and Snyder, P. K. (2005). Global consequences of land use. *Science* 309(5734): 570–574.

Fukui, Y., Nakanishi, K. and Okamura, S. (1980). Computer evaluation of sprinkler irrigation uniformity. *Irrigation Science* 2(1): 23–32.

González Perea, R., Daccache, A., Rodríguez Díaz, J. A., Camacho Poyato, E. and Knox, J. W. (2017). Modelling impacts of precision irrigation on crop yield and in-field water management. *Precision Agriculture* 19(5): 1–16.

Goodwin, I. and O'Connell, M. G. (2008). The future of irrigated production horticulture – world and Australian perspective. *Acta Horticulturae* 792: 449–458.

Han, S., Evans, R. G. and Kroeger, M. W. (1994). Sprinkler distribution patterns in windy conditions. *Transactions of the American Society of Agricultural Engineers* 37(5): 1481–1489. https://doi.org/10.13031/2013.28231.

Hedley, C. B., Bradbury, S., Watson, E., Dalrymple, H. and Wright, J. (2011). Farm-scale trials of variable rate irrigation to assess the benefits of modifying existing sprinkler systems for precision application. *International Journal of Agricultural Management* 1(2): 51–55.

Hedley, C. B. and Yule, I. J. (2009). Soil water status mapping and two variable-rate irrigation scenarios. *Precision Agriculture* 10(4): 342–355.

Hedley, C. B., Yule, I. J., Eastwood, C. R., Shepherd, T. G. and Arnold, G. (2004). Rapid identification of soil textural and management zones using electromagnetic induction sensing of soils. *Australian Journal of Soil Research* 42(4): 389–400.

Jones, J. W., Hoogenboom, G., Porter, C. H., Boote, K. J., Batchelor, W. D., Hunt, L. A., Wilkens, P. W., Singh, U., Gijsman, A. J. and Ritchie, J. T. (2003). DSSAT cropping system model. *European Journal of Agronomy* 18(3–4): 235–265.

Knox, J. W., Kay, M. G. and Weatherhead, E. K. (2012). Water regulation, crop production and agricultural water management – understanding farmer perspectives on irrigation efficiency. *Agricultural Water Management* 108: 3–8.

Lacey, T. R. and Ober, E. (2011). Impact of irrigation practices on Rijnsburger bulb onion husbandry, quality and storability—II FV 362a. Annual Report, Year 1.

Lacey, T. R. and Ober, E. (2012). Impact of irrigation practices on Rijnsburger bulb onion husbandry, quality and storability—II FV 362a. Annual Report, Year 2.

Lal, R. (2009). Soils and world food security. *Soil and Tillage Research* 102(1): 1–4.

Lankford, B., Closas, A., Dalton, J., López Gunn, E., Hess, T., Knox, J. W., van der Kooij, S., Lautze, J., Molden, D., Orr, S., Pittock, J., Richter, B., Riddell, P. J., Scott, C. A., Venot, J., Vos, J. and Zwarteveen, M. (2020). A scale-based framework to understand the promises, pitfalls and paradoxes of irrigation efficiency to meet major water challenges. *Global Environmental Change*. https://doi.org/10.1016/j.gloenvcha.2020.102182.

McCarthy, A. C., Hancock, N. H. and Raine, S. R. (2010). VARIwise: a general-purpose adaptive control framework for spatially and temporally varied irrigation at sub-field level. *Computers and Electronics in Agriculture* 70(1): 117–128.

Meier, J., Zabel, F. and Mauser, W. (2018). A global approach to estimate irrigated areas – a comparison between different data and statistics. *Hydrology and Earth System Sciences* 22(2): 1119–1133.

Molle, B. and Gat, Y. (2000). Model of water application under pivot sprinkler. II: calibration and results. *Journal of Irrigation and Drainage Engineering* 126(6): 348–354.

Monaghan, J. M., Daccache, A., Vickers, L. H., Hess, T. M., Weatherhead, E. K., Grove, I. G. and Knox, J. W. (2013). More 'crop per drop' – constraints and opportunities for precision irrigation in European agriculture. *Journal of the Science of Food and Agriculture* 93(5): 977–980.

Montero, J., Tarjuelo, J. M. and Carrion, P. (2001). SIRIAS: a simulation model for sprinkler irrigation: II. Calibration and validation of the model. *Irrigation Science* 20(2): 85–98.

Perez-Ortola, M., Daccache, A., Hess, T. M. and Knox, J. W. (2015). Simulating impacts of irrigation heterogeneity on onion (*Allium cepa* L.) yield in a humid climate. *Irrigation Science* 33(1): 1–14.

Playan, E., Zapata, N., Faci, J. M., Tolosa, D., Lacueva, J. L., Pellegrin, J., Salavador, R., Sanchez, I. and Lafia, A. (2006). Assessing sprinkler irrigation uniformity using a ballistic simulation model. *Agricultural Water Management* 84(1-2): 89-100.

Raes, D., Steduto, P., Hsiao, T. C. and Fereres, E. (2009). AquaCrop–the FAO crop model to simulate yield response to water: II. Main algorithms and software description. *Agronomy Journal* 101(3): 438-447.

Richards, P. J. and Weatherhead, E. K. (1993). Prediction of raingun application patterns in windy conditions. *Journal of Agricultural Engineering Research* 54(4): 281-291.

Ruelle, P., Mailhol, J. C., Quinones, H. and Granier, J. (2003). Using NIWASAVE to simulate impacts of irrigation heterogeneity on yield and nitrate leaching when using a travelling rain gun system in a shallow soil context in Charente (France). *Agricultural Water Management* 63(1): 15-35.

Sadler, E. J., Evans, R. G., Buchleiter, G. W., King, B. A. and Camp, C. R. (2000). Design considerations for site specific irrigation. In:Evans, R. G., Benham, B. L., and Trooien, T. P. (Eds.), Proceedings of the 4th Decennial National Irrigation Symposium, pp.304-315.

Sadler, E. J., Evans, R. G., Stone, K. C. and Camp, C. R. (2005). Opportunities for conservation with precision irrigation. *Journal of Soil and Water Conservation* 60(6): 371-378.

Salemi, H., Soom, M. A. M., Lee, T. S., Mousavi, S. F., Ganji, A. and KamilYusoff, M. (2011). Application of AquaCrop model in deficit irrigation management of winter wheat in arid region. *African Journal of Agricultural Research* 6(10): 2204-2215.

Seginer, I., Kantz, D. and Nir, D. (1991). The distortion by wind of the distribution patterns of single sprinklers. *Agricultural Water Management* 19(4): 341-359.

Siebert, S., Burke, J., Faures, J. M., Frenken, K., Hoogeveen, J., Ḋoll, P. and Portmann, F. T. (2010). Groundwater use for irrigation - a global inventory. *Hydrological Earth Systems Sciences* 7: 3977-4021.

Siebert, S., Henrich, V., Frenken, K. and Burke, J. (2013). *Global Map of Irrigation Areas Version 5*. Rheinische Friedrich-Wilhelms-University, Bonn, Germany / Food and Agriculture Organization of the United Nations, Rome, Italy.

Smith, R. and Baillie, J. (2009). Defining precision irrigation: a new approach to irrigation management. In: Irrigation Australia 2009: Irrigation Australia Irrigation and Drainage Conference: Irrigation Today - Meeting the Challenge, Irrigation Australia Ltd, Swan Hill, Vic, Australia, 18-21. Swan Hill, Australia.

Smith, R. J. (1989). Uniformity of applications from lateral move irrigation machines. CIGR 11th International Congress on Agricultural Engineering, Dublin. In: Dodd, V. and Grace, P. (Eds.), *Agricultural Engineering, Volume 1: Land and Water Use*. AA Balkema, Rotterdam.

Smith, R. J., Baillie, J. N., McCarthy, A. C., Raine, S. R. and Baillie, C. P. (2010). *Review of Precision Irrigation Technologies and Their Application*. National Centre for Engineering in Agriculture. Publication 1003017/1. USQ, Toowoomba.

Smith, R. J., Gillies, M. H., Newell, G. and Foley, J. P. (2008). A decision support model for travelling gun irrigation machines. *Biosystems Engineering* 100(1): 126-136.

Steduto, P., Hsiao, T. C., Raes, D. and Fereres, E. (2009). AquaCrop–the FAO crop model to simulate yield response to water: I. Concepts and underlying principles. *Agronomy Journal* 101(3): 426-437.

Stricevic, R., Cosic, M., Djurovic, N., Pejic, B. and Maksimovic, L. (2011). Assessment of the FAO AquaCrop model in the simulation of rainfed and supplementally irrigated maize, sugar beet and sunflower. *Agricultural Water Management* 98(10): 1615–1621.

Tarjuelo, J. M., Carrion, P. and Valiente, M. (1994). Simulaciòn de la distribuciòn del riego por aspersiòn en condiciones de viento. *Investigación Agraria:* Producción y Protección *Vegetales* 9(2): 255–271.

Thenkabail, P. S., Knox, J. W., Ozdogan, M., Gumma, M. K., Congalton, R., Wu, Z., You, S., Milesi, C., Giri, C., Nagler, P., Finkral, A., Marshall, M. and Mariotto, I. (2012). Assessing future risks to agricultural productivity, water resources and food security: how can remote sensing help? *Photogrammetric Engineering and Remote Sensing* August 2012: 773–782.

Thorp, K. R, Batchelor, W. D., Paz, J. O., Steward, B. L. and Caragea P. C. (2006). Methodology to link production and environmental risks of precision nitrogen management strategies in corn. *Agricultural Systems* 89(2–3): 272–298. https://doi.org/10.1016/j.agsy.2005.09.005.

Thorp, K. R., DeJonge, K. C., Kaleita, A. L., Batchelor, W. D. and Paz, J. O. (2008). Methodology for the use of DSSAT models for precision agriculture decision support. *Computers and Electronics in Agriculture* 64(2): 276–285.

Tilman, D. (1999). Global environmental impacts of agricultural expansion: the need for sustainable and efficient practices. *Proceedings of the National Academy of Sciences of the United States of America* 96(11): 5995–6000.

Von Bernuth, R. D. (1988). Effect of trajectory angle on performance of sprinklers in wind. *Journal of Irrigation and Drainage Engineering* 114(4): 579–587. https://doi.org/10.1061/(ASCE)0733-9437(1988)114:4(579).

Von Bernuth, R. D. and Gilley, J. R. (1984). Sprinkler droplet size distribution estimation from single leg test data. *Transactions of the American Society of Agricultural Engineers* 27: 1435–1441. https://doi.org/10.13031/2013.32984.

Von Bernuth, R. D. and Seginer, I. (1990). Wind considerations in sprinkler system design. In: Elliot, R. L. (Ed.), Visions of the Future: Proceedings of the 3rd International Irrigation Symposium. ASAE, St. Joseph, MI, pp. 334–339.

Vories, E. D., Von Bernuth, R. D. and Mickelson, R. H. (1987). Simulating sprinkler performance in wind. *Journal of Irrigation and Drainage Engineering* 113(1): 119–130.

White, J. W., Hoogenboom, G., Kimball, B. A. and Wall, G. W. (2011). Methodologies for simulating impacts of climate change on crop production. *Field Crops Research* 124(3): 357–368.

Zeleke, K. T., Luckett, D. and Cowley, R. (2011). Calibration and testing of the FAO AquaCrop model for canola. *Agronomy Journal* 103(6): 1610–1618.

Zhang, N., Wang, M. and Wang, N. (2002). Precision agriculture – a worldwide overview. *Computers and Electronics in Agriculture* 36: 113–132.

Chapter 15

Managing the impacts of climate change, droughts and water scarcity on agriculture

Ray-Shyan Wu, National Central University, Taiwan; and Fiaz Hussain, PMAS-Arid Agriculture University Rawalpindi, Pakistan*

1 Introduction

Climate change is a long-term change or shift in average weather (temperatures and precipitation patterns) conditions. The world is getting warmer as the land and ocean average (1850–2020) annual temperature is 1.1°C higher than it was in the nineteenth century (IPCC, 2021). The observed temperature change is unequivocal. It is mainly due to greenhouse gas emissions from human-influenced anthropogenic activities. Climate change is widespread, rapid, and intensifying according to the Intergovernmental Panel on Climate Change (IPCC) 6th Assessment Report (AR6) (IPCC, 2021). In interpreting

* Corresponding author: raywu@ncu.edu.tw.

http://dx.doi.org/10.19103/AS.2023.0123.21

the findings from AR6, we are in a climate emergency, so there is a need to respond now in terms of climate resilience and not wait for the end of twenty-first century. Climate change is one of the most important non-traditional security challenges (Hussain et al., 2022). There is a wake-up call to act in a short timeframe to prepare for, recover from, and adapt to climate change impacts and to prevent or reduce its effects. The impacts of climate change are pervasive, unprecedented, and intensified in terms of floods, heatwaves, hurricanes, drought, wildfires, and glacial melting. It is recommended to take measures at national, regional, and as well as at local scales by calculating and predicting changes in average conditions of temperatures and weather patterns (Aktaş, 2020).

Climate change syndromes, i.e. high temperature and low rainfall, are known to have significant impacts on water-dependent sectors such as agriculture which accounts around 70% of worldwide freshwater use. Agriculture lies at the heart of many fundamental global challenges including food security, climate change, environmental degradation, and economic development (Fig. 1).

Agriculture and climate change are directly and indirectly linked. Studies reported in the literature indicate that agriculture directly contributes about 20% of global greenhouse gas emissions (carbon dioxide (CO_2), methane

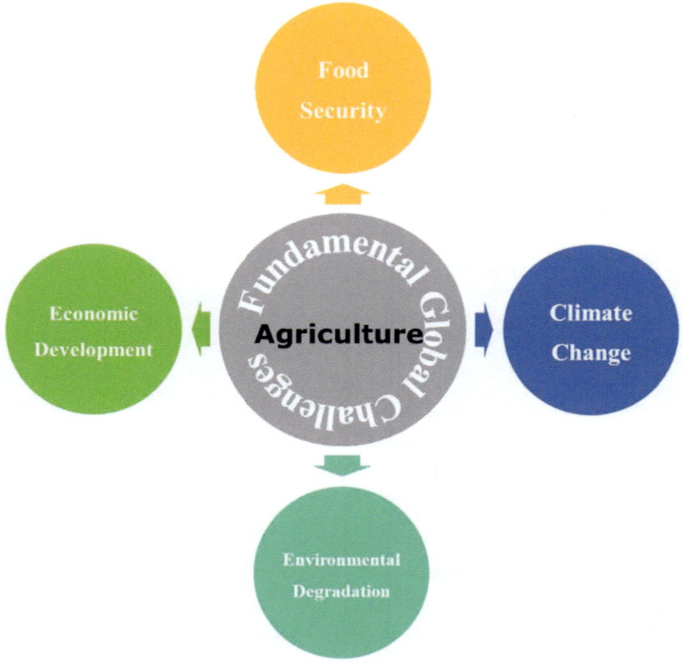

Figure 1 Interlinkages in the agricultural sector with other fundamental global challenges.

(CH_4), nitrous oxide (N_2O)); as discussed in Section 3. Crop growth is also sensitive to temperature as increased temperatures may accelerate the CO_2 fertilization rate during the respiration process of plants and disrupt optimal net growth with consequent reductions in yield (Gardner et al., 1985). High temperatures also accelerate plant physiological development, resulting in faster maturation and reduced yields (Vu et al., 1997). High temperatures will also increase evapotranspiration (ET) rates by reducing available moisture for plants and creating early soil moisture stress conditions that ultimately leads to higher irrigation water demand (Bhatt and Hossain, 2019). The available water resources are also shrinking due to changing patterns and decreasing amounts of rainfall that will ultimately affect agricultural productivity. High temperatures and low rainfall syndromes are the main drivers that weaken national security in a connectivity of water and food security (Fig. 2), as any decline in water resources (both quantity and quality) raise non-traditional security challenges such as water security, food security, and national security.

Climate change, water, and agriculture are inextricably linked. With worsening climatic conditions, water availability is either becoming scarce, unpredictable, polluted, or all three together directly affecting quality and quantity of crops and create food insecurity. Drought and water scarcity challenge water security, agriculture, and food security globally and know no national boundaries and affect all socioeconomic sectors of society. Water security is directly linked with agriculture security, while climate change is boosting these key variables to insecurity. Droughts affect rainfed agriculture, while water scarcity affects irrigated agriculture. There are also multiple drought-induced factors such as heat stress, water deficiency, and insects attacks that harm agricultural productivity. For instance, crop productivity in Central and Eastern Europe usually drops due to insect attacks during drought events (Eitzinger et al., 2012) and the severity and frequency of droughts is

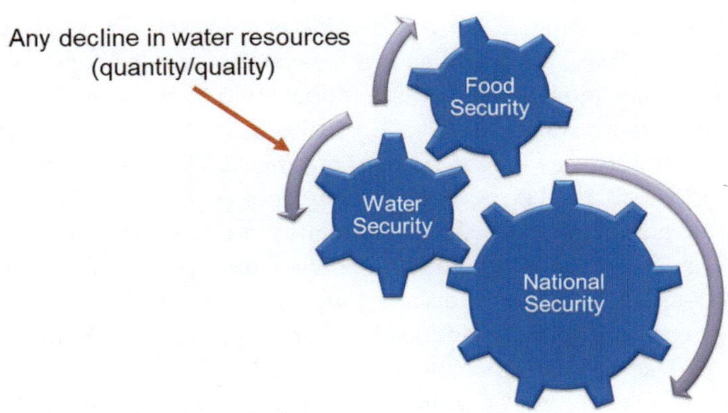

Figure 2 Inextricable links between water, food, and national securities.

reported to be increasing in many regions of the world due to climate change as observed in the past few decades (Mishra and Singh, 2010).

In this context, this chapter focusses on how to manage climate change, droughts, and water scarcity and their combined impacts on agriculture. The scope of this chapter covers the context of climate change and global warming (Section 2), climate change and agriculture (Section 3), droughts and water scarcity (Section 4), climate change impacts on agricultural production and water use (Section 5), and drought and water scarcity impacts on agriculture under a changing climate (Section 6). The evidence from relevant literature is discussed to propose relevant strategies at global, national, regional, and as well as local scales for managing the impacts of climate change, droughts, and water scarcity on agriculture (Section 7).

2 Climate change and global warming

Mainly there are two climate change syndromes i.e. high temperature and low rainfall and the third one is the extreme weather events. The current and projected warming of the Earth is unequivocal, and this is due to significant increase (47%) in the greenhouse gas (CO_2) concentration (412 ppm (parts per million) in March 2019), compared to 280 ppm at the beginning of the industrial age (Buis, 2019). Industrialization, burning fossil fuels, deforestation, and some agricultural practices have contributed to this increase in greenhouse gases (CO_2, CH_4, sulphur hexafluoride (SF_6), N_2O, nitrogen trifluoride, hydrofluorocarbons (HFCs), perfluorocarbons (PFCs) and chlorofluorocarbons (CFCs)) concentrations into the atmosphere. According to the Climate Analysis Indicators Tool report (CAIT, 2011), CO_2 alone is responsible for 77% of global warming. Methane global warming potential is almost 20 times higher than from carbon dioxide. Water vapours are also considered as a greenhouse gas.

Although the impacts of climate change are global, Asia, Africa, and the arid and semi-arid regions are more vulnerable. According to Assessment Reports (ARs) of IPCC, the mean land surface temperature has been increased by 0.60°C in the twentieth century at a global scale as reported in AR3, while in AR4 this increase was 0.74°C (IPCC, 2007). In AR5, the mean (2006–2015) global temperature was 0.86°C higher than the 1850–1900 duration termed the pre-industrial baseline. Climate change projections are likely to reach +1.5°C between 2030 and 2052 if it continues to increase at the current rate (IPCC, 2018). There is overall consistency between the projections from climate models in AR4 and AR5 for large-scale patterns of change.

Changes in precipitation and temperature extremes have been observed globally and some will continue in the future with intensification, e.g. in Africa (Aguilar et al., 2009), in India (Dash and Mamgain, 2011), in China (Zhou et al., 2012), and the rest of the world (IPCC, 2012). According to the

Global Environmental Risk Index, Pakistan's average annual precipitation has increased by 25% and the mean annual temperature has increased by 0.57°C compared to South Asia (0.75°C) over the last century (Eckstein et al., 2021). In Europe, the warming rate is about 1°C during the last century (Shrestha et al., 2013). The IPCC used different Representative Concentration Pathways (RCPs) for developing their climate change projections. In AR5, the predicted increase in global mean surface temperature relative to 1986–2005 period for the mid (2046–2056) and late (2081–2100) twenty-first century was reported. By the mid-century, the projected increase would be in the range of 0.4–1.6°C (under RCP2.6), 0.9–2.0°C (under RCP4.5), 0.8–1.8°C (under RCP6.0), 1.4–2.6°C (under RCP8.5), while for the late century, the projected increase would be in the range of 0.3–1.7°C (under RCP2.6), 1.1–2.6°C (under RCP4.5), 1.4–3.1°C (under RCP6.0), and 2.6–4.8°C (under RCP8.5) based on the Coupled Model Intercomparison Project 5 (CMIP5; IPCC, 2014). The warming projection for the African continent would be higher than the global average under RCP8.5 as indicated by average temperature increase rate of 3°C to 6°C by the end of the century (Ofori et al., 2021). According to climate model simulations for the twenty-first century, there is a consistent pattern of water availability increase in high latitudes and parts of the tropics and a decrease in some subtropical and lower mid-latitude regions. This is primarily because of the changes in precipitation patterns, with an increase in heavy precipitation events and more frequent and severe droughts in some regions. The water supplies stored in glaciers and snow cover are also projected to decline, which could lead to a reduction in water availability during warm and dry periods in regions supplied by meltwater from major mountain ranges.

All the emission scenarios show projected increases in surface temperature, with heat waves becoming more frequent and lasting longer, and extreme precipitation events becoming more frequent and intense in many regions over the twenty-first century. In general, these projections specified climate change which is accelerating the water cycle via more frequent and intense rainfall episodes. However, there are differential spatio-temporal impacts in different regions i.e. an increase in precipitation especially at high latitudes, the equatorial Pacific, and parts of monsoon regions, whereas in sub-tropics parts and limited areas of tropics, there are projected to be decreasing patterns in precipitation. The increasing temperatures, uncertain rainfall patterns, and climatic variability (including extreme climate events) indicate that climate change will have a dramatic effect on crop productivity with severe impacts on food security (Porter et al., 2014). Climate change and its impact on agriculture is the most serious environmental threat to water, food and national securities. The increased probability of frequent extreme weather events creates havoc and high-risk for agriculture. Short and slow maturing crops are more susceptible to extreme heat and drought, while longer growing slowly maturing crops may be more resilient because maturing more slowly allows crops to absorb and transform sunlight into growth and yield stages. For instance, in the US southern states, longer growing season varieties with shorter maturing times are selected to save the crops from late summer heat. Similar vulnerable impacts have been observed in sub-Saharan African regions where the onset and length of growing seasons governed by rainfall patterns could decrease crop yields by 6–24% depending on the climate scenario and management strategy (Waha et al., 2013).

Although the variability and trends in climate change will vary between regions, global agricultural water demand is projected to decline slightly despite the projected warming in climate. For example, in China, Africa, and Australia, crop water demand is expected to be lower than usual in future, while other regions (including North India, Europe, eastern South America, and eastern United States) also showed similar trends with uncertainty. Climate change has a direct impact on water resources, and crop yields is directly associated with this impact. Water will therefore become a major limiting factor for crop production in the context of climate change. There may or may not be a direct impact of climate change on crop production as explained in Fig. 3. The direct impact on crop suitability and yield may be due to three driving forces including CO_2 concentration changes, mean and extreme temperatures, and the amount and timing of rainfall. These driving factors also change water availability patterns

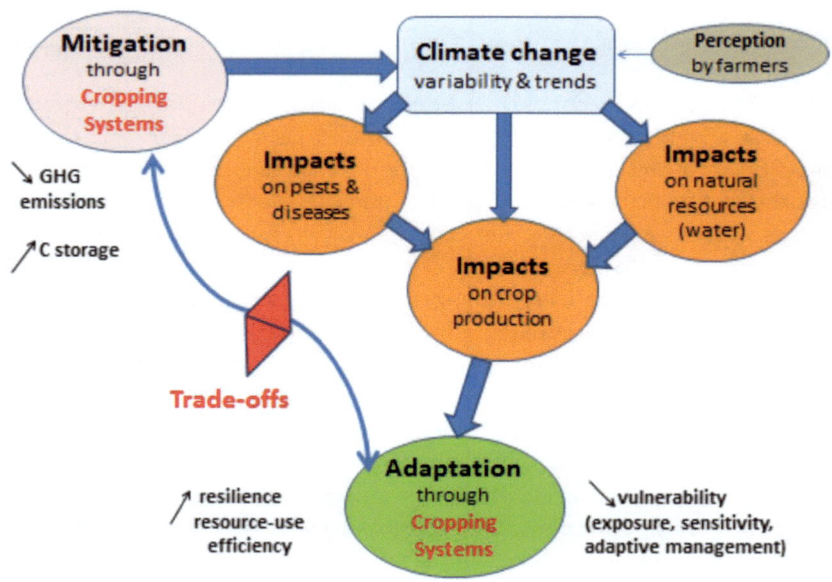

Figure 3 Schematic representation of climate change impacts on crop production and the trade-off measures.

and crop water requirements that subsequently affect agricultural productivity (Anwar et al., 2012). In addition, climate change also imposes indirect effects (negative impacts) on crop production, e.g. via an increase in pests and disease risks, and new invasive species may result in dramatic crop damages (Gregory et al., 2009).

The Paris Agreement on Climate Change (United Nations Framework Convention on Climate Change (UNFCCC) was the first universal, legally binding global deal to combat climate change and adapt to its effects (UNFCCC, 2015). According to the National Oceanic and Atmospheric Administration (NOAA), there are 10 main indicators of climate change. Seven of these indicators (tropospheric temperature, temperature over oceans, sea surface temperature, oceanic heat content, temperature over land, sea level, and humidity) are expected to increase, and three indicators (glaciers, snow cover, and sea ice) show a declining trend. The literature reveals the impact of extreme climate events on agriculture in terms of damage and yield losses. Further research is needed to understand and assess the impacts of climate change on agriculture according to local and regional scales, including future extremes (frequency and magnitude) and the actions needed to support sustainable agriculture.

3 Climate change and agriculture

Agriculture is considered most vulnerable to global climate change (Wang et al., 2009) and adversely affected the production of crops and contributing to the food security issues (Kirby et al., 2016). There may be direct or indirect impact of climate change on crop production. Direct impacts mean changes in cropping calendars and crop phenology, soil loss and shifting of cultivation areas, changes in crop water demand and water supply, and direct effects of increased levels of CO_2 on crop growth. Indirect effects have further negative impacts on crop production, e.g. increase in crop diseases and pests. Crop growth is sensitive to optimal temperature as increased temperatures may accelerate the CO_2 rate during respiration process of plant that disturb optimal net growth and drop in yield. High temperature accelerates plant physiological development, resulting in hastened maturation and reduced yield. The high temperature increases ET rates by reducing available moisture for plants and creating early soil moisture stress conditions that ultimately lead to higher water demand and an increase in energy demand to pump water.

A changing climate means changes in temperature, rainfall and intensity, and frequency of extreme events viz., extreme temperatures, drought, flood, and intense radiation. Both crop productivity and quality are greatly influenced by a changing climate. For example, in sub-Saharan Africa, a changing climate is shortening the crop growing period in Southern and Western Africa at an average rate of 20% causing a 40% decrease in crop yields by 2050; on the other hand, warming at high altitudes in Eastern Africa is promoting maize production (Niang et al., 2014). Arid and semi-arid crops of Pakistan are more vulnerable to climatic change due to the shifting of cropping seasons and expanding of summer. In dry years, there is merging of spring with summer or winter in wet years. Wheat crop is at greater risk from such climatic change in these regions because of possible drought at the time of sowing or heavy rains during grain formation (Khan and Khan, 2018). It was estimated that crop yields will decrease by about 6% to 18% in arid and semi-arid regions due to a shortening of the growing season, with an increase in crop water requirements and heat stress at critical reproductive stages (Saif Ullah, 2017). It is also estimated that agricultural productivity will decrease by 8% to 10% if the temperature rises between 0.5°C and 2°C by 2040 (Dehlavi et al., 2015). Crop growth and production is directly correlated with climatic variables such as temperature and precipitation and it is evident that climate change altered water availability, crop water requirements, and water quality that negatively affected agricultural productivity and ultimately threatened global food security.

There is a two-way link between climate change and agriculture as it is estimated that agriculture is the fifth largest contributor (i.e. 20% of greenhouse gas emissions) in global climate change (Dickie et al., 2014). Direct agriculture production contribution is about 40% in this emission, agricultural supply chain and on-farm machinery contribute nearly 20% and ~40% from deforestation, agriculture fires, and peat loss. Different organizations such as the CDP and FAO stated similar contributions as given in Fig. 4 and Table 1. According to the CDP (2015) agriculture and land use currently contribute about 24% of global GHG emissions and about half of that (12%) coming from agriculture.

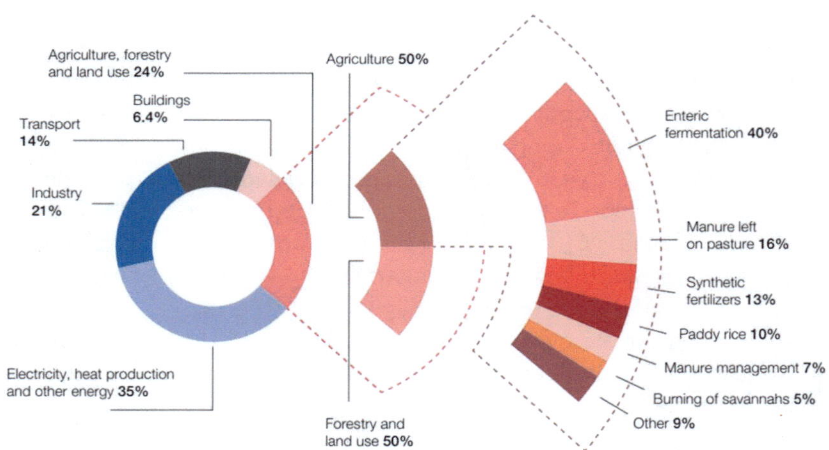

Figure 4 Distribution of global GHG emissions and contribution from different sectors. Source: Adapted from: CDP (2015).

Table 1 Reported sources of direct agricultural emissions

Source of emission	Percentage
Enteric fermentation (CH_4 emissions from ruminants (cattle, sheep, goats, buffalo, etc.) during digestion system.	43%
Manure deposition on grazing/pasture lands (N_2O emissions from manure and urine left on grazing lands)	16%
Synthetic fertilizers (N_2O emissions from nitrogen fertilizer application to crops)	15%
Rice production (CH_4 emissions from flooded paddy fields due to anaerobic decomposition)	11%
Stored manure (CH_4 and N_2O emissions through livestock manure and urine decomposition in wet and dry storage systems, respectively)	7%
Crop residues (N_2O emissions from crop residues on agricultural lands)	3%
Manure deposited on croplands (N_2O emissions from manure, the source of nitrogen fertilizer)	2%
Cultivation of organic soils (N_2O emissions from drained organic soils)	2%

Source: FAO (2013b).

Agriculture is a fundamental part of the solution to boost resilience and combat climate change impacts, especially in developing countries where agriculture is often the backbone of the economy. Globally, agriculture has to take sustainable steps to reduce its carbon footprint because of being partly responsible of global GHG emissions (Smith et al., 2014). In the European Union (EU), agriculture contributes about 10% of all GHG from 1990 to 2017 and this sector already reduced its GHG emissions by 19%. To accomplish net zero GHG emissions by 2050, an increase in carbon sequestration is promoted in the agriculture sector in EU (Jacobs et al., 2019). Agriculture can help in three ways to reduce GHG emissions (i) via a reduction in CO_2, N_2O, and CH_4 emissions, (ii) additional carbon storage in biomass and soils, and (iii) bio-energy production to replace fossil energies.

4 Droughts and water scarcity

Drought is a natural hazard where deficits in available water supplies may be due to below average precipitation for an extended period of months or years (hydrological drought), whilst water scarcity represents the long-term imbalance between water demand and supply to satisfy the long-term average needs in a region. Droughts can trigger water scarcity and water scarcity can aggravate droughts in a way that both are interrelated, whilst droughts affect rainfed agriculture, while water scarcity affects irrigated agriculture. These climatic features have substantial impacts on agriculture and ecosystems. Drought is a hydro-meteorological hazard, and its identification (onset and recovery) is a complex phenomenon that makes it one of worst natural hazards (Mishra and Desai, 2005; Adnan et al., 2017). Drought prediction is a relatively slow and complex process due to understanding its causes, extent, and timing for different spatio-temporal scales (Ndehedehe et al., 2020). There are a number of different drought indices that are commonly used to measure the severity and spatio-temporal extent of droughts. These include, for example,

- standardized precipitation index (SPI);
- weighted anomaly standardized precipitation index;
- percent of normal precipitation index;
- rainfall variability index;
- standardized precipitation temperature index;
- standardized precipitation evapotranspiration index (SPEI);
- standardized soil moisture anomaly index;
- self-calibrated Palmer drought severity index;
- reconnaissance drought index (RDI);
- deciles index;
- China Z-index;
- modified CZI;
- Z-score;
- composite index; and
- percentage area weighted departure.

Each of these indices uses different data and methods to calculate drought conditions, and they can provide different perspectives on the severity of drought and its spatial and temporal extent. Some of these indices are based on precipitation data, while others consider temperature, ET, and soil moisture. Some are based on a single parameter such as precipitation, while others use composite parameters to provide a more comprehensive understanding of drought conditions.

Drought is a deficit in water availability to satisfy long-term average water requirements, caused by a lack of precipitation or an increase in ET (Poljanšek et al., 2017). There are different types of droughts such as meteorological, agricultural, hydrological, and socioeconomical. Meteorological drought is defined as the amount of dryness and the duration of the dry period. Agricultural drought refers to a deficit of soil water that affects crops, and hydrological drought refers to a shortage of water supply. Socioeconomic drought occurs when the demand for an economic good exceeds supply as a result of a weather-related shortfall in water supply. According to the IPCC (2012), there is 'medium confidence' that droughts will intensify in some seasons and areas in the twenty-first century, due to a combination of more variable precipitation and/or increased ET. This includes regions such as Southern Africa, Central and Southern Europe, Central North America, Central America, Mexico, Northeast

Brazil, and the Mediterranean. Drought intensity and variability is high in the African continent and studies reported that how climate change, El Niño-Southern Oscillation (ENSO), and sea surface temperatures (SSTs) have led to mechanisms inducing droughts (Masih et al., 2014; Bhaga et al., 2020). El Niño conditions tend to coincide with a period of weak monsoon and rising temperatures in South Asian countries, which can increase the probability of drought occurrence and disturb crop production and water supply. Additionally, El Niño-related deficit rainfall often lowers the summer crop production such as rice, sugarcane, cotton, and oilseeds, which can result in high inflation rates and a lower GDP due to the high contribution of agriculture to the economy of South Asian countries.

According to Malik et al. (2013), the most severe and longest drought episode experienced over most parts of Asia occurred between 1999 and 2002. This was considered to be one of the most prolonged and intense drought events in recent history, affected large areas of Southeast and South Asia and had significant impacts on water resources, agriculture, and food security. This event was likely caused by a combination of factors, including the El Niño-Southern Oscillation (ENSO) phenomenon and human-induced land-use and water management changes. The ENSO, which is a natural climate pattern that affects the sea-surface temperatures and atmospheric circulation over the Pacific Ocean, has a large impact on the Asian monsoon, which provides the water needed for agriculture and to meet societal needs. The impacts of this event were felt over a wide area, including India, Pakistan, Bangladesh, and parts of Southeast Asia and it resulted in water scarcity, crop failures, and food shortages and even a humanitarian crisis in the region.

Water scarcity is a major global challenge in the twenty-first century. With an increasing population and the demand for water resources from various sectors such as agriculture, industry, and the domestic sector, many regions around the world are facing a growing water scarcity problem. Climate change, urbanization, and industrialization are also putting pressure on water resources. Water scarcity is intensifying with climate change – from California to China's eastern provinces and from Jordan to the southern tip of Africa, almost 4 billion people live with severe water shortages for at least some of the year. There are many regions in the world already experiencing a water scarcity situation that is not only a problem of quantity but also of quality. According to the Ministry of Irrigation and Water Resources in Egypt, the country is projected to face absolute water scarcity, defined as less than 500 m^3 water per capita per year, by 2025. In 2017, the per capita water availability in Pakistan was 900 m^3 per year, indicating water scarcity. The threshold for water-stressed conditions, which is when the water resources available to a population are limited and the demand is high, is generally considered to be 1700 m^3 per capita per year. With a per capita water availability of 900 m^3 per year, Pakistan is below this threshold and is facing water scarcity. Water storage capacity is alarmingly low at only 132 m^3 per person and available only for 30 days, while water storage capacity in the USA and Australia is 6150 m^3 per person and 5000 m^3 per person, respectively (Saeed et al., 2018). Sub-Saharan Africa is particularly vulnerable to the impacts of climate change, due to its low capacity to adapt or mitigate its effects. One of the main areas of vulnerability is water resources, which are being negatively impacted by the changes in precipitation and evaporation due to climate change. According to Bates et al. (2008), the water capacity in the region is getting reduced due to a decrease in precipitation and an increase in evaporation. Climate change is expected to exacerbate water scarcity in the region through changes in temperature, precipitation patterns and variability, and evaporation rates, which would lead to changes in surface water and groundwater availability. Different studies (Niang et al., 2014) have also suggested that the effect of climate change on future water scarcity in the region is uncertain. While some studies suggest that climate change will exacerbate water scarcity, others suggest that it could lead to an increase in water resources in some areas. This uncertainty is due to the complex nature of the interactions between precipitation, evaporation, and other factors, as well as the lack of data and understanding of these interactions in the region. Overall, the region is extremely vulnerable to climate change, and it is imperative that adequate measures are taken to adapt to and mitigate its effects, particularly in the areas of water resources management, to ensure that the water resources that are critical for the survival of the population and economy of the region are adequately protected.

Increasing water scarcity under climate change is a major challenge for adaptation in many regions of the world. Under conditions of water scarcity, it is important to use irrigation water efficiently to maintain or increase agricultural production while reducing the risk of water shortages. This can be achieved through the implementation of modern, climate-smart agricultural management practices that focus on the rational use of irrigation water. Climate smart agriculture, when implemented in an integrated and holistic manner, can help to increase water and land productivity under water scarcity. Therefore, it is important for countries to implement adaptation strategies that focus on water conservation, efficient water use, infrastructure development and management, and drought management to overcome the challenges of water scarcity under climate change.

5 Climate change impacts on agricultural production and agricultural water use

In agriculture, the terms 'green water' and 'blue water' are used to describe different sources of water available for crop growth. Green water refers to water stored in soil that is available for rainfed crops, i.e. the water that infiltrates into the soil from precipitation and can be used for crop growth. Blue water, on the other hand, refers to water available for irrigated crops, which includes both green water and water from surface and groundwater sources, such as rivers, lakes, and aquifers (Falkenmark, 2013). The green water sources are essential in drought-prone regions, and they play a critical role in the availability of water to support crop growth. Blue water can be managed through infrastructure like dams and canals to provide water to crops during the dry seasons. Climate change is altering crop water requirement and water availability significantly due to changing temperatures and precipitation at the regional and local scales. Figure 5 shows how climate change affects all the elements of the water cycle and its impact on agriculture (FAO, 2013a). There is some uncertainty in predictions regarding the impact of climate change on irrigation water requirements at the global scale. According to Döll (2002), irrigation requirements are expected to slightly increase globally due to rising temperatures and changes in precipitation patterns. However, Zhang and Cai (2013) found that irrigation water requirements may decline slightly despite the projected warmer climate, a phenomenon which is not fully understood. The uncertainty in these predictions is due to the complex interactions between temperature, precipitation, and ET, as well as the lack of data and understanding of these interactions. Climate change also has different effects

Figure 5 Climate change effects on the water cycle and its impacts on agriculture Source: Adapted from: FAO (2013a).

on different regions and on different types of crops, making it difficult to make general predictions on global scale.

Climate change can have significant impacts on the timing of sowing and harvesting of crops, as well as the length of growing seasons. There are reports of shifts in the sowing time and length of growing seasons geographically, as a result of changes in temperature and precipitation patterns. This could alter planting and harvesting dates of crops and varieties currently used in a particular area (Mahdi, 2019). Additionally, warmer temperatures can negatively affect crop yields. For example, studies have found that the yield of wheat may decline up to 10% for each 1°C increase in mean seasonal temperature (Craufurd and Wheeler, 2009). This decline in yield is due to heat stress and water scarcity. Agriculture is one of the sectors that is particularly vulnerable to the impacts of extreme weather events such as heatwaves, droughts, floods, and storms. These events can significantly damage crops and infrastructure, reduce yields and quality, and increase production costs, ultimately leading to losses for farmers and food insecurity for communities (Turral et al., 2011; Nelson et al., 2009).

It has been projected that the impacts of climate change could lead to significant losses in agriculture income in the EU by 2050. According to a study by Jacobs et al. (2019), it was projected that there could be up to a 16% loss in EU agriculture income by 2050, with large regional variations. It is important to note that this estimate is a projection and does not account for potential adaptation measures that might help to mitigate the impacts of climate change on agriculture. This also is a general projection, so the actual loss may vary depending on the region and on the specific agricultural activity. The EU is working on measures to mitigate the negative impacts of climate change on agriculture, such as the Common Agricultural Policy (CAP) and the Green Deal for Europe. Climate change has impacting on global crop production, and many regions have reported negative impacts on yields and quality as a result of changes in temperature and precipitation patterns. According to Porter et al. (2014), the observed effects of past climate trends on crop production are evident in several regions of the world, with negative impacts more common than positive ones, including several periods of price spikes following climate extremes in key- producing regions. There is evidence that climate change has already negatively affected wheat and maize yields in many regions, as well as globally. This is primarily due to the increase in temperature, extreme weather events, and changes in precipitation patterns, which have led to heat stress and water scarcity. According to Lobell et al. (2011), the increased frequency of unusually hot nights, particularly in rice- producing regions, is damaging for most crops and has observed negative impacts on rice yields and quality. The above-mentioned impacts are likely to worsen in the future if we do not take the necessary steps to mitigate the effects of climate change on agriculture. Therefore, it is important to continue researching and monitoring the effects of climate change on crop production and to implement adaptation measures such as crop management practices and irrigation that can help to reduce the negative impacts of climate change on agriculture and maintain food security.

Climate change fundamentally altered global food production patterns and several methods and crop models can be used to estimate how future climate change will affect crop production (Frieler et al., 2015; Rosenzweig et al., 2014). According to a study conducted by the Agricultural Model Intercomparison and Improvement Project (AgMIP) and Inter-Sectoral Impact Model Intercomparison Project (ISI-MIP), by 2100, the impacts of climate change on crop yields are expected to be severe. That study reported that, under a high-emissions climate change scenario, crop yields for maize could decline by as much as 45%, for wheat by 50%, for rice by 30%, and for soybean by 60%. The study also showed that the impact of climate change on crop yields may be somewhat mitigated by the CO_2 fertilization effect, which can increase crop growth and yields under increased atmospheric CO_2 concentrations. However, if nitrogen limitations are explicitly considered, the benefit from CO_2 fertilization is reduced and the negative impacts of climate change on crop yields are amplified (Müller and Elliott, 2015).

Financiers can provide capital for water productivity investments, and engineers can develop technologies to improve agricultural efficiency and consumers can reduce food loss and waste. Different water saving measures in irrigated agriculture can save a handsome amount of water such as no-till farming and plastic mulching can save up to 50% irrigation water as studies conducted in China and the USA (van Donk et al., 2010; Wu et al., 2015). Switching from flood (furrow) irrigation to drip irrigation can save water up to 32–57% as studies conducted in Turkey, Pakistan, and the USA (Cetin and Bilgel, 2002; Tagar et al., 2012; Enciso et al., 2015). Regulated deficit irrigation can save up to 80% water as studies conducted in the USA, Turkey, Spain, Bolivia, and China have demonstrated (Stewart et al., 2011; Ünlü et al., 2014; Lozano et al., 2016; Geerts et al., 2008; Zhou et al., 2012). Alternate wetting and drying of rice can save water up to 31%, according to the study of the USA (Linquist et al., 2015). Irrigation scheduling based up soil moisture metering can save up to 22–42% in different countries such as India, China, France, and the USA (Perveen et al., 2012; Zhang, 2002; Molle et al., 2012; Gleason, 2013). Crop shifting (different comparison) such as rice to cereal crops in Australia can save potential water by up to 70–85% (ABS, 2022) and in the USA by 60–80% (USDA, 2013). Sugar cane to sunflower shifting can save 74% water in Egypt (Dawoud, 2014). Therefore, farmers and agricultural extension services need to adapt to the changing climate by selecting or developing crops that are better suited to the new conditions and using new crop management strategies such as irrigation and changing the planting dates. This can help to maintain or even increase agricultural productivity in the face of climate change. In addition, research and development in climate-resilient agriculture and precision agriculture may help to mitigate the negative impacts of climate change on agriculture.

Crop models can be an effective tool for assessing the impacts of future climate change on agriculture and for suggesting suitable mitigation and adaptation strategies. Crop models are mathematical representations of the growth and development of crops, and they can simulate the interactions between crops and their environment, including temperature, precipitation, CO_2 levels, and soil conditions. Crop models can effectively integrate the direct and indirect impacts of climate change, including changes in temperature, precipitation, CO_2 levels, soil health, and pest and disease infestation. The results from these models can provide valuable information on the potential impacts of future climate change on crop productivity and can help in the development of suitable mitigation and adaptation strategies. However, it is important to note that there are a number of uncertainties that need to be considered when using crop models for assessing the impacts of future climate change. These include uncertainties in climate change projections, intra-seasonal variability at the local scale, nonlinear impacts with climate variables, and socioeconomic factors such as changes in population dynamics, crop prices, and political structures (Schlenker and Roberts, 2006; Wang, 2011a,b; Parry et al., 2004). Accordingly, it's important to note that crop models are based on certain assumptions and many factors can affect the outcome like the quality of input data, the assumptions used in the model and the complexity of the model. Therefore, it is important to consider and account for these uncertainties when interpreting the results from crop models and when developing and implementing mitigation and adaptation strategies. Several simulation models including INFOCROP, WTGROWS, ORYZA, DSSAT, and APSIM are all based on the principles of crop physiology. These models can be used for different purposes including resource and input management, plant ideotype designs, climate change and variability impact evaluation, yield forecasting and addressing extreme and episodic events. The INFOCROP model is used for crop productivity forecasting, WTGROWS model is used to simulate crop growth and yield under different weather conditions and management practices, the ORYZA model is used to simulate rice crop growth and development under various environmental conditions, the DSSAT model is used to simulate crop growth and development under different soil and weather conditions, and the APSIM model is used to simulate the crop growth, water, nutrient, and pest management.

6 Droughts and water scarcity impacts on agriculture under climate change

Climate change, water, and agriculture are inextricably linked. With worsening climatic conditions, water availability is either becoming scarce, unpredictable, polluted, or all three together directly affecting quality and quantity of crops and create food insecurity. Global and regional studies have consistently projected that droughts and water scarcity affected areas will likely increase during the twenty-first century as a result of a combination of more variable precipitation and increased ET. A study by Berry et al. (2006), e.g., projected that global warming will increase the frequency and severity of droughts in many regions, particularly in semi-arid and arid regions. Drought and water scarcity challenge water, agriculture, and food security globally and know no national boundaries and affect all socioeconomic sectors of society. Water security is directly linked with agriculture security, while climate change is boosting these key variables to insecurity.

Drought is a climate extreme associated with dry and hot weather conditions (Mishra and Singh, 2010) that can have serious impacts on water quantity and quality, agricultural production, food security and economy at local, regional, and national scales. Droughts can cause losses in agriculture through multiple factors, including heat stress, insects, as well as water deficit. The severity and frequency of droughts have been increasing in many regions of the world in recent decades, which is consistent with climate change projections. Droughts propagate from meteorological to agricultural systems and can greatly impair agricultural production as well as food security. Droughts can cause crop failures, reduced yields, and complete crop loss. This can have significant impacts on food security, particularly in developing countries where agriculture is a major source of income and livelihood. In addition, droughts can cause serious problems for water resources, such as lowering water levels in rivers, lakes, and aquifers, and reducing the availability of water for irrigation, drinking, and other uses. In addition to the direct impacts on crop yields, droughts can also have other indirect impacts on agriculture, such as increasing the spread of insects or increasing the risk of wildfire. For example, in Central and Eastern Europe, crop productivity usually drops due to spreading of insects during droughts (Eitzinger et al., 2013). Furthermore, droughts can produce adverse impact on farmers' income (Burke et al., 2006) and have negative impacts on the economy.

Drought results in a variety of negative impacts on plants and crops, including wilting, reduction in photosynthesis, disturbances in physiological processes, cessation of growth, or even death of the plant or plant parts, which ultimately reduce the potential yield. Drought stress during the production phase of crops can affect their physiology and morphology in such a way as to influence susceptibility to weight loss in storage (Kumar et al., 2006). One of the negative effects associated with water deficit is that pre-harvest water stress (watering to 25–75% of soil water field capacity) can weaken the cells, resulting in higher membrane leakage (i.e. cell damage) and consequently greater weight loss in storage. Many aspects of plant growth are affected by moisture stress, including leaf expansion which is reduced due to sensitivity of cell growth to water stress. The reduction in leaf area reduces crop growth and thus biomass production (Brown et al., 1985). Different crops are affected differently by drought stress, and the timing of the stress is also an important factor. For example, in sugarcane, drought stress occurring when the leaf canopy is already well established has more serious impacts on final yield in terms of total biomass, stalk biomass, and stalk sucrose than when drought occurs earlier during the season (Robertson et al., 1999). It's important to note that, although drought stress has severe negative impacts on crop yield and quality, it's also important to consider other factors such as the crop's physiological characteristics and its stage of growth, as well as the duration and severity of the drought stress. Additionally, adapting crop management practices to suit the water availability and crop characteristics can help to mitigate the impacts of droughts on crops.

The drought is an extreme event and its impact under climate change is the biggest concern of agricultural communities for agriculture water management around the world (Cai et al., 2015). It is acknowledged that it is difficult to conclude how drought has changed on a global scale due to the lack of adequate evidence and the complex nature of drought analysis (Trenberth et al., 2014). However, it is agreed that there are regions where drought trend is increasing and regions where it is decreasing. Regions where drought trend is increasing include Mediterranean and West Africa, southern and central Europe, Alpine region, central North America, the Mediterranean region, Australia, southern Africa, Mexico, and northeast Brazil (IPCC, 2014; Wang et al., 2011a,b; Planton et al., 2008). Europe is also facing the challenges of increasing drought risk, which will reduce productivity through negative impacts on grassland productivity. Historical trends in droughts in Europe report that drought frequency and severity increased over the Mediterranean area and in eastern Europe over the period 1950–2015 (Spinoni et al., 2017). Future projections report an increase in drought frequency and intensity in the Mediterranean area, western Europe and northern Scandinavia by the end of the twenty-first century (under the moderate climate scenario RCP 4.5) (Spinoni et al., 2018), as well as a greater increase in the length of meteorological dry spells, mainly in southern Europe (Kovats et al., 2014; IPCC, 2019). Droughts will be prolonged, with 1–2 days per year in the Nordic region (all of Denmark, Norway, and the very south of Sweden) (Wiréhn, 2018). In the Mediterranean area, including most of Portugal, Galicia in Spain, northern Scandinavia, and

Mediterranean Turkey, the increase in drought severity will be more pronounced (Spinoni et al., 2018).

Droughts affect water quantity and quality that can drastically decrease agricultural production (Fraser et al., 2013). In South Asian countries, major droughts have occurred during El Niño events (Rojas et al., 2014). Many regions around the world have faced more frequent and severe droughts over the past decades, resulting in significant damages to agriculture and other sectors. Climate change is expected to exacerbate this trend, leading to more intense droughts that occur more frequently in many regions around the world (Mishra and Singh, 2010). The Alpine region, the USA, the Mediterranean basin, Australia, South Africa, among others, are all expected to be impacted by more intense and frequent droughts in the future (Wang et al., 2011; Mpelasoka et al., 2008). In addition to these regions, some areas in the Amazon and West Africa are also expected to experience drier soils and more frequent droughts during the June–August season. Similarly, the Asian monsoon region is also expected to experience drier conditions and more frequent droughts during the December–February season (Fraser et al., 2013). Drought events can have a significant impact on crop production in Europe, and in recent years, many farmers in northern and central Europe have faced crop failures due to the intensity of regional droughts. In addition to yield reduction, droughts can also increase the demand for irrigation water. Studies have projected that irrigation demand for water in the Mediterranean area is likely to increase by 4–18% by the end of the century under different climate change scenarios (RCPs 4.5 and 8.5). This will have implications for water resources management and the use of water for irrigation in the region (Cramer et al., 2018).

Climate change is also creating worse water crises as indicated by the World Resources Institute (WRI) who reported that 17 countries (Qatar, Israel, Lebanon, Iran, Jordan, Libya, Kuwait, Saudi Arabia, Eritrea, United Arab Emirates, San Marino, Bahrain, India, Pakistan, Turkmenistan, Oman, and Botswana) will face 'extremely high' water stress within 20 years where more than 80% of their available supply is withdrawn by irrigated agriculture, industries, and municipalities every year (Hofste et al., 2019). Agricultural output is highly vulnerable to climate change in these countries and depends significantly on climate change variables, e.g. wheat sowing already delayed across the Pakistan at least 14 days because of changes in precipitation patterns (Ahmad et al., 2014), this will have adverse effect on wheat production. The predicted impacts of climate change to agriculture sustainability are at the forefront, particularly the projected impacts of climate change on agriculture in Asia are concerning. According to a study by the Ministry of the Environment of Japan, Asian countries are projected to lose about 50% of their wheat production by 2100. Similar projections have been made for other cereal crops in South Asian countries. Net cereal production in South Asian countries is projected to decline by 4–10% by the end of this century under the most conservative climate change scenarios, as reported by Alam (2007) and Sathar and Khan (2019). According to a study by Nelson et al. (2009), the impacts of climate change on irrigation wheat and maize yields in developing countries are projected to be significant. The study projects that by the 2050s, under various climate change projection scenarios, irrigated wheat and maize yields could decline by 20–28%. Lobell and Field (2007) found that six most widely grown crops (wheat, rice, maize, soybeans, barley, and sorghum) of the world showed more than 30% variations per year in global average yields due to variations in optimum temperature and precipitation required for crop growth.

Due to increasing temperature trends, the crop yields in sub-Saharan Africa will be negatively affected (Schlenker and Lobell, 2010). In summary, climate change affects crop yield. Evidence of the impacts have been found using historical literature studies. A 0.1°C to 0.5°C increase in temperature can reduce rice yield by approximately 1–5% in Sri Lanka, while by 2050, average paddy irrigation water requirement during the wet season will increase up to 23% due to reductions in average rainfall, increase in potential ET, and early ending of rainfall in Sri Lanka (Eriyagama et al., 2010). Increase of temperature at 1% could result in 3.44% drop in current rice yield, while a 1% increase of rainfall could reduce rice yield by 0.12% in Malaysia (Alam et al., 2014). According to the projection under high-end emission scenario, yields of non-irrigated crops like wheat, corn, and sugar beet are projected to decrease in southern Europe by up to 50% by 2050 (Iglesias et al., 2009). This is shown in Fig. 6 where the

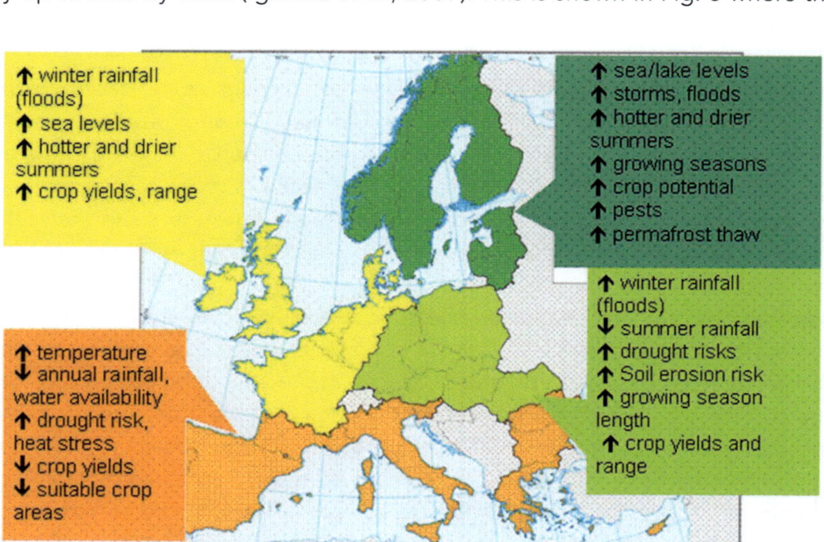

Figure 6 Climate change impacts in different EU regions. Source: Adapted from: Sabrina and Donatella (2014).

adverse effects of climate change are already affecting agricultural production in southern Europe (Sabrina and Donatella, 2014).

Impacts of climate change on agriculture maybe categorized into three areas (i) impact of temperature on crop growth; (ii) impact of CO_2 on crop growth; and (iii) impact of precipitation on crop growth. Agricultural productivity can be affected by climate change in a number of ways. Climate change can impact crop growth, development, and yield by altering temperature, precipitation, and CO_2 levels, either directly or indirectly. Direct impacts include changes in temperature, precipitation, and CO_2 levels. Rising temperatures can lead to heat stress and reduced crop growth, especially during critical stages of development. Changes in precipitation patterns can lead to water scarcity, and in some cases, flooding, both of which can affect crop growth and yield. The CO_2 fertilization effect, in which higher CO_2 levels can increase crop growth and yield, is also considered as direct impact. Indirect impacts include changes in soil health and infestation by pests and disease, which can further reduce crop productivity. Increased temperatures can accelerate soil organic matter decomposition, which can lead to soil degradation and reduced soil fertility. Moreover, shifts in pest and disease ranges and increased frequency of extreme weather events can also have negative impacts on crop productivity. It's worth mentioning that these impacts are not mutually exclusive and often interact with each other. For example, water scarcity can exacerbate heat stress, and changes in temperature and precipitation can affect soil health and infestation. Therefore, climate change can have complex and multi-dimensional impacts on agriculture, making it challenging to both predict and adapt to.

It's important to note that while the link between climate change and regional droughts has been established in many studies, the actual impact of droughts on agriculture and water resources will depend on a variety of factors such as precipitation, temperature, and soil type. Additionally, because droughts are a complex natural hazard with multiple drivers, the relationship between drought and climate change can be complex, with multiple pathways to consider. It's worth noting that, despite the impacts of droughts on agriculture, the relationship between droughts and climate change is complex, involving multiple pathways and factors. The impacts also depend on the region, type of crops and other characteristics, thus it's important to consider these factors in assessing the specific impacts of droughts on agriculture in specific regions.

7 Management of climate-induced drought and water scarcity

The main consequences of changes in water resources due to climate change induced draught and water scarcity for agricultural production include the following:

1. Reduced crop yields: changes in temperature and precipitation patterns, drought and water scarcity can lead to water stress for crops, which can reduce growth and yield. This can impact food security and the livelihoods of farmers.
2. Changes in crop selection: climate change, drought, and water scarcity can make it difficult to grow certain crops in certain regions, leading to changes in the types of crops grown and shifts in agricultural production.
3. Increased irrigation requirements: climate change can exacerbate water scarcity and in regions where water is already scarce, drought, and water scarcity can exacerbate the need for irrigation and lead to increased water withdrawals from groundwater or surface water sources.
4. Soil degradation: reduced water availability and increased temperature can lead to soil erosion and a decline in soil quality, which can negatively impact crop yields and future crop productivity.
5. Reduced water quality: drought and water scarcity can lead to lower water levels in rivers and lakes and can result in increased water temperatures and salinity, which can affect the quality of water used for irrigation.
6. Economic impacts: climate change, drought, and water scarcity can lead to reduced agricultural production and revenues, which can have negative economic impacts on farmers, agricultural businesses, and entire communities.
7. Water conflicts: Increased competition for water resources can lead to conflicts between different water users and can be difficult to resolve.
8. Impact on downstream ecosystems: changes in water flow and water availability due to climate change, drought, and water scarcity can lead to reduced water flow in rivers which can have negative impacts on aquatic life and ecosystems.
9. Increased pest and disease pressure: climate change can increase the range and severity of certain pests and diseases, which can negatively impact crop yields.
10. Flooding: rising sea levels and increased heavy rainfall events can increase the risk of flooding in coastal and low-lying areas, negatively impacting crop production and infrastructure.

Water scarcity and drought are major challenges for agriculture, and climate change is expected to exacerbate these issues in many regions. Agriculture is the largest consumer of water worldwide and it is a critical component for crop growth and food security. In areas where water resources are already limited, drought and water scarcity can lead to crop failures and reduced yields, which in turn can impact food security and the livelihoods of farmers. Droughts and water scarcity management strategies involve proactive structured approaches to minimize adverse impact on agriculture such as drought monitoring and early warning system. There are a variety of management strategies that can be used to mitigate the effects of drought and water scarcity on agriculture, including:

1. *Drought monitoring*: continuous observation of rainfall levels and comparisons with current usage levels can help prevent man-made drought.
2. *Irrigation management:* this includes techniques like precision irrigation, where water is applied only where and when it is needed, and deficit irrigation, where crops are grown with less water than normal in order to conserve water.
3. *Crop selection and management:* drought-resistant crops can be grown in regions where water is scarce, and certain management practices like mulching and crop rotation can help to reduce water loss from the soil.
4. *Soil moisture monitoring:* this involves measuring the amount of water in the soil and using that information to optimize irrigation schedules.
5. *Water harvesting and storage:* this includes techniques like rainwater harvesting, where water is collected and stored for later use, and the construction of dams and other water storage structures.
6. *Water conservation and reuse:* this involves reducing water usage through measures like efficient irrigation systems, water recycling, and reducing water leaks.
7. *Climate information services:* this involves obtaining and using weather and climate forecast information to plan and make informed decisions on irrigation, drought, and water scarcity.
8. *Watershed management:* this means managing natural resources that affect water availability, such as through reforestation and conservation of wetlands.
9. *Community-based approaches:* involving local communities in water management and helping them to develop the necessary capacity to implement water conservation and management measures.
10. *Adaptation of soil and moisture conservation techniques*: adaptation of agro-mechanical measures such as contour cultivation, contour bunding, contour trip cropping, graded bunding, mixed cropping, bench terracing, tillage, curved land ploughing, broad bed furrow system, mulching, vertical mulching, and zero tillage for the *in situ* soil and moisture conservation.

11. *Enrichment of soil organic carbon content*: mixing of bio-fertilizers, green manuring, crop residues and farmyard manure to the agricultural soil enhance organic matter, improves soil structure and soil moisture holding capacity. The adaptation of integrated bio-composting techniques are helpful to conserve soil moisture during droughts and water stress conditions.
12. *Hydroponic and aquaponics techniques*: hydroponics is a method of growing plants in a nutrient-rich water solution instead of soil and aquaponics is a combination of aquaculture (raising fish) and hydroponics. Hydroponic agriculture is a sustainable farming to the challenge posed by drought and climate change.

Climate change is expected to intensify the existing risks, particularly in regions where water scarcity is already a concern, as well as create new opportunities in some areas. It's important to note that the specific consequences of drought and water scarcity on agricultural production can vary depending on factors like the region, the type of crops grown, and the specific agricultural practices used. It's also crucial to note that these consequences can also be compounded by other factors such as rising temperature, increased frequency of extreme weather events, and other climate change impacts. Effective management of water resources in agriculture, such as implementing irrigation management practices and crop selection, can help to mitigate the effects of drought and water scarcity on agricultural production.

8 Managing climate change, droughts and water scarcity affecting agriculture

Mitigation and adaptation are the two key solutions to manage climate change, droughts and water scarcity affecting agriculture. Generally, mitigation is an intervention to reduce the 'causes' of climate changes while adaptation addresses the 'impacts'. Mitigation is decisive to limit changes in climate system by reducing the sources or enhancing the sinks of greenhouse gases. Adaptation mean adjustment in that system in response to actual or expected climatic changes or their effects. For example, crop production of southern Europe and Mediterranean regions is projected to decrease due to climate change and European Union's agriculture sector must adapting to climate change at top priority to improve resilience against extreme events such as droughts, heatwaves and floods. Based on this extensive literature review we characterize the effort and benefit of a number of agronomic and policy measures, aiming to develop concrete adaptation plans and responding to concrete regional challenges. although the type and scope of adaptation strategies vary from region to region, and socio-economic and agro-ecological environments are constantly changing (Abid et al., 2015).

The following cutting-edge solutions can be adopted to reduce water stress in any geographical area:

1. *Increase agricultural water productivity:* implementation of more crop per drop strategy by growing low delta high cash crops and applying smart irrigation techniques by using precision water metering rather than flooding.
2. *Investment in grey and green infrastructure*: Built grey infrastructure (like pipes and treatment plants) and green infrastructure (like wetlands and healthy watersheds) can work in tandem to tackle issues of both water supply and water quality.
3. *Treat, reuse and recycle*: Treat and reuse wastewater as a 'new' water source and adopt 'a resource conserved is a resource generated approach'.

Climate change impacts can also be reduced by the adaptation of following options in conjunction with each other:

1. *Climate solutions:* acting now to address the issue and changing its course using science and technology and positive leadership.
2. *Cut emissions:* 'net zero' carbon emissions by 2050 or sooner by adaptation and implementations of green energy sources, better transportation system, stop deforestation, and climate-friendly agricultural system.
3. *Remove carbon dioxide:* planting new forests (afforestation) or restoring old ones (reforestation), enhanced land management practices, new technologies that suck CO_2 out of the air ('direct air capture'), or prevent it from leaving smokestacks ('carbon capture and storage').
4. *Fight disinformation:* disinformation misleads and confuses the public about the growing consequences of global warming and makes it more difficult to implement the solutions.
5. *Prepare and adapt:* cutting carbon is the only long-term solution for avoiding climate impacts. In the short-term, we need to adapt. That means everything from discouraging development in high-risk areas to planning for water scarcity and to building more resilient cities and communities.
6. *Act:* The best policy ideas in the world aren't worth much if we don't have activists, experts, and everyday people fighting for change. From school groups to churches and from corporate boardrooms to mayors and local leaders, we need collective and concerted action.

9 Climate change mitigation and adaptation strategies

9.1 Adaptation strategies at the national level

There are several mitigation and adaptation strategies that can be implemented at the national level to address the impacts of climate change on water resources and agricultural production:

- *National water resource management*: Developing and implementing national policies and plans for managing water resources can help to ensure the sustainable use of water for agricultural production and other sectors.
- *Climate-smart agriculture*: This approach integrates the use of climate information and forecasting, crop management, and conservation practices to help farmers adapt to the impacts of climate change and increase their resilience.
- *Irrigation modernization*: Investing in efficient irrigation systems and technologies can help to reduce water loss and increase water use efficiency in agriculture.
- *Water storage and harvesting*: Building dams and other water storage structures, as well as promoting rainwater harvesting and other forms of water storage can help to ensure water availability during the periods of water scarcity.
- *Research and development*: Investing in research and development of crop varieties that are more drought- and heat-tolerant and that can adapt to changing climate conditions.
- *Water pricing reform*: Encouraging water pricing reform to reflect the true cost of water, as well as providing subsidies for water-saving technologies and practices, can help to promote water conservation.
- *Capacity building*: Developing the capacity of farmers, extension workers, and water managers to implement water conservation and management measures can help to ensure the effective use of water resources.
- *National climate information services*: Developing and providing access to reliable and actionable climate information services can help farmers, water managers, and other decision-makers plan and make informed decisions on water resources and agriculture.
- *Climate-resilient infrastructure*: Investing in infrastructure that can withstand the impacts of climate change, such as flood protection measures, can help to reduce the risk of crop losses and protect agricultural production.
- *Carbon sequestration*: Encouraging the use of farming practices that sequester carbon, such as conservation agriculture, agroforestry, and wetlands restoration can help to mitigate climate change.

It's important to note that the specific strategies that are most appropriate will depend on the country's specific context and the particular challenges it faces. Additionally, implementing these strategies often requires collaboration between different sectors and stakeholders, including government, private sector, civil society, and local communities.

9.2 Adaptation strategies at the local level

In addition to national-level strategies, there are also several mitigation and adaptation strategies that can be implemented at the local level to address the impacts of climate change on water resources and agricultural production:

- *Community-based water management*: Encouraging the participation of local communities in water management can help to ensure the sustainable use of water resources and promote conservation;
- *Farmer field schools*: Providing training and education to farmers on water conservation and management practices can help to increase their knowledge and skills in adapting to the impacts of climate change;
- *Small-scale irrigation systems*: Investing in small-scale irrigation systems, such as drip irrigation and sprinkler systems, can help to increase water use efficiency and reduce water loss;
- *Water storage systems*: Building small-scale water storage systems, such as rainwater harvesting systems and small dams, can help to ensure water availability during periods of water scarcity;
- *Crop diversification*: Encouraging farmers to diversify their crop production can help to reduce the risk of crop failures due to climate change;
- *Agroforestry*: Incorporating trees into farming systems can help to increase water availability and reduce soil erosion, as well as sequester carbon and provide other ecosystem services;
- *Soil conservation*: Implementing soil conservation measures, such as terracing, contouring, and mulching, can help to reduce soil erosion and increase water retention in the soil;
- *Micro-irrigation*: Encouraging the use of micro-irrigation systems, such as drip irrigation, can help to increase water use efficiency and reduce water loss;
- *Water quality monitoring*: Monitoring water quality at the local level can help to identify potential water quality issues and take appropriate action to address them; and
- *Climate-smart villages*: Encouraging the adoption of climate-smart village concept, which integrate different aspects of sustainable development, including climate change adaptation and mitigation, can help to increase resilience of local communities to climate change.

Table 2 Climate change adaptation measures in the agricultural sector

Adaptation needs	Potential adaptation measures
Adaptation focussed on average impacts	• Changes in crops and cropping patterns • Changes in cultivation practices • Increased input of agro-chemicals • Introduce new irrigation areas • Develop climate change resilient crops • Diversify livelihood • Relocate farm processing industry
Adaptation focussed on reducing variability	• Promote insurance • Provide supplemental irrigation • Shift crops from vulnerable areas • Improve soil moisture retention capacity
Adaptation focussed on both changes in the mean and the variability	• Implement regional adaptation plans • Provide advisory services • Promote research on technology and biotechnology • Promote research on water use efficiency • Promote research on management and planning
Adaptation focussed on eliminating barriers to potential agriculture opportunities	• Develop adaptation plans to maintain optimal farming conditions and increased crop productivity • Provide advice based on expert judgement

Source: Adapted from González-Zeas et al. (2013).

It's important to note that these strategies are often most effective when they are tailored to the specific needs of the local community and when they involve the active participation of local farmers and other stakeholders. There are many strategies that could be considered for climate change adaptation and adoption of mitigation measures in agriculture sector (Table 2). In addition, coordination and collaboration between different levels of government, as well as with the private sector and civil society organizations, can help to ensure successful implementation.

9.3 Greenhouse gas mitigation in the agricultural sector

According to FAO estimate there is a total of 5.4–6.3 Gt CO_2e of mitigation potential in the agricultural sector through a combination of emissions reduction, sequestration of carbon in agricultural systems, and major shifts in consumption patterns (FAO, 2013b). Figure 7 provides a useful schematic to highlight the GHG mitigation options at the farm level.

Shifting dietary trends: shift meat consumptions to lower carbon alternatives (e.g. from beef to poultry).

Soil carbon sequestration: from biochar applied to cropland through use of rice residue and sugar cane residues.

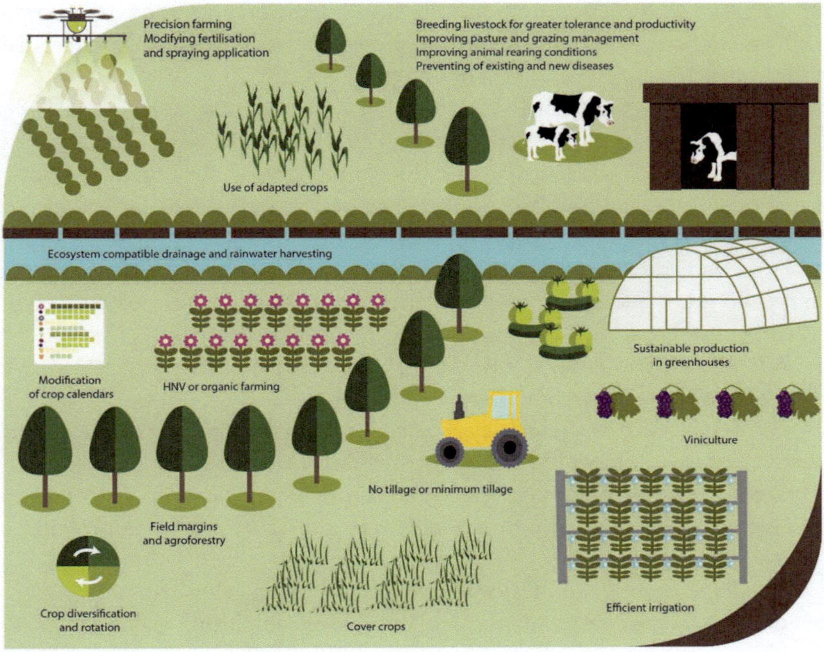

Figure 7 GHG mitigation measures at the farm level. Source: Adapted from: Jacobs et al. (2019).

Enteric fermentation: i.e. reducing ruminant population by increasing their productivity.

Reduce food wastage: for cost savings and emission reduction via transportation.

Grazing land soil carbon sequestration: in large areas of grazing land, e.g. Brazil, China, Kenya, Ethiopia, and Mongolia.

Improved nutrient management: i.e. in China farmers reduce their fertilizer application rates by 30% to 60% without harming yields whilst also reducing emissions (Zhang et al., 2013; Ju et al., 2009).

Emissions: reduction from stored manure and from industrial fertilizer production.

Water and straw management: rice has highest carbon footprints of any crop because of methane generation from cultivating in wet system. Mid-season drainage in irrigated systems and improved management of rice straws can provide significant emission reductions and are cost effective where water is expensive.

Carbon sequestration from agroforestry: Incorporating trees into farming systems can help to sequester carbon, provide other ecosystem services, and reduce emissions from land use change (van Vark, 2013).

Management of soil carbon in cropping systems: There are two main ways to increase carbon stocks in cropland soils: (1) reduced tillage practices can help to increase soil organic matter, reduce emissions of nitrous oxide, a potent greenhouse gas, and help to sequester carbon in the soil; (2) retaining crop residues on the croplands, applying biochar, etc.

Improve carbon storage in grazing lands: via a variety of measures such as timing and rotation of livestock, application of biochar and compost to increase soil carbon storage.

Adapted crops: Using adapted crops (new crops) has synergies to reduce the impact of extreme weather and climate events (e.g. droughts).

Use of cover crops and artificial soil covers: Cover crops can help to reduce erosion, improve soil health, and sequester carbon in the soil.

Crop diversification and rotation: Crop diversification and rotation improve the resilience of crops against climate change and improved soil quality.

No tillage or minimum tillage: Reduced tillage practices can help increase soil organic matter content, reduce emissions of nitrous oxide, a potent greenhouse gas, and help sequester carbon in the soil.

Modifying cropping calendars: Modifying the timing of sowing and harvesting to changed climatic regimes can make use of better soil moisture conditions and improves the crop yields.

Precision agriculture: The use of precision agriculture technologies, such as precision irrigation and nutrient management, can help to increase crop yields and reduce the use of inputs like fertilizer, which in turn can reduce emissions.

Biogas production: Capturing methane from livestock manure and other organic waste materials and converting it into biogas can be used as an energy source and can help to reduce emissions.

Sustainable land use practices: Adopting sustainable land use practices, such as conservation agriculture, agroforestry, and wetlands restoration, can help to sequester carbon in the soil and vegetation and reduce emissions from land use change.

Agroforestry: Incorporating trees into farming systems can help to sequester carbon, provide other ecosystem services, and reduce emissions from land use change.

Manure management: Implementing best practices in manure management can help to reduce emissions of methane and nitrous oxide.

Conservation of wetlands: wetlands can act as carbon sinks and help to reduce emissions.

There may be two strategies that should be considered for climate change adaptation and adoption of mitigation measures in the agriculture sector (Stokes and Howden, 2010), including (i) incremental changes and autonomous farmer's responses for short-term adaptation (e.g. shifting the planting date, switching crop varieties) and (ii) planned and transformational changes (e.g. development of new crop varieties, expanding irrigation infrastructure, introducing agroforestry) for medium- and long-term adaptations planned by the agricultural sector and policy makers. According to Pellerin et al. (2014), following agricultural practices can also be adopted:

1. N_2O can be reduced by more legumes in crop rotations and a better adjustment of N fertilization;
2. C storage can be increased by no-till, residue management, introduction of cover crops, and more trees in agro-ecosystems; and
3. CH_4 can be reduced by paddy rice management (e.g. AWD technique, reducing depth of paddy fields).

9.4 Adaptation of cropping systems to a changing climate

9.4.1 Short-term adaptation at the farm level

Adjustments in crop management, production technology and cropping system composition to adapt to climate change at farm level (Anwar et al., 2012) include, for example:

Shifting cropping patterns and calendars: Shifting planting dates according to precipitation and temperature patterns is low-cost simplest adaptation strategy. For example, earlier sowing of spring crops with early maturing varieties at higher latitudes in southern Europe is an efficient drought-escaping strategy (Moradi et al., 2013). In wheat–maize system 'double-delay' technology (delaying sowing time of wheat and the harvesting time of maize in autumn) is used in the North China Plain under climate change leading to an increase of 4–6% total grain yield (Wang et al., 2012). In the Loess Plateau of China and similar semiarid regions, the sowing date of wheat is delayed by 20–25 days and 10–20 days in the dry and wet years, respectively, compared to usual sowing date for optimum soil water condition to phenological development. Sunflower sowing in late autumn or winter instead of spring is common strategy in some Mediterranean regions to escape water stress at flowering and grain-filling stage (Soriano et al., 2004).

Choosing new crops and cultivars: Using new crops and cultivars combined with change of planting date have synergies to reduce climate extreme vulnerability. For example, the shorter crop duration with increasing temperature could be compensated by using long cycle cultivars combined with early sowing dates. Heat and drought tolerant varieties should be adopted in water scarce areas with increased temperature and heat stress in phenological phases. For instance, earlier flowering varieties could be adopted to allow grain filling in the cooler and wetter parts of the year (Debaeke, 2004). It is recommended to substitute irrigated maize by rainfed crops (e.g. sunflower, sorghum) in western and southern Europe to handle increasing water problems (Debaeke et al., 2008).

Modifying crop rotations: Crop diversification and rotation as a self-insurance measure improve the resilience of crops against infrequent and fluctuant climate change, pest and disease infestations, and price context (Bradshaw et al., 2004). In Europe, shifting of heat-requiring crops toward northern regions diversifying winter-based crop rotations (Tuck et al., 2006). In sub-Saharan Africa, sequential cropping systems including double cropping is a way to maintain overall crop production under changing climate (Waha et al., 2013).

Increasing water availability and water use efficiency by irrigation, crop management, and cropping systems: Supplemental or deficit irrigation is an effective way to maintain or increase grain yield in dry condition. Extra precipitation in winter could however be stored for securing summer irrigation when possible (Zhao et al., 2015). Crop management opportunities could be optimized to cope with drought-prone conditions and changing climate in rainfed agriculture (Debaeke, 2004). In semi-arid regions the use of soil and water conservation techniques as mulching in no-till systems (Aboudrare et al., 2006) are likely to maintain soil quality, protecting against erosion, and facilitating water infiltration (Lal et al., 2011). In Australia, adoption of zero tillage, stubble retention, early sowing, enhanced weed control, and perennial crops and pastures are all suggested to farmers for adapting to reduced precipitation under rainfed conditions (Howden et al., 2007).

9.4.2 Long-term adaptation in the agricultural sector

Following transformational changes in land allocation and management, farming systems and plant breeding should be adjusted to consider possible disruptions due to climate changes that are anticipated by the end of this century(Kates et al., 2012; Anwar et al., 2012).

Breeding new crop varieties that are better adapted to changing climate conditions is a long-term adaptation strategy that can help to increase resilience to climate change. The new crop varieties should be adapted to thermal shocks, such as heat and cold, drought, waterlogging, and higher atmospheric CO_2 concentration, as the current cultivars were selected for different targets (Ziska et al., 2012). This strategy is often suggested as an effective way to adapt to changing climate conditions, as it can help to improve crop yields and reduce the risk of crop failures. However, it's important to note that breeding new crop varieties is a slow process and it can take up to 10 years to develop a new crop cultivar to the point of commercial production. Therefore, it's crucial to begin the breeding process well before it's expected that new cultivars will be needed (Asseng and Pannell, 2012). Additionally, there is a high level of uncertainty about the extent of future change for rainfall, moderately high for temperature, and relatively low for CO_2 concentration, which can make it challenging to breed crops that are well adapted to future climate conditions. One way to overcome these challenges is to use a combination of breeding and genetic engineering techniques, as well as to focus on specific traits that are more likely to be affected by climate change, such as drought and heat tolerance. Additionally, the use of precision breeding techniques, such as marker-assisted selection and genome editing, can help to speed up the breeding process and improve the efficiency of the selection process. Moreover, it's important to work with farmers and other stakeholders to identify the most pressing needs and to evaluate the performance of new crop varieties under different climate conditions and farming systems.

Model-based tools, climate forecasting, and site-specific technology can play an important role in helping farmers to optimize, support, and secure their decisions in adapting to changing climate conditions. These tools can help to provide farmers with the information and guidance they need to make informed decisions on crop management, water use, and other agricultural practices. Remote sensing technologies, such as drones and satellites, can provide detailed information on plant canopies and other crop-related data, which can be used to increase the precision of crop management and maximize resource-use efficiency. The use of precision farming technologies, such as precision irrigation and nutrient management, can also help to improve crop yields and reduce the risk of crop failures. Adaptation to climate change can range from tactical fine-tuning of existing cropping systems to more fundamental changes in cropping systems, such as diversifying crops and varieties to increase resilience at the farm or territorial level. This can have downstream impacts on land use, agricultural sector activity, such as machinery, inputs, and market, and it can change deeply the socio-technical system, such as cooperatives, consultants, and seed companies. Additionally, the extent of numerical agriculture and the willingness to diversify crops and varieties will play a role in determining the success of the adaptation. However, it is important to note that the adoption of these innovations can be hindered by lock-ins, which might occur in the existing system.

10 Conclusion

This chapter has highlighted the challenges faced by agriculture worldwide due to climate change, droughts, and water scarcity. It also provides an overview of the various strategies that can be implemented to address these challenges, including adaptation, mitigation, water management, government policies, and international cooperation. The IPCC's Fifth Assessment Report (AR5) states that crop yields could decline by up to 2% per decade due to climate change, with the potential for larger declines in some regions. Additionally, the report states that water scarcity caused by climate change could lead to reductions in crop yields of up to 30% in some regions. According to the FAO, climate change is projected to lead to a decline in crop yields of up to 30% in some regions, with the potential for even larger declines in developing countries. In addition, the FAO estimates that by 2050, up to 600 million people could be at risk of hunger as a result of climate change. In terms of water scarcity, the World Bank estimates that by 2050, water scarcity could affect up to 5 billion people, with the potential to reduce crop yields by up to 50%. Drought, which is becoming more frequent and severe in many regions as a result of climate change, is also having a significant impact on agriculture, with the potential to reduce crop yields by up to 30%.

Climate change, droughts, and water scarcity are having a significant impact on agriculture around the world. In many regions, rising temperatures and changing rainfall patterns are leading to more frequent and severe droughts, which are reducing crop yields and making it more difficult to grow certain crops. For example, in Africa, the Sahel region is facing increasing water stress due to climate change, which is making it difficult for farmers to grow crops such as millet and sorghum. In Central Asia, the increasing frequency and severity of droughts is having a major impact on irrigation-dependent crops such as cotton and rice. Water scarcity is also a major concern, particularly in regions where water resources are already limited. In many parts of the Middle East and North Africa, e.g. water scarcity is a major constraint on agriculture, and farmers are increasingly turning to irrigation technologies such as drip irrigation and desalination to try to overcome this constraint. In India, water scarcity is a major problem, particularly in the states of Gujarat, Maharashtra, and Rajasthan, where farmers are facing increasing difficulty in growing crops such as cotton, sugarcane, and rice due to water scarcity. The effects of climate change, droughts, and water scarcity are also having a major impact on the livelihoods of smallholder farmers, particularly in developing countries. Many smallholder farmers are already facing significant challenges due to poverty, lack of access to markets, and limited access to technology and information. Climate change, droughts, and water scarcity are making it even more difficult for these farmers to make a living, and many are being forced to leave their land in search of other livelihoods.

Adaptation and mitigation measures can play an important role in reducing the impacts of climate change, water scarcity, and drought on agriculture. Adaptation measures such as crop diversification, irrigation efficiency, and water storage can help farmers to become more resilient to changes in water availability at the global, regional and local levels. Mitigation measures such as sustainable agricultural practices and water-saving technologies can help to reduce greenhouse gas emissions and promote sustainable agriculture at the global, regional, and local levels. Government policies such as subsidies, tax incentives, and regulations can encourage the adoption of sustainable agricultural practices and water-saving technologies at the global, regional, and local levels. In order to address these challenges, it is important to implement a combination of strategies that are tailored to the specific context of each region and country. This may include investing in irrigation infrastructure, promoting sustainable agricultural practices, and providing smallholder farmers with access to information, training, and technology. Governments, international organizations, and the private sector also have important roles to play in addressing these challenges, by providing financial resources, technical expertise, and policy support.

Future policies on managing climate change, droughts, and water scarcity affecting agriculture should include the following key elements:

- Recognition of the significant impacts of climate change, droughts, and water scarcity on agriculture and the need for immediate action to address these challenges.
- Identification of the most vulnerable regions and populations, such as smallholder farmers in developing countries, and the need to prioritize support for these groups.
- Emphasis on the need for a multifaceted approach that includes both mitigation and adaptation measures, tailored to the specific context of each region and country.
- Investment in irrigation infrastructure, water-saving technologies, and sustainable land management practices to increase the resilience of agricultural systems to changes in water availability.
- Promoting sustainable agricultural practices such as agroforestry, conservation tillage, and crop diversification to reduce greenhouse gas emissions and increase the resilience of the agricultural sector to climate change.
- Support for smallholder farmers through access to information, training, and technology to promote sustainable and climate-resilient agricultural practices.
- Encouraging the private sector to invest in sustainable and climate-resilient agricultural practices, through the use of subsidies, tax incentives, and regulations.
- Encouraging international cooperation to address the common challenge of climate change through reducing greenhouse gas emissions, sharing information, and technology.

- Monitoring and evaluation of the effectiveness of the policies and programmes implemented to address the challenges of climate change, droughts, and water scarcity affecting agriculture.
- Regularly updating policies and programmes to reflect the latest scientific and technical knowledge on the impacts of climate change, droughts, and water scarcity on agriculture.

Overall, this chapter highlights the need for a multifaceted approach to managing climate change, droughts, and water scarcity affecting agriculture. It is crucial to adopt a combination of strategies that are tailored to the specific context of each region and country. This will help to promote sustainable agriculture and ensure food security in the face of these challenges.

11 References

Abid, M., Scheffran, J., Schneider, U. A. and Ashfaq, M. (2015). Farmers' perceptions of and adaptation strategies to climate change and their determinants: the case of Punjab Province, Pakistan. *Earth System Dynamics* 6(1), 225–243. https://doi.org/10.5194/esd-6-225-2015.

Aboudrare, A., Debaeke, P., Bouaziz, A. and Chekli, H. (2006). Effects of soil tillage and fallow management on soil water storage and sunflower production in a semi-arid Mediterranean climate. *Agricultural Water Management* 83(3), 183–196. https://doi.org/10.1016/j.agwat.2005.12.001.

Adnan, S., Ullah, K., Shuanglin, L., Gao, S., Khan, A. H. and Mahmood, R. (2017). Comparison of various drought indices to monitor drought status in Pakistan. *Climate Dynamics* 51(5–6), 1885–1899. https://doi.org/10.1007/s00382-017-3987-0.

Aguilar, E., Aziz Barry, A., Brunet, M., Ekang, L., Fernandes, A., Massoukina, M., Mbah, J., Mhanda, A., do Nascimento, D. J., Peterson, T. C., Thamba Umba, O., Tomou, M. and Zhang, X. (2009). Changes in temperature and precipitation extremes in western central Africa, Guinea Conakry, and Zimbabwe, 1955–2006. *Journal of Geophysical Research* 114(D2). https://doi.org/10.1029/2008JD011010.

Ahmad, M., Siftain, H. and Iqbal, M. A. (2014). Impact of climate change on wheat productivity in Pakistan: a district level analysis. MPRA Paper, Pakistan Institute of Development Economics.

Aktaş, B. (2020). *Possible Changes in Some Climate Parameters and Climate Types in Konya Depending on Global Warming*. Kastamonu University, Institute of Science, Department of Sustainable Agriculture and Natural Plant Resources, Kastamonu.

Alam, M. (2007). Impacts, vulnerability and adaptation to climate change in Asia. UNFCCC Asia Regional Workshop on Climate Change Adaptation, Beijing, 11–13 April. Available at: https://unfccc.int/sites/default/files/alam_back_paper.pdf (Accessed 21 August 2023).

Alam, Md. M., Siwar, C., Talib, B. and bin Toriman, M. E. (2014). Impacts of climatic changes on paddy production in Malaysia: micro study on IADA at North West Selangor. *Research Journal of Environmental and Earth Sciences* 6(5), 251–258. https://doi.org/10.19026/rjees.6.5767.

Anwar, M. R., Liu, D. L., Macadam, I. and Kelly, G. (2012). Adapting agriculture to climate change: a review. *Theoretical and Applied Climatology* 113(1-2), 225–245. https://doi.org/10.1007/s00704-012-0780-1.

Asseng, S. and Pannell, D. J. (2012). Adapting dryland agriculture to climate change: farming implications and research and development needs in Western Australia. *Climatic Change* 118(2), 167–181. https://doi.org/10.1007/s10584-012-0623-1.

Australian Bureau of Statistics (ABS) (2022). Statistics | Australian Bureau of Statistics. www.abs.gov.au. Available at: https://www.abs.gov.au/statistics.

Bates, B. C., Kundzewicz, Z. W., Wu, S. and Palutikof, J. P. (2008). Climate Change and Water.*Technical Paper of the Intergovernmental Panel on Climate Change* 210 pp. IPCC Secretariat, Geneva.

Berry, P. M., Rounsevell, M. D. A., Harrison, P. A. and Audsley, E. (2006). Assessing the vulnerability of agricultural land use and species to climate change and the role of policy in facilitating adaptation. *Environmental Science and Policy* 9(2), 189–204. https://doi.org/10.1016/j.envsci.2005.11.004.

Bhaga, T. D., Dube, T., Shekede, M. D. and Shoko, C. (2020). Impacts of climate variability and drought on surface water resources in sub-Saharan Africa using remote sensing: a review. *Remote Sensing* 12(24), 4184. https://doi.org/10.3390/rs12244184.

Bhatt, R. and Hossain, A. (2019). Concept and consequence of evapotranspiration for sustainable crop production in the era of climate change. *Advanced Evapotranspiration Methods and Applications* https://doi.org/10.5772/intechopen.83707.

Bradshaw, B., Dolan, H. and Smit, B. (2004). Farm-level adaptation to climatic variability and change: crop diversification in the Canadian prairies. *Climatic Change* 67(1), 119–141. https://doi.org/10.1007/s10584-004-0710-z.

Brown, E. A., Caviness, C. E. and Brown, D. A. (1985). Response of selected soybean cultivars to soil moisture deficit. *Agronomy Journal* 77(2), 274–278. https://doi.org/10.2134/agronj1985.00021962007700020022x.

Buis, A. (2019). The Atmosphere: Getting a Handle on Carbon. *Sizing Up Humanity's Impacts on Earth's Changing Atmosphere: A Five-Part Series*. NASA: Global Climate Change: Vital Signs of the Planet. Available at: https://climate.nasa.gov/news/2915/the-atmosphere-getting-a-handle-on-carbon-dioxide/.

Cai, X., Zhang, X., Noël, P. H. and Shafiee-Jood, M. (2015). Impacts of climate change on agricultural water management: a review. *Wiley Interdisciplinary Reviews: Water* 2(5), 439–455. https://doi.org/10.1002/wat2.1089.

CDP (2015). Global climate change report 2015 – CDP. Available at: https://www.cdp.net/en/research/global-reports/global-climate-change-report-2015.

Cetin, O. and Bilgel, L. (2002). Effects of different irrigation methods on shedding and yield of cotton. *Agricultural Water Management* 54(1), 1–15.

CIAT (2011). Climate analysis indicators tool. Available at: http://cait.wri.org Accessed 10 December 2022.

Cramer, W., Guiot, J., Fader, M., Garrabou, J., Gattuso, J.-P., Iglesias, A., Lange, M. A., Lionello, P., Llasat, M. C., Paz, S., Peñuelas, J., Snoussi, M., Toreti, A., Tsimplis, M. N. and Xoplaki, E. (2018). Climate change and interconnected risks to sustainable development in the Mediterranean. *Nature Climate Change* 8(11), 972–980. https://doi.org/10.1038/s41558-018-0299-2.

Craufurd, P. Q. and Wheeler, T. R. (2009). Climate change and the flowering time of annual crops. *Journal of Experimental Botany* 60(9), 2529–2539. https://doi.org/10.1093/jxb/erp196.

Dash, S. K. and Mamgain, A. (2011). Changes in the frequency of different categories of temperature extremes in India. *Journal of Applied Meteorology and Climatology* 50(9), 1842–1858. https://doi.org/10.1175/2011JAMC2687.1.

Dawoud, Z. (2014). Economic optimal allocation of irrigation water in Egypt. *Journal of Development and Agricultural Economics* 6(12), 472–480. https://doi.org/10.5897/jdae2014.0609.

Debaeke, P. (2004). Scenario analysis for cereal management in water limited conditions by the means of a crop simulation model (STICS). *Agronomie* 24(6-7), 315–326. https://doi.org/10.1051/agro:2004035.

Debaeke, P., Bergez, J.-E. and Leenhardt, D. (2008). Perspectives agronomiques et génétiques pour limiter ou réguler la demande en eau d'irrigation. *La Houille Blanche* 94(6), 17–25. https://doi.org/10.1051/lhb:2008067.

Dehlavi, A., Groom, B., Grost, A. and Zaman, F. (2015). *Climate Change Adaptation in the Indus Ecoregion: A Microeconometric Study of the Determinants, Impacts, and Cost Effectiveness of Adaptation Strategies*. WWF-Pakistan, Karachi.

Dickie, A., Streck, C., Roe, S., Zurek, M., Haupt, F. and Dolginow, A. (2014). Strategies for mitigating climate change in agriculture: recommendations for philanthropy – executive summary. Climate Focus and California Environmental Associates, prepared with the support of the Climate and Land Use Alliance. Report and supplementary materials available at: www.agriculturalmitigation.org.

Döll, P. (2002). Impact of climate change and variability on irrigation requirements: a global perspective. *Climatic Change* 54(3), 269–293. https://doi.org/10.1023/A:1016124032231.

Eckstein, D., Künzel, V. and Schäfer, L. (2021). Global climate risk index 2021. Who suffers most from extreme weather events? Weather related loss events in 2019 and 2000–2019. German Watch, p. 15.

Eitzinger, J., Trnka, M., Semerádová, D., Thaler, S., Svobodová, E., Hlavinka, P., Šiška, B., Takáč, J., Malatinská, L., Nováková, M., Dubrovský, M. and Žalud, Z. (2012). Regional climate change impacts on agricultural crop production in Central and Eastern Europe – hotspots, regional differences and common trends. *The Journal of Agricultural Science* 151(6), 787–812. https://doi.org/10.1017/S0021859612000767.

Enciso, J., Jifon, J., Anciso, J. and Ribera, L. (2015). Productivity of onions using subsurface drip irrigation versus furrow irrigation systems with an internet based irrigation scheduling program. *International Journal of Agronomy* 2015, 1–6. https://doi.org/10.1155/2015/178180.

Eriyagama, N., Smakhtin, V., Chandrapala, L. and Fernando, K.(2010). *Impacts of Climate Change on Water Resources and Agriculture in Sri Lanka: A Review and Preliminary Vulnerability Mapping*. International Water Management Institute, Colombo. 51 p. IWMI Research Report 135. https://doi.org/10.3910/2010.211.

Falkenmark, M. (2013). Growing water scarcity in agriculture: future challenge to global water security. *Philosophical Transactions of the Royal Society A: Mathematical, Physical and Engineering Sciences* 371(2002), 20120410. https://doi.org/10.1098/rsta.2012.0410.

FAO (2013a). Climate-smart agriculture sourcebook. Available at: http://www.fao.org/3/i3325e.pdf.

FAO (2013b). *Food and Agriculture Organization of the United Nations*. FAOSTAT. Retrieved 2013-2014, from http://faostat.fao.org.

Fraser, E. D. G., Simelton, E., Termansen, M., Gosling, S. N. and South, A. (2013). Vulnerability hotspots': integrating socio-economic and hydrological models to identify where cereal production may decline in the future due to climate change induced drought. *Agricultural and Forest Meteorology* 170, 195–205. https://doi.org/10.1016/j.agrformet.2012.04.008.

Frieler, K., Levermann, A., Elliott, J., Heinke, J., Arneth, A., Bierkens, M. F. P., Ciais, P., Clark, D. B., Deryng, D., Döll, P., Falloon, P., Fekete, B., Folberth, C., Friend, A. D., Gellhorn, C., Gosling, S. N., Haddeland, I., Khabarov, N., Lomas, M., Masaki, Y., Nishina, K., Neumann, K., Oki, T., Pavlick, R., Ruane, A. C., Schmid, E., Schmitz, C., Stacke, T., Stehfest, E., Tang, Q., Wisser, D., Huber, V., Piontek, F., Warszawski, L., Schewe, J., Lotze-Campen, H. and Schellnhuber, H. J. (2015). A framework for the cross-sectoral integration of multi-model impact projections: land use decisions under climate impacts uncertainties. *Earth System Dynamics* 6(2), 447–460. https://doi.org/10.5194/esd-6-447-2015.

Geerts, S., Raes, D., Garcia, M., Vacher, J., Mamani, R., Mendoza, J., Huanca, R., Morales, B., Miranda, R., Cusicanqui, J. and Taboada, C. (2008). Introducing deficit irrigation to stabilize yields of quinoa (*Chenopodium quinoa* Willd.). *European Journal of Agronomy* 28(3), 427–436. https://doi.org/10.1016/j.eja.2007.11.008.

Gleason, D. J. (2013). Evapotranspiration-based irrigation scheduling tools for use in eastern Colorado. MS Thesis, Colorado State University, Fort Collins, CO, p. 225.

González-Zeas, D., Quiroga, S., Iglesias, A. and Garrote, L. (2013). Looking beyond the average agricultural impacts in defining adaptation needs in Europe. *Regional Environmental Change* 14(5), 1983–1993. https://doi.org/10.1007/s10113-012-0388-0.

Gregory, P. J., Johnson, S. N., Newton, A. C. and Ingram, J. S. I. (2009). Integrating pests and pathogens into the climate change/food security debate. *Journal of Experimental Botany* 60(10), 2827–2838.

Hofste, R. W., Reig, P. and Schleifer, L. (2019). 17 countries, home to one-quarter of the world's population, face extremely high water stress. Available at: https://www.wri.org/insights/17-countries-home-one-quarter-worlds-population-face-extremely-high-water-stress.

Howden, S. M., Soussana, J. F., Tubiello, F. N., Chhetri, N., Dunlop, M. and Meinke, H. (2007). Adapting agriculture to climate change. *Proceedings of the National Academy of Sciences of the United States of America* 104(50), 19691–19696.

Hussain, F., Wu, R.-S., Nabi, G., Noor, R. S., Anjum, M. N., Azam, M. and Afzal, A. (2022). Analysis of temperature data using innovative trend pivot analysis method and trend polygon star concept: a case study of Soan River Basin, Pothohar, Pakistan. *Pure and Applied Geophysics*180, 475–507.

Iglesias, A., Garrote, L., Quiroga, S. and Moneo, M. (2009). "Impacts of climate change in agriculture in Europe" PESETA-agriculture study. JRC Scientific and Technical Research Series EUR 24107 EN, European Commission, Joint Research Centre.

IPCC (2007). Climate change 2007, impacts, adaptation and vulnerability. Cambridge University Press, Cambridge/New York.

IPCC (2012). In: Field, C. B., Barros, C., Stocker, T. F., Qin, D., Dokken, D. J., Ebi, K. L., Mastrandrea, M. D., Mach, K. J., Plattner, G.-K., Allen, S. K., Tignor, M. and Midgley, P. M. (Eds), *Managing the Risks of Extreme Events and Disasters to Advance Climate Change Adaptation*. Cambridge University Press, Cambridge, 582 p.

IPCC (2014). Climate Change 2014: synthesis report. In: Core Writing Team, Pachauri, R. K. and Meyer, L. A. (Eds), Contribution of Working Groups I, II and III to the Fifth Assessment Report of the Intergovernmental Panel on Climate Change. Geneva, IPCC, p. 151.

IPCC (2018). Summary for policymakers. Global Warming of 1.5°C. An IPCC special report on the impacts of global warming of 1.5°C above pre-industrial levels and related global greenhouse gas emission pathways, in the context of strengthening the global response to the threat of climate change, sustainable development, and efforts to eradicate poverty. IPCC, Geneva. Available at: www.ipcc.ch/sr15/chapter/spm/.

IPCC (2019). Climate change and land. Summary for policy makers, Intergovernmental Panel on Climate Change, Geneva, Switzerland. Available at: https://www.ipcc.ch/srccl-reportdownload-page/ Accessed 8 August 2019.

IPCC (2021). Summary for policymakers. In: Masson-Delmotte, V., Zhai, P., Pirani, A., Connors, S. L., Péan, C., Berger, S., Caud, N., Chen, Y., Goldfarb, L., Gomis, M. I., Huang, M., Leitzell, K., Lonnoy, E., Matthews, J. B. R., Maycock, T. K., Waterfield, T., Yelekçi, O., Yu, R. and Zhou, B. (Eds), Climate Change 2021: The Physical Science Basis. Contribution of Working Group I to the Sixth Assessment Report of the Intergovernmental Panel on Climate Change. Cambridge University Press, Cambridge, UK and New York, pp. 3–32. https://doi.org/10.1017/9781009157896.001.

Jacobs, C., Berglund, M., Dworak, T., Marras, S., Mereu, V. and Michetti, M. (2019). Climate change adaptation in the agriculture sector in Europe. EEA Report No. 04/2019, European Environment Agency, Kongens Nytorv, Copenhagen K, Denmark.

Ju, X. T., Xing, G. X., Chen, X. P., Zhang, S. L., Zhang, L. J., Liu, X. J., Cui, Z. L., Yin, B., Christie, P., Zhu, Z. L. and Zhang, F. S. (2009). Reducing environmental risk by improving N management in intensive Chinese agricultural systems. *Proceedings of the National Academy of Sciences of the United States of America* 106(9), 3041–3046.

Kates, R. W., Travis, W. R. and Wilbanks, T. J. (2012). Transformational adaptation when incremental adaptations to climate change are insufficient. *Proceedings of the National Academy of Sciences of the United States of America* 109(19), 7156–7161.

Khan, M. J. and Khan, I. (2018). Impact of climate change on key crops of Khyber Pakhtunkhwa. ADP Project Financed by the Government of Khyber Pakhtunkhwa. Climate Change Center, The University of Agriculture, Peshawar.

Kirby, J. M., Mainuddin, M., Mpelasoka, F., Ahmad, M. D., Palash, W., Quadir, M. E., Shah-Newaz, S. M. and Hossain, M. M. (2016). The impact of climate change on regional water balances in Bangladesh. *Climatic Change* 135(3–4), 481–491.

Kovats, R. S. and Valentini, R. (2014). Europe. In: *Climate Change 2014: Impacts, Adaptation, and Vulnerability. Part, B: Regional Aspects*. Contribution of Working Group II to the Fifth Assessment Report of the Intergovernmental Panel on Climate Change. Cambridge University Press, Cambridge, pp. 1267–1326.

Kumar, K. K., Rajagopalan, B., Hoerling, M., Bates, G. and Cane, M. (2006). Unraveling the mystery of Indian monsoon failure during El Niño. *Science* 314(5796), 115–119.

Lal, R., Delgado, J. A., Groffman, P. M., Millar, N., Dell, C. and Rotz, A. (2011). Management to mitigate and adapt to climate change. *Journal of Soil and Water Conservation* 66(4), 276–285.

Linquist, B. A., Anders, M. M., Adviento-Borbe, M. A. A., Chaney, R. L., Nalley, L. L., Da Rosa, E. F. F. and Van Kessel, C. (2015). Reducing greenhouse gas emissions, water use, and grain arsenic levels in rice systems. *Global Change Biology* 21(1), 407–417.

Lobell, D. B. and Field, C. B. (2007). Global scale climate-crop yield relationships and the impacts of recent warming. *Environmental Research Letters* 2(1), 014002.

Lobell, D. B., Schlenker, W. and Costa-Roberts, J. (2011). Climate trends and global crop production since 1980. *Science* 333(6042), 616–620.

Lozano, D., Ruiz, N. and Gavilán, P. (2016). Consumptive water use and irrigation performance of strawberries. *Agricultural Water Management* 169, 44–51.

Malik, K. M., Taylor, P. A., Szeto, K. and Khan, A. H. (2013). Characteristics of central southwest Asian water budgets and their impacts on regional climate. *Atmospheric and Climate Sciences* 03(2), 259–268.

Masih, I., Maskey, S., Mussá, F. E. F. and Trambauer, P. (2014). A review of droughts on the African continent: a geospatial and long-term perspective. *Hydrology and Earth System Sciences* 18(9), 3635–3649.

Mishra, A. K. and Desai, V. R. (2005). Drought forecasting using stochastic models. *Stochastic Environmental Research and Risk Assessment* 19(5), 326–339.

Mishra, A. K. and Singh, V. P. (2010). A review of drought concepts. *Journal of Hydrology* 391(1–2), 202–216.

Molle, B., Tomas, S., Hendawi, M. and Granier, J. (2012). Evaporation and wind drift losses during sprinkler irrigation influenced by droplet size distribution. *Irrigation and Drainage* 61(2), 240–250.

Moradi, R., Koocheki, A., Mahallati, M. N. and Mansoori, H. (2013). Adaptation strategies for maize cultivation under climate change in Iran: irrigation and planting date management. *Mitigation and Adaptation Strategies for Global Change* 18(2), 265–284.

Mpelasoka, F., Hennessy, K., Jones, R. and Bates, B. (2008). Comparison of suitable drought indices for climate change impacts assessment over Australia towards resource management. *International Journal of Climatology* 28(10), 1283–1292.

Müller, C. and Elliott, J. (2015). The Global Gridded Crop Model intercomparison: approaches, insights and caveats for modelling climate change impacts on agriculture at the global scale. In: Elbehri, A. (Ed.), *Climate Cchange and Food Systems: Global Assessments and Implications for Food Security and Trade*. FAO, Rome.

Ndehedehe, C. E., Agutu, N. O., Ferreira, V. G. and Getirana, A. (2020). Evolutionary drought patterns over the Sahel and their teleconnections with low frequency climate oscillations. *Atmospheric Research* 233, 104700.

Nelson, G. C., Rosegrant, M. W., Koo, J., Robertson, R., Sulser, T., Zhu, T., Ringler, C., Msangi, S., Palazzo, A., Batka, M., Magalhaes, M., Valmonte-Santos, R., Ewing, M. and Lee, D. (2009). *Climate Change Impacts on Agriculture and Costs of Adaptation*. Washington, DC, International Food Policy Research Institute.

Niang, I., Ruppel, O. C., Abdrabo, M. A., Essel, A., Lennard, C., Padgham, J. and Urquhart, P. (2014). Africa. In: Barros, V. R., Field, C. B., Dokken, D. J., Mastrandrea, M. D. and Mach, K. J. (Eds), *Climate Change 2014: Impacts, Adaptation, and Vulnerability. Part B: Regional Aspects*. Contribution of Working Group II to the Fifth Assessment Report of the Intergovernmental Panel on Climate Change. Cambridge University Press, Cambridge, UKand New York.

Ofori, S. A., Cobbina, S. J. and Obiri, S. (2021). Climate change, land, water, and food security: perspectives from sub-Saharan Africa. *Frontiers in Sustainable Food Systems* 5, 680924. https://doi.org/10.3389/fsufs.2021.680924.

Parry, M. L., Rosenzweig, C., Iglesias, A., Livermore, M. and Fischer, G. (2004). Effects of climate change on global food production under SRES emissions and socio-economic scenarios. *Global Environmental Change* 14(1), 53–67.

Perveen, S., Krishnamurthy, C. K., Sidhu, R. S., Vatta, K., Kaur, B., Modi, V., Fishman, R., Polycarpou, L. and Lall, U. (2012). *Restoring Groundwater in Punjab, India's Breadbasket: Finding Agricultural Solutions for Water Sustainability*. Columbia Water Center White Paper. Columbia University, New York, p. 28.

Planton, S., Déqué, M., Chauvin, F. and Terray, L. (2008). Expected impacts of climate change on extreme climate events. *Comptes Rendus Geoscience* 340(9–10), 564–574.

Poljanšek, K., Montserrat, M. F., Clark, I. and De Groeve, T. (2017). *Science for Disaster Risk Management 2017: Knowing Better and Losing Less*. Publications Office of the European Union, Luxembourg.

Porter, J. R., Xie, L., Challinor, V., Cochrane, K., Howden, S. M., Iqbal, M. M., Lobell, D. B. and Travasso, M. I. (2014). Food security and food production systems. In: *Climate Change 2014: Impacts, Adaptation, and Vulnerability*. Cambridge University Press, Cambridge, UK and New York, pp. 485–533.

Robertson, M. J., Inmam-Bamber, N. G., Muchow, R. C. and Wood, A. W. (1999). Physiology and productivity of sugarcane with early and mid-season water deficit. *Field Crops Research* 64(3), 211–227.

Rojas, N., Yanyun Li, E. and Renato Cumani, N. (2014). *Understanding the Drought Impact of El Niño on the Global Agricultural Areas: An Assessment Using FAO's Agricultural Stress Index (ASI), Environment and Natural Resources Management Series*. Food and Agriculture Organization of the United Nations, Rome.

Rosenzweig, C., Elliott, J., Deryng, D., Ruane, A. C., Müller, C., Arneth, A., Boote, K. J., Folberth, C., Glotter, M., Khabarov, N., Neumann, K., Piontek, F., Pugh, T. A. M., Schmid, E., Stehfest, E., Yang, H. and Jones, J. W. (2014). Assessing agricultural risks of climate change in the 21st century in a global gridded crop model intercomparison. *Proceedings of the National Academy of Sciences of the United States of America* 111(9), 3268–3273. https://doi.org/10.1073/pnas.1222463110.

Sabrina, A. and Donatella, V. (2014). Climate change effects and agriculture in Italy: a stochastic frontier analysis at regional level. Available at: https://mpra.ub.uni-muenchen.de/53500/ MPRA Paper No. 53500.

Saeed, M., Sheikh, R. U. and Shoaib, M. (2018). A review on: water saving techniques for domestic, agricultural and industrial water usage. World Water Day 2018, 75th Session. Pakistan Engineering Congress, Lahore.

Saif Ullah (2017). Climate change impact on agriculture of Pakistan – a leading agent to food security. *International Journal of Environmental Science and Natural Resources* 6(3), 1–4.

Sathar, Z. A. and Khan, K. (Eds) (2019). *Climate, Population, and Vulnerability in Pakistan: Exploring Evidence of Linkages for Adaptation*. Population Council, Islamabad.

Schlenker, W. and Lobell, D. B. (2010). Robust negative impacts of climate change on African agriculture. *Environmental Research Letters* 5(1), 014010.

Schlenker, W. and Roberts, M. J. (2006). Nonlinear effects of weather on crop yields: implications for climate change. *Review of Agricultural Economics* 28(3), 391–398.

Shrestha, S., Ciaian, P., Himics, M. and Van Doorslaer, B. (2013). Impacts of climate change on EU agriculture. *Review of Agricultural and Applied Economics* XVI(2), 24–39.

Smith, P., Bustamante, M., Ahammad, H., Clark, H., Dong, H., Elsiddig, E. A., et al. (2014). Agriculture, forestry and other land use (AFOLU). In: Edenhofer, O., Pichs-Madruga, R., Sokona, Y., Farahani, E., Kadner, S., Seyboth, K., Haberl, H., Harper, R., House, J., Jafari, M., Masera, O., Mbow, C., Ravindranath, N. H., Rice, C. W., Abad, C. R., Romanovskaya, A., Sperling, F. and Tubiello, F. N. (Eds), *Climate Change 2014: Mitigation of Climate Change Contribution of Working Group III to the Fifth Assessment Report of the Intergovernmental Panel on Climate Change*. Cambridge University Press, Cambridge, UK and New York.

Soriano, M. A., Orgaz, F., Villalobos, F. J. and Fereres, E. (2004). Efficiency of water use of early plantings of sunflower. *European Journal of Agronomy* 21(4), 465–476.

Spinoni, J., Naumann, G. and Vogt, J. V. (2017). Pan-European seasonal trends and recent changes of drought frequency and severity. *Global and Planetary Change* 148, 113–130.

Spinoni, J., Vogt, J. V., Naumann, G., Barbosa, P. and Dosio, A. (2018). Will drought events become more frequent and severe in Europe? *International Journal of Climatology* 38(4), 1718–1736. (https://doi.org/10.1002/joc.5291).

Stewart, W. L., Fulton, A. E., Krueger, W. H., Lampinen, B. D. and Shackel, K. A. (2011). Regulated deficit irrigation reduces water use of almonds without affecting yield. *California Agriculture* 65(2), 90–95.

Stokes, C. J. and Howden, M. (2010). *Adapting Agriculture to Climate Change: Preparing Australian Agriculture,Forestry and Fisheries for the Future*, CSIRO Publishing, Collingwood.

Tagar, A., Chandio, F. A., Mari, I. A. and Wagan, B. (2012). Comparative study of drip and furrow irrigation methods at farmer's field in Umarkot. *World Academy of Science, Engineering and Technology* 69, 863–867.

Trenberth, K. E., Dai, A., Van der Schrier, G., Jones, P. D., Barichivich, J., Briffa, K. R. and Sheffield, J. (2014). Global warming and changes in drought. *Nature Climate Change* 4(1), 17–22. https://doi.org/10.1038/nclimate2067.

Tuck, G., Glendining, M. J., Smith, P., House, J. I. and Wattenbach, M. (2006). The potential distribution of bioenergy crops in Europe under present and future climate. *Biomass and Bioenergy* 30(3), 183–197.

Turral, H., Burke, J. and Faurès, J. (2011). *Climate Change, Water and Food Security*. Food and Agricultural Organization of the United Nations, Rome.

UNFCCC (2015). *Adoption of the Paris Agreement*. United Nations Framework Convention on Climate Change, Bonn. Available at: https://unfccc.int/resource/docs/2015/cop21/eng/l09r01.pdf.

Ünlü, M., Kanber, R., Koç, D. L., Özekİcİ, B., Kekeç, U., Yeşİloğlu, T., Ortaş, İ., Ünlü, F., Kapur, B., Tekİn, S., Käthner, J., Gebbers, R., Zude, M., Peeters, A. and Ben-gal, A. (2014). Irrigation scheduling of grapefruit trees in a Mediterranean environment throughout evaluation of plant water status and evapotranspiration. *Turkish Journal of Agriculture and Forestry* 38, 908–915.

US Department of Agriculture (USDA) (2013). *2012 Census of Agriculture. Crops Grown in the Open and Pasture Data*. USDA, Washington, DC.

van Donk, S. J., Martin, D. L., Irmak, S., Melvin, S. R., Petersen, J. L. and Davison, D. R. (2010). Crop residue cover effects on evaporation, soil water content, and yield of deficit-irrigated corn in west-central Nebraska. *Transactions of the American Society of Agricultural and Biological Engineers* 53, 1787–1797.

van Vark, C. (2013). How agroforestry systems can improve food security in developing countries. *The Guardian*. Available at: https://www.theguardian.com/global-development-professionals-network/2013/feb/26/agroforestry-farming-food-security.

Vu, J. C. V., Allen, L. H. Jr., Boote, K. J. and Bowes, G. (1997). Effects of elevated CO_2 and temperature on photosynthesis and RuBisCO in rice and soybean. *Plant, Cell and Environment* 20(1), 68–76.

Waha, K., Müller, C., Bondeau, A., Dietrich, J. P., Kurukulasuriya, P., Heinke, J. and Lotze-Campen, H. (2013). Adaptation to climate change through the choice of cropping system and sowing date in sub-Saharan Africa. *Global Environmental Change* 23(1), 130–143.

Wang, A., Lettenmaier, D. P. and Sheffield, J. (2011a). Soil moisture drought in China, 1950–2006. *Journal of Climate* 24(13), 3257–3271. https://doi.org/10.1175/2011JCLI3733.1.

Wang, D., Hejazi, M., Cai, X. and Valocchi, A. J. (2011b). Climate change impact on meteorological, agricultural, and hydrological drought in central Illinois. *Water Resources Research* 47(9).

Wang, J., Mendelsohn, R., Dinar, A., Huang, J., Rozelle, S. and Zhang, L. (2009). The impact of climate change on China's agriculture. *Agricultural Economics* 40(3), 323–337.

Wang, J., Wang, E., Yang, X., Zhang, F. and Yin, H. (2012). Increased yield potential of wheat-maize cropping system in the North China Plain by climate change adaptation. *Climatic Change* 113(3–4), 825–840.

Wiréhn, L. (2018). Nordic agriculture under climate change: a systematic review of challenges, opportunities and adaptation strategies for crop production. *Land Use Policy* 77, 63–74.

Wu, Y., Jia, Z., Ren, X., Zhang, Y., Chen, X., Bing, H. and Zhang, P. (2015). Effects of ridge and furrow rainwater harvesting system combined with irrigation on improving water use efficiency of maize (*Zea mays* L.) in semi-humid area of China. *Agricultural Water Management* 158, 1–9.

Zhang, F., Chen, X. and Vitousek, P. (2013). An experiment for the world. *Nature* 497, 33–35. https://doi.org/10.1038/497033a.

Zhang, X. (2002). Linking water balance to irrigation scheduling: a case study in the piedmont of Mount Taihang. In: McVicar, T. R., Li, R., Walker, J., Fitzpatrick, R. W. and Liu, C. (Eds), *Regional Water and Soil Assessment for Managing Sustainable Agriculture in China and Australia*. ACIAR Monograph No. 84. Australian Centre for International Agricultural Research, Canberra, pp. 57–69.

Zhang, X. and Cai, X. (2013). Climate change impacts on global agricultural water deficit. *Geophysical Research Letters* 40(6), 1111–1117. https://doi.org/10.1002/grl.50279.

Zhao, G., Webber, H., Hoffmann, H., Wolf, J., Siebert, S. and Ewert, F. (2015). The implication of irrigation in climate change impact assessment: a European-wide study. *Global Change Biology* 21(11), 4031–4048.

Zhou, G., Wan, S., Feng, G. and He, W. (2012). Effects of regional warming on extreme monthly low temperatures distribution in China. *International Journal of Climatology* 32(3), 387–391.

Ziska, L. H., Bunce, J. A., Shimono, H., Gealy, D. R., Baker, J. T., Newton, P. C. D., Reynolds, M. P., Jagadish, K. S. V., Zhu, C., Howden, M. and Wilson, L. T. (2012). Food security and climate change on the potential to adapt global crop production by active selection to rising atmospheric carbon dioxide. *Proceedings of the Royal Society B: Biological Sciences* 279(1745), 4097–4105.

Chapter 16

The water-energy-food nexus (WEF)

Oscar Melo[1], Sadurní Vendrell and William Foster, Pontificia Universidad Católica de Chile, Chile

1 Introduction

Water management and water-related policy design often do not consider the implications they may have on other aspects not directly central to water-related objectives. Water-related research also often ignores its connections to other social, environmental, and economic aspects. Since elements such as energy and food are often intrinsically connected with water, the decisions on the latter impinge upon the former, whether knowingly or not. The nexus approach addresses the potential synergies and trade-offs between these three elements: water, energy, and food (WEF). This approach offers a path to sustainable development through research, management, and policy but with an integrated view of these elements.

The WEF nexus represents an interconnected system wherein water, energy, and food form a cohesive unit characterised by complexity and feedback. When one component of this nexus faces pressure, it can lead to consequences and trade-offs in other aspects. The uncertainties and apprehensions surrounding water, energy, and food security have highlighted the WEF nexus approach (Zhang et al., 2018). Although the nexus approach is multicentric (Liu et al., 2017) and addresses the three elements in a balanced way, in many cases, water is at the centre of the interconnectedness and plays a crucial role in balancing the other elements.

[1] Corresponding author: omelo@uc.cl.

http://dx.doi.org/10.19103/AS.2023.0123.22

This interconnection or nexus between the elements can take different forms and directions. For example, increasing irrigation withdrawals in agriculture may require more energy to pump water from underground aquifers. Also, using more water for agriculture may reduce the water available for hydroelectricity generation, or negatively impact on environmental flows. Increasing food security in a particular area may require more water for agriculture, threatening water security. Understanding and accounting for the trade-offs and potential synergies in using the resources and achieving water, energy, and food security is at the centre of the nexus approach.

2 The nexus and its elements: a global perspective

The nexus concept's origin is linked to the World Economic Forum's response to a concern regarding global water mismanagement and its impact on the global economy and food security under a changing climate (The World Economic Forum, 2011). Hoff (2011) also produced a background paper for the 2011 Bonn Conference, 'The Water, Energy and Food Security Nexus, Solutions for the Green Economy' in preparation for the Rio+20 conferences. These origins gave rise to extensive discussion in the science literature regarding the nexus concept and its applications. In many cases, they extend the concept to include other elements including the environment, climate, and other social needs. We now turn to a brief presentation of the main elements of the nexus from a global perspective, and then discuss their interlinkages.

The concerns relating to water security are global and reflect two main issues: changes in the availability of water (supply) and changes in the use of water (demand), and the use of other resources that affect the availability of clean freshwater. For the first issue, there is growing evidence that a changing climate is altering water availability across the globe, reducing and/or increasing supplies but also changing, for example, the intensity of precipitation, snowfall accumulation, frequency of extreme events, including flooding and droughts, and atmospheric humidity (Konapala et al., 2020; Srinivasan et al., 2017). The second issue is related to population and economic growth, land use change, dietary changes, and inadequate governance and regulation. This concern has raised new interest in balancing water supplies with demand. One approach has been using concepts such as water footprinting (Hoekstra, 2005), including water as part of the United Nations Sustainable Development Goals (SDG),[2] or establishing planetary boundaries for water (Gleeson et al., 2020).

On the other hand, food security has been a permanent global concern, primarily focussing on increasing food availability in low-income countries. However, the increasing demand for food due to population and economic

[2] https://sdgs.un.org/goals/goal6.

growth coupled with dietary changes has raised concerns about the food systems' sustainability. Global food demand is expected to increase by 70% by 2050 (Schütz et al., 2009). To meet this demand, agricultural productivity must similarly increase substantially, especially in developing countries, and the global yield gap must be reduced, especially if proposed planetary boundaries are not to be exceeded. However, agricultural land is not only dedicated to food production but it is also used for energy (biofuel) cropping. If additional energy crops are to be produced to meet expected demand increases for 'clean' bioenergy, significant cropland expansion will be required, suggesting a trade-off between land used for food production, land allocated for energy crops, and land that should be preserved for ecosystems.

As countries develop, their energy consumption also grows. Since most of the energy produced worldwide still relies on fossil fuels, this has meant increasing emissions of greenhouse gases (GHGs) (U.S. Energy Information Administration, 2023). Asia stands out as a continent where population and economic growth have driven a significant increase in energy consumption (Energy Institute, 2023). However, for the countries in this region to continue developing and reducing their emissions requires a more sustainable approach that reduces dependence on fossil fuels and instead relies on either nuclear (Verbruggen, 2008) or renewable energy sources, including wind, solar, tidal, and geothermal energy (Brouwer, 2022). In the case of energy, the planetary boundaries are directly linked to climate change and the goals from the Paris Agreement that aim for energy production to avoid 'the increase in the global average temperature to well below 2°C above pre-industrial levels' and to pursue efforts 'to limit the temperature increase to 1.5°C above pre-industrial levels'. To limit global warming to 1.5°C, GHG emissions must peak before 2025 at the latest and decline by 43% by 2030 (United Nations, 2015).

3 Understanding nexus elements and their interlinkages

This section briefly discusses the linkages between the various nexus elements using examples to illustrate them. The nexus approach is represented in Fig. 1, showing each element's interdependencies. The three main elements of the system, water, energy, and food, have two tags; one refers to the SDG goal that is focussed on the specific element, and the other tag is focussed on the planetary boundary that each one is limited to maintain the goals for the fight against climate change and Agenda 2030 (United Nations, 2015). Between each element, there are two arrows representing the bidirectional nature of the connection. Trade-offs and synergies and their effects on stakeholders are also represented, with some examples provided. The specific examples are summarised in Fig. 1.

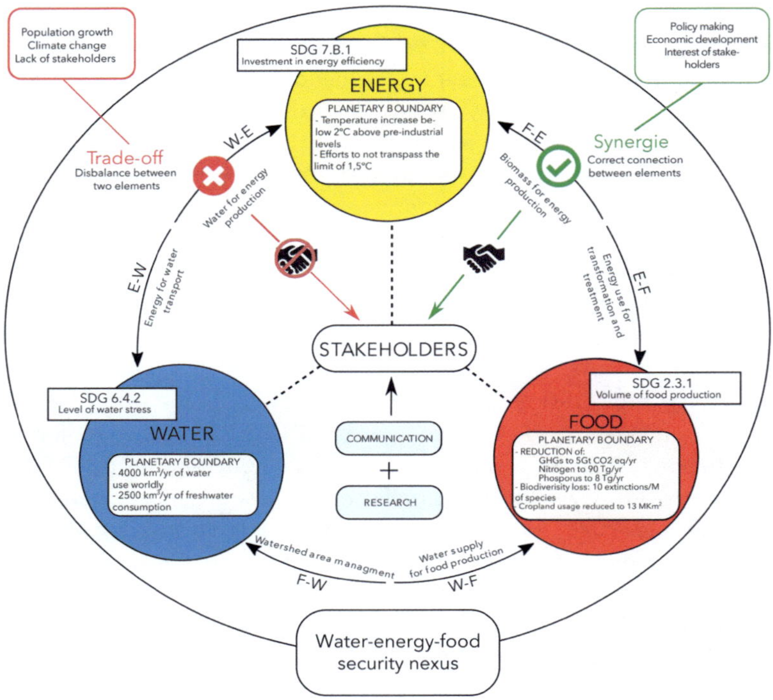

Figure 1 The WEF nexus linkages. Source: Modified from: Cremades et al. (2019).

3.1 Water-energy, energy-water

Water is an essential resource in the production of energy. Since the industrial revolution, it has been used as one of the most valuable resources for its relatively easy access and for its physical and chemical properties. Water is required for hydroelectricity generation and in many processes for extracting raw materials. For example, fracking consumes water to fracture porous geology to extract oil, which is needed for energy production. Transitioning towards renewable energy generation, mainly through the utilisation of wind and solar power, while reducing reliance on biofuels, will play a significant role in decreasing the water demand associated with energy production. Nevertheless, green hydrogen production from wind or solar requires water as the base material for the electrolysis, albeit substantially less than hydrogen from thermoelectrically powered electrolysis (Beswick et al., 2021; Webber, 2007). Another effective measure is retrofitting existing power plants to enhance water efficiency by adopting water-recirculating technologies. These combined efforts will contribute to a noticeable reduction in overall water consumption within the energy sector (Susnik, 2022). Examples where water is needed for energy are for biofuel production, electricity generation, and urbanisation.

Water is also essential in biofuel crop production (e.g. sugarcane, maize, or sugar beet). The amount of water used varies significantly, influenced by the specific crop type, agricultural techniques employed, prevailing climate conditions, soil characteristics, and local topography. These multifaceted factors collectively determine the water requirements for successful crop growth and biofuel production (Mielke et al., 2020).

Hydropower energy plays a crucial role in the fight against climate change, having the advantage of the potential to store energy and the flexibility of the services (Energy Efficiency and Renewable Energy Office, 2022). This flexibility represents the greatest advantage of hydropower, having the capability of providing energy to the grid immediately and the possibility of being a backup power supply during periods of peak demand. The counterpart is the water losses associated with storage because of water evaporation and seepage (deep percolation), which translates directly into an energy loss (Hoekstra et al., 2012).

According to Susnik (2022), increasing urbanisation has also led to a significant interdependence between water and electricity. In urban environments, a notable trend has emerged where the waste heat generated by wastewater treatment plants is being harnessed for energy generation purposes. This innovative approach allows for previously wasted heat energy utilisation, thereby contributing to more sustainable and efficient energy production within urban areas (Cipolla and Maglionico, 2014; Meggers and Leibundgut, 2011). Furthermore, energy can be generated by implementing in-stream micro-generation techniques within the water supply system. These innovative methods harness the power of flowing water to produce electricity. Integrating such techniques into the water supply infrastructure makes it possible to generate renewable energy from the natural movement of water, presenting a promising avenue for sustainable energy production (Du et al., 2017). Energy use is implicit in providing water security to satisfy the daily needs of households and businesses and is also necessary for the treatment of wastewater, especially if it is to be reused. The treatment of wastewater involves high amounts of energy use, more than that needed for raw water treatment. There is also significant variability in the energy needs depending on the level of treatment required, depending on the initial quality of the wastewater source (Susnik, 2022).

Since water is relatively heavy, its provision requires energy for transport (pumping), especially if the use of gravity for conveyance is impossible. Historically, settlements and communities relied on natural sources to obtain water. Supplying large quantities of water for fountains was a luxury few communities and states could afford before the Roman era (Mays, 2016).

Nowadays, electricity and motors allow pumps to move water without using gravity, but the energy cost and associated GHG emissions cannot be ignored.

One of the significant transformative advancements in agriculture, beyond the influential Green Revolution that defied earlier predictions of food scarcity, was the adoption of modern irrigation techniques in farming practices, leading to significant crop productivity increases (Pingali, 2012). Scientific studies have provided evidence indicating that irrigated agriculture exhibits higher crop yields compared to rainfed cultivation. Moreover, irrigated farming systems offer enhanced reliability regarding water availability for crop irrigation. These findings underscore the importance of irrigation in achieving higher agricultural productivity and ensuring a more consistent and reliable food supply (Pei et al., 2017).

Transforming salt water from oceans and seas into freshwater takes advantage of the water potential we have on our planet. However, one key limitation is the energy to convert saline water to fresh water. According to fundamental principles based on the second law of thermodynamics, the minimum energy required to produce 1 m^3 of freshwater from seawater (at 20°C) is 0.79 kWh m^{-3}. Currently, reverse osmosis desalination plants represent an efficient technological solution for water purification. These plants typically consume approximately 2–4 kWh/m^3 of water processed (Voutchkov, 2018). The trade-offs are therefore the very high energy costs and GHG emissions associated with desalinisation unless the energy can be produced from renewable sources. Another key aspect of desalination is disposing of residual saline waters without affecting terrestrial or marine ecosystems.

An often overlooked aspect of energy consumption lies in the heating of water, which accounts for a substantial proportion, ranging from 80% to 90%, of the total energy demand in the water sector (Susnik, 2022). This energy demand is due to water heating for hot tap water and home heating.

3.2 Water–food

Without water, producing the food required to meet global demands for a growing population is not currently possible, nor in the foreseeable future. Sufficient water resources are essential for primary food production, as substantial quantities of water are required. It is worth noting that more than two-thirds of global freshwater abstraction worldwide (and potentially up to 90% in certain countries) is dedicated to fulfilling irrigated agriculture's water demands to sustain food production (Kirby et al., 2003). At the same time, food production needs to grow substantially to meet the needs of the future population (Zhang et al., 2018).

Brouwer (2022) reported that the supply- and demand-side factors influence the amount of water required for food. The supply factor in food production is influenced by various elements such as local climatic conditions, soil characteristics (Allen et al., 1998; Hoekstra, 2005; IAASTD, 2009), irrigation efficiency (WWAP, 2012), and water losses within irrigation distribution networks. These factors play significant roles in determining the gross volumes of water required to meet food demand in different regions. There is also concern about food waste because it is directly connected with energy (and water) losses. According to the FAO, in 2011, nearly one-third of all food was lost annually; this food waste occurs along the entire food supply chain, starting during the harvest on the fields and ending with consumption and food waste in individual households (de Nobile et al., 2012). This concern has triggered literature and policy recommendations to reduce food waste in different countries. For example, in Europe, the Farm to Fork proposal is integrated into the Common Agricultural Policy (CAP).

3.3 Energy-food, food-energy

The interlinkages between energy and food are usually the nexus element that is most overlooked. Since the Industrial Revolution, energy has been vital to producing, maintaining, preserving, and transporting food. On the other hand, organic matter has a huge potential for energy production.

Energy is a critical requirement for ensuring reliable access to food at the household level. Modern food production involves extensive processing to meet society's quality and variety demands. The industrialisation of food production, shifts in consumption patterns, increased dietary diversity, and changes in agricultural practices are a few examples highlighting the crucial connection between energy and food. Sustaining this system necessitates the availability of energy. The following two examples illustrate significant interconnections between energy and food.

In Western diets, there is a tendency towards higher food processing levels, which significantly contributes to the overall energy demand. Apart from the energy required for processing, there is also a significant energy expenditure associated with the transportation of food (Pimentel and Pimentel, 2007). Considering the energy perspective, it is crucial to recognise that food loss also translates into a significant waste of energy, resulting in inefficiencies within the food system. This losses imply the unnecessary utilisation of energy and, consequently, the potential reduction of GHG emissions if processes are made more efficient, decreasing food losses and waste (Roidt and Avellán, 2019).

Current agricultural systems are heavily dependent on the use of inorganic fertilisers and pesticides with the aim of intensifying production to generate greater yields with fewer inputs. However, despite ongoing research to make processes and field production more efficient, energy is still vital given that fertilisers require high levels of energy to be produced (Gellings and Parmenter, 2004).

Food serves as both a source of energy and a consumer of fossil energy throughout its lifecycle, encompassing production, transportation, and preparation stages (Cuéllar and Webber, 2010). This embedded energy can be recycled by some transformations which are explained in the following examples.

Sugars, primarily derived from sugarcane and digested starchy grains, are currently used in fermentation to produce ethanol and butanol, serving as potential substitutes for gasoline (petrol). The fermentation of sugars to produce ethanol as a fuel source has been recognised since prehistoric times. Although its transportation and calorific value as an automotive fuel are relatively less favourable than butanol, an entire industry was established around this fermentation process (Jones and Woods, 1986). The concern relating to crop production is the move away from crops being grown for food production to those for energy production, thus shifting the crop's target market and the pattern of land use (Henry et al., 2018). However, the advantage of biofuel for energy production and transport is that they have lower GHG impacts than fossil fuels (Gerbens-Leenes et al., 2009).

The anaerobic digestion of biomass wastes could significantly impact on renewable energy requirements. The conversion of organic waste derived from agricultural practices, livestock rearing, industrial operations, municipal waste, and various human activities offers a well-suited approach for generating alternate sources of energy and fertiliser (Manyi-Loh et al., 2013). The source of this energy production is methane produced by the decomposition of organic matter from microorganisms. The second end product is a residual solid nutrient-rich substance that can be used as an organic fertiliser, reducing the need for energy-intensive synthetic fertilisers and representing a synergistic nexus connection that should be exploited. In the near future, the practice of biodigesters as decentralised power infrastructure will bring better capabilities than the actual model of large, centralised power stations (Brouwer, 2022).

4 Case studies of the nexus in Andean countries

This section presents three case studies from Andean countries that apply the nexus approach to analyse sustainable development strategies. The first focusses on the application of a water–energy–food–environment modelling toolkit for agricultural development which is applied it to the Ica Valley in Peru and based on the work of Melo et al. (2022) and Correa-Cano et al. (2022). The second offers a case study of irrigation expansion in Santa Elena, Ecuador, and draws from Melo et al. (2022) and Naranjo et al. (2023). The third shows a water–food–energy model that connects agriculture and hydroelectric production to evaluate climate change adaptation strategies and is primarily based on Jander et al. (2023).

A wide variety of modelling methods have been developed to analyse the different aspects and components of the WEF nexus. However, most of these methods only consider one or two nexus elements. A thorough examination of the literature shows that existing nexus tools seldom capture the full range of interactions that exists between individual nexus components (Albrecht et al., 2018). Furthermore, nexus research rarely considers environmental and economic aspects crucial for policy recommendations (referring to WEFE, rather than WEF). The work by Correa-Cano et al. (2022) developed a modelling toolkit that integrated the biophysical and socio-economic aspects by linking three modelling tools: an irrigation simulation module (Hess et al., 2000), an environmental module based on a life cycle assessment (LCA) model and a socio-economic module. The irrigation module assessed productivity measures in terms of net income per unit of energy and per unit of water for different crops, farm sizes, and irrigation methods (Daccache et al., 2014). These measures were then used in the socio-economic and environmental modules to assess the impacts on three scenarios: (i) a decrease in surface water availability, (ii) a decrease in underground water availability, and (iii) an increase in prices. Given the productive and economic implications of scenarios analysed, the socio-economic module was based on a model of farmer behaviour that represents land use decisions (Melo and Foster, 2021). The environmental module used an LCA model implemented in OpenLCA (www.openlca.org) to assess the environmental impacts of activities from field to farm gate, focussing on water consumption and global warming potential.

This nexus analysis allowed stakeholders to review the synergies and trade-offs between local impacts, such as socio-economics and water availability, and more global impacts, such as climate change, to promote more sustainable agreements regarding future development pathways. This toolkit can also aid more integrated water resource management through cross-sectoral collaboration, especially for sectors that must identify shared natural resource investment options to build resilience to future climate and water risks. Applying this toolkit to two major agricultural export crops (asparagus and table grapes) in Peru's Ica Valley provided field-level details on different farming systems' water and energy efficiency under future scenarios for socio-economic outcomes (food prices, employment, income) and regional (air and water pollution) and global (global warming potential) environmental impacts.

The study in Ecuador by Naranjo et al. (2023) used the WEF nexus modelling toolkit to evaluate two participatory policy scenarios of irrigation expansion in a semi-arid region. The first one was an expansion of the irrigated area but maintaining the current mix and proportion of crops cultivated; the second scenario proposed a more limited expansion and only of export-oriented

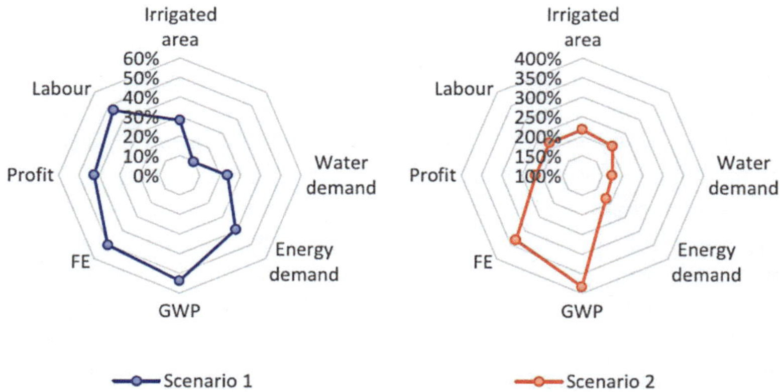

Figure 2 Synergies and trade-offs: proportional change in each dimension between each scenario and the baseline in Santa Elena, Ecuador. GWP, global warming potential; FE, freshwater eutrophication. Source: Adapted from: Naranjo et al. (2023, p. 10).

crops. Export-oriented crops have a different impact in water and energy use, the environment (global warming potential, freshwater eutrophication, soil acidification, and particulate matter formation), and employment and income, compared to domestic market-oriented crops. In the study area, farming is water-restricted and dependent on supply infrastructure and transfers from an adjacent watershed. The results illustrate the trade-offs between water, energy, the environment, and socio-economic outcomes of possible paths of irrigation expansion represented by the two scenarios. Figure 2 presents changes in each dimension with respect to the base case. For example, scenario 2 has a larger impact on employment and in global warming potential than scenario 1.

Jander et al. (2023) used an WEF nexus model to link agricultural production with hydropower generation with distribution rules to project climate change effects of different adaption strategies in the Maule basin in Chile. The two strategies considered were to set out linear hedging rules for a reservoir and to establish adaptive water rights. Simulations using 14 different climate scenarios and water availability were run. Using a multi-objective optimisation of environmental, agricultural, and energy outcome indicators, non-inferior portfolios are generated incorporating the proposed strategies. The results show how these strategies can increase the system's performance and resilience, illustrating the trade-offs between vulnerability and environmental consequences of productive activities.

5 Conclusion

This chapter presents an outline description and example applications of the WEF nexus approach to support more integrated and holistic frameworks to inform water research and policy. The concept's origin has been presented in a global context and more specific examples of the different trade-offs and synergies between the elements. We have illustrated how this concept can be applied to research informing policy decisions in three case studies in Andean countries. Although research on the nexus is considerably broader than what this chapter has briefly reviewed, the central ideas that underpin this approach have been presented. Several other topics related to the nexus approach are presented extensively in the literature. These include the legal aspects, knowledge and communication, policy integration, and governance. The nexus approach has also been applied more specifically to varying geographical areas and scales including cities, marine areas, and different agricultural systems.

6 Where to look for further information

The seminal work that introduced the concept is the work by Hoff: https://mediamanager.sei.org/documents/Publications/SEI-Paper-Hoff-UnderstandingTheNexus-2011.pdf and the book by the World Economic Forum https://link.springer.com/book/10.5822/978-1-61091-026-2.

A recent and complete exposition of the nexus can be found in the book by Brouwer: https://www.elgaronline.com/edcollbook/book/9781839100550/9781839100550.xml.

Albert et al. (2018) present an excellent review of different methods used in nexus analysis. Cremades et al. (2019) propose ten principles to integrate the nexus and climate services for integrated assessments. Zhang et al. (2018) also review concepts, research questions, and methods.

Some brief presentations of the subject are: https://www.fao.org/land-water/water/watergovernance/waterfoodenergynexus/en/. https://www.unwater.org/water-facts/water-food-and-energy. https://international-partnerships.ec.europa.eu/policies/climate-environment-and-energy/water-energy-food-ecosystem-nexus_en.

A web page on the subject, with training material and a link to projects and events, is the https://www.water-energy-food.org/.

A global national-level WEF nexus index is presented at: https://wefnexusindex.org/.

7 References

Albrecht, T. R., Crootof, A. and Scott, C. A. (2018). The water-energy-food nexus: a systematic review of methods for nexus assessment. *Environmental Research Letters* 13(4). https://doi.org/10.1088/1748-9326/aaa9c6.

Allen, R. G., Pereira, L. S., Raes, D. and Smith, M. (1998). Crop evapotranspiration: guidelines for computing crop water requirements. *FAO Irrigation and Drainage Paper* 56, 328.

Beswick, R. R., Oliveira, A. M. and Yan, Y. (2021). Does the green hydrogen economy have a water problem? *ACS Energy Letters* 6(9), 3167–3169. https://doi.org/10.1021/ACSENERGYLETT.1C01375.

Brouwer, F. (2022). *Handbook on the Water-Energy-Food Nexus*. Edward Elgar Publishing Limitedhttps://doi.org/10.4337/9781839100550.

Cipolla, S. S. and Maglionico, M. (2014). Heat recovery from urban wastewater: analysis of the variability of flow rate and temperature in the sewer of Bologna, Italy. *Energy Procedia* 45, 288–297. https://doi.org/10.1016/j.egypro.2014.01.031.

Correa-Cano, M. E., Salmoral, G., Rey, D., Knox, J. W., Graves, A., Melo, O., Foster, W., Naranjo, L., Zegarra, E., Johnson, C., Viteri-Salazar, O. and Yan, X. (2022). A novel modelling toolkit for unpacking the water-energy-food-environment (WEFE) nexus of agricultural development. *Renewable and Sustainable Energy Reviews* 159, 112182. https://doi.org/10.1016/J.RSER.2022.112182.

Cremades, R., Mitter, H., Tudose, N. C., Sanchez-Plaza, A., Graves, A., Broekman, A., Bender, S., Giupponi, C., Koundouri, P., Bahri, M., Cheval, S., Cortekar, J., Moreno, Y., Melo, O., Karner, K., Ungurean, C., Davidescu, S. O., Kropf, B., Brouwer, F. and Marin, M. (2019). Ten principles to integrate the water-energy-land nexus with climate services for co-producing local and regional integrated assessments. *Science of the Total Environment* 693, 133662. https://doi.org/10.1016/j.scitotenv.2019.133662.

Cuéllar, A. D. and Webber, M. E. (2010). Wasted food, wasted energy: the embedded energy in food waste in the United States. *Environmental Science and Technology* 44(16), 6464–6469. https://doi.org/10.1021/es100310d.

Daccache, A., Ciurana, J. S., Rodriguez Diaz, J. A. and Knox, J. W. (2014). Water and energy footprint of irrigated agriculture in the Mediterranean region. *Environmental Research Letters* 9(12). https://doi.org/10.1088/1748-9326/9/12/124014.

de Nobile, F. O., Galbiatti, J. A., Muraishi, R. I. and Spadoni, T. B. (2012). *Pérdidas y desperdicio de alimentos en el mundo*. Food and Agriculture Organization (Vol. 9, Issue 1). https://doi.org/10.3738/1982.2278.562.

Du, J., Yang, H., Shen, Z. and Chen, J. (2017). Micro hydro power generation from water supply system in high rise buildings using pump as turbines. *Energy* 137, 431–440. https://doi.org/10.1016/j.energy.2017.03.023.

Energy Efficiency and Renewable Energy Office. (2022). *Benefits of Hydropower*. https://www.energy.gov/eere/water/benefits-hydropower#:$~$:text=Because.

Energy Institute. (2023). *Primary Energy Consumption by World Region*. Our World in Data.

Gellings, C. W. and Parmenter, K. E. (2004). Energy efficiency in fertilizer production and use. *Efficient Use and Conservation of Energy* II. 123–136.

Gerbens-Leenes, W., Hoekstra, A. Y. and Van Der Meer, T. H. (2009). The water footprint of bioenergy. *Proceedings of the National Academy of Sciences of the United States of America* 106(25), 10219–10223. https://doi.org/10.1073/pnas.0812619106.

Gleeson, T., Wang-Erlandsson, L., Zipper, S. C., Porkka, M., Jaramillo, F., Gerten, D., Fetzer, I., Cornell, S. E., Piemontese, L., Gordon, L. J., Rockstro, J., Oki, T., Sivapalan, M., Wada, Y., Brauman, K. A., Martina Flörke, M. and Famiglietti, J. S. (2020). The water planetary boundary: interrogation and revision. *One Earth* 12. https://doi.org/10.1016/j.oneear.2020.02.009.

Henry, R. C., Engström, K., Olin, S., Alexander, P., Arneth, A. and Rounsevell, M. D. A. (2018). Food supply and bioenergy production within the global cropland planetary boundary. *PLoS ONE* 13(3), e0194695. https://doi.org/10.1371/journal.pone.0194695.

Hess, T., Leeds-Harrison, P. and Counsell, C. (2000). *IWASRI WaSim Technical Manual*. HR Wallingford and Cranfield University. Available at: http://www.silsoe.cranfield.ac.uk/.

Hoekstra, A. Y. (2005). Globalization of water. *Water Encyclopedia*. https://doi.org/10.1002/047147844X.wr51.

Hoekstra, A. Y., Mekonnen, M. M., Chapagain, A. K., Mathews, R. E. and Richter, B. D. (2012). Global monthly water scarcity: blue water footprints versus blue water availability. *PLoS ONE* 7(2). https://doi.org/10.1371/journal.pone.0032688.

Hoff, H. (2011). Understanding the nexus. In: Background Paper for the Bonn 2011 Conference: The Water, Energy and Food Security Nexus. Stockholm Environment Institute.

IAASTD. (2009). Agriculture at a crossroads: the global report. *International Assessment of Agricultural Knowledge, Science, and Technology*. Washington, DC: Island Press.

Jander, V., Vicuña, S., Melo, O. and Lorca, Á. (2023). Adaptation to climate change in basins within the context of the water-energy-food nexus. *Journal of Water Resources Planning and Management* 149(11), 04023060. https://doi.org/10.1061/JWRMD5.WRENG-5566.

Jones, D. T. and Woods, D. R. (1986). Acetone-butanol fermentation revisited. *Microbiological Reviews* 50(4), 484–524. https://doi.org/10.1128/mr.50.4.484-524.1986.

Kirby, R. M., Bartram, J. and Carr, R. (2003). Water in food production and processing: quantity and quality concerns. *Food Control* 14(5), 283–299. https://doi.org/10.1016/S0956-7135(02)00090-7.

Konapala, G., Mishra, A. K., Wada, Y. and Mann, M. E. (2020). Climate change will affect global water availability through compounding changes in seasonal precipitation and evaporation. *Nature Communications* 11(1), 1–10. https://doi.org/10.1038/s41467-020-16757-w.

Liu, J., Yang, H., Cudennec, C., Gain, A. K., Hoff, H., Lawford, R., Qi, J., de Strasser, L., Yillia, P. T. and Zheng, C. (2017). Challenges in operationalizing the water–energy–food nexus. *Hydrological Sciences Journal* 62(11), 1714–1720. https://doi.org/10.1080/02626667.2017.1353695.

Manyi-Loh, C. E., Mamphweli, S. N., Meyer, E. L., Okoh, A. I., Makaka, G. and Simon, M. (2013). Microbial anaerobic digestion (bio-digesters) as an approach to the decontamination of animal wastes in pollution control and the generation of renewable energy. *International Journal of Environmental Research and Public Health* 10(9), 4390–4417. https://doi.org/10.3390/ijerph10094390.

Mays, L. W. (2016). *Ancient Water Technologies*. Springer Dordrecht.

Meggers, F. and Leibundgut, H. (2011). The potential of wastewater heat and exergy: decentralized high-temperature recovery with a heat pump. *Energy and Buildings* 43(4), 879–886. https://doi.org/10.1016/j.enbuild.2010.12.008.

Melo, O. and Foster, W. (2021). Agricultural and forestry land and labor use under long-term climate change in Chile. *Atmosphere* 12(3), 305. https://doi.org/10.3390/atmos12030305.

Melo, O., Naranjo, L., Rey, D., Salmoral, G., Viteri-Salazar, O. and Zegarra, E. (2022). *Agricultural Development in the Andean Countries and the Nexus*. Edward Elgar Publishing. Available at: https://www.elgaronline.com/display/book/9781839100550/book-part-9781839100550-20.xml.

Naranjo, L., Correa-Cano, M. E., Rey, D., Chengot, R., España, F., Sactic, M., Knox, J. W., Yan, X., Viteri-Salazar, O., Foster, W. and Melo, O. (2023). A scenario-specific nexus modelling toolkit to identify trade-offs in the promotion of sustainable irrigated agriculture in Ecuador, a Belt and Road country. *Journal of Cleaner Production* 413, 137350. https://doi.org/10.1016/J.JCLEPRO.2023.137350.

Nations, U. (2015). The Paris agreement. In: *The Paris Agreement*. UNFCCC. Available at: https://unfccc.int/process-and-meetings/the-paris-agreement.

Pei, H., Min, L., Qi, Y., Liu, X., Jia, Y., Shen, Y. and Liu, C. (2017). Impacts of varied irrigation on field water budgets and crop yields in the North China Plain: rainfed vs. irrigated double cropping system. *Agricultural Water Management* 190, 42–54. https://doi.org/10.1016/j.agwat.2017.05.007.

Pimentel, D. and Pimentel, M. H. (2007). *Food, Energy, and Society*. CRC Press.

Pingali, P. L. (2012). Green revolution: impacts, limits, and the path ahead. *Proceedings of the National Academy of Sciences of the United States of America* 109(31), 12302–12308. https://doi.org/10.1073/PNAS.0912953109.

Roidt, M. and Avellán, T. (2019). Learning from integrated management approaches to implement the nexus. *Journal of Environmental Management* 237(March), 609–616. https://doi.org/10.1016/j.jenvman.2019.02.106.

Schütz, H., Jansen, M. and Verhoff, M. A. (2009). How to feed the world in 2050. *FAO* 228(5–6), 151–159.

Srinivasan, V., Konar, M. and Sivapalan, M. (2017). A dynamic framework for water security. *Water Security* 1, 12–20. https://doi.org/10.1016/J.WASEC.2017.03.001.

Susnik, J. (2022). Global nexus relationships and trends. In: *Handbook on the Water-Energy-Food Nexus* (p. 430). https://doi.org/10.4337/9781839100550.

The World Economic Forum. (2011). *Water Security the Water-Food-Energy-Climate Nexus* (1st edn.). Island Press/Center for Resource Economics. https://doi.org/10.5822/978-1-61091-026-2.

U.S. Energy Information Administration. (2023). *Primary Energy Consumption Worldwide*. Our World in Data.

United Nations. (2015). Transforming our world: the 2030 agenda for sustainable development. Available at: https://sdgs.un.org/sites/default/files/publications/21252030%20Agenda%20for%20Sustainable%20Development%20web.pdf.

Verbruggen, A. (2008). Renewable and nuclear power: a common future? *Energy Policy* 36(11), 4036–4047. https://doi.org/10.1016/j.enpol.2008.06.024.

Voutchkov, N. (2018). Energy use for membrane seawater desalination – current status and trends. *Desalination* 431(October 2017), 2–14. https://doi.org/10.1016/j.desal.2017.10.033.

Webber, M. E. (2007). The water intensity of the transitional hydrogen economy. *Environmental Research Letters* 2(3), 034007. https://doi.org/10.1088/1748-9326/2/3/034007.

WWAP. (2012). *UNESCO World Water Assessment Programme*. World Water Development Report 4. Paris: UNESCO Publishing.

Zhang, C., Chen, X., Li, Y., Ding, W. and Fu, G. (2018). Water-energy-food nexus: concepts, questions and methodologies. *Journal of Cleaner Production*. https://doi.org/10.1016/j.jclepro.2018.05.194.

Index